EPSTEIN-BARR VIRUS
AND ASSOCIATED DISEASES

DEVELOPMENTS IN MEDICAL VIROLOGY

Yechiel Becker, Series Editor
Julia Hadar, Managing Editor

Levine, et al (eds.), *Epstein-Barr Virus and Associated Diseases* (1985)

EPSTEIN-BARR VIRUS AND ASSOCIATED DISEASES

Proceedings of the First International Symposium on Epstein-Barr Virus-Associated Malignant Diseases (Loutraki, Greece—September 24-28, 1984)

edited by

P.H. Levine
D.V. Ablashi
G.R. Pearson
S.D. Kottaridis

Martinus Nijhoff Publishing
a member of the Kluwer Academic Publishers Group
Boston/Dordrecht/Lancaster

Distributors for North America:
Kluwer Academic Publishers
190 Old Derby Street
Hingham, MA 02043, USA

Distributors for the UK and Ireland:
Kluwer Academic Publishers
MTP Press Limited
Falcon House, Queen Square
Lancaster LA1 1RN, UK

Distributors for all other countries:
Kluwer Academic Publishers Group
Distribution Centre
P.O. Box 322
3300 AH Dordrecht, THE NETHERLANDS

*The figure on the cover is from Proc Natl Acad
Sci USA, 81:4652, 1984, and is used with permission. It also
appears on page 447 of this book.*

Library of Congress Cataloging in Publication Data
International Symposium on Epstein-Barr Virus-
    Associated Malignant Diseases (1st : 1984 :
        Loutraki, Greece)

    Epstein-Barr virus and associated diseases.

    (Developments in medical virology)
    Includes bibliographies.
    1. Epstein-Barr virus—Congresses.   2. Nasopharynx—
Cancer—Congresses.   3. Oncologic viruses—Congresses.
4. Virus diseases—Congresses.   I. Levine, P.H.
(Paul H.) II. Title. III. Series. [DNLM: 1. Epstein-
Barr Virus—immunology—congresses.   2. Epstein-Barr
Virus—pathogenicity—congresses.   3. Immunologic
Deficiency Syndromes—congresses.   4. Medical Oncology—
congresses.   5. Nasopharyngeal Neoplasms—congresses.
6. Tumor Virus Infections—congresses.
QW 165.5.H3 I616 1984e]
QR400.2.E68I57   1984   576'.6484   85–15484

ISBN-13:978-1-4612-9641-6   e-ISBN-13:978-1-4613-2625-0
DOI:10.1007/978-1-4613-2625-0

Copyright © 1985 by Martinus Nijhoff Publishing, Boston
Softcover reprint of the hardcover 1st edition 1985

CONTENTS

SERIES PREFACE   xi
PREFACE   xiii
ACKNOWLEDGMENTS   xv

IMMUNODEFICIENCY DISEASES

1. Association of Epstein-Barr Virus and Lymphoproliferative   3
   Diseases in Immune Deficient Persons *D.T. Purtilo*
2. Relapsing, Recurrent and Chronic Infectious Mononucleosis   18
   in the Normal Host *S.E. Straus*
3. Immune Assessment of Patients with Chronic Active EBV   34
   Infection (CAEBV) *J.F. Jones, S.E. Straus, G. Tosato,
   R.A. Kibler, M. Hicks, D. Lucas, O. Preble, T. Lawley,
   G. Armstrong, G. Pearson and R.M. Blaese*
4. Reversal of Common Variable Hypogammaglobulinemia-   43
   Associated Suppressor Cell Activity by Specific
   Carbohydrates *G. Tosato, S.E. Pike and R.M. Blaese*
5. Unusual Primary Tumors of Brain and Lungs Associated   53
   with Epstein-Barr Virus (EBV) *J. Joncas, F. Ghibu,
   N. Lapointe, L. Begin, J. Michaud, P. Simard, P. Benoit,
   G. Rivard, J. Demers and R. Raymond*
6. Lytic, Non-Transforming Epstein-Barr Virus (EBV) from   63
   Two Patients with Chronic Active EBV Infection *C. Alfieri,
   F. Ghibu and J. Joncas*
7. Brief Communication: The Significance of Antibodies   73
   Against the Epstein-Barr Virus-Specific Membrane Antigen
   gp250 in Acute and Latent EBV Infections *W. Jilg and
   H. Wolfe*

EBV-ASSOCIATED MALIGNANCIES

8. The Epidemiology of Epstein-Barr Virus-Associated   81
   Malignancies *P.H. Levine*
9. Genetic Aspects of EBV-Associated Malignancies *M. Simons*   90
10. Pathology of Epstein-Barr Virus (EBV)-Associated Disease   106
    (the Lymphatic System) *G.R.F. Krueger*
11. The "Family Study" Approach to Investigating the Role of   131
    Genetic Factors in Nasopharyngeal Carcinoma *S.J. Bale,
    A.E. Bale and P.H. Levine*

v

12. An EBV-Associated Salivary Gland Cancer *A.P. Lanier,*   145
S. Krishnamurthy, S.E. Clift, K.T. Kline, G.W. Bornkamm,
W. Henle, A. Gown and D. Thorning

13. Nasopharyngeal Carcinoma: Early Detection and IgA-   151
Related pre-NPC Condition. Achievements and Prospectives
Y. Zeng and G. de The

14. Evaluation of Epstein-Barr Virus Serologic Analysis in North   164
American Patients with Nasopharyngeal Carcinoma and in
Comparison Groups *H.B. Neel, III, G.R. Pearson and*
W.F. Taylor

15. Use of Epstein-Barr Virus Serology in the Diagnosis of   180
Nasopharyngeal Carcinoma in Malaysia *M. Yadav, N.*
Malliha, A.W. Norhanom and U. Prasad

16. An Analysis of the Relationship Between Clinical Pathology   193
and Serological Level of EB Virus VCA-IgA Antibody in
Nasopharyngeal Carcinoma *P.S. Huang, Y.D. Chen and*
Y.Y. Shen

17. Brief Communication: Fossa of Rosenmuller: The Site for   200
Initial Development of Carcinoma of the Nasopharynx
U. Prasad, J. Singh and R. Patmanathan

18. Brief Communication: Carcino-Embryonic Antigen (CEA) in   207
Nasopharyngeal Carcinoma and Chronic Nasopharyngitis
Y.Y. Shen and P.S. Huang

19. Brief Communication: HLA-Antigens in NPC Patients from   211
a Low-Risk Area (Cologne, FRG) *G. Bertram, J. Kruger*
and K. Sesterhenn

MOLECULAR BIOLOGY OF EBV

20. Persistence and Expression of the Epstein-Barr Virus   221
Genome in Latent Infection and Growth Transformation of
Lymphocytes *E. Kieff, K. Hennessy, T. Dambaugh,*
T. Matsuo, S. Fennewald, M. Heller, L. Petti and
M. Hummel

21. An Epstein-Barr Virus-Determined Nuclear Antigen Encoded   248
by a Region within the EcoRI A Fragment of the Viral
Genome *L. Rymo and G. Klein*

22. An EBV RNA with a Repetitive Spliced Structure   256
M. Bodescot, B. Chambraud and M. Perricaudet

23. Characterization of an EBV-Associated Protein Kinase   267
E. Fowler

24. Characterization of the Genes within the BamHl Fragment   278
of Epstein-Barr Virus DNA that may Determine the Fate of
Viral Infection *J. Sample, A. Tanaka, G. Lancz and*
M. Nonoyama

25. Effects of Tunicamycin on Binding of Epstein-Barr Virus   289
    N. Balachandran and L.M. Hutt-Fletcher
26. Brief Communication: Structure and Expression of the   299
    Epstein-Barr Virus Genome P.J. Farrell, J. Dyson,
    P. Tuffnell, M. Biggin, T. Gibson, A. Bankier, G. Hudson,
    G. Hatfull, S. Satchwell and B. Barrell
27. Brief Communication: Antibody Response to Epstein-Barr   307
    Virus-Specific DNase in Thirteen Patients with Naso-
    pharyngeal Carcinoma J.Y. Chen, R.P. Beasley, C.S. Chien
    and C.S. Yang
28. Brief Communication: Recent Developments in Nucleic Acid   311
    Hybridization H. Wolf, S. Gu, M. Haus and U. Leser

MECHANISMS OF EBV TRANSFORMATION

29. The Host-Cell Range of the Epstein-Barr Virus R. Glaser   319
30. Epstein-Barr Virus (EBV) Growth Transformation is   334
    Associated with an Alteration in C-MYC Chromatin
    Structure W.H. Schubach, B.H. Steiner and M. Birkenbach
31. Study of Nucleosomal Organization of Chromatin in EBV   345
    Producer and Non-Producer Cells V. Zongza and
    S.D. Kottaridis
32. Novel Biological Functions Associated with Epstein-Barr   355
    Virus DNA B.E. Griffin, D. King and L. Karran
33. EBV DNA Content and Expression in Nasopharyngeal   362
    Carcinoma N. Raab-Traub, D. Huang, C.S. Yang and
    G. Pearson
34. Transformation of Human Lymphocytes by Coinfection with   373
    EBV DNA and Transformation-Defective Virions
    D.J. Volsky, B. Volsky, M. Hedeskog, F. Sinangil and
    T.G. Gross
35. Epstein-Barr Virus-Activating Substance(s) from Soil Y. Ito,   383
    H. Tokuda, H. Ohigashi, K. Koshimizu and Y. Zeng
36. Hydrocortisone Enhancement of both EBV Replication and   392
    Transformation of Human Cord Lymphocytes D.V. Ablashi,
    J. Whitman, J. Dahlberg, G. Armstrong and J. Rhim
37. Brief Communication: Detection of EBNA and Rescue of   402
    Transforming EBV in Megakaryocyte Cells Established in
    Culture D. Morgan and D.V. Ablashi

EBV PROTEINS

38. Advances in the Identification of EBV-Specific Proteins: 411
    An Overview G.R. Pearson
39. Bacterially Synthesized EBNA as a Reagent for Enzyme   426

Linked Immunosorbent Assays *G. Milman, D.K. Ades, M-S Cho, S.C. Hartman, G.S. Hayward, A.L. Scott and S.D. Hayward*

40. Epstein-Barr Virus Nuclear Antigen (EBNA): Antigenicity of the Molecule Encoded by the BamHl K Fragment of the EBV Genome *M.J. Allday and A.J. MacGillivray*   436

41. The Use of Antibodies Against Synthetic Peptides for Studying the EBV Nuclear Antigen *J. Dillner, L. Eliasson, L. Sternås, B. Kallin, G. Klein and R.A. Lerner*   446

42. Characterization of Two Forms of the 72,000 MW EBNA and a Cross-Reacting Cellular Protein *J. Luka, T. Kreofsky, T.C. Spelsberg, G.R. Pearson, K. Hennessey and E. Kieff*   456

43. Identification of EBV-Specific Antigens Following Micro-injection of Subgenomic DNA Fragments *A. Boyd, J. Stoerker, J. Holliday and R. Glaser*   466

44. Localization of Epstein-Barr Virus Early Antigen (EA) by Electron Microscopy *M. Kishishita, Y. Ito, J. Luka and G.R. Pearson*   477

45. Selection and Production by Genetechnological Methods of Medically Relevant EBV-related Antigens *H. Wolf, M. Motz, R. Kühbeck, R. Seibl, G.J. Bayliss, S. Modrow and J. Fan*   485

46. Identification of Multiple Epstein-Barr Virus Nuclear Antigens *T.B. Sculley, P.J. Walker, D.J. Moss and J.H. Pope*   495

47. Brief Communication: Production of Human Monoclonal Antibodies by EBV Immortalization *J. Paire and C. Desgranges*   504

IMMUNOLOGY

48. Cellular Immunity in EBV Infections *J.H. Pope*   511

49. T Cell Responses to Epstein-Barr Virus Infection *A.B. Rickinson*   525

50. Epstein-Barr Virus and Immunosuppression *J. Menezes and S.K. Sundar*   535

51. *In Vitro* Immunogenicity of Human Lymphoid Tumour Cell Lines *D.J. Moss, J.A. Staples, S.R. Burrows, J. Ryan and T.B. Sculley*   553

52. Monocyte Contrasuppression of EBV-Immune Regulatory T Cells *G. Tosato, S.E. Pike and R.M. Blaese*   562

53. Analysis of Intratumoral Lymphocyte Subsets in Patients with Undifferentiated Nasopharyngeal Carcinoma *T. Tursz, P. Herait, M. Lipinski, G. Ganem, M.C. Dokhelar, C. Carlu, C. Micheau and G. de The*   572

54. Potential Usefulness of Isoprinosine as an Immunostimulating   580
    Agent in EBV-Associated Disorders: In Vitro Studies
    S.K. Sundar, G. Barile and J. Menezes
55. Mechanisms of Expression of a Burkitt Lymphoma-Associated   588
    Antigen (Globotriaosylceramide) in Burkitt Lymphoma and
    Lymphoblastoid Cell Lines *J. Wiels, E.H. Holmes,*
    N. Cochran, S.I. Hakomori and T. Tursz

CONTROL OF EBV-ASSOCIATED MALIGNANT DISEASE

56. Prevention of EB Virus-Associated Malignant Diseases   603
    M.A. Epstein
57. A Perspective on Treatment of EBV Infection States   619
    J.S. Pagano
58. Clinical and Pathobiological Features of Burkitt's   631
    Lymphoma and their Relevance to Treatment *I. Magrath*
59. Management of Nasopharyngeal Carcinoma *A.T. Huang,*   644
    I.R. Crocker, S.R. Fisher and M.J. Wallman
60. Treatment of Nasopharyngeal Carcinoma with the Antiviral   660
    Drug 9-[(2-Hydroxyethoxymethyl)] Guanine: A Case Report
    J.W. Sixbey, E. Thompson and E.C. Douglass
61. EBV-specific Transfer Factor in the Treatment of African   666
    Burkitt's Lymphoma: A Pilot Study *F.K. Nkrumah,*
    G. Pizza, D. Viza, J. Neequaye, C. DeVinci and P.H. Levine
62. Brief Communication: The Teatment of Nasopharyngeal   673
    Carcinoma (NPC) *J.H. Ho*

SPECIAL LECTURE

63. Epstein-Barr Virus: Past, Present and Future *G. Henle and*   677
    W. Henle

LIST OF PARTICIPANTS   687

1

ASSOCIATION OF EPSTEIN-BARR VIRUS AND LYMPHOPROLIFERATIVE

DISEASES IN IMMUNE DEFICIENT PERSONS

David T. Purtilo, M.D.

Department of Pathology and Laboratory
Medicine, Pediatrics, and Eppley Institute
University of Nebraska Medical Center
Omaha, Nebraska 68105

SUMMARY

Multiple immune responses ordinarily provide tight
security against life-threatening Epstein-Barr virus (EBV)-
induced diseases. However, studies performed predominantly
during the recent decade have demonstrated that individuals
with acquired or inherited immune deficiency disorders are
subject to life-threatening diseases related to EBV. The
diseases seem to result depending on the type and degree of
the immune deficiency and when the immune deficiency occurs
with respect to primary infection by the virus. The X-
linked lymphoproliferative syndrome (XLP) serves as a model
demonstrating that immune deficient individuals can develop
a spectrum of diseases including acquired agammaglobulinemia, aplastic anemia, red cell aplasia, or proliferative disorders such as chronic or fatal infectious
mononucleosis, pseudolymphoma or malignant B cell lymphoma.
Similarly, renal transplant recipients can develop a fatal
infectious mononucleosis-like disease in young seronegative
patients, whereas older individuals tend to show reactivation of virus and develop malignant lymphomas. Patients
with AIDS have also developed EBV-carrying malignant lymphomas. The serological findings in immune-deficient patients
usually reveal excessively high or low antibody titers. The
conversion from polyclonal B cell proliferation to monoclonal B cell malignancies probably occurs as a result of cyto-

genetic and/or molecular changes involving immunoglobulin
gene loci and oncogenes, such as c-myc. Recognition that
EBV can induce life-threatening diseases can lead to development of rational strategies for preventing immune deficiency and also for treating patients before they develop
the life-threatening diseases.

INTRODUCTION

The immune competence and age of a person determine
the clinical consequence when primary infection by
Epstein-Barr virus (EBV) occurs. Although mainly malignant
B cell disorders will be dealt with, a variety of
EBV-associated diseases which affect human beings will also
be summarized. Thereby, I intend to illustrate the biological bridges between benign disorders and lymphoma which
are found in an individual or within families or in communities and are associated with immune deficiency and
viral infection (Purtilo and Sakamoto, 1982; Purtilo,
1984a).

PREGNANCY, THE FETUS, EBV, AND BIRTH DEFECTS

Owing to physiological immune suppression accompanying
normal pregnancy (Purtilo et al., 1972), the virus frequently becomes reactivated (Sakamoto et al., 1982).
Reactivation of EBV in pregnant women has been linked with
an increased frequency of pathologic births (Icart and
Didier, 1981). Others (Goldberg et al., 1981) have noted
that congenital heart disease can arise when the mother
becomes infected by EBV early in pregnancy. Maternal antibodies to the virus increase during pregnancy. They are
passed transplacentally to the fetus (Sakamoto et al.,
1982).

PASSIVE AND ACTIVE PERIODS OF PROTECTION AGAINST EBV

The transplacental passage of neutralizing EBV antibodies protect the child from primary infection for periods up
to 10 months. Seldom does an infant develop infectious
mononucleosis or Burkitt's lymphoma prior to approximately
four months, nor do phenotypes of the X-linked lymphoproliferative syndrome (XLP) appear. Children, until approxima-

tely 10 years of age, silently convert following primary
infection. This protection is due to their efficient
immune responses against the virus.

IMMUNE DEFICIENT CHILDREN, EBV, AND DISEASES

In contrast, children with immune deficiency disorders
can exhibit aberrant responses to the virus. The clinical
outcome depends on the child's type and degree of immune
incompetence. XLP is a model for investigating these phenomena (Purtilo et al., 1975). Due to immune regulatory/
immune deficiency, males with the syndrome variously develop a variety of phenotypes following infection by the
virus. The predominant phenotype is fatal infectious mononucleosis. It results from uncontrolled polyclonal B cell
proliferation which infiltrates vital organs such as the
heart, brain, and liver. Massive liver necrosis associated
with lymphoid cellular infiltration is the cause of death.
Concomitant or evolving phenotypes include aplastic anemia,
hypo- or agammaglobulinemia, and malignant B cell lymphoma
(Purtilo, 1984a). Characteristically, infected males are
unable to mount normal cellular and humoral immune responses to viral antigens. Nearly 100% of the males possessing
the defective XLP lymphoproliferative control locus on the
X chromosome, manifest these phenotypes and the mortality
approaches 85% by 20 years of age.

Other immune deficiency disorders showing aberrant
immune responses to the virus and lymphoproliferative disorders include ataxia-telangiectasia, severe combined immune
deficiency, Wiskott-Aldrich syndrome, common variable immune
deficiency, and Chediak-Higashi syndrome (Purtilo, 1984a).
Gatti and Good (1971) had noted an increased frequency of
lymphoma in immune deficient children. EBV is likely to be
responsible for some of these malignancies. Curiously, only
patients with XLP have manifested fatal infectious mononucleosis. We think that the clinical outcome of EBV
infection is determined by the immune capacity of the
infected child. Alternative explanations include remote
possibilities that fragments of genome or defective virus
may be responsible for individual phenotypes. Rarely,
sporadic fatal infectious mononucleosis and non-X-linked
familial cases are identified. Chronic, active EBV infections or recurrent infectious mononucleosis have also been
documented (Purtilo et al., 1981; Ballow et al., 1982).

RECURRENT AND CHRONIC ACTIVE EBV INFECTION

Currently, textbooks state that infectious mononucleosis is not a recurrent or chronic disease. Data is
mounting that rarely, individuals suffer prolonged and
recurrent untoward consequences of infection by EBV. These
individuals are predominantly females who exhibit weakness,
malaise, fever, and, occasionally, lymphadenomegaly. They
variously exhibit immune deficient responses to EBV
including persistent anti-early antigen antibodies, defective regression of autologous lymphoblastoid cell proliferation in the presence of their T cells, and lymphokine
production may be deficient (DuBois et al., 1984). Straus
discusses this in detail elsewhere in this volume. This
syndrome arises usually during the late teen years or up to
the fourth decade. The advent of organ allografting,
abnormal social practices and immune suppression secondary
to malignancy and use of cytotoxic drugs has produced
another group of patients vulnerable to EBV.

ACQUIRED IMMUNE DEFICIENCY DISORDERS AND EBV

Various cultural and medical practices such as use of
immunosuppressive measures to transplant organs has rendered
such patients vulnerable to the virus. Summarized here are
studies on allograft recipients, male homosexuals, and
cancer patients regarding EBV and lymphoproliferative
malignancies.

Organ Transplantation and Epstein-Barr Virus

Penn and Starzl (1972) have noted an increased frequency of malignancies in renal allograft recipients, especially, nonHodgkin's lymphoma. Collaborative studies
performed during the period 1979-1981 by investigators in
Minneapolis, Stockholm, and Massachusetts had revealed a
fatal infectious mononucleosis-like illness in young renal
allograft recipients and polyclonal B cell lymphoma-like
proliferative lesions in elderly recipients. These lesions
have all contained EBV genome (Hanto et al., 1984).
Similarly, cardiac and hepatic allograft recipients have
developed fatal EBV-carrying lymphoproliferative diseases
(Cleary et al., 1984; Starzl et al., 1984). Only rarely

have bone marrow transplant recipients died of EBV-induced
malignancies: Neudorf et al. (1984) have demonstrated that
HLA-matched bone marrow allografting of immune deficient
patients prevents them from developing lymphoma. In
contrast, nonHLA matched allograft recipients may acquire
EBV-carrying lymphoma (Purtilo, 1984b). Removal of cytotoxic T cells by plant lectins or monoclonal antibodies
potentially removes the immunological surveillance of the
patient.

Acquired Immune Deficiency Syndrome (AIDS)

In 1981, an epidemic emerged among male homosexuals,
intravenous drug users, Haitians, and hemophiliacs which
was manifested by unusual infectious agents such as
Pneumocystis carinii and opportunistic malignancies
including Kaposi's sarcoma, squamous cell carcinoma and B
cell lymphomas which are also encountered in renal
transplant recipients (Sonnabend et al., 1983). Pre-AIDS
lymphadenomegaly has subsequently been identified.
NonHodgkin's lymphoma has been found in approximately 90
male homosexuals since 1981 in large cities where AIDS
occurs (Ziegler et al., 1984). We (Petersen et al., 1984)
and others (Ziegler et al., 1982) have demonstrated
EBV-carrying malignant lymphomas in such patients. The
agent responsible for AIDS may be a retrovirus variously
termed the lymphadenopathy-associated virus (LAV) or human
T cell lymphotropic virus-III (HTLV-III). This virus can
infect EBV-infected Raji or lymphoblastoid cell lines and
become very productive in vitro (Montagnier et al., 1984).
Perhaps in vivo, interaction between EBV and HTLV-III
which simultaneously infects the immune system may be
responsible for the progressive, irreversible immune deficiency of the patients.

For many years, individuals in Subsaharan Africa have
manifested an excess incidence of Burkitt's lymphoma,
Kaposi's sarcoma, squamous cell carcinoma and opportunistic
infections which have recently developed in patients with
AIDS and renal transplant recipients (Purtilo, 1976;
Purtilo, 1984c).

EBV-ASSOCIATED BURKITT'S LYMPHOMA IN AFRICANS

The role of immunodeficiency and EBV in the pathogenesis of Burkitt's lymphoma in the African setting is substantial. Likely, a subtle immune deficiency arises from the immunosuppressive impact of malnutrition, holoendemic malaria, measles, and parasitism which would allow primary EBV infection to induce a smoldering B cell proliferative process. Moreover, males who show lesser immunocompetence than do females are chiefly involved. Monoclonal malignancy likely develops vis-a-vis from molecular attentions and/or a reciprocal translocation involving sites where c-myc resides on chromosome 8 and immunoglobulin gene loci on chromosomes 14, 22, and 2. The prospective study in the West Nile by DeThe and others (1975) have demonstrated that approximately 1 to 2 1/2 years transpire from the time of primary infection to the appearance of Burkitt's lymphoma. The patients manifest a much higher VCA titer than do those free of tumor. Henle and Henle, 1981 and Sakamoto et al., 1981 have demonstrated that immune deficient patients exhibit either too high or too low antibody titers to the viral antigens. Defective antibody responses can be due to immune suppression accompanying malignancy.

CANCER PATIENTS AND EBV

Although fatal EBV-induced lymphoproliferative disease in a patient with malignancy is rare (Hardy et al., 1984), reactivation of the virus is common in individuals with hematologic malignancies. For example, some individuals with Hodgkin's disease show elevated EBV antibody titers (Masucci et al., 1981). Similarly, patients with hairy cell leukemia exhibit markedly elevated titers indicative of reactivation of primary infection (Sakamoto et al., 1981). Recently, we have demonstrated that reactivation of EBV in patients with nonHodgkin's lymphoma correlates with the prognosis of the patient: patients showing reactivation patterns have a worse prognosis than those who do not (Lipscomb and Purtilo - unpublished observations). Normal immunological responses to EBV are dealt with by Rickinson elsewhere in this volume, however, a brief synopsis is provided to compare with the defects found in the immune deficient patients.

IMMUNE SURVEILLANCE TO EBV

Numerous mechanisms have evolved to protect against the virus. Briefly, these include inhibition of virus by dilution with saliva and mucus; mucosal barriers; cell barriers due to lack of viral receptors; suppression of EBV infected B cell outgrowth by interferon; natural killer cell activity; non-HLA cytotoxic T cell activity responses to the lymphocyte-defined membrane antigen; HLA-restricted cytotoxic T cells; antibody-dependent cellular cytotoxicity; antibodies to membrane, viral capsid, early and EB nuclear-associated antigens. A patient will not become vulnerable to a life-threatening EBV infection if they have only limited defective immune responses to the virus. Therefore, multiple failsafe-type mechanisms have evolved to prevent lethal disease.

The defective immune responses to the virus can be categorized in two groups: qualitative defects such as deficient interferon production in a French girl (Virelizier et al., 1978), and natural killer cell defects in patients with Chediak-Higashi disease (Merino, 1984) are seldom reported. Quantitative immune defects to EBV are more frequently encountered. For example, in XLP, the patients show defective antibody responses to EBV nuclear-associated antigen and antibody-dependent cellular cyto-toxicity; natural killer cell defects are acquired. Cytotoxic T cell, suppressor and helper T cell function are also deficient. T cell regulation of the complex immune responses required to subdue the virus are defective in XLP. Hence the extreme vulnerability to this virus. Diagnosis of infectious mononucleosis has traditionally been based on demonstrating a triad of clinical, hematologic, and serological findings (Hoagland, 1967). In immune-deficient patients with EBV infection, other diagnostic strategies are required.

DIAGNOSIS OF EBV INFECTION IN IMMUNE DEFICIENT PATIENTS

Suspicion that an unusual EBV-induced disease is present is usually based on the history of inherited or acquired immune deficiency. Often the patients manifest fever, lymphadenomegaly, and other findings seen in non-immune deficient patients with infectious mononucleosis. Peripheral blood smears may reveal an unusually large

number of plasmacytoid lymphocytes. EBV has the capacity
to drive B cells to form end-stage plasma cells. Serum Ig,
especially IgM, usually becomes elevated because of the
capacity of the virus to evoke polyclonal B cell activation.

The antibody responses to EBV typically show too high,
too low, or lack of complete spectrum of antibodies to the
viral antigens. Throat washings containing EBV can transform cord lymphocytes; lymphoblastoid cell lines can be
established from peripheral blood lymphocytes or biopsied
specimens; EBNA can be demonstrated in touch imprints of
infected organs; virally infected lymphoid cells can be
detected by monoclonal antibodies in flow cytometry; and
employment of hybridization techniques can demonstrate the
viral genome (Purtilo, 1980). The immune-deficient patient
offers an opportunity to define normal immune responses to
EBV and the steps in a multistep process of lymphomagenesis
initiated by immune deficiency and promoted by the virus.

MOLECULAR AND CYTOGENETIC ASPECTS OF LYMPHOMAGENESIS

Defective immune surveillance to EB viral antigens can
allow the B cell proliferation to persist. If the immune
surveillance defects are profound, the individual will
likely succumb to a polyclonal B cell proliferation reminiscent of infectious mononucleosis. Patients with more
subtle immune deficiencies appear to develop a smoldering B
cell proliferation that converts to monoclonality (Klein,
1979). We (Hanto et al., 1982) have found conversion from
polyclonal to monoclonal EBV carrying B cell proliferation
in a renal transplant recipient. The demonstration by
Manolov and Manolova (1971) that a 14q+ aberration was present in Burkitt's lymphoma has provided a glimpse of mechanisms of lymphomagenesis. Characteristically, reciprocal
translocation is found in 80% of Burkitt's lymphoma lines
between 8q24 where the c-myc oncogene resides and the heavy
chain immunoglobulin locus at 14q32. In 10% of cases the
exchange with 8 occurs at the lambda locus on 22, and the
remaining cases involve the kappa gene locus on chromosome
2. Investigators postulate (Leder et al., 1983) that the
approximation of the active immunoglobulin gene with the
suppressed c-myc oncogene promotes translocation of message
of c-myc, thereby promoting growth of the B cell. Thus the
altered cell acquires growth advantages and monoclonality

PREFACE TO THE SERIES, DEVELOPMENTS IN MEDICAL VIROLOGY

It is my pleasure to introduce the first volume in our new series, Developments in Medical Virology. This series is planned to add another dimension to our understanding of the processes involving viruses as pathogens of cells and organisms. Our series Developments in Molecular Virology is devoted to the basic molecular aspects of virus replication, while Developments in Veterinary Virology deals with viral disease processes in domestic animals.

It is hoped that Developments in Medical Virology will provide advanced information on the understanding of virus diseases of man. The current volume deals in depth with various aspects of Epstein-Barr virus. Forthcoming volumes will be devoted to viruses affecting various organs and will also focus on different illnesses caused by a specific family of viruses.

Certain volumes in the current series are intended to complement their counterparts in Developments in Molecular Virology, so as to update our knowledge of human virus diseases and the practices used in their control.

I would like to express my appreciation to the editors of this book and to all those already engaged in the preparation of volumes to appear in the future.

Yechiel Becker
Series Editor

PREFACE

It has been slightly more than two decades since the Epstein-Barr virus (EBV) was discovered by Prof. M.A. Epstein and his colleagues at the University of Bristol in their search for the causative agent of Burkitt's lymphoma. For several years EBV was a "virus in search of a disease." The first documentation that EBV was pathogenic for humans was in 1969 when Drs. Gertrude and Werner Henle identified it as the causative agent for infectious mononucleosis. Seroepidemiologic and biochemical studies subsequently linked EBV to Burkitt's lymphoma (BL), nasopharyngeal carcinoma (NPC), and more recently to the X-linked lymphoproliferative syndrome. With its widespread pattern of infection and a predilection for producing clinical signs and symptoms in only certain individuals, EBV has provided a model for many other candidate oncogenic viruses, including papilloma viruses, herpes simplex, and HTLV/LAV.

In 1975, an international workshop was sponsored by the National Cancer Institute to address the problem of EBV production, thus facilitating basic research on the virus. This proved to be the last international meeting on EBV for almost a decade. In the past, progress in both clinical and basic research on EBV has been presented in two types of international meetings, the international herpesvirus workshops devoted primarily to basic research on both human and animal herpesviruses, and the international symposia on NPC, in which EBV-related studies were interspersed with clinical, epidemiologic and other etiologic aspects of this important human neoplasm. Because of the rapid advances in both basic and applied research on this widely investigated virus, scientists in several countries initiated a meeting in September 1985 to review the current status of the field and to propose new areas of research. Attracting clinicians, epidemiologists, immuno-virologists, geneticists and workers in other areas of carcinogenesis, the meeting proved to be a stimulating forum for fruitful discussion.

The continual progress in understanding the biochemistry as well as the biology of EBV is apparent from the manuscripts presented at this symposium and appearing in this book. EBV, a ubiquitous virus which may cause many illnesses still unidentified, has long merited the attention and concentrated discussion permitted by

this First International Symposium. It is our hope that
the documentation of the recent progress in studies of
this virus and the observation of areas still needing
concentrated attention (such as the laboratory
investigation of nasopharyngeal carcinoma, which has still
resisted attempts to grow the undifferentiated tumor cells
in tissue culture), will provide a guideline as well as a
status report to new investigators interested in pursuing
the enigma of why a ubiquitous virus can produce so many
different clinical outcomes. It is apparent from the
diverse problems being studied and the many unanswered
questions still being raised that there are still many
opportunities for clinicians and basic scientists to find
new leads regarding the pathogenesis and control of this
important human virus.

ACKNOWLEDGMENTS

The Editors wish to thank the following for their support and cooperation which helped immeasurably in making this symposium a success.

ABBOT LABORATORIES U.S.A.
ADVANCED BIOTECHNOLOGY U.S.A.
ADAMS-CHILCOTT S.A.
AGMARTIN S.A.
AMERICAN KIDNEY FUND U.S.A.
BACAKOS S.A.
BANTIN AND KINGMAN LTD.
BEHRINGWERKE A G
B.I.B.E. ANAPLIOTIS S.A.
BAYER RESEARCH LABORATORIES W. GERMANY
BIOTECH RESEARCH LABORATORIES, INC. U.S.A.
BRISTOL HELLAS
BURROUGHS WELCOME CO. U.S.A.
ELI-LILLY HELLAS
ESSEX HELLAS
GREEK ANTICANCER INSTITUTE
GREEK ANTICANCER SOCIETY
HARTELLAS S.A.
JANSSEN PHARMACEUTICA A.E.B.E.
KALIFRONAS S.A.
KYRIAKIDIS S.A.
LITTON BIONETICS INC.
MERCK, SHARPE AND DOHME RESEARCH LABORATORIES U.S.A.
MINISTRY OF CULTURE AND SCIENCES (GREECE)
MINISTRY OF HEALTH (GREECE)
N.D. VARELAS S.A.
NATIONAL CANCER INSTITUTE (USA)
NATIONAL TOURIST ORGANIZATION OF GREECE
PRINIOTAKIS S.A.
SHOWA UNIVERSITY RESEARCH INSTITUTE FOR
    BIOMEDICINE IN FLORIDA
SMITH, KLINE AND FRENCH U.S.A.
SQUIBB S.A.
VIANEX S.A.

Credits

We also thank Ms. Adriana Cabrales, Ms. Stephanie Coleman, Mrs. Kristine L. Ablashi and Mrs. Harriet Huebner, for their assistance in preparing manuscripts for publication.

IMMUNODEFICIENCY DISEASES

supervenes. Given the requirement that immunoglobulin genes become rearranged during normal immune responses, the immunoglobulin loci in chromosomes 14, 22 and 2 are vulnerable to cleavage when B cells are stimulated by EBV or other B cell-activating agents. Perhaps molecular affinities between c-myc and the immunoglobulin loci are important in aligning chromosomes for cleavage and translocation.

During this decade, investigators have developed methods for identifying EBV infection in immune-deficient patients and have documented the untoward clinical consequences of uncontrolled immune responses to the virus. We now seek measures to prevent and intervene in these life-threatening diseases.

PREVENTION AND INTERVENTION OF EBV INFECTION

Standard methods for preventing a viral infection include vaccination and use of gammaglobulin and antivirals. Presently, a vaccine for EBV is being developed, however, it is not yet available. A practical approach to preventing primary infection in children with XLP is the use of intravenous immunoglobulin (Purtilo et al., 1985). This prophylaxis provides neutralizing antibodies, potentially opsonizes infected EBV cells for destruction, and arms antibody-dependent cellular cytotoxicity. Correction of the immune deficiency is possible by bone marrow transplantation (Neudorf et al., 1984; Filipovich et al., to be published). We have restored immune competence in a patient with XLP who had received HLA-matched transplant from his sister.

Therapy for individuals who become infected with EBV can be approached rationally by improving immunity and providing antiviral therapy. The antiviral therapy available today includes interferon and acycloguanosine. The limited studies attempted so far have been disappointing. Immunomodulating substances to correct quantitative immune defects such as interleukin-2 or thymosin are being considered for use in clinical trials.

SUMMARY

A spectrum of untoward consequences of EBV infections in immune-deficient patients has been presented in patients at various stages in the life cycle. Immunologic shielding of the fetus and infant from primary infection is provided by maternal antibodies. Children with severe primary infection are vulnerable to life-threatening diseases following dissipation of maternal antibodies beginning about four to 10 months of age. Recurrent and chronic active EBV infections occur in rare incidences, predominantly in fe-males. A large group of patients with acquired immune deficiency owing to allograft transplantation, unusual lifestyles, geographical locale, and the development of malignancies are at risk for fatal lymphoproliferative diseases. The immune surveillance apparatus that has evolved in concert with the virus provides substantial defense against the virus. But rarely, both qualitative and quantitative immune defects are encountered in persons. No single immune defect is responsible for vulnerability to EBV. Conversion from polyclonal to monoclonal proliferation probably occurs due to molecular and/or cytogenetic events which activate one or more cellular oncogenes. These agent endow the altered cell with survival advantages. Finally, prevention of and intervention in life-threatening EBV infections in immune deficient patients can be potentially achieved by improving the patient's immune responses and by providing antiviral therapy.

ACKNOWLEDGEMENTS

This work was supported in part by grant PHS CA30196, awarded by the National Cancer Institute DHHS, the Nebraska State Department of Health LB506, and the Lymphoproliferative Research Fund. Space has precluded review of work by many authors who have made valuable contributions to understanding relationships between immune deficiency and EBV.

REFERENCES

Ballow, M., Seeley, K., Sakamoto, K., St. Onge, S., Rickles, F.R., and Purtilo, D.T., Familial chronic mononucleosis. Ann. Int. Med., 97, 821-825 (1982).

Cleary, M.L., Warnke, R., and Sklar, J., Monoclonality of
lymphoproliferative lesions in cardiac transplant recipients. N. Engl. J. Med., 310, 477-523 (1984).

DeThe, G., Day, N.E., Geser, A., Lavoue, M.F., Ho, J.H.C.,
Simons, M.J., Sohier, R., Tukei, P., Vonka, V., and
Zavadova, H., Seroepidemiology of the Epstein-Barr virus:
preliminary analysis of an international study. A review.
In: G. DeThe, M.A. Epstein, and H. ZurHausen (eds.),
Epidemiology, Host Response, and Control, Part 2, pp. 3-16,
International Agency for Research on Cancer, Lyon (1975).

DuBois, R., Seeley, J., Harada, S., Ballow, M., and Purtilo,
D.T., Chronic Epstein-Barr virus infection and mononucleosis. Southern Med. J., 77, 1376-1382 (1984).

Filipovich, A.H., Blazar, B.R., Ramsay, N.K.C., Kersey,
J.H., Zelkowitz, L., Harada, S., and Purtilo, D.T.,
Allogeneic bone marrow transplantation for X-linked
lymphoproliferative syndrome. Submitted for publication.

Gatti, R.A., and Good, R.A., Occurrence of malignancy and
immunodeficiency diseases. Cancer, 28, 89-98 (1971).

Goldberg, G.N., Fulginiti, V.A., Ray, C.G., Ferry, P.,
Jones, J.F., Cross, H., Minnich, L., In utero Epstein-Barr
virus (infectious mononucleosis) infection. J. Am. Med.
Assoc., 246, 1579-1581 (1981).

Hanto, D.W., Frizzera, G., Gajl-Peczalska, K.J., Purtilo,
D.T., and Simmons, R.L., Lymphoproliferative diseases in
renal allograft recipients. In: D.T. Purtilo (ed.),
Immune Deficiency and Cancer: Epstein-Barr Virus and
Lymphoproliferative Malignancies, pp. 321-347, Plenum
Press, New York (1984).

Hanto, D.W., Frizzera, G., Gajl-Peczalska, K.J., Sakamoto,
K., Purtilo, D.T., Balfour, H.H., Simmons, R.L., and
Najarian, J.S., Epstein-Barr virus-induced B-cell lymphoma
after renal transplantation: Acyclovir therapy and tran-
sition from polyclonal to monoclonal B-cell proliferation.
N. Eng. J. Med., 306, 913-918 (1982).

Hardy, C., Harada, S., Lubin, B., Sanger, W., von Schmidt,
B., Yetz, J., Saemundsen, A., Feusner, J., Lennette, E.,

14

Linder, J., Seeley, J.K., and Purtilo, D.T., Fatal
Epstein-Barr virus induced lymphoproliferation complicating
acute lymphoblastic leukemia. J. Pediat., 105, 64-67
(1984).

Henle, W., and Henle, G., Epstein-Barr virus-specific serology in immunologically compromised individuals. Cancer
Res., 41, 4222-4225 (1981).

Hoaglund, R.J., Infectious Mononucleosis, Grune & Stratton,
New York, pp. 1-50 (1967).

Icart, J., and Didier, J., Infections due to Epstein-Barr
virus during pregnancy. J. Infect. Dis., 143, 499 (1981).

Klein, G., Lymphoma development in mice and humans: diver-
sity of initiation is followed by convergent cytogenetic
evolution. Proc. Natl. Acad. Sci., 76, 2442-2446 (1979).

Leder, P., Battey, J., Lenoir, G., Moulding, C., Murphy, W.,
Potter, H., Stewart, T., and Taub, R., Translocations among
antibody genes in human cancer. Science, 222, 765-771,
1983.

Manolov, G., and Manolova, Y., Marker band in one chromosome
No. 14 in Burkitt's lymphomas. Hereditas, 69, 300-302
(1971).

Masucci, M.G., Szigeti, R., Ernberg, I., Masucci, G., Klein,
G., Chessel, J., Sieff, C., Lie, S., Glomstein, A., Businco,
L., Henle, W., Henle, G., Pearson, G., Sakamoto, K., and
Purtilo, D.T., Cellular immune defects to Epstein-Barr
virus-determined antigens in young males. Cancer Res., 41,
4284-4291 (1981).

Merino, F., Immunodeficiency to Epstein-Barr virus in
Chediak-Higashi syndrome. In: D.T. Purtilo (ed.), Immune
Deficiency and Cancer: Epstein-Barr Virus and
Lymphoproliferative Malignancies, pp. 143-164, Plenum Press,
New York (1984).

Montagnier, L., Guest, J., Chamaret, S., Dauguet, C., Axler,
C., Guetard, D., Nugeyre, M.T., Bare-Sinoussi, F., Chermann,
J-C., Brunet, J.B., Klatzmann, D., and Gluckman, J.C.,
Adaptation of lymphadenopathy asociated virus (LAV) to
replication in EBV-tranformed B lymphoblastoid cell lines.
Science, 225, 63-66 (1984).

Neudorf, S.L., Filipovich, A.H., and Kersey, J.H., Immunoreconstitution by bone marrow. In: D.T. Purtilo (ed.), Immune Deficiency and Cancer: Epstein-Barr Virus and Lymphoproliferative Malignancies, pp. 471-480, Plenum Press, New York (1984).

Penn, I., and Starzl, T.E., Malignant tumors arising de novo in immunosuppressed organ transplant recipients. Transplantation, 14, 407-417 (1972).

Petersen, J.M., Tubbs, R.R., Savage, R.A., Calabrese, L.C., Proffitt, M.R., Manolova, Y., Manolov, G., Schumaker, A., Tatsumi, E., McClain, K., and Purtilo, D.T., Small noncleaved B cell Burkitt-type lymphoma with chromosome t(8;14) translocation and carrying Epstein-Barr virus in a male homosexual with the acquired immune deficiency syndrome. Am. J. Med., in press.

Purtilo, D.T., Malignancies in the tropics. In: D.H. Connor, and C. Binford (eds.), Pathology of Tropical and Extraordinary Diseases, pp. 647-660, Armed Forces Institute of Pathology, Washington, D.C. (1976).

Purtilo, D.T., Epstein-Barr-virus-induced oncogenesis in immune deficient individuals. Lancet, 1, 300-303 (1980).

Purtilo, D.T., Immune Deficiency and Cancer: Epstein-Barr Virus and Lymphoproliferative Malignancies, Plenum Pess, New York, pp. 1-481 (1984a).

Purtilo, D.T., Squamous cell carcinomas, Kaposi's sarcoma, and Burkitt's lymphoma are consequences of impaired immune surveillance to ubiquitous viruses in acquired immune deficiency syndrome, allograft recipients, and tropical African patients. In: A.O. Williams, and G.T. O'Conor (eds.), Viruses and Tumors in Africans, IARC Scientific Publications, Geneva, in press (1984b).

Purtilo, D.T., Clonality of EBV-induced lymphoproliferative diseases in immune deficient patients. N. Engl. J. Med., 311, 191 (1984c).

Purtilo, D.T., Hallgren, H., and Yunis, E.J., Depressed maternal lymphocyte response to phytohemagglutinin in human pregnancy. Lancet, 1, 769-771 (1972).

Purtilo, D.T., and Sakamoto, K., Reactivation of Epstein-Barr virus in pregnant women, social factors, and

immune competence as determinants of lymphoproliferative
disease. Med. Hypothesis, 8, 401-408 (1982).

Purtilo, D.T., Sakamoto, K., Saemundsen, A.K., Sullivan,
J.L., Synnerholm, A.C., Andersson-Anvret, M., Pritchard, J.,
Sloper, C., Sieff, C., Pincott, J., Pachman, L., Rich, K.,
Cruzi, F., Cornet, J., Collins, R., Barnes, N., Knight, J.,
Sandstedt, B., and Klein, G., Documentation of Epstein-Barr
virus infection in immunodeficient patients with life-
threatening lymphoproliferative disease by clinicial, virological and immunopathological studies. Cancer Res., 41,
4226-4235 (1981).

Purtilo, D.T., Yang, J.P.S., Cassel, C.K., Harper, P.,
Stephenson, S.R., Landing, B.H., and Vawter, G.F., X-linked
recessive progressive combined variable immunodeficiency
(Duncan's Disease). Lancet, 1, 935-950 (1975).

Purtilo, D.T., Harada, S., Ochs, H., Sakamoto, K.,
Tatsumi, E., and Davis, J., Epstein-Barr virus specific
antibodies pre- and post-intravenous gammaglobulin in primary immune deficiency syndromes. Submitted for publication.

Sakamoto, K., Aiba, M., Katayama, I., Sullivan, J.L.,
Humphreys, R.E, and Purtilo, D.T., Antibodies to
Epstein-Barr virus-specific antigens in patients with Hairy
Cell Leukemia. Int. J. Cancer, 27, 453-458 (1981).

Sakamoto, K., Greally, J., Gilfillan, R., Sexton, J.,
Seeley, J., Barnabei, V., O'Dwyer, E., Bechtold, T., and
Purtilo, D.T., Epstein-Barr virus in normal pregnant women.
Am. J. Reprod. Immunology, 2, 217-221 (1982).

Sonnabend, J., Witkin, S., and Purtilo, D.T., Acquired immunodeficiency syndrome, opportunistic infections, and
malignancies in male homosexuals. A hypothesis of etiologic
factors in pathogenesis. J. Am. Med. Assoc., 249,
2370-2374 (1983).

Starzl, T.E., Porter, K.A., Iwatsuki, S., Rosenthal, J.T.,
Shaw, B.W., Atchison, R.W., Nalesnik, M.A., Ho, M.,
Griffith, B.P., Hakala, T.R., Hardesty, R.L., and Jaffe, R.,
Reversibility of lymphomas and lymphoproliferative lesions
developing under cyclosporin-steroid therapy. Lancet, 1,
583-587 (1984).

Virelizier, J.-L., Lenoir, G., and Griscelli, C., Persistent
Epstein-Barr virus infection in a child with hypergam-
maglobulinemia and immunoblastic proliferation associated

17

Lancet, 2, 231-234 (1978).

Ziegler, J.L., Beckstead, J.A., Volberding, P.A., Abrams,
D.I., Levine, A.M., Lukes, R.J., Gill, P.A., Burkes, R.L.,
Riggs, S.A., Butler, J.J., Cabanillas, F.C., Hersh, E.,
Newell, G.R., Laubenstein, L.J., Knowles, D., Odajnyk, C.,
Raphael, B., Koziner, B., Urmacher, C., and Clarkson, B.D.,
Non-Hodgkin's lymphoma in 90 homosexual men. Relation to
generalized lymphadenopathy and acquired immunodeficiency
syndrome. N. Engl. J. Med., 310, 565-570 (1984).

Ziegler, J.L., Miner, R., Rosenbaum, E., Lennette, E.T.,
Shilitoe, E., Casavant, C., Drew, W.L., Mintz, L., Gershow,
J., Greenspan, J., Beckstead, J., Yamamot, K., Outbreak of
Burkitt's-like lymphoma in homosexual men. Lancet, 631-633
(1982).

2

RELAPSING, RECURRENT, AND CHRONIC INFECTIOUS MONONUCLEOSIS

IN THE NORMAL HOST

Stephen E. Straus, M.D.

Medical Virology Section, Laboratory of Clinical
Investigation, National Institute of Allergy and Infectious
Diseases, National Institutes of Health, Bethesda, Maryland
20205 U.S.A.

Infectious mononucleosis is a well described disorder,
so there is no need for me to review the features of the
classic syndrome or the extensive evidence for Epstein-Barr
virus (EBV) as its major cause. I will, instead, review a
number of selected aspects of infectious mononucleosis
that have occupied my attention for the past few years,
but which are incompletely addressed in the literature.
These include the rather provocative issues of relapse,
recurrence, and chronicity of EBV infection in the normal
host.

Forty to sixty years ago, these issues were addressed
quite seriously by a number of investigators who made the
now uncommon effort to precisely observe and record the
features and sequelae of illness in hundreds of sequential
patients. Unfortunately, those observations were dependent
upon what we understand today to be inadequate diagnostic
tools. Nonetheless, they are of heuristic value.

We, who are armed today with more specific and
sensitive means of documenting and following EBV infection,
find ourselves confronted on occasion by patients with serologically proven infections who appear to be improving and
then suffer an exacerbation of selected symptoms. Others
present with histories of infectious mononucleosis on more

than one occasion. We offhandedly dismiss most such cases as inadequately documented. Then, to our greatest consternation, we are beset by individuals who claim continued problems dating back to a remote episode of a mononucleosis-like illness. It is because of such patients as these that I have attempted in an open-minded fashion to consider whether relapses, recurrences, or chronic illness follow acute EBV infection. During the course of this review, I will attempt to suggest how these manifestations of EBV infection should be defined and documented and discuss what pathophysiologic processes may underlie them.

RELAPSE

To begin with, I'd like to propose an operational definition for a relapse of infectious mononucleosis, namely, the return of selected symptoms associated with the initial illness within one month of their abatement. Classical infectious mononucleosis is quite protean in its manifestations and, as with many diseases, the expression of the illness may fluctuate, with periods of greater and lesser severity.

In classic reviews of over 800 cases of infectious mononucleosis published in the 1940s, Bernstein (1940), Contratto (1944), and Wechsler et al. (1946) all stated that in their experiences, relapses are not uncommon, occurring in 4%-9% of their patients. Symptoms, including fever, returned within 1-27 days after having disappeared. Contratto (1944) was impressed that these relapses occurred only in individuals who prematurely resumed their normal level of activity. He observed no relapses in individuals whom he felt had a complete recovery, as manifested by a total disappearance of all symptoms and abnormal blood findings.

Obviously, since the concept of an exacerbation or relapse is in accord with common clinical experience, it is most palatable, and serves as the basis for our standard advice to patients: activity must be resumed gradually. To my knowledge, however, there are no studies that address this issue in current terms. For example, we have no incidence figures for individuals who fulfill EBV-specific serologic criteria for acute infection. Do individuals who relapse have sustained elevations of selected antibody

titers such as to the diffuse component of early antigen
(EA-D)? Are they more likely to display antibodies to the
restricted rather than the diffuse component of EA, as the
serologic data of Horwitz et al. (1975) appear to. suggest?
Is it possible that their development of EB nuclear-
associated (EBNA) antibodies is delayed? These are questions that the current technology permits us to address. I
believe that our understanding of the immune responses to
EBV remains too primitive, but some of the technologies,
which the other discussants in this session and I will
address, may also be suitable for examining patients who
suffer from relapses or possible recurrences.

RECURRENCE

Evidence for recurrent infectious mononucleosis, like
that for relapsing infection, largely resides within the
realm of "old wives' tales" or, at best, clinical anecdote,
but "old wives' tales" and clinical anecdote often have
merit. Therefore, since all other herpesviruses are
capable of inducing clinically apparent recurrences, it is
reasonable to consider that this might extend to EBV.

As before, I'll begin with an operational definition.
Recurrent infectious mononucleosis involves the reappearance of a typical mononucleosis-like illness in an individual who has been free of such symptoms for, let us say,
2 months or longer.

Hoagland (1967), in a summary of 500 consecutive cases
of infectious mononucleosis studied over 20 years, stated
that none of the cases had provided evidence of recurrent
infection. Citing several cases reported in the 1920s to
1940s, however, Bernstein (1940) concluded that recurrences
do exist, but are rare. To aid in an objective discussion
of recurrent mononucleosis, I have collected reports that
have been published in the English language over the past
40 years and that purported to demonstrate recurrent
infections confirmed by clinical, hematologic and serologic
data. These are displayed below in two tables, according
to whether or not EBV-specific serologies were performed.
The cases in the first table were reported as examples of
recurrent infection, depending largely on clinical and
epidemiologic criteria. An illness was considered to be

infectious mononucleosis, even in the absence of an atypical lymphocytosis and/or heterophile antibodies, if a close contact had recently experienced an illness with these hematologic or serologic abnormalities.

By these criteria, I found 15 cases that could be considered to be recurrent infectious mononucleosis (Table I). Only 10 of the case reports indicated the presence of atypical lymphocytes on more than 1 occasion. Only 6 patients (Nos. 5, 6, 7, 13, 14, and 15) had heterophile responses and atypical lymphocytes in each episode. These latter cases are more intriguing, but atypical lymphocytosis and heterophile antibodies are not the current sine qua non of EBV infection. There are numerous causes of atypical lymphocytosis, although high percentages (>50%) of atypical lymphocytes are unusual except in EBV infection. Moreover, it is not always clear in these reports whether the heterophile tests incorporated proper absorption studies to enhance their specificity; and even when performed properly, these tests still yield false positives (Ginsburg et al., 1977). Thus, under the scrutiny of the contemporary critical eye, the cases in Table I are suggestive but can be dismissed.

Table II summarizes data from 7 reported cases in which EBV-specific serologies had been included. Unfortunately, we don't know what EBV-specific serologic patterns a recurrent infection should exhibit. A reappearance of VCA-IgM or EA antibodies would be desirable, as would a 4-fold or greater rise in VCA-IgG titers. Are these reasonable criteria? On the surface they would appear so, but recurrent infections caused by herpes simplex virus and varicella-zoster virus are often not accompanied by reappearance or a rise in specific antibody titers.

Of the 7 patients described in Table II, only Patient 7 had a recurrent mononucleosis-like illness with atypical lymphocytes and heterophile antibodies on each occasion. EBV serologies were not done during the first episode, making this case no more impressive than those of Table I. Patients 1-4 had positive but low EBV VCA-IgG titers 10 months to 3 years before presenting with a heterophile-positive mononucleosis-like illness. If we assumed the first titers to date from a prior EBV infection and the

TABLE I. REPORTS OF RECURRENT INFECTIOUS MONONUCLEOSIS

NO.	AGE	SEX	DOCUMENTATION OF FIRST ATTACK			INTERVAL BETWEEN ATTACKS	DOCUMENTATION OF SECOND ATTACK			REFERENCES
			CLINICAL PICTURE*	ATYPICAL LYMPHS	HETEROPHILE†		CLINICAL PICTURE	ATYPICAL LYMPHS	HETEROPHILE†	
1	25	M	+	22%	112	8 yr	+	ND††	ND	Kaufman, 1950
2	3½	F	+	37%	ND	3 mo	+	ND	ND	"
3	8	F	+	57%	ND	2½ mo / 2 mo **	+	2% / ND	ND / ND	"
4	7	F	+	+	ND	3 mo	+	+	ND	"
5	32	M	+	+	32	8½ yr	+	48%	28	"
6	11	F	+	+	224	8 yr / 10 yr **	+	79% / 27%	56 / -	"
7	20	F	+	+	320	2½ yr	+	34%	28	"
8	20	F	+	+	ND	2 yr	+	12%	56	"
9	11	F	+	ND	ND	4 yr	+	ND	ND	"
10	37	F	+	ND	ND	6 mo	+	23%	ND	"
11	18	F	+	ND	ND	10 mo	+	53%	56	"
12	20	F	+	+	ND	12 yr	+	43%	ND	"
13	14	M	+	+	+	4 yr / 7 yr **	+	+ / +	+ / 512	Paterson and Pinninger, 1955
14	23	M	+	+	112	5 yr	+	50%	896	Bender, 1962
15	22	M	+	40%	+	3 yr	+	32%	112	Graves, 1970

* = presence or absence of a mononucleosis-like illness with fever and adenopathy.
† = reciprocal serologic titers.
†† = ND, not done. ** = third episodes.

22

23

TABLE II. REPORTS OF RECURRENT EBV INFECTION

NO.	AGE	SEX	DOCUMENTATION OF FIRST ATTACK				INTERVAL BETWEEN ATTACKS	DOCUMENTATION OF SECOND ATTACK				REFERENCES
			CLIN.*	ATYP. LYMPHS	HETERO-PHILE**	EBV** TITER		CLIN.*	ATYP. LYMPHS	HETERO-PHILE**	EBV** TITER	
1	?	?	+	ND†	ND	VCA-IgG 20	1182 d.	+	ND	≥80	VCA-IgG 160	Chang, 1975
2	?	M	+	ND	ND	VCA-IgG 40	497 d.	+	ND	20	VCA-IgG 160	"
3	?	?	+	ND	ND	VCA-IgG 10	1128 d.	+	ND	≥80	VCA-IgG 640	"
4	?	?	-	ND	ND	VCA-IgG 10	290 d.	+	ND	+	VCA-IgG 40	Chang & Maddock,1980
5	19	M	+	+	14,336	VCA-IgM 320 VCA-IgG 80 EA-D 20 EBNA <2	8 mo.	+	-	112	VCA-IgM ND VCA-IgG 160 EA-D <10 EA-R 40 EBNA 10	Horwitz et al., 1975
6	21	F	+	+	3,584	VCA-IgM ND VCA-IgG 320 EA-D 80 EBNA <2	29 mo.	+	ND	-	VCA-IgM ND VCA-IgG 160 EA-D <10 EA-R 40 EBNA 80	"
7	29	F	+	+	+	ND	10 yr.	+	+	+	VCA 32	Stevens et al., 1970

* = presence or absence of a clinical history of a mononucleosis-like illness with fever and adenopathy.
** = reciprocal serologic titer.
? = data not given.
† = ND, not done.

heterophile responses in the second attack to be nonspecific, then all 4 cases could be dismissed. But then it would be hard to explain the 4-fold and greater rises in EBV titers between attacks. Thus, these cases are interesting, particularly that of Patient 2, who experienced recurrent illness and had substantial titers on each occasion.

The most complete and provocative data are those of Horwitz et al. (1975), who followed the patients who reacted to their primary infection with an EA-R rather than just an EA-D response. Antibodies to EA-R are typical of immune-deficient patients in whom reactivation infections are presumed (Henle and Henle, 1981). Thus, their presence during clinical recurrences in these presumably normal hosts, is reasonable. Two patients (Nos. 5 and 6) (Horwitz et al., 1975) were found to have substantial EA-R titers during their recurrences. Unfortunately, none was proven to be devoid of antibodies to EA prior to the documentation of his EA-R titers.

A final issue, which must be addressed when considering recurrent infection, is that of reactivation as opposed to exogenous reinfection. Exogenous reinfections have been proven to occur in herpes simplex and varicella-zoster virus infections. Theoretically, these could happen with EBV as well, but none of the clinical, hematologic, or serologic studies would alert us to the occurrence of such events. Restriction endonuclease analysis of isolates from sequential infections would need to be performed, but in EBV, the tedium and expense involved in the procedure will preclude its use until there are major technological advances.

CHRONIC MONONUCLEOSIS

As indicated in my introductory remarks, the concept of chronic mononucleosis is most disconcerting, and investigators interested in EBV appear most inclined to dismiss it, except as it may relate to the immune-compromised host. Because Dr. Purtilo discusses this subject elsewhere in this book, I will not attempt to review in any detail the evidence for chronic EBV disease in patients with

congenital or acquired immune disorders. Suffice it to say
here that a combination of methodologies including serologies, fluorescence microscopy, cell-mediated immune studies,
nucleic acid hybridizations, cytogenetics, and others have
led over the past several years to the recognition of
chronic proliferative and aproliferative sequelae of EBV.
These sequelae have appeared in young boys with the X-
linked lymphoproliferative syndrome, transplant recipients,
other sporadic cases, and possibly AIDS patients as well.
Studies of these model disorders have allowed us to develop
a series of criteria for recognition of chronic EBV disease.

Having recognized the serious form of chronic EBV
disease as manifested in the immune-deficient host, we turn
now to the question of whether a milder spectrum of illness
may develop in individuals who are, at least superficially,
normal from a host-defense standpoint. Unfortunately,
several of the methodologies relevant to the documentation
of chronic infection in the immune-deficient patients may
not be sufficiently sensitive or appropriate to the investigation of persisting illness in normal patients. For the
present, I would propose to define chronic infection in the
normal host as that in which symptoms persist for 1 year or
longer following acute symptomatic or initially asymptomatic infection. It is this group of individuals that I
will consider for the remainder of this review.

As with the reports suggesting relapsing and recurrent
infectious mononucleosis, many cases in which chronic
infections were reported had preceded the development of
EBV-specific serologic studies. An oft-cited example is
the study of Isaacs (1948) in which he summarized his experience with prolonged symptomatology following infectious
mononucleosis. Of 206 patients, 53 had protracted symptoms
for 3 months or longer, and 25 of the 53 were ill for over
1 year. All 25 continued to have atypical lymphocytes but
in small numbers (1-7%), and none had persisting heterophile titers above 1:64. Most patients were in their 3rd
to 5th decades of life, two-thirds were women, and fatigue
was the predominant complaint. The report is noteworthy
for its subjectivity but nonetheless presages a number of
more recent studies.

Table III displays a summary of 10 cases reported during the last dozen years in which persistent EBV infections were suggested. Banatlava et al. (1972) demonstrated continued IgM-VCA antibodies for over 6 months in an individual (Patient 1) with persisting lassitude and lymphadenopathy. Askinazi et al. (1976) reported recurrent positive heterophile titers, but unremarkable EBV serologies for 4 years in another individual (Patient 2). Purtilo et al. (1980) reported persistent transfusion-acquired infectious mononucleosis and transient immune deficiency in a young man. The EBV-specific serologic studies demonstrated positive but falling EA titers for a year. More recently, Tobi et al. (1982) described 7 patients with prolonged vague illnesses, of whom 4 had persistent EA titers. Curiously, all 7 had positive IgM-VCA titers, but these could not be confirmed in other laboratories. In a subsequent report on these same patients, Morag et al., (1982) documented elevated levels of [2'-5']-oligo adenylate synthetase. This is an enzyme activated by interferon, and elevated levels have been observed in a number of acute infections including acute infectious mononucleosis.

There are now a number of other groups, including my own, that are engaged in investigations of patients similar to those of Tobi. With their permission, I have excerpted data from a report by DuBois et al. (1984). Table IV lists a series of findings and their frequency in 14 of DuBois' patients. Their illnesses, like those of the patients of both Tobi et al. (1982) and Isaacs (1948), are quite debilitating in the apparent absence of remarkable abnormal findings on physical and routine laboratory examinations. Table V summarizes the first and last serologies recorded by DuBois for these patients. The VCA-IgM titers are generally negative, while the VCA-IgG and EA titers are higher than one would expect.

Elsewhere in this volume, Jones reviews selected immunologic findings in patients that we have studied in Washington, Tucson, and now Denver, so I won't dwell on their observations except to state that we find moderately abnormal EBV serologies and a series of immunologic abnormalities in individuals with prolonged lethargy and other constitutional and emotional problems.

TABLE III. REPORTS OF CHRONIC MONONUCLEOSIS IN NORMAL PATIENTS

NO.	AGE	SEX	INITIAL ILLNESS				CHRONIC ILLNESS				DURATION (months)	REFERENCES
			CLINICAL FEATURES*	ATYPICAL LYMPHS	HETERO-PHILE	EBV[†] TITER	CLINICAL FEATURES	ATYPICAL LYMPHS	HETERO-PHILE	EBV[†] TITER		
1	?**	F	+	ND*	+	VCA-IgM 10	fatigue adenopathy	ND	-	VCA-IgM 10	6	Banatvala et al.,1972
2	21	M	+	ND	+	ND	fatigue adenopathy	-	1:14	VCA-IgG 80 EA-D <10 EA-R <10	45	Askinazi et al., 1976
3	21	M	+	74%	-	VCA-IgG 640 EA 320	?	+	-	VCA-IgG 320 EA 20	12	Purtilo et al., 1980
4	25	F	fever malaise adenopathy	ND	-	VCA-IgM 32 VCA-IgG 256 EA-D <8 EA-R 8 EBNA 40	fever adenopathy	ND	ND	VCA-IgM 64 VCA-IgG 256 EA-D <8 EA-R 8 EBNA 80	15	Tobi et al., 1982
5	19	M	fever adenopathy	ND	-	VCA-IgM 32 VCA-IgG 512 EA-D ND EA-R ND EBNA ND	fever adenopathy malaise	ND	ND	VCA-IgM 32 VCA-IgG 128 EA-D <8 EA-R 8 EBNA 20	18	"
6	28	F	fever malaise	ND	-	VCA-IgM 32 VCA-IgG 256 EA-D <8 EA-R <8 EBNA 80	fever malaise	ND	ND	VCA-IgM 32 VCA-IgG 256 EA-D <8 EA-R <8 EBNA ND	12	"

28

	Age	Sex	Symptoms			Serology		Symptoms		Serology		
7	28	F	fever adenopathy	ND	+	VCA-IgM 16 VCA-IgG 256 EA-D <8 EA-R 32 EBNA 160	ND	fever adenopathy	ND	VCA-IgM 32 VCA-IgG 256 EA-D <8 EA-R 32 EBNA 80	19	"
8	25	F	fever fatigue	ND	ND	VCA-IgM 32 VCA-IgG 512 EA-D <8 EA-R <8 EBNA 40	ND	fever fatigue adenopathy	ND	VCA-IgM 32 VCA-IgG 128 EA-D <8 EA-R <8 EBNA 80	15	"
9	33	F	fever malaise	ND	ND	VCA-IgM 64 VCA-IgG 512 EA-D <8 EA-R <8 EBNA 80	ND	fever malaise	ND	VCA-IgM 16 VCA-IgG 128 EA-D <8 EA-R <8 EBNA 320	11	"
10	19	F	adenopathy	ND	ND	VCA-IgM 64 VCA-IgG 512 EA-D ND EA-R ND EBNA ND	ND	adenopathy	ND	VCA-IgM 32 VCA-IgG 128 EA-D <8 EA-R 8 EBNA 160	22	"

* = history of an acute mononucleosis-like illness or symptoms as listed.

† = reciprocal serologic titer.

ND = not done.

** = data not given.

29

TABLE IV. FEATURES OF 14 PATIENTS WITH CHRONIC
MONONUCLEOSIS SYNDROME[*]

Symptom	Number of Patients	Percent
Weakness and Fatigue	14	100
Fever	13	93
Myalgias, arthralgias	13	93
Depression	10	71
Recurrent pharyngitis	7	50
Lymphadenopathy	6	43
Hepatic tenderness	4	29
Splenomegaly	1	7
Lymphocytosis, atypical lymphocytes	6	43
Partial hypogammaglobulinemia	10	71
Prior infectious mononucleosis	6	43

*Modified from DuBois et al. with permission.

Studies done to date on these patients have failed to document the kinds of EBV-associated lymphoproliferative disorders seen in immune-deficient patients. Thus, we are deprived of one important marker for chronic EBV disease and are left to piece together a series of still isolated observations. The challenge will be to develop additional tools to confirm or refute the notion that EBV underlies the abnormalities documented in these individuals. We have yet to enumerate sufficiently precise criteria for diagnosis. We have only a few cases to suggest spontaneous resolution of the illness, and we know nothing of the long-term consequences of this syndrome in terms of risk for development of lymphoproliferative malignancies or other serious immune disorders.

There is also no real sense of how to treat these patients. We and others have attempted steroidal and non-steroidal anti-inflammatory agents, immunoglobulins, interferon, and antiviral drugs. There are anecdotal reports of

TABLE V. SEROLOGIC FINDINGS IN 14 PATIENTS WITH CHRONIC MONONUCLEOSIS*

NO.	AGE	SEX	INITIAL OBSERVATION						FINAL OBSERVATION						STUDY INTERVAL (months)
			HETERO-PHILE	VCA IgM	VCA IgG	EA D	EA DR	EBNA	HETERO-PHILE	VCA IgM	VCA IgG	EA D	EA DR	EBNA	
1	14	F	+	< 2**	1280	80	80	80	ND†	ND	1280	<10	40	40	30
2	19	M	+	<10	>640	ND	ND	5	+	ND	320	ND	40	ND	16
3	31	M	+	<10	160	20	20	≥5	+	ND	160	<10	10	>5	37
4	35	F	-	<10	>640	320	320	≥5	ND	ND	1280	80	10	80	26
5	33	F	-	<10	>640	10	10	≥5	ND	ND	160	<10	20	40	25
6	32	F	-	<10	40	20	20	≥5	ND	ND	160	<10	20	40	26
7	15	F	-	< 5	160	40	40	≥5	ND	ND	320	<10	40	80	22
8	36	F	-	<10	>640	40	40	≥5	ND	<2	80	< 5	20	80	7
9	39	F	-	< 2	640	40	160	160	ND	<2	≥1280	<10	80	80	18
10	37	F	-	< 2	320	40	40	80	ND	<2	640	<10	<10	40	18
11	36	M	-	< 2	1280	160	320	40	-	<2	2560	20	80	40	7
12	20	F	-	< 2	320	ND	80	ND	+	<2	160	10	20	5	27
13	39	M	-	<10	640	20	20	≥5	ND	ND	640	<10	40	80	22
14	28	M	-	< 2	640	<5	40	20	ND	<2	160	< 5	< 5	40	17

* = modified from DuBois et al. (1984), with permission.

** = reciprocal EBV-specific titers.

† = ND, not done.

improvement and failures with each of these modalities. We have initiated a placebo-controlled trial of intravenous and oral acyclovir. There is no strong belief that this treatment will be beneficial even transiently. Our intent is to learn how to prospectively approach this syndrome to facilitate later studies, should more potent EBV-inhibiting or immunomodulatory agents become available.

ACKNOWLEDGMENTS

I thank M. Tobi for helpful discussions, J. Wilkinson for assisting in the literature review, C. Crout and K. Leighty for manuscript preparation and editing.

REFERENCES

ASKINAZI, C., COLE, F.S., and BRUSCH, J.L., Positive differential heterophile antibody test. J.A.M.A., 236, 1492-1493 (1976).

BANATVALA, J.E., BEST, J.M., and WALLER, D.K., Epstein-Barr virus-specific IgM in infectious mononucleosis, Burkitt lymphoma, and nasopharyngeal carcinoma. Lancet, 1, 1205-1208 (1972).

BENDER, C.E., Recurrent mononucleosis. J.A.M.A., 182, 954-956 (1962).

BERNSTEIN, A., Infectious mononucleosis. Medicine, 19, 85-159 (1940).

CHANG, R.S., EB-virus-seropositive person is susceptible to infectious mononucleosis. N. Engl. J. Med., 292, 925 (1975).

CHANG, R.S., and MADDOCK, R., Recurrence of infectious mononucleosis. Lancet, 1, 704 (1980).

CONTRATTO, A.W., Infectious mononucleosis. A study of one hundred and ninety-six cases. Arch. Int. Med. 73, 449-459 (1944).

DUBOIS, R.E., SEELEY, J.K., BRUS, I., SAKAMOTO, K., BALLOW, M., HARADA, S., BECHTOLD, T.A., PEARSON, G., and PURTILO, D.T., Chronic mononucleosis syndrome. South. Med. J., 77, 1376-1382 (1984).

GINSBURG, C.M., HENLE, W., HENLE, G., and HORWITZ, C.A., Infectious mononucleosis in children. J.A.M.A. 237, 781-785 (1977).

GRAVES, S., JR., Recurrent infectious mononucleosis. J. Kentucky Med. Assoc. December issue, pp. 790-792 (1970).

HENLE, W., and HENLE, G., Epstein-Barr virus-spe fic serology in immunologically compromised individuals. Cancer Res., 41, 4222-4225 (1981).

HOAGLAND, R., Infectious Mononucleosis, Grune and Stratten, Inc., New York (1967).

HORWITZ C.A., HENLE, W., HENLE, G., and SCHMITZ, H., Clinical evaluation of patients with infectious mononucleosis and development of antibodies to the R component of the Epstein-Barr virus-induced early antigen complex. Am. J. Med., 58, 330-338 (1975).

ISAACS, R., Chronic infectious mononucleosis. Blood, 3, 858-861 (1948).

KAUFMAN, R.E., Recurrences in infectious mononucleosis. Amer. Practit. 1, 673-676 (1950).

MORAG, A., TOBI, M., RAVID, Z., SCHATTNER, A., and PATERSON, J.K., and PINNINGER, J.L., A case of recurrent infectious mononucleosis. Br. Med. J., 2, 476 (1955).

PATERSON, J.K., and PINNINGER, J.L., A case of recurrent infectious mononucleosis. Br. Med. J., 2, 476 (1955).

PURTILO, D.T., PAQUIN, L.A., SAKAMOTO, K., HUTT, L.M., YANG, J.P.S., SPARLING, S., BEBERMAN, N., and McAULEY, R.A., Persistent transfusion-associated infectious mononucleosis with transient acquired immunodeficiency. Am. J. Med., 68, 437-440 (1980).

REVEL, M., Elevated [2'-5']-oligo A synthetase activity in patients with prolonged illness associated with serologic evidence of persistent EBV infection. Lancet, 1, 744-746 (1982).

STEVENS, D.A., PRY, T.W., and MANAKER, R.A., Infectious mononucleosis—Always a primary infection with herpes-type virus? J. Natl. Cancer Inst., 44, 533-537 (1970).

TOBI, M., RAVID, Z., FELDMAN-WEISS, V., BEN-CHETRIT, E., MORAG, A., CHOWERS, I., MICHAELI, Y., SHALIT, M., and KNOBLER, H., Prolonged atypical illness associated with serological evidence of persistent Epstein-Barr virus infection. Lancet, 1, 61-64 (1982).

WECHSLER, H.F., ROSENBLUM, A.H., and SILLS, C.T., Infectious mononucleosis: Report of an epidemic in an army post. Part II. Ann. Int. Med., 25, 236-265 (1946).

3

IMMUNE ASSESSMENT OF PATIENTS WITH CHRONIC ACTIVE EBV INFECTION (CAEBV)

J.F. Jones,[1] S.E. Straus,[2,3] G. Tosato,[5] R.A. Kibler,[1] M. Hicks,[1] D. Lucas,[1] O. Preble,[6] T. Lawley,[3,4] G. Armstrong,[5] G. Pearson,[7] and R.M. Blaese.[3,4]

The University of Arizona,[1] Tucson, AZ, NIAID,[2] NCI,[3] NIH,[4] FDA[5] and USUHS,[6] Bethesda, MD, Georgetown University,[7] Washington, D.C.

SUMMARY

Screening and disease - specific immune responses were evaluated in patients with chronic illness and serologic evidence of active EBV infection. CBC, differentials and sedimentation rates were normal as were T and B cell numbers, T cell subsets, mitogen responses and NK and mononuclear cell numbers. Circulating immune complexes were present in 50% of patients studied. T-cell suppression of Ig production persisted in patients. Production of IL-2, IFN γ and NK activity were diminished in patients. Whether these latter differences are due to or result in the ongoing disease is not understood.

INTRODUCTION

Seventy-five patients with similar, undefined but debilitating illnesses lasting for over one year were evaluated. Symptoms began either as classical infectious mononucleosis or as an illness characterized by fatigue, myalgia, neuralgia, arthralgia, depression and dyslogia with varying degrees of fever, pharyngitis and lymphadenopathy. These latter symptoms then either persisted or recurred multiple times in all patients. Sixty-two (83%) of the patients (39 in Arizona and 23 in Bethesda) had serologic profiles suggesting active EBV infection. Similar patients have been previously described (Tobi et al., 1982; DuBois et al., 1984). The thirteen remaining patients were EBV seronegative and/or had alternative diagnoses, including Sjogren's syndrome, systemic lupus erythematosis, Mycoplasma pneumoniae infection, multiple sclerosis and lymphoma.

Previous reports associating elevated EBV-specific antibody titers with immune disorders and pregnancy (Fleisher et al., 1983; Joncas et al., 1977; Lange et al., 1980; Sumaya, 1977) led to the analysis of serological findings and immune function testing reported here.

METHODS
Forty-four patients (18 children under 15 years of age and 26 adults) from Arizona and 31 adults followed at the NIH had been ill for greater than one year. EBV serologies were performed by Dr. Henle or Dr. George Ray, by reported methods (Henle et al., 1974; Ray et al., 1982). Immune function tests included: phenotyping of lymphoid cells by OKT and Leu series monoclonal antibodies; surface IgM-bearing cells; E-rosette forming cells; natural killer cells (leu 7 and morphology); mononuclear cell enumeration; HLA and DR typing and mitogen (PHA, ConA and PWM stimulation) (Jones et al., in press). Levels of circulating immune complexes, serum interferon, leukocyte 2'-5' A synthetase activity, and suppression of Ig secretion were performed as previously described (Straus et al., in press). Induction by ConA-PMA of interleukin 2 (IL-2) and interferon γ (IFN-γ) and measurement of natural killer (NK) cell activity were performed as described (Kibler et al., in press). Measurement of the antibody mediating antibody dependent cell cytotoxicity was performed by Dr. Gary Pearson (Pearson et al., 1978).

RESULTS

Clinical Findings
Table 1 depicts the symptoms of the EBV-seropositive patients. Table 2 compares selected demographic features of the patients and shows the remarkable similarity between groups of patients at each center.

Table 1: CLINICAL FEATURES OF CHRONIC ACTIVE EBV INFECTION

Symptom	Percent of Patients
Fatigue	100
Fever	80
Allergy	75
Pharyngitis	61
Neurologic	58
Lymphadenopathy	53
Psychologic	50
Arthritis/Arthralgia	45
Myalgia	43
Weight loss	22
Rash	10
Hepatomegaly	7

Table 2: DEMOGRAPHIC FEATURES OF 62 PATIENTS WITH CHRONIC ACTIVE EBV INFECTION

	Arizona	NIH
Sex (M/F)	13/26	11/12
Age (years)	24 (3 - 54)	35 (21 - 48)
Duration of symptoms (years)	6 (1 - 38)	5.7 (1.8 - 17)
Heterophile +	10 (25%)	11 (48%)

Serological Findings

In Arizona the geometric mean titer of IgG antibodies to the EBV capsid antigen (VCA) was 258 in patients versus 32 in age-matched controls (30 of 33 were EBV positive) ($p < 0.001$). The geometric mean titer of antibodies to early antigens (EA) was 113 versus 22 in controls ($p < 0.001$). Although serologies were performed in 2 different laboratories, analysis of the distribution of various serological findings (Table 3) also demonstrated striking similarities between the patient groups. Anti-EA titers could not be compared in detail due to the different cell lines used for assay and differing starting dilutions. Beside the high anti-VCA and anti-EA titers, the lack of production of anti-EBNA and antibodies defected by the ADCC assay in 15% and 12% of all patients, respectively, is remarkable.

Table 3: NUMBER AND PERCENT OF 62 PATIENTS
WITH SELECTED EBV-SPECIFIC TITERS

Antibody	Reciprocal of Titers	Arizona	NIH	Combined
IgM-VCA	10	3 (8)[1]	5 (22)	8 (13)
IgG-VCA	320	36 (92)	19 (83)	55 (88)
	640	18 (46)	14 (61)	32 (52)
EA - D	10	37 (95)	9 (39)	---
EA - R	10	2 (5)	18 (78)	---
EA (D or R)	10	---	19 (83)	---
	20	39 (100)	16 (70)	55 (88)
	40	36 (92)	---	---
EBNA	5	8 (21)	7 (30)	15 (24)
ADCC	240	8 (22)[2]	4 (36)[3]	12 (33)

1   percent of patients studied
2   n = 36
3   n = 11

Immune Function

Phenotypic analyses demonstrated no lasting deficits among or between the patients and normal laboratory control values. The percentage of OKT_4-positive cells in 10 NIH patients was elevated (57.1 ± 2.8% vs 42 ± 1.1) whereas in Arizona the values were 46 ± 7 vs 44 ± 8, respectively. One Arizona patient had an inversion of the T helper/suppressor ratio (0.7) during one of multiple exacerbations. Mitogen responses were likewise normal to both B and T cell mitogen in 32 and 13 patients tested, respectively. Again, the same Tucson patient had diminished response to PHA only on one occasion. No HLA or DR restriction appeared to be present in the 21 cases examined.

Circulating immune complexes were studied by CIq and Raji cell assays in 15 of the NIH patients. Eleven had elevated levels in one or the other assay. Six patients had CIq levels greater than normal (10%); these values were 12, 13, 14, 18, 19 and 24% binding activity. Eight had elevated Ti/Ui ratios (greater than 1.0) in the Raji cell assay; values here ranged 1.1 to 1.6. These values, although elevated, are less than usually seen in diseases in which circulating immune complexes are associated with clinical disease.

Despite previous reports of high levels of circulating interferon (IFN) in diseases with symptoms similar to those of these patients, i.e., chronic infection and immunodeficiency states, circulating IFN α was not present in 50 of 51 patients studied. In one patient and in controls with acute infectious mononucleosis (IM) levels of IFN α were detected (Table 4). The presence of 2'-5' oligoadenylate (A) synthetase activity in cells of all 5 patients studied, however, suggests that IFN-associated pathways may have been active in these individuals. Since induction of 2-5 A synthetase is not wholly dependent on IFN, this matter remains unsettled.

Table 4: INTERFERON STUDIES

Serum IFN Levels

CAEBV	Acute IM
n = 51	n = 7
Positive: 1/51	Positive: 7/7
Range: 17-81 i.u./ml	Range: 9-27 i.u./ml

2-5 Synthetase

CAEBV	Normals
n = 5	n = 9
Range = 38 (26 - 54) units	Range = 16 (8 - 31) units

T cell suppression of PWM induced Ig production occurs during the acute stages of IM with asymptomatic seropositive individuals demonstrating less than 10% suppression. T cells of eighteen of nineteen CAEBV patients showed significant suppressive (mean ± SEM = 52 ± 6.8%) activity suggesting active disease.

Control of EBV infection is at least in part dependent upon lymphokine-mediated T and NK cell regulation of proliferating B cells. Therefore we examined in vitro production of IL2 and IFN γ as well as NK activity in 16 patients. Table 5 shows decreased ConA-PMA induced IL-2 and IFN γ production in CAEBV patients. PHA-PMA induced production of these lymphokines was also decreased. A strong correlation between decreased IL-2 and IFN-γ production was present (R = 0.75, p < 0.001) when controls were compared to patients.

Table 5: MEAN (± S.E.M.) IN VITRO IL-2
AND IFN γ PRODUCTION

	IL-2 units ConA/PMA	IFN γ (log U/ml) ConA/PMA	PHA
Patient	$1.7 \pm 0.69(16)^1$	$3^{2.77}(13)$	$3^{2.25}(8)$
Control	$2.58 \pm 1.28(19)^2$	$3^{5.75}(4)$	$3^{4.4}(15)^3$

1 number of subjects studied
2 significant differences between controls and patients ($p < 0.001$)
3 significant differences between controls and patients ($p < 0.05$)

NK activity was also decreased in patients versus controls when whole mononuclear cell preparations were tested (Table 6). The reduced responses were seen at 3 effector to target ratios and when the data were analyzed as percent specific release or lytic units per 10^6 cells. When the preparation was enriched for large granular lymphocytes (LGL) by Percoll gradient centrifugation, NK activity was comparable between patient and controls.

Table 6: MEAN (± S.E.M.) NATURAL KILLER CELL ACTIVITY

	CAEBV (n = 11)	Controls (n = 33)[1]	
Unfractionated	16 ± 9.6	43 ± 31.8	$P < .01$
Fractionated	110 ± 64.7	162 ± 60.7	NS

[1] in lytic units per 10^6 cells

		% SPECIFIC RELEASE (n = 11)	(n = 27)	
Unfractionated	10:1	18 ± 8	33 ± 12	$P < .005$
	5:1	11 ± 6	23 ± 10	$P < .005$
	25:1	6 ± 4	14 ± 7	$P < .01$
Fractionated	10:1	54 ± 16	61 ± 13	NS
	5:1	41 ± 16	51 ± 17	NS
	2.5:1	29 ± 4	42 ± 21	NS

DISCUSSION

In contrast to reported findings in patients with acute infectious mononucleosis, patients with chronic active EBV infection had normal (or elevated) T cell subset numbers and ratios and mitogen responses and essentially no circulating interferon. 2-5 A synthetase activity was higher in patients than in controls, in agreement with findings in similar patients (Morag et al., 1982).

The EBV-VCA IgG and EA titers were elevated, but 12 - 15% of patients failed to produce antibodies to EBNA or antibody capable of supporting ADCC activity. Whether this is a failure of the immune system to properly control the infection or due to variable viral gene product expression can only be speculated at this time.

Although seemingly divergent, four apparent alterations in immune function seen here may be linked and if so could help us understand the pathophysiology of the illness observed in these patients. The persistence of T cell-mediated suppression of immunoglobulin production and decreased IL-2 and IFN- γ production may be attributed to the same T-cell population (Papermaster et al., 1983). Patient NK activity was restored to normal levels when preparations enriched for large granular cells were used; control of NK activity has also been linked to T-cell regulation (Tarkkanen et al., 1983). Since each of these parameters has been shown to be important in control of acute EBV infections, their persistence adds support to antibody data that an ongoing infection may be responsible for the symptoms displayed by the patients.

Future studies will require longitudinal analysis of these immune factors, virus excretion, identification of virus strain, determination of mediators which are known to elicit these symptoms, and evaluation of the role allergy may play in these processes.

REFERENCES

DuBois, R.E., Seeley, J., Brus, I., Sakamoto, K., Ballow, M., Harato, S., Bechtold, T., Pearson, G., Purtillo, D. Chronic mononucleosis syndrome. South. Med. J. 77:1376-1381, (1984).

Fleisher, G., Bolognese, R. Persistent Epstein-Barr virus infection and pregnancy. J. Infect. Dis. 147:982-986 (1983).

Henle, W., Henle, G., Horowitz, C.A. Epstein-Barr virus specific diagnostic tests in infectious mononucleosis. Hum. Path. 5:551-565 (1974).

Joncas, J., Lapointe, N., Gernais, F., Leyritz, M., Wills, A. Unusual prevalence of antibodies to Epstein-Barr virus early antigen in ataxia telangiectasia. Lancet 1:1160 (1977).

Jones, J.F., Ray, C.G., Minnich, L.L., Hicks, M.J., Kibler, R.A., Lucas, D.O. Evidence for active Epstein-Barr virus infection in patients with persistent, unexplained illnesses: Elevated anti-early antigen antibodies. Ann. Intern. Med. 102:1-7 (1985).

Kibler, R., Lucas, D.O., Hicks, M.J., Poulos, B.T., Jones, J.F. Immune function in persistent Epstein-Barr virus infection. J. Clin. Immunol. In Press.

Lange, B., Henle, W., Meyers, J.D., Yang, L.C., August C., Koch, P., Arbeter, A., Henle, G. Epstein-Barr virus related serology in marrow transplant recipients. Int. J. Cancer 26:151-157 (1980).

Morag, A., Tobi, M., Ravid, Z., Revel, M., Schattner, A. Increased (2' - 5')-oligo A synthetase activity in patient with prolonged illness associated with persistant Epstein-Barr virus infection. Lancet 1:744 (1982).

Papermaster, V., Torres, B.A., Johnson, H.M. Evidence for suppressor T-cell regulation of human gamma interferon production. Cell Immunol. 79:279-287 (1983).

Pearson, G.R., Johansson, B., Klein, G. Antibody dependent cellular cytotoxicity against Epstein-Barr virus associated antigens in African patients with nasopharyngeal carcinoma. Int. J. Cancer 22:120-125 (1978).

Ray, C.G., Gall, E.P., Minnich, L.L., Roediger, J., DeBenedetti, C., Corrigan, J.J. Acute polyarthritis associated with active Epstein-Barr virus infections. JAMA 248:2990-2993 (1982).

Straus, S.E., Tosato, G., Armstrong, G., Lawley, T., Preble, O.T., Henle, W., Davey, R., Pearson, G., Epstein, J., Brus, I., Blaese, R.M. Persisting illness and fatigue in 23 adults: serologic and immunologic evidence for chronic active Epstein-Barr virus infection. Ann. Intern. Med. 102:7-16, (1985).

Sumaya, C.V.   Endogenous reactivation of Epstein-Barr virus infections. J. Infect. Dis. 135:374-379 (1977).

Tarkkanen, J., Saksela, E., von Willebrand, E., Lehtonen, E. Suppressor cells of the human NK activity; characterization of the cells and mechanism of action. Cell Immunol. 79:265-278 (1983).

Tobi, M., Morag, A., Ravid, Z., Chowess, I., Feldman-Weiss, V., Michaeli, Y., Ben-Chetrit, E., Shalit, M., Knobler, H. Prolonged atypical illness associated with serological evidence of persistent Epstein-Barr virus infection. Lancet 1:61-63 (1982).

4

REVERSAL OF COMMON VARIABLE HYPOGAMMAGLOBULINEMIA-

ASSOCIATED SUPPRESSOR CELL ACTIVITY BY SPECIFIC

CARBOHYDRATES

G. Tosato[1], S.E. Pike[1], and R.M. Blaese[2]

FDA, Bldg. 29-520[1] and NIH Bldg. 10-6B05[2]

Bethesda, MD 20205

Summary

Patients with common variable hypogammaglobulinemia (CVH) often have circulating suppressor cells that profoundly inhibit normal immunoglobulin (Ig) production in vitro. We have examined the nature of signals operating in the interaction between CVH-associated suppressor cells and their targets, and explored the possibility that lectin-like receptor molecules and their specific sugars might contribute to the specificity of these interactions. When D-mannose was added to suppressed cocultures of normal PWM-activated mononuclear cells and patient T cells a significant enhancement of Ig production was observed. N-acetyl-D-glucosamine had a similar enhancing effect when added to suppressed cocultures of normal mononuclear cells and patient non-T cells. Since D-mannose and N-acetyl-D-glucosamine did not enhance Ig production by normal cells when cultured alone, these sugars were interfering with the process of suppression. In contrast, a number of other saccharides had no effect on suppression. These results suggest that selected saccharides may represent critical components in the cellular receptors involved in suppressor cell interactions.

Introduction

Little is known about the molecular nature of signals operating in the interaction between suppressor cells and their targets. In other systems it has been shown that specific cell surface carbohydrates may serve as recognition and interaction structures (Shen et al., 1968; Reisner et al., 1977; Ofek et al., 1977; Muramatsu et al., 1979; Sharon, 1983). We have recently reported that D-mannose and some of its derivatives can significantly reverse inhibition of immunoglobulin production mediated by suppressor T cells activated during the course of acute EBV-induced infectious mononucleosis (Tosato et al., 1983). This finding indicated that D-mannose and selected mannose derivatives were interfering with the process of suppression, and suggested that certain carbohydrates may represent critical components involved in physiologic suppressor cell interactions.

In the present study we have tested the hypothesis that specific carbohydrate molecules might be involved in other suppressor cell interactions, and looked at the in vitro effects of a panel of sugars on suppression mediated by T cells as well as non-T cells from a selected group of patients with common variable hypogammaglobulinemia (CVH).

Patients and Methods

Mononuclear cells were obtained from peripheral blood of normal individuals and 11 patients with CVH. These patients were known to have circulating suppressor cells (Waldmann et al., 1974). Selected clinical and laboratory data relating to these patients are shown in Table I. Of interest, patient no. 2 developed hypogammaglobulinemia 1 year earlier, following an illness clinically and serologically defined as acute Epstein-Barr virus (EBV)-induced infectious mononucleosis (Henle et al., 1974). B cell-enriched and T cell-enriched cell subsets were obtained by incubating the mononuclear cells with AET-sensitized sheep red blood cells and separating the rosette forming cells on Ficoll-Hypaque gradients, as described (Tosato et al., 1983). Normal and patient mononuclear cells (1×10^6) were cultured alone in the presence of pokeweed mitogen (PWM) or EBV. In addition, normal mononuclear cells (0.5×10^6) were cultured in the presence of PWM either alone or mixed in culture with patient T (0.5×10^6) or non-T cells (0.25×10^6). At the end of a 6-8 day culture period the number of immunoglobulin (Ig) secreting cells was determined.

TABLE I

Selected clinical and laboratory data on the patient
population

	Age	Sex	Serum Immunoglobulin (I.U./ml)			Duration of Illness
			IgG	IgA	IgM	(yrs)
1	37	M	23	<20	65	33
2	6	M	24	<20	<20	5
3	66	F	30	37	<20	46
4	23	M	<20	<20	22	18
5	26	F	35	<20	<20	1
6	57	M	<20	<10	76	40
7	39	F	22	<20	<20	1
8	50	M	22	<20	<20	27
9	61	M	25	<20	26	30
10	62	F	34	<20	58	11
11	57	F	32	<20	<20	5

Normal range   72–204   30–261   36–266

Results

Cultures of patient mononuclear cells ($1x10^6$)
produced significantly lower numbers of Ig secreting cells
than did cultures of normal cells. The geometric mean
response (x/÷ S.E.M.) for the 11 patients was 111 (1.56)
Ig secreting cells in the presence of PWM, and 334 (2.25)
in the presence of EBV.   Control values were 13,857 (1.13)
Ig secreting cells with PWM and 5,392 (1.14) with EBV.
    Ten of the patients had circulating T cells that
profoundly inhibited Ig production by normal cocultured
mononuclear cells.   Thus, while normal PWM activated
mononuclear cells ($0.5x10^6$) produced an average of 13,680
(1.14) Ig secreting cells when cultured alone, cocultures
of the same mononuclear cells with patient T cells (0.5 x
10^6) produced an average of 1,139 (1.72) Ig secreting
cells (91 percent suppression).   This T cell inhibition
was radiosensitive, since cocultures of the normal
mononuclear cells with patient T cells irradiated with
2000 R prior to culture produced 11,272 (1.24) Ig
secreting cells.

Addition of a variety of carbohydrates to the
suppressed cocultures, including L-rhamnose; D-galactose,
L-fucose, gentiobiose, cellobiose, N-acetyl-D-glucosamine,
mannosamine and L-mannose, had no significant effect on
the Ig-secreting cell response (data not shown). By
contrast, as shown in Fig. 1, D-mannose (25 mM final
concentration) markedly reversed T cell suppression. In
the same experiments, D-mannose had little or no effect on
the normal Ig secreting cell response, indicating that
this sugar was affecting the process of suppression rather
than directly stimulating the responder mononuclear cells.
Thus, similar to the results previously obtained with
infectious mononucleosis-associated suppression (Tosato et
al., 1983), D-mannose consistently reversed inhibition
mediated by CVH T cells.

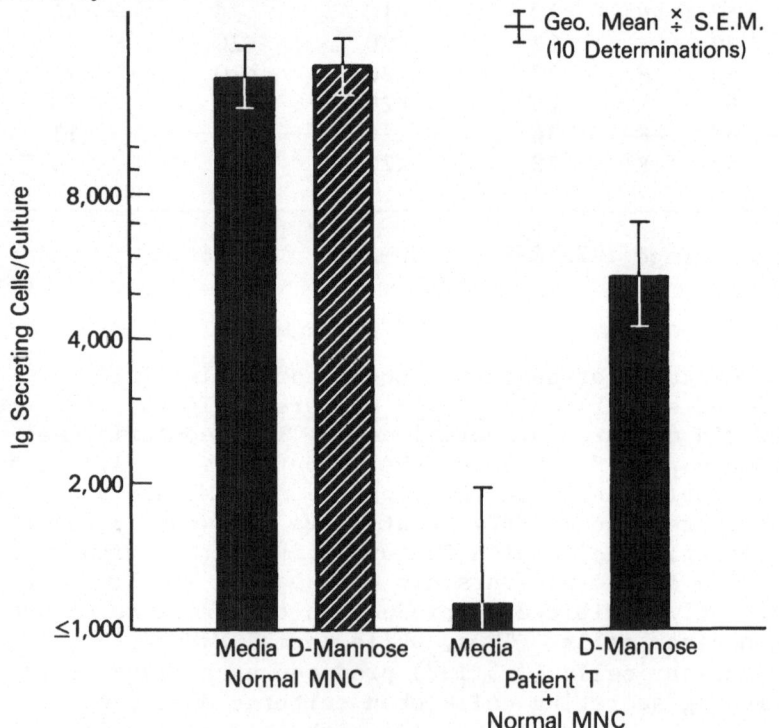

Fig 1. Reversal of CVH-associated T cell suppression by
D-mannose. Mononuclear cells (0.5×10^6) from 10
normal individuals were cultured alone or mixed
with T cells (0.5×10^6) from 10 patients with CVH,
in the presence of PWM only or PWM and
D-mannose (25 mM).

TABLE II

Non-T cell suppression in patients with common variable
hypogammaglobulinemia

Cultures*	Ig secreting cells/culture		
	1.	2.	3.
Normal MNC alone	23315	11100	8190
Cocultures of normal MNC and patient non-T cells	3003	1950	40

*Normal mononuclear cells (MNC) were cultured either alone
$(0.5x10^6)$ in the presence of PWM or mixed in coculture
with patient non-T cells $(0.25x10^6)$.

     Three of the 11 patients (no. 1, 5 and 6) had in
their peripheral blood non-T cells capable of inhibiting
normal Ig production by over 75% (Table II). This was
associated with T cell suppression in two of these
patients. We first examined simultaneously the effects of
a panel of monosaccharides on T cell and non-T
cell-mediated suppression, both present in the same
patient. As shown in Fig. 2, both patient cell
populations, T and non-T, suppressed markedly the normal
mononuclear cell response to PWM. The addition of
D-mannose significantly reversed T cell suppression but
had no significant effect on non-T cell suppression. By
contrast, N-acetyl-D-glucosamine (25 mM) had no
significant effect on T cell mediated suppression, but
substantially reversed inhibition mediated by non-T cells.
Thus, N-acetyl-D-glucosamine appears to selectively
inhibit the process of suppression mediated by non-T
cells. This was a consistent finding in all 3 patients
who had evidence of a marked non-T cell mediated
suppression (Fig. 3). Also, since N-acetyl-D-glucosamine
had little or no effect on the normal response to PWM,
this monosaccharide most likely interferes with the
process of non-T cell suppression.

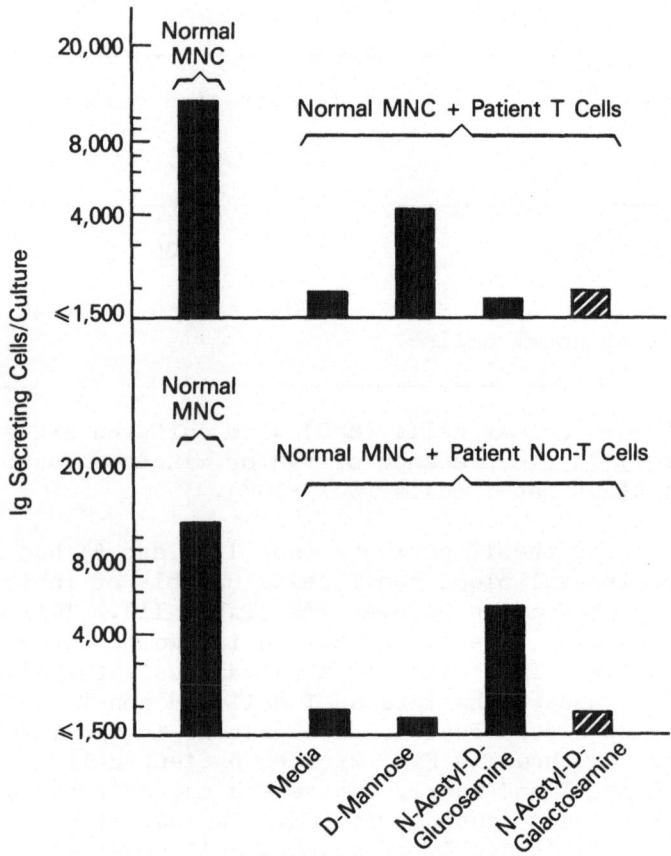

Fig 2. Carbohydrate specificity of T cell and non-T cell suppression associated with CVH patients. Normal mononuclear cells (0.5×10^6) were cultured either alone in the presence of PWM or mixed with either T cells (0.5×10^6) or non-T cells (0.25×10^6 from a patient with CVH. Selected monosaccharides were added to individual cocultures at the final concentration of 25 mM.

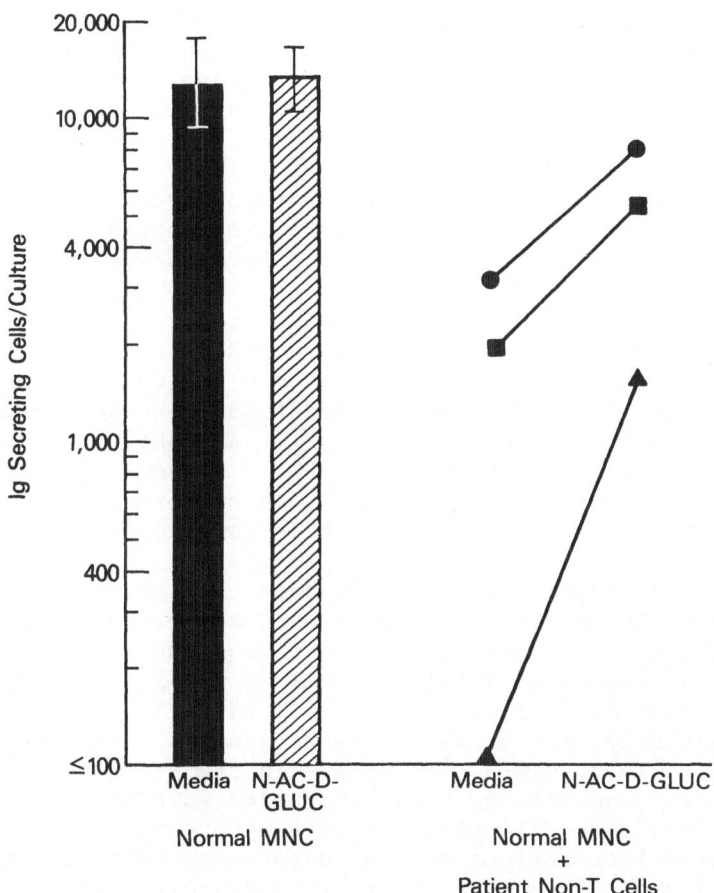

Fig 3. Reversal of CVH–associated non–T cell suppression by N–acetyl–D–glucosamine. Mononuclear cells (0.5×10^6) from 3 normal individuals were cultured alone or mixed with non–T cells (0.25×10^6) from 3 patients with CVH, in the presence of PWM only or PWM and N–acetyl–D–glucosamine (25 mM).

Discussion

Common variable hypogammaglobulinemia (CVH) is an
acquired disorder of largely unknown etiology
characterized by recurrent bacterial infections and low
serum immunoglobulin levels. Recent reports have linked
certain cases of acquired hypogammaglobulinemia with
EBV-induced infectious mononucleosis (Provisor et al.,
1975; Purtilo, 1981; Greally et al., 1983). During the
acute phase of this viral illness suppressor T cells
become activated that profoundly inhibit normal Ig
production (Tosato et al., 1979). Persistence of
abnormally elevated numbers of suppressor T cells beyond
the acute disease might be responsible for the occurrence
of hypogammaglobulinemia (Tosato and Blaese, 1984). We
have recently reported that D-mannose and certain mannose
derivatives markedly reverse infectious mononucleosis-
associated T cell suppression, while a number of other
sugars do not (Tosato et al., 1983). Since D-mannose has
little or no effect on Ig production by normal
lymphocytes, we concluded that this monosaccharide was
interfering with the process of suppression.

To investigate further the nature of signals
operating in the interaction between suppressor cells and
their targets we have extended our studies to examine the
effect of sugars on T cell as well as non-T cell
suppression associated with certain patients with CVH.
Interestingly, one of the patients studied here developed
hypogammaglobulinemia following an acute viral illness
clinically and serologically diagnosed as EBV-induced
infectious mononucleosis. We have hypothesized that
selective interactions between suppressor cells and their
targets might depend upon the presence of a specific
lectin-like structure on one cell surface and a
correspondent carbohydrate molecule on the other
interacting cell. If this was the case, addition of large
quantities of a specific sugar to the suppressed cultures
should competitively saturate the lectin's binding
capacity, and could thus interfere with suppression.
One would also expect that suppression involving
different types of interacting cells would have different
sugar specificity. We have shown that D-mannose
consistently reverses T cell mediated suppression but has
no effect on suppression mediated by non-T cells. In
contrast, N-acetyl-D-glucosamine always markedly reversed
non-T cell mediated suppression but had no effect on T

cell mediated suppression. These results suggest that
certain carbohydrates may represent critical components in
the receptor structures involved in suppressor cell
interactions. These findings also suggest that different
types of suppression may differ in their sugar
specificity.

It is not known what structures bear the
sugar-specific receptors. These endogenous lectins could
be located either on the target cells for suppression, or
on the suppressor cells, or possibly on the soluble
mediators they may produce.

Further study will be necessary to define the role of
saccharides in immune cell interactions, but our studies
clearly indicate that selected carbohydrates consistently
counteract specific types of suppression.

References

Grabel, L.B., Rosen, S.D., and Martin, G.R.,
Teratocarcinoma stem cells have a cell surface
carbohydrate-binding component implicated in cell-cell
adhesion. Cell 17, 477-484 (1979).

Greally, J.F., Lowlor, E., Lyons, J., Rickinson, A.,
Sakamoto, K., and Purtilo, D., Acquired
hypogammaglobulinemia following infectious mononucleosis.
J. Clin. Immunol. 3, 207-211 (1983).

Henle, W., Henle, G.E., and Horowitz., Epstein-Barr virus
specific diagnostic tests in infectious mononucleosis.
Hum. Pathol. 5, 551-565 (1974).

Muramatsu, T., Gachelin, G., Damonneville, M., Delabre,
C., and Jacob, F., Cell surface carbohydrate of embrional
carcinoma cells: polysaccharidic side chains of F9
antigens and of receptors to two lectins FBP and PNA.
Cell 18, 183-191 (1979).

Ofek, I., Mirelman, D., and Sharon, N., Adherence of
Escherichia coli to human mucosal cells mediated by
mannose receptors. Nature 265, 623-625 (1977).

Provisor, A.J., Lacuone, J.J., Chilcote, R.R., Neiburger, R.G., Crussi, F.G., and Baener, R.L., Acquired agammaglobulinemia after a life-threatening illness with clinical and laboratory features of infectious mononucleosis in three related male children. N. Engl. J. Med. 293, 62-65 (1975).

Purtilo, D.T., X-linked lymphoproliferative syndrome. An immunodeficiency disorder with acquired agammaglobulinemia, fatal infectious mononucleosis or malignant lymphoma. Arch. Pathol. Lab. Med. 105, 119-121 (1981).

Reisner, Y., Gachelin, G., Dubois, P., Nicholas, J.F., Sharon, N., and Jacob, F., Interaction of peanut agglutinin, a lectin-specific for non-reducing terminal D-galactosyl residues, with embrional carcinoma cells. Dev. Biol. 61, 20-27 (1977).

Sharon, N., Lectin receptors as lymphocyte surface markers. Adv. Immunol. 34, 213-298 (1983).

Shen, L., Grollman, E.F., and Ginsburg, V., An enzymatic basis for secretor status and blood group substance specificity in humans. Proc. Natl. Acad.Sci. USA 59, 224-237 (1968).

Tosato, G., Magrath, I., Koski, I., Dooley, N., and Blaese, R.M., Activation of suppressor T cells during Epstein-Barr-virus-induced infectious mononucleosis. N. Engl. J. Med. 301, 1133-1137 (1979).

Tosato, G., Pike, S.E., and Blaese, R.M., Reversal of infectious-mononucleosis-associated suppressor T cell activity by D-mannose. J. Exp. Med. 158, 1048-1060 (1983).

Tosato, G., and Blaese, R.M., Epstein-Barr virus infection and immunoregulation in man. Adv. Immunol. In press, 1984.

Waldmann, T.A., Durm, M., Broder, S., Blackman, M., Blaese, R.M., and Strober, W., Role of suppressor T cells in pathogenesis of common variable hypogammaglobulinemia. Lancet 2, 609-613 (1974).

5

UNUSUAL PRIMARY TUMORS OF BRAIN AND LUNGS ASSOCIATED

WITH EPSTEIN-BARR VIRUS (EBV)

J. Joncas, F. Ghibu, N. Lapointe, L. Begin, J. Michaud,
P. Simard, P. Benoit, G. Rivard, J. Demers and R. Raymond
Departments of Microbiology & Immunology, Pediatrics and
Pathology, Sainte-Justine Hospital and Jewish General
Hospital, Montreal, Canada H3T 1C5

The association of African Burkitt's lymphoma (BL) and
nasopharyngeal carcinoma (NPC) with EBV is well known.
Opportunistic EBV associated polymorphic B cell lymphoma
have also been observed in renal transplant patients (Hanto
et al., 1983) and in bone marrow transplant patients
(Starzl et al., 1984). Similar tumors in congenital or acquired immunodeficiency syndromes have rarely been reported
(Saemundsen et al., 1981; Reece et al., 1981; Blanc et al.,
1984). Such tumors have also been reported very rarely in
apparently normal individuals (Hochberg et al., 1983). We
have recently seen a primary monoclonal polymorphic B cell
lymphoma of the brain in a 3 year old girl with congenital
combined immunodeficiency associated with adenosine deaminase deficiency, a similar lung lymphoma in a 4 year old
girl with acute lymphocytic leukemia in remission and a
primary non keratinizing squamous cell carcinoma of the
lingula in a 43 year old women, all associated with EBV.

MATERIAL AND METHODS

Serology: Serum titres of IgM and IgG antibodies to VCA
and IgG antibodies to EA (both diffuse and restricted components) were measured by indirect immunofluorescence, as
previously described (Henle and Henle, 1966). Antibodies to
EBNA were assayed by anticomplement immunofluorescence
(Reedman and Klein, 1973). Cytomegalovirus antibodies were
measured by the complement fixation test.
Immunofluorescence for viral antigens in tissue: We

tested for EBNA antigen on imprints and frozen sections of
biopsy or autopsy specimens by anticomplement immunofluorescence with control sera known to be EBNA-positive and
EBNA-negative. The reference sera were negative for anti-
nuclear antibodies on Molt-4 cells.

Establishment of lymphoid cell lines: Bone marrow cells
or tumor cell homogenates were centrifuged on Ficoll-Hypaque gradients. The separated lymphocytes were grown in
suspension cultures in Falcon plastic flasks using RPMI 1640
medium supplemented with 15% foetal calf serum, penicillin
100 units/ml and gentamycin 50 µg/ml or cocultivated with
cord blood lymphocytes.

Isolation of virus: Human fibroblasts WI-38 (ATCC, CCL-
75) and HEL, another human diploid fibroblast cell strain
established in our laboratory were used to propagate cytomegalovirus from the urine,throat and lung biopsy. These
cells were maintained in Eagle's minimal essential medium
(MEM) with 10% foetal calf serum and antibiotics in the
concentration mentioned above. After reaching confluence,
the amount of foetal calf serum was reduced to 4%. The
identification of CMV was done by anticomplement immunofluorescence using a reference serum positive for CMV antibodies but negative for Herpes simplex and Varicella-Zoster
antibodies on coverslips of human embryo fibroblasts infected with the isolated viruses.

Cell surface and cytoplasmic immunoglobulin detection:
The detection of cytoplasmic or surface immunoglobulin was
done on acetone-fixed cells, both by direct immunofluorescence using Hyland fluorescein-conjugated anti-human IgM
(µ-chain specific) and anti-human IgG (γ-chain specific)
and anti K, anti λ light chain specific antisera.

RESULTS

CASE 1
A Caucasian girl with congenital combined immunodeficiency associated with adenosine deaminase deficiency died
at the age of 3 years from a monoclonal (IgM and λ light
chain positive) primary B cell lymphoma of the brain. The
tumor was detected by CAT scan and identified by brain
biopsy. In the 9 months preceding death this child had a
primary cytomegalovirus (CMV) and Epstein-Barr virus (EBV)
infection which were confirmed by virus isolation and seroconversion to both viruses (Table I). These infections
were probably acquired from blood and blood products

TABLE I - SUMMARY OF PATHOLOGICAL AND SEROLOGICAL FINDINGS

	CASE 1	CASE 2	CASE 3
EBNA in TUMOR CELLS	+	+	Unsuitable material
EBV SEROLOGY			
VCA-IgG	10	320	10,240
IgA	ND		2,560
EA (D or R)	<5	80(R)	1,280(D)
EBNA	±5	<5	80
TYPE of TUMOR	Polymorphic B cell lymphoma	Polymorphic B cell lymphoma	Undifferentiated non keratinizing squammous cell carcinoma
CLONALITY	Monoclonal	Monoclonal	ND
OUTCOME	Death	Remission	Death

EBNA: Epstein-Barr virus (EBV) nuclear antigen.
VCA-IgG: EBV viral capsid antigen (Immunoglobulin G anti-
          body against VCA).
EA-D/R: EBV early antigen diffuse/restricted.
ND: Not done.

repeatedly given to this patient to correct the adenosine deaminase deficiency and to prevent opportunistic infections. CMV was first isolated from urine 9 months before death and then from a lung biopsy 6 months later. Seroconversion to CMV occurred at or before the age of 2 years and seroconversion to EBV 3 to 4 months before death. A EB nuclear antigen (EBNA) positive cell line was established spontaneously from the brain tumor biopsy obtained 2 months before death. The fresh tumor cells (imprints and frozen sections) were also EBNA positive by anticomplement immunofluorescence (ACIF). In addition several endothelial cells lining blood vessels within the tumor were positive for CMV by ACIF immunofluorescence using a CMV positive EBV negative, antinuclear antibody negative human serum. There was no response of the tumor to Acyclovir 1500 mg/M2/day given IV for 15 days nor to local radiotherapy both initiated 3 weeks before death. At autopsy multiple lymphomatous masses were found in brain, lungs, liver, mesenteric and mediastinal lymph nodes, spleen, kidney and adrenals (Fig. 1,2).

CASE 2

A 4 year-old Caucasian girl with acute lymphocytic leukemia in remission developed, while on maintenance anti-leukemic therapy, progressively enlarging lung infiltrates and masses which failed to respond to antibacterial and antifungal therapy. An open lung biopsy, eventually done (Fig. 3), revealed a polymorphic B cell monoclonal lymphoma (K light chain). The lymphoma was EBNA positive by anticomplement immunofluorescence. Serology revealed a EBV-VCA antibody titer of 1/320, a EBV-EA titer of the restricted type (EA-R) only, of 1/80 but no detectable EBNA antibodies. Maintenance anti-leukemic therapy was discontinued and 2 successive courses of IV Acyclovir 1500 mg/M2/day for 15 days at 2 month interval were given. The lung masses have decreased in size on X-ray and CAT Scan (Fig. 4,5).

CASE 3

A nodule measuring approximately 2 cm in diameter was first detected in the lingula in 1978 on routine chest X-ray of a 38 year old female nurse born in the Philippines. The patient was a non smoker and had no contributory family or personal past history. The nodule remained stable in size for 2 years but then suddenly doubled in size. A needle biopsy revealed malignant cells and extensive resections of the left upper lobe, lingula and segments of the left lower lobe were done. Pathological examination revealed a poorly differentiated non keratinizing squamous cell carcinoma by light and electron microscopy (Fig. 6,7). The

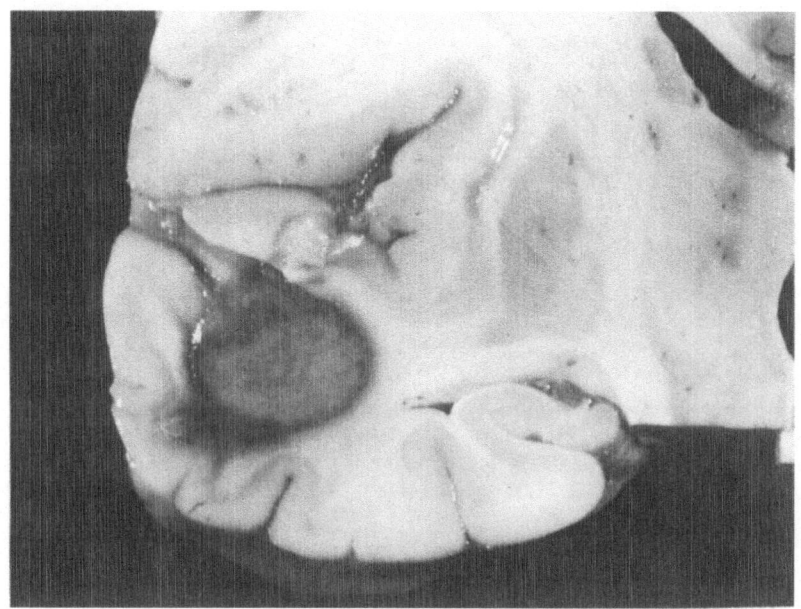

Fig. 1: Macroscopic appearance of brain tumor at autopsy (Case 1)

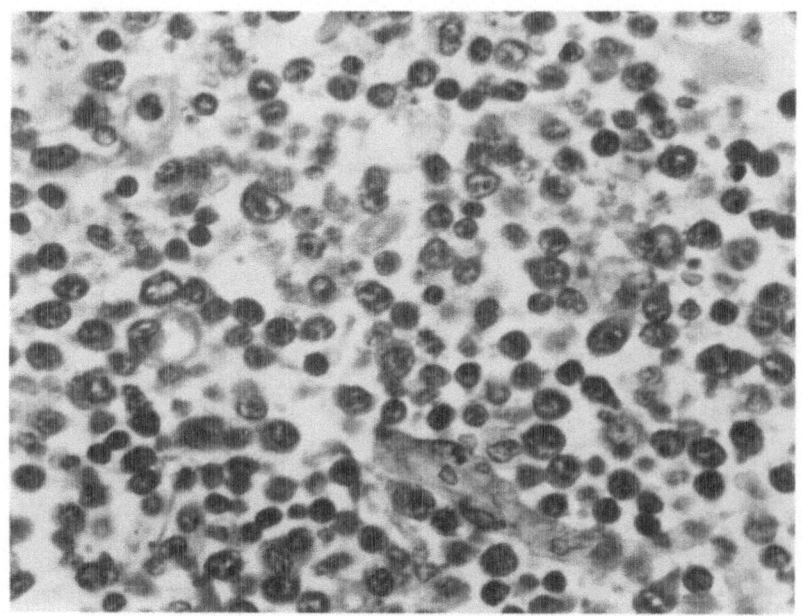

Fig. 2: Polymorphic B cell lymphoma of brain. Original Magnification: 630X (Case 1)

58

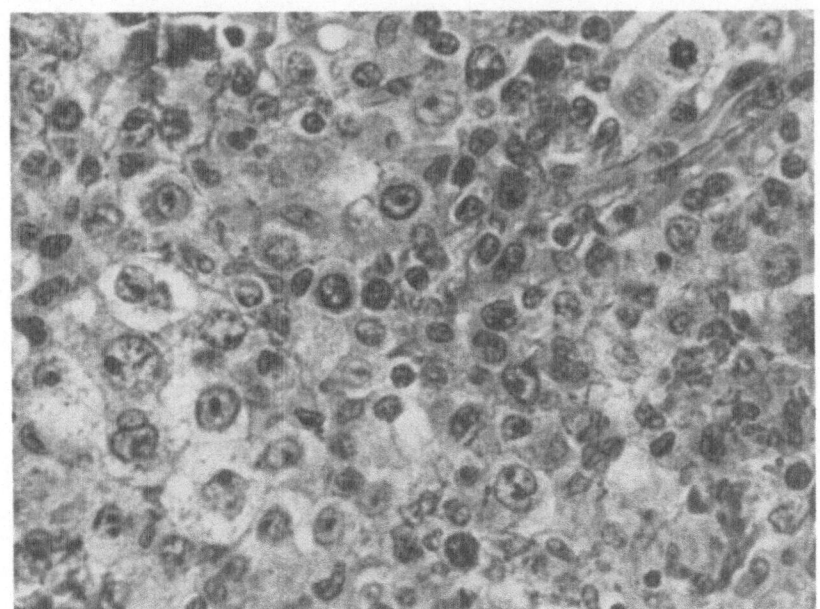

Fig. 3:   Lung lymphoma (biopsy).   Original magnification
          630X (Case 2)

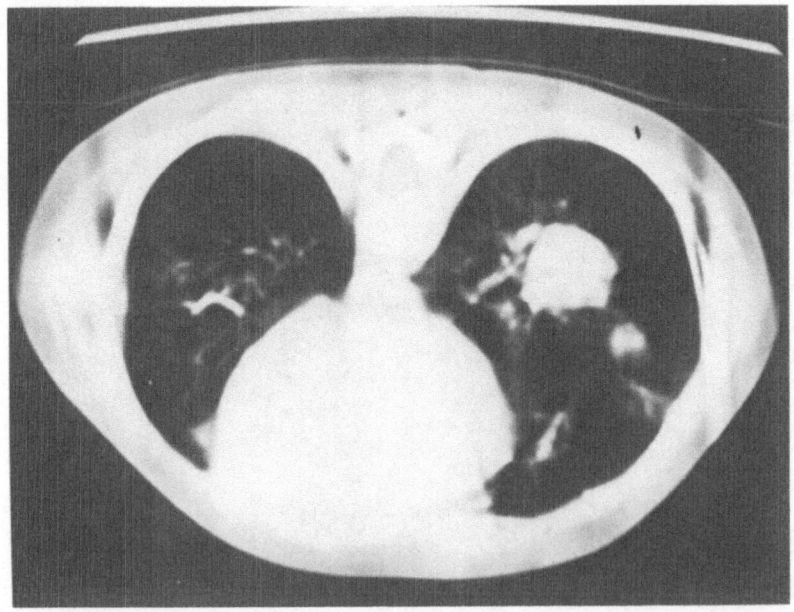

Fig. 4:   Large tumor mass on Cat Scan in right lung before
          therapy (Case 2)

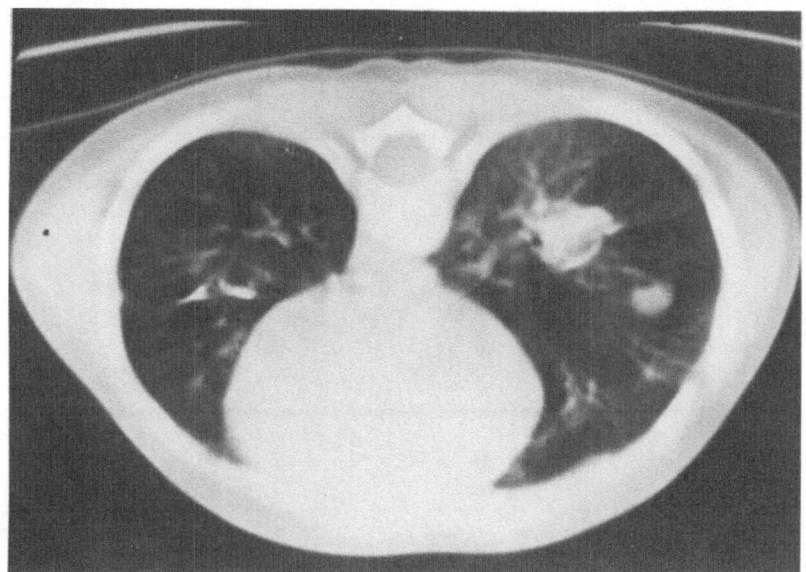

Fig. 5: Reduced size of tumor mass 6 weeks later, after
two, 2 week   courses of Acyclovir (Case 2)

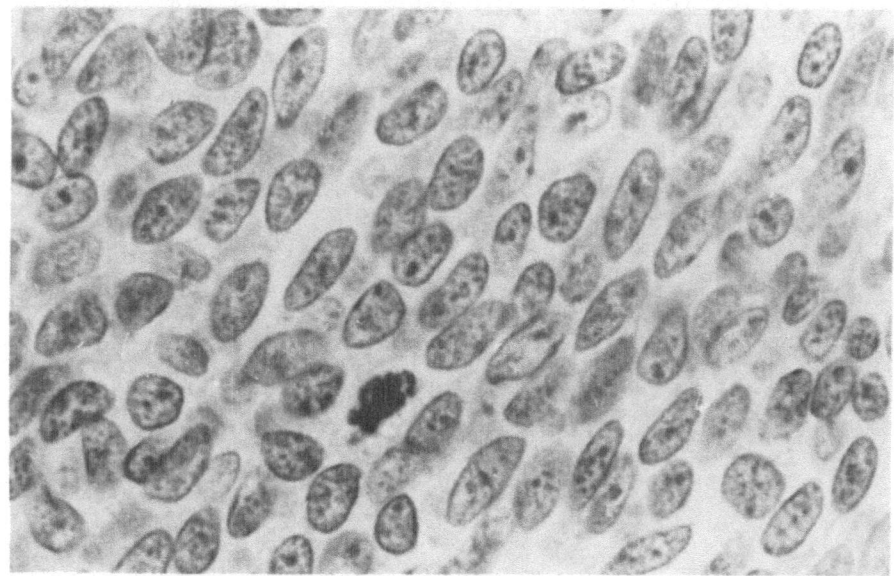

Fig. 6: Light microscopy; the tumor is characterized by a
syncytial arrangement of short spindle cells without differentiation. The chromatin is stipled and
one mitotic figure is present (630X) orig. magnif.

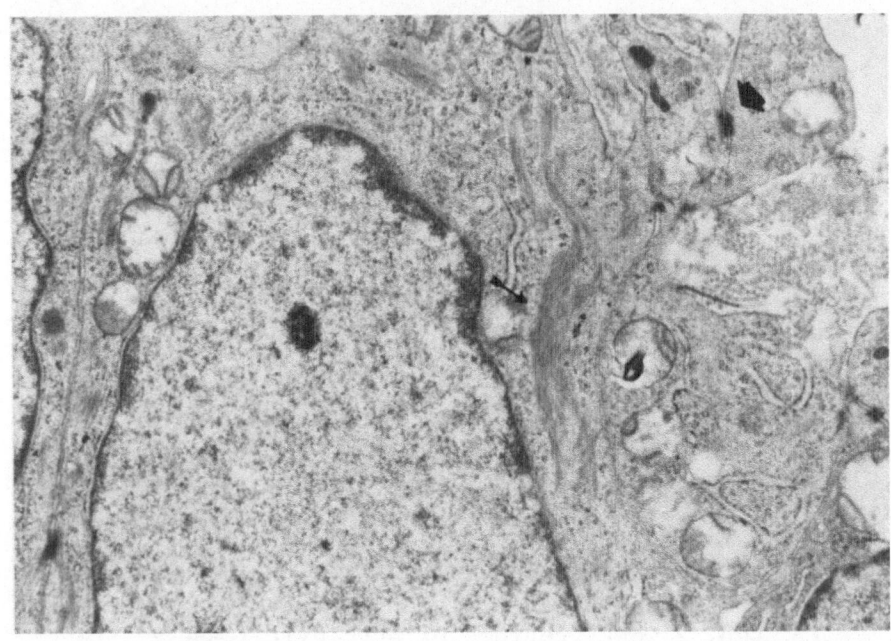

Fig. 7:   Electron microscopy; ultrastructural examination
          revealed multiple desmosomes and intracytoplasmic
          bundles of intermediate tonofilaments, confirming
          the squammous nature of this neoplasm (17100X)
          Arrow: tonofilaments; Arrowhead: desmosomes

patient did well for about 1 year but chest X-ray eventually
disclosed a developing mass in the remaining left lower
lobe. Surgical exploration revealed several pleural nodules
with extension to the pericardium. The pathological findings were the same as those of the original tumor. Surgical
resection was followed by radiotherapy. The patient did
well for approximately 8 months but metastases were noted in
the supraclavicular nodes. These were resected and the area
subsequently irradiated. A careful surgical exploration of
the rhinopharynx failed to disclose any tumor at this site.
Within another 6 months the patient complained of abdominal
pain and a CAT scan revealed metastases in the retroperito-
neal region. Radiotherapy and chemotherapy with cyclophosphamide, adriamycin and vincristine were unsuccesful and
the patient died in 1983 approximately 5 years after the
first nodule had been detected in her lung. Autopsy was
refused. The EBV-VCA IgG antibody titer of 3 sera taken
over the course of approximately one year from July 1982 to
May 1983 rose from 1/2560 to 1/10240 while the corresponding
IgA titer averaged 1/2560. The EBV-EA-D titer remained stable at 1/1280 while the EBNA titer did not exceed 1/80. The
formaldehyde fixed surgical specimen unfortunately did not
allow visualization of EBNA.

DISCUSSION

The first 2 cases represent well documented examples of
EBV associated lymphomas in children with congenital and
acquired immunodeficiency respectively. Both of these
children received blood in therapy of their disease. The
occurrence of these tumors stresses the hazards of transfusion in young EBV seronegative immunocompromised children.
Clinical trials of EBV-CMV hyperimmune immunoglobulin are
presently being considered for immunocompromised children to
whom blood has to be given in an attempt to prevent or minimize the consequences of primary infection by these viruses
in such children. The association of EBV with the tumor in
the 3rd case is only supported by serology since suitable
tissue for specific antigen detection was not available.
The NPC like histology of the tumor however and the high
EBV-VCA IgG and IgA and high EBV-EA-D antibody titers seen
in this patient strongly suggest an association with EBV.

REFERENCES

Blanc, A.P., Routy, J.P., Pedinielli, F.J., Xau, P.F., Miletto, G., Chardon, H., Alliez, B., Payan, M.J., Gambarelli, D., and Jauffret, P., Lymphome cérébral, complication d'un syndrome d'immuno-déficit acquis. La Presse Médicale,14, 884 (1984).

Hanto, D.W., Gajl-Peczalska, J., Frizzera, G., Arthur, D.C., Balfour, H.H., McClain, K., Simmons, R.L., and Najarian, J.S., Epstein-Barr virus (EBV) induced polyclonal and monoclonal B-cell lymphoproliferative diseases occurring after renal transplantation. Ann. Surg., 198, 356-69 (1983).

Henle, G., and Henle, W., Immunofluorescence in cells derived from Burkitt's lymphoma. J. Bacteriol., 91, 1248-56 (1966).

Hochberg, F.H., Miller, G., Schooley, R.T., Hirsch, M.S., Feorino, P., and Henle, W., Central-nervous-system lymphoma related to Epstein-Barr virus. N. Engl. J. Med., 309, 745-48 (1983).

Reece, E.R., Gartner, J.G., Seemayer, T.A., Joncas, J.H., and Pagano, S., Epstein-Barr virus in a malignant lymphoproliferative disorder of B-cells occurring after thymic epithelial transplantation for combined immunodeficiency. Cancer Res., 41, 4243-47 (1981).

Reedman, B.M., and Klein, G., Cellular localization of an Epstein-Barr virus (EBV)-associated complement-fixing antigen in producer and non-producer lymphoblastoid cell lines. Int. J. Cancer, 11, 499-520 (1973).

Saemundsen, A.K., Izzet Berkel, A., Henle, W., Henle, G., Anvret, M., Sanal, O., Ersoy, F., Caglar, M., and Klein, G., Epstein-Barr virus carrying lymphoma in a patient with ataxia-telangiectasia. Br. Med. J., 282, 425-27 (1981).

Starzl, T.E., Nalesnik, M.A., Porter, K.A., Ho., M., Iwatsuki, S., Griffith, B.P., Rosenthal, J.T., Hakala, T.R., Shaw, B.W., Hardesty, R.L., Atchison, R.W., Jaffe, R., and Bahnson, H.T., Reversibility of lymphomas and lymphoproliferative lesions developing under cyclosporin-steroid therapy. The Lancet 1, 583-87 (1984).

6

LYTIC, NON-TRANSFORMING EPSTEIN-BARR VIRUS (EBV) FROM TWO
PATIENTS WITH CHRONIC ACTIVE EBV INFECTION

Carolina Alfieri, Félicia Ghibu, Jean Joncas

Department of Microbiology and Immunology,

Sainte-Justine Hospital, Montreal, Canada

INTRODUCTION

The Epstein-Barr virus (EBV) is the causative agent
of infectious mononucleosis. EBV is also associated with
Burkitt's lymphoma and nasopharyngeal carcinoma, and is
implicated in other disorders such as the X-linked
lymphoproliferative syndrome (Purtilo et al., 1982).

There are two known laboratory strains of EBV
(Menezes et al., 1975; Miller et al., 1975), one designated as the B95-8, or transforming strain, because of
its capacity to transform EBV-negative lymphocytes into
permanently established EBV-positive lymphoblastoid cell
lines. The other is the P3HR-1 or lytic strain which
does not have transforming potential, but will instead
cause lysis of cord blood lymphocytes (CBL) and growth
inhibition of Raji cells. This latter strain also has
the property of inducing the Epstein-Barr virus early
antigen (EA) in Raji cells.

All wild-type strains of EBV characterized to date
are transforming viruses capable of immortalizing CBL and
are therefore akin to the B95-8 biotype. There have, as
yet, been no reports of lytic activity associated with a
wild-type EBV isolate. The study reported here is a
follow-up of a case of chronic active EBV infection in an
adolescent girl and her father with persistent spleno-
megaly, whose throat washings were unable to transform

CBL but which contain an agent capable of inducing EA in Raji cells.

MATERIALS AND METHODS

Patients

Blood and throat washings were obtained periodically from our patient - and from her father, when possible - over a period of five years. At these times the patient also underwent a physical examination. The girl's spleen was invariably palpable 3 to 5 cm below the left costal margin. Recurrent proliferative lesions undergoing ulceration were frequently seen in her mouth. Other clinical data pertinent to these two patients were published elsewhere (Joncas et al., 1984). Particularly interesting was the consistently low T4/T8 ratio, low NK activity and decreased lymphocyte transformation index in response to stimulation by mitogens. However, despite these immunological abnormalities presumed to be acquired, the patient and her father are not prone to opportunistic infections and enjoy relatively normal lives.

Establishment of the Lymphoblastoid Cell Lines

Separation of mononuclear cells from whole, fresh blood was done essentially as previously described (Joncas et al., 1977). After separation on a Ficoll-hypaque density gradient the mononuclear cell fraction was washed three times in Medium RPMI-1640, aliquoted in test tubes in volumes of 1.0 ml (10^6 cells/ml) with RPMI-1640 medium containing 16% fetal calf serum (RPMI-16) and incubated at 37°C in a CO_2 humidified incubator.

A second series of washed lymphocytes was also aliquoted as above but the cells were sedimented and then resuspended in 0.2 ml of a 100-fold dilution of the B95-8 strain of EBV (titre = $10^{4.5}$ TD_{50}%/ml). After incubating at 37°C for one hour the volume was adjusted to 1.0 ml (10^6 cells/ml) with RPMI-16. Cultures were observed twice a week at which time the medium was also changed. The presence of growing clumps concomitant with significant acidification of the medium indicated that the cells had been transformed. Once transformation was evident, the cells were placed in flasks and fed twice a week. At no time after transformation was the cell density allowed to exceed 10^6 cells/ml.

Immunofluorescence for the Epstein-Barr Nuclear Antigen (EBNA), the Viral Capsid Antigen (VCA) and EA

Immunofluorescent staining for EBNA was done not only on the patient's established lines (with B95-8 virus) but also on her fresh leukocytes (without addition of B95-8 virus) during the first week of culture. Smears were made, fixed, stained and examined for anticomplement immunofluorescence as described previously (Reedman and Klein, 1973). Smears of the patient's cell line (and that of her father) were also prepared for detection of VCA and EA by indirect immunofluorescence (Henle and Henle, 1966).

Cocultivation of Fresh CBL with the Patient's Cell Line

Cord blood mononuclear cells were obtained and separated essentially as described above. Cocultivation experiments were performed using the technique described by Henle et al. (1967) with minor modifications: 10^6 male CBL suspended in RPMI-16 were combined in a 50 ml conical centrifuge tube with a) 10^6 fresh mononuclear leukocytes from the female patient, or with b) 10^6 cells obtained from her previously transformed (with B95-8 virus) lymphoblastoid cell line, and spun at 300 g for 10 minutes. Excess medium was removed and the cells were incubated for two to 24 hours. They were then resuspended in RPMI-16 and aliquoted in one millilitre volumes in Falcon culture tubes at a concentration of 500,000 cells/ml. Acidification of cultures was monitored daily and the cells were transferred to larger containers when the cell density exceeded 10^6 cells/ml.

Cytogenetic Analysis

The cocultivated cells were cultured for three months before cytogenetic analyses were performed. This was only possible with (b). Cocultures (a) failed to transform. Karyotypes of the resulting cell line (b) were obtained using the C-banding technique (Summer, 1972). Presence of the male Y chromosome was confirmed by quinacrine labelling (QFQ-banding) (Caspersson et al., 1970).

Isolation of Virus and Induction of EA in Raji Cells by Throat Washings and Cell Line Supernatants

Throat washings and cell line supernatants were first concentrated by high speed centrifugation. They were then resuspended in approximately 1/50 of the original volume and 0.1 ml of this suspension was incubated with 200,000 Raji cells in order to test for EA induction. The test was otherwise performed as described for the P3HR-1 virus.

The efficiency of EA induction is expressed as EA-inducing units/1.0 ml and is calculated by the following formula:

$$\text{EA-inducing units} = \frac{\% \text{ antigen-positive cells x initial no. of cells infected}}{100} \times \text{Dil. factor}$$

Use of this formula gave a titre of 2×10^6 EA-inducing units/ml for the control P3HR-1 virus.

Throat washings were also inoculated onto CBL separated as previously described (Joncas et al., 1977) and the cultures were observed during a two-month period for evidence of transformation.

Standard procedures were used for isolation of herpes simplex virus, cytomegalovirus and varicella-zoster virus (Lennette and Schmidt, 1979), and for the electron microscopy studies.

RESULTS

Unusual recurring ulcers were observed in the mouth of the patient (daughter). Throat washings were repeatedly negative for herpes simplex virus, cytomegalovirus and varicella-zoster virus. All throat washings from this patient (more than ten were tested) and from her father failed to transform CBL.

Permanent lymphoblastoid cell lines were obtained from the daughter and her father's peripheral blood, but only by B95-8 infection of their leukocytes. In a total of at least 30 trials, a cell line was never obtained spontaneously. On two occasions smears were made from short-term cultures of the daughter's leukocytes before total loss of viability and these were stained for EBNA; both showed EBNA positivity. The indirect immunofluorescence technique showed that the B95-8 induced father's and daughter's cell lines both produced EA and VCA (Table 1). A nonenvelopped herpesvirus particle was found in the concentrated supernatant of the daughter's cell line obtained by in vitro transformation of her lymphocytes with the B95-8 strain of EBV. This supernatant (as well as that obtained from the father's cell line) failed to transform CBL. However, when the daughter's cell line was cocultivated with CBL obtained from a male donor, an EBNA-positive cell line exhibiting the male karyotype was obtained (see Table 1 for summary of studies performed on peripheral blood leukocytes of father and daughter).

However, supernatants from father's and daughter's cell lines were capable of inducing EA in Raji cells (Table 1). Even more interesting was the demonstration that seven out of fourteen of the daughter's throat washings tested and two out of ten of the father's also induced EA in Raji cells (Table 2).

Furthermore, when two of the daughter's seven throat washings positive for EA induction were inoculated onto EBV-negative peripheral blood lymphocytes, EBV antigens (EA and/or VCA) could be detected within three days of culture using an EBV-positive serum; the test was negative with an EBV-negative control serum. Uninoculated lymphocytes remained negative. Finally, two of three smears obtained from the daughter's buccal lesions were also positive for EBV antigens by the same procedure.

Table 1

SUMMARY OF STUDIES PERFORMED ON FATHER AND DAUGHTER OF AFFECTED FAMILY
A. Peripheral blood leukocytes

Patient/Control	Peripheral blood leukocytes			
	Spontaneous transformation	B95-8 induced transformation	EBV antigens in transformed cells	EA induction in Raji[1]
Father	-	+	EBNA (> 80%) EA (5-20%) VCA (0.1-10%)	3.2×10^3
Daughter	-	+	EBNA (> 80%) EA (5%) VCA (0.1-1%)	7.2×10^2
Controls[2]	+(1)	+(1)	EBNA (> 80%) EA (0%) VCA (0%)	0

[1] Expressed as EA-inducing units/ml

[2] Spontaneously established (1) and B95-8-induced transformation (1) of peripheral blood leukocytes from normal individuals of similar age groups.

Table 2

SUMMARY OF STUDIES PERFORMED ON FATHER AND DAUGHTER OF
AFFECTED FAMILY
B. Throat washings

Patient/Control	CBL transformation	Throat washings EA induction in Raji[1]
Father	−	2/10 (1.2×10^3)
Daughter	−	7/14 (3.2×10^3)
Controls[2]	+(5) −(2)	0/7

[1]No. of samples positive/no. of samples tested
(EA-inducing units/ml)

[2]Five transforming throat washings from two patients with
infectious mononucleosis and two from patients with the
acquired immunodeficiency syndrome.

DISCUSSION

The literature describes two distinct biotypes of
EBV, the transforming B95-8 strain and the non-transforming (lytic) P3HR-1 strain (Menezes et al., 1975; Miller
et al., 1975). These two biotypes differ at the genomic
level: hybridization experiments show that the viral
genome of P3HR-1 lacks a segment of about 3000 base pairs
in the U2-IR2 region, whose role in transformation is
still unknown (Kieff et al., 1982).

The virus present in cell lines and throat washings
of our two patients failed to transform CBL but did
induce EA in Raji cells, and may therefore be described
as P3HR-1-like. The ability of the concentrated cell
line extracts and throat washings to induce EA in Raji
cells, observed concurrently with their inability to

transform CBL, seems to suggest that the majority of viral particles produced are defective, exhibiting P3HR-1-like properties. The presence in very small numbers of transforming virus in the overall population cannot be excluded. Nonetheless, because only very minute quantities of the latter may be produced, or perhaps due to an inability of the transforming virus to bind to its target cell, it may not be able to effectively transform unless special conditions such as cell-cell contact are allowed as in cocultivation experiments.

Whether the patients were originally infected with this mutant strain or whether such a mutant appeared subsequently cannot be determined. However, an increased permissiveness of the host target cell, documented by the productive properties (EA and VCA antigen-positivity) of the B95-8 induced cell lines of these patients, may have significantly contributed to the appearance of a mutant virus with the properties of the P3HR-1 strain. Such increased production of EBV, and/or its antigens, presumably led to the extremely high EA and VCA antibody titres observed. Further work is in progress to better characterize the virus and its interaction with the host cell.

ACKNOWLEDGEMENTS

We are grateful to Ms Louise Robillard for her assistance with the electron microscopy, to Dr Claude-Lise Richer and Ms Elise Ménard for performing the cytogenetic analyses, and to Ms Sylvie Tassé for typing this manuscript. Tables 1 and 2 are reprinted with the permission of the Can Med Assoc J.

REFERENCES

Caspersson, T., Zech, L., Johansson, C., and Modest, E.J., Identification of human chromosomes by DNA-binding fluorescent agents. Chromosoma, 30, 215-227 (1970).

Henle, G., and Henle, W., Immunofluorescence in cells derived from Burkitt's lymphoma. J. Bacteriol., 91, 1248-1256 (1966).

Henle, W., Diehl, V., Kohn, G., zur Hausen, H., and Henle, G., Herpes-type virus and chromosome marker in normal leukocytes after growth with irradiated Burkitt cells. Science, 157, 1064-1065 (1967).

Joncas, J.H., Ghibu, F., Blagdon, M., Montplaisir, S., Stefanescu, I., and Menezes, J., A familial syndrome of susceptibility to chronic active Epstein-Barr virus infection. Can. Med. Assoc. J., 130, 280-285 (1984).

Joncas, J., Menezes, J., Patel, P., Wills, A., Gervais, F., and Leyritz, M., The Epstein-Barr virus infection in the immunosuppressed host. In: Proceedings of the 9th International Course on Transplantation and Clinical Immunology, Lyon, France, June 6-8, 1977. Amsterdam: Excerpta Medica 1977: 41-47 (International Conress Series no. 423).

Kieff, E., Dambaugh, T., Heller, M., King, W., Cheung, A., van Santen, V., Hummel, M., Beisel, C., Fennewald, S., Hennessy, K., and Heineman, T., The biology and chemistry of Epstein-Barr virus. J. Infect. Dis., 146, 506-517 (1982).

Lennette, E.H., and Schmidt, N.J. (ed.), Diagnostic procedures for viral, rickettsial and chlamydial infections, chapters 11-13, American Public Health Association, Washington (1979).

Menezes, J., Leibold, W., and Klein, G., Biological differences between Epstein-Barr virus (EBV) strains with regard to lymphocyte transforming ability, superinfection and antigen induction. Exp. Cell Res., 92, 478-484 (1975).

Miller, G., Robinson, J., Heston, L., and Lipman, M., Differences between laboratory strains of Epstein-Barr virus based on immortalization, abortive infection and interference. In: G. de-Thé, M.A. Epstein and H. zur Hausen (ed.), Oncogenesis and Herpesvirus II, Part 1, pp. 395-408, IARC Scientific Publications no. 11, Lyon, France (1975).

Purtilo, D.T., Sakamoto, K., Barnabei, V., Seeley, J., Bechtold, T., Rogers, G., Yetz, J., and Harada, S., Epstein-Barr virus-induced diseases in boys with the X-linked lymphoproliferative syndrome (XLP): update on studies of the registry. Am. J. Med., 73, 49-56 (1982).

Reedman, B.M., and Klein, G., Cellular localization of an Epstein-Barr (EBV)-associated complement-fixing antigen in producer and non-producer lymphoblastoid cell lines. Int. J. Cancer, 11, 499-520 (1973).

Sumner, A.T., A simple technique for demonstrating centromeric heterochromatin. Exp. Cell Res., 75, 304-306 (1972).

7

BRIEF COMMUNCATION

The Significance of Antibodies Against the Epstein-Barr Virus-Specific Membrane Antigen gp 250 in Acute and Latent EBV Infections

Wolfgang Jilg and Hans Wolf

Max-von-Pettenkofer-Institut, Munich, FRG

INTRODUCTION

The serodiagnosis of Epstein-Barr Virus (EBV) infections is based on the demonstration of antibodies against three different antigens of EBV: the virus capsid antigen (VCA), the EBV-related early antigen (EA), and the EBV nuclear antigen (EBNA). Antibodies against a fourth antigen complex, the membrane antigen (MA) (Qualtiere and Pearson, 1979; North et al, 1980), are not determined routinely becuase of the complexities of the available assays (Klein et al, 1968). These antibodies, however, should be of special interest as they include neutralising antibodies, which are directed against the two closely related EBV membrane proteins gp 350 and gp 250 (Thorley-Lawson and Geilinger, 1980; Thorley-Lawson and Poodry, 1982). Usually, anti-EBV antibodies are detected by indirect immunofluorescence (Henle and Henle, 1966). With this method, however, antigen complexes rather than single proteins are detected. In the present study, we examined sera of EBV-infected persons for antibodies against the membrane protein gp 250 using a radioimmunoprecipitation assay.

METHODS

Antibodies against the membrane antigen gp 250 were determined by immunoprecipitation of detergent extracts of ^{125}iodine surface-labelled P3HR1 cells. For induction of

EBV antigens, cells were treated with phorbol-ester (Bayliss and Wolf, 1981) three days before surface labeling. Iodination of cells and extraction of labelled proteins using 0.5% NP-40 was done as described by Jilg and Hannig (1981). To increase the sensitivity of the assay, cell extracts were precleared with an antiserum against EBV-negative lymphoid cells prepared in rabbits. To reduce background activity sera were preincubated with an extract from EBV-negative lymphoid cells (BJAB). Immunocomplexes were isolated using protein A covalently bound to sepharose, and analysed by SDS-page followed by autoradiography (Jilg and Hannig, 1981).

RESULTS AND DISCUSSION

Sera of 108 patients with acute or latent EBV infection or nasopharyngeal carcinoma (NPC) were examined for antibodies against the EBV-specific membrane protein gp 250. These antibodies were absent in most patients with serologically proven infectious mononucleosis; in a small number of patients, anti-gp 250 was found in traces. They were clearly detectable in most persons a few months after acute EBV infection (Table 1, Fig. 1).

Table 1. ANTIBODIES AGAINST GP 250 IN ACUTE, RECENT AND PAST EBV-INFECTION

	ANTI-VCA IGG IGM	ANTI-EBNA	TOTAL NO.	ANTI-GP 250 NO.POS.	NO.NEG.	%POS.
ACUTE EBV-INFECTION	+ +	Ø	26	2*	24	7.7*
RECENT EBV-INFECTION	+ +/Ø	Ø/+	6	2*	4	33.3*
PAST EBV-INFECTION	+ Ø	+	52	49	3	94.0*

Figure 1.

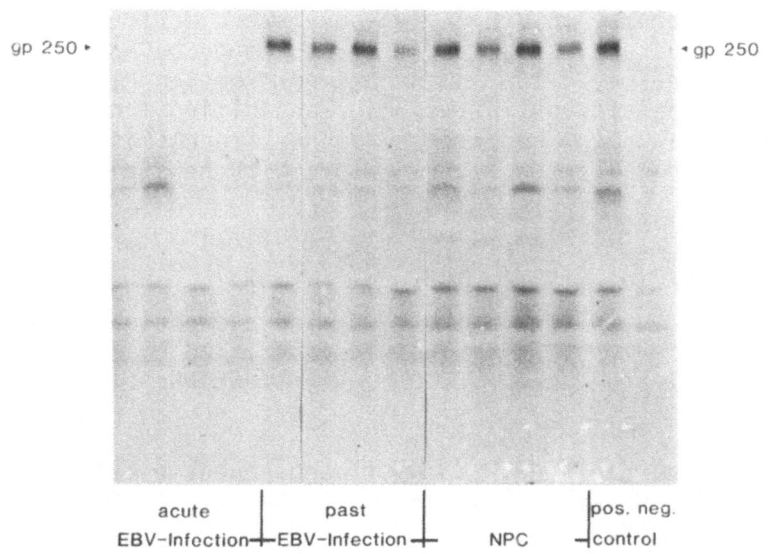

Examination of Patient's Sera for Antibodies Against gp 250 by Immunoprecipitation.

Three out of 49 individuals with past EBV infection were negative for anti-gp 250; two of them showed signs of persistent active EBV infection (lymph node enlargement, recurrent low-grade fever, headache, hepatosplenomegaly). A third woman did not show any clinical symptoms, but gave birth to a child with hepatosplenomegaly and minimal congenital defects. Although the child's serology was consistent with an infection during the first months of life, an intrauterine infection can not be excluded. In the latter case there may be a correlation between the EBV infection, and the hepatosplenomegaly and the minimal malformations.

Of 24 anti-VCA and anti-EBNA positive patients who were suspected to have persistent active EBV infections according to their clinical picture, 3 (12.5%) lacked detectable antibodies against gp 250.

In 14 patients with NPC, anti-gp 250 could be demonstrated in high concentrations.

The above results suggest that anti-gp 250 is a useful marker for past EBV infections; lack of this antibody in the presence of other markers of EBV indicates acute EBV infection, or may be a hint for persistent active infection. As antibodies against gp 250 are assumed to be neutralising, their absence may allow endogenous reinfection giving rise to chronic persistent infection. In addition intrauterine infection may occur under these circumstances, as may be suggested in the case of the newborn with hepatosplenomegaly.

REFERENCES

Bayliss, G.J., Wolf, H. The regulated expression of Epstein-Barr virus. III. Proteins specified by EBV during the lytic cycle. J. Gen. Virol. 56:105-118 (1981).

Henle, W. and Henle, G. Immunofluorescence in cells derived from Burkitt's lymphoma. J. Bacteriol. 91: 1248-1256 (1966).

Jilg, W. and Hannig, K. Lymphocyte surface proteins recognized by an anti-thymocyte-globulin. Hoppe-Seyler's Z. Physiol. Chem. 362:1474-1485 (1981).

Klein, G., Pearson, G., Nadkami, J.S., Nadkami, J.J., Klein, E., Henle, G., Henle, W. and Clifford, P. Relation between Epstein-Barr virus and cell membrane immunofluorescence on presence of EB virus. J. Exp. Med. 128:1011-1020 (1968).

North, J., Morgan, A.J. and Epstein, M.A. Observations on the EB virus envelope and virus-determined membrane antigen (MA) polypeptides. Int. J. Cancer 26:231-240 (1980).

Qualtiere, L.F. and Pearson, G.R. Epstein-Barr virus-induced membrane antigens: immunochemical characterization of Triton X-100 solubilized viral membrane antigens from EBV-superinfected Raji cells. Int. J. Cancer 23:808-817 (1979).

Thorley-Lawson, D.A. and Poodry, C.A. Identification and isolation of the main component (gp 350-gp 220) of Epstein-Barr virus responsible for generating neutralizing antibodies <u>in vivo</u>. J. Virol. 43:730-736 (1982).

Thorley-Lawson, D.A. and Geilinger, K. Monoclonal antibodies against the major glycoprotein (gp 350/220) of Epstein-Barr virus neutralize infectivity. Proc. Natl. Acd. Sci. U.S.A. 77:5307-5311 (1980).

EBV-ASSOCIATED MALIGNANCIES

8

THE EPIDEMIOLOGY OF EPSTEIN-BARR VIRUS-ASSOCIATED MALIGNANCIES

P.H. LEVINE

Clinical Epidemiology Branch
National Cancer Institute
Bethesda, Maryland

SUMMARY

Epidemiologic studies of Epstein-Barr Virus (EBV)-associated malignancies continue to demonstrate the interrelationship of multiple etiologic factors, thereby allowing several opportunities for intervention with the disease process. For the two malignancies most closely linked with EBV, Burkitt's lymphoma (BL) and nasopharyngeal carcinoma (NPC), the risk factors are more readily identified in high-incidence areas than in low-incidence areas. The report of an apparent decline in BL incidence in Africa is in contrast with a reported increase in incidence among young white males in the United States. The search for etiologic factors in BL, in addition to EBV and malaria, is currently focusing on retroviruses. For NPC, evidence continues to support the major roles of genetics and salted fish in southern Chinese, but other etiologic factors appear to be important in non-Chinese. In areas of low incidence for BL and NPC, less homogeneous patterns are seen, suggesting the inclusion of unrelated cases. Precise definition of study groups may be aided by new laboratory assays. The further development and utilization of cancer registry data throughout the world will also increase our knowledge of these diseases.

As noted in the many recent reviews concerning the epidemiology of nasopharyngeal carcinoma (NPC)

(Shanmugaratnam, 1982; Simons and Shanmugaratnam, 1982; Levine and Connelly, 1985) and Burkitt's lymphoma (BL) (Lenoir et al, 1985), the two malignancies most closely linked with the Epstein-Barr virus (EBV), the interrelationship among a number of etiologic co-factors continues to be an area of active investigation. In the context of prevention, knowledge of multiple etiologic factors offers several approaches to intervening in the chain of events leading to these malignancies. The lack of agreement on disease classification has continued to plague epidemiologic studies, however, and it is to this problem that we and others in this symposium (Magrath, this volume; Bale et al, this volume) will give particular attention. The tendency to consider only tumors which have detectable EBV genomes may be inappropriate since it is not technically feasible to assay more than a small percentage of neoplasms. In addition, it appears that as more sensitive assays are developed the percentage of apparently genome-negative cases declines. The appropriate approach to studying the epidemiology of BL and NPC at the present time, therefore, is to define the disease by clinical/pathologic manifestations and to turn to the laboratory only in specific substudies where a high percentage of cases can be evaluated by standardized laboratory assays.

Burkitt's Lymphoma

Definition of BL continues to be a challenging and controversial problem. The majority of cases of BL seen in equatorial Africa appear to be biologically different from most cases seen in low-incidence areas, although there is considerable overlap (Klein, 1974; Lenoir et al, 1985; Levine et al, 1982; Magrath, this volume). At the present time, therefore, investigators should base the diagnosis of BL on the histopathologic criteria as defined by the World Health Organization (Berard et al, 1969). Difficulties encountered in making the diagnosis of BL histopathologically, even for experienced hematopathologists, must be addressed in any studies of BL, particularly in non-endemic areas. The rapid growth of this tumor makes prompt and careful fixation of biopsy material mandatory. The opportunity for a firm diagnosis may be hampered by improper processing of the biopsy. A second problem occurs in the nature of the tumor itself; most cases have the typical morphology allowing full agreement among experienced pathologists but some have atypical features that require a consensus diagnosis. The American Burkitt Lymphoma Registry (ABLR) initially addressed this problem by classifying the biopsies according to the identification of typical features. The presence of such features was found to correlate with an elevated EBV antibody titer (Levine et al, 1972), suggesting a correlation between histologic pattern and EBV-association. Some investigators question whether these minor differences among cases are biologically meaningful (Magrath, this volume), and subsequent studies have not used such analyses (Levine et al, 1982). However, the addition of new tools to measure cell maturation and other biologic properties may permit more precise categorization of tumor types and biological insights.

Despite the problem of classification, many similarities can be observed between BL occurring in high-incidence and low-incidence areas. In both, the identification of translocations involving chromosome 8 and the identification of the myc oncogene are regularly found (Lenoir et al, 1985). EBV genomes can be detected in tumors from both areas, although more frequently in high incidence areas. Thus far, no other viral footprint has been reported for the non-EBV-associated cases. The clinical presentation differs somewhat between high- and low-incidence area cases (Magrath, this volume), but in both, jaw tumors are more frequent in young children and abdominal tumors more frequent in older patients (Burkitt, 1962; Biggar et al, 1981; Levine et al, 1982). In general, there appears to be a predilection for the disease to appear in organs which are undergoing rapid growth. Time-space clustering has been reported in both endemic (Pike et al, 1967) and non-endemic BL (Judson et al, 1977; Levine et al, 1973); however the significance of these reports are unclear since no clustering has been noted in several studies and controversy continues over the statistical evaluation of such clusters as well as their biologic significance.

Among the more important recent observations relating to African BL is the apparent decline in the disease incidence. This observation is difficult to document because of the absence of population-based tumor registries in the areas of interest, but at the Lyon conference on Burkitt's lymphoma in December 1983, clinicians managing patients in Tanzania, Ghana and Uganda agreed that the decline in new cases apparently exceeded any decline that could be explained by transportation difficulties or other logistic problems. Supporting the validity of this observation is a trend noticed in a study of the patient population at the Burkitt Tumor Project in Accra, which revealed that patients were presenting at an

Accra, which revealed that patients were presenting at an older age and there was more abdominal and less jaw involvement (Biggar et al, 1981). This changing pattern was consistent with a later age of onset and perhaps a decline in incidence. One plausible explanation for such a trend is a decline in the incidence of malaria, a possible cofactor for BL, perhaps due to the wider availability of antimalarial drugs.

EBV and malaria are insufficient cofactors for BL because of the relatively low incidence in comparison with the ubiquity of EBV and malaria in endemic BL areas. Although it has been 13 years since Pike and Morrow (1972) reviewed the need to find the third etiologic factor(s) for BL, little progress has been made in identifying what separates the BL patient from his neighbors. Other infectious agents that should be given further attention are the newly discovered human retroviruses, since they appear to be more widespread in Africa than previously suspected and their pathogenicity has not yet been completely defined. In one recent study, HTLV-1 was shown to infect approximately 10% of Ghanaians and 60% of Ugandans (Saxinger et al, 1984) whereas HTLV-III apparently infected even a higher percentage of these populations (Saxinger et al, submitted). Perhaps co-infection by multiple viruses in a specific sequence is necessary for the production of a malignancy.

Regarding the epidemiology of BL in the United States, a finding that exemplifies the problem in dealing with BL outside of Africa is the report of an increasing incidence of BL in the young white male population (Levine et al, in press). This increase has been identified in three independent data sources: 1) the Surveillance, Epidemiology and End Results (SEER) Program, which monitors cancer occurrence in approximately 10% of the US population; 2) the National Center for Health Statistics which records mortality data on the entire US population; and 3) the American Burkitt's Lymphoma Registry, a voluntary reporting system. This increase is confined only to white males and not females or non-whites of either sex. The possibility exists that a more particular subtype of lymphoma resembling BL is occurring selectively in this group. The precise classification of these cases awaits a more detailed clinical/pathologic/ laboratory study with consistent evaluation of all suspected cases.

Nasopharyngeal carcinoma

The evidence supports a multifactorial etiology for NPC, as with BL, and many of the specific co-factors suggested in the past continue to be identified in recent studies. The strong association of EBV with undifferentiated NPC is well documented but it is premature to exclude EBV from having a role in well-differentiated NPC. Some cases of well-differentiated NPC (WHO I histologic type) have been shown to have EBV genome in the tumor cells (Raab-Traub et al, this volume) and as more sensitive techniques are developed, the percentage of genome-positive well-differentiated tumors may increase. It is quite likely, however, that the individual co-factors are not the same in every individual or group of NPC patients. In addition to EBV, consumption of salted fish is a prominent risk factor in Cantonese Chinese. First suggested as an important potential carcinogen by Ho (1972) and supported by laboratory (Huang et al, 1978) as well as epidemiologic observations (Ho, 1972; Shanmugaratnam, 1982; Simons and Shanmugaratnam, 1982), salted fish has been found to be associated with NPC in Malaysians with a dose-response curve emphasizing the likelihood of the significance (Armstrong et al, 1983). As reviewed recently (Levine and Connelly, 1985), respiratory carcinogens such as smoke and soot, and natural tumor promoters may play a role in determining the disease pattern (Ito et al, 1981). In some groups, particularly Cantonese Chinese, genetics may play an important role as summarized elsewhere in this symposium (Simons, this volume).

Several intriguing epidemiologic patterns have emerged from studies of NPC in the United States. First, there appears to be a decreasing incidence of NPC among Chinese living in the US. This was first reflected in mortality data reported by Fraumeni and Mason (1974) and is supported by recent SEER data (Levine and Connelly, 1985). There are several possible explanations, including changes in dietary habits or other manifestations of a different life style, a greater proportion of immigrants from Northern China, where NPC is less common, or to an improvement in therapy. Other findings requiring additional study are the declining incidence in young white males (Levine et al, submitted) and the identification of a coastal pattern for nasopharyngeal cancer in the U.S. white population (Levine et al, 1983) with a predominance along the southeast Atlantic and Gulf coasts. The continued monitoring of new cases through the SEER program will continue to add to our understanding of the patterns and perhaps obtain additional clues to the etiology of NPC in the US.

As discussed elsewhere in this symposium (Simons, 1985), one of the most important areas of research is the role of genetics, which appears to be particularly important in the Chinese population. There are a number of approaches to studying the role of genetics in human cancer, however, and few of them have been systematically applied to NPC patients. Among these approaches, recently reviewed by Miller (in press), are studies of twins, inbred populations and cancer families. One of these, the "Family Study" approach, is discussed elsewhere in this symposium. It is quite apparent that in the area of genetics, perhaps more than any other area, the clinician has a key role to play in uncovering clues as to the etiology of NPC.

The development of collaborative projects among investigators in different parts of the world, fostered by meetings such as this one, has provided continuing improvement in our knowledge of EBV-associated malignancies. While particularly striking advances have been made in the laboratory, newer techniques are also being utilized by epidemiologists as well. Integration of standardized laboratory assays, population-based registries, and newly developed computer programs with sophisticated methods of analysis provide an excellent opportunity to define the diseases under study more precisely, thereby leading to a greater opportunity for progress in etiologic studies.

ACKNOWLEDGEMENT

The comments of Dr. Robert Biggar and Mr. Roger Connelly, and the preparation of the manuscript by Mrs. Barbara Salins and Mary Ann Abraham are gratefully acknowledged.

REFERENCES

ARMSTRONG, R.W., ARMSTRONG, M.J., YU, M.C., and HENDERSON, B.E., Salted fish and inhalants as risk factors for nasopharyngeal carcinoma. Cancer, **43**, 2967-2970 (1983)

BALE, S., BALE, A.E., and LEVINE, P.H., The "Family Study" approach to investigating the role of genetic factors in nasopharyngeal carcinoma (this volume).

BERARD, C.W., O'CONOR, G.T., THOMAS, L.B., and TORLONI, H., Histopathological Definition of Burkitt's Tumor, World Health Organization, New York (1969).

BIGGAR, R.J., NKRUMAH, F.K., NEEQUAYE, J., and LEVINE, P.H., Changes in presenting tumor site of Burkitt's lymphoma in Ghana, West Africa, 1975-1978. Br. J. Cancer, 46, 632-636 (1981).

BURKITT, D., A lymphoma syndrome in African children. Ann. Royal Coll. Surg. Eng., 30, 211-219 (1962).

FRAUMENI, J.F., Jr., and MASON, T.J., Cancer mortality among Chinese Americans, 1950-1969. JNCI, 52, 659-665 (1974).

HO., J.H.C., Nasopharyngeal carcinoma (NPC). In: G. Klein, S. Weinhouse, and A. Haddow (eds.), Advances in Cancer Research, pp. 57-92, Academic Press, New York (1972).

HUANG, D.P., SAW, D., TEOH, T.B., AND HO, J.H.C., Carcinoma of the nasal and paranasal regions in rats fed Cantonese salted marine fish. In: G. de-The, and Y. Ito (eds.), Nasopharyngeal Carcinoma: Etiology and Control, pp. 315-328, International Agency for Research in Cancer, Lyon (1978).

ITO, Y., KISHISHITA, M., MORIGAKI, T., YANASE, S., and HIRAYAMA, T., Induction and intervention of Epstein-Barr virus expression in human lymphoblastoid cell lines: a simulation model for study of cause and prevention of nasopharyngeal carcinoma and Burkitt's lymphoma. In: E. Grundmann, G.R.F. Krueger, and D.V. Ablashi (eds.), Nasopharyngeal Carcinoma, pp. 255-262, Gustav Fischer Verlag, Stuttgart/New York (1981).

JUDSON, S.C., HENLE, W., and HENLE, G., A cluster of Epstein-Barr virus-associated American Burkitt's lymphomas. N. Engl. J. Med., 297, 464-468 (1977).

KLEIN, G., Studies on the Epstein-Barr virus genome and the EBV-determined nuclear antigen in human malignant disease in tumor viruses. Cold Spring Harbor Symp. Quant. Biol., 39, 783-796 (1974).

LENOIR, G., O'CONOR, G.T., and OLWENY, C. (eds.), Burkitt's Lymphoma: A Human Tumor Model, International Agency for Research on Cancer, Lyon (1985).

LEVINE, P.H., and CONNELLY, R.R., Epidemiology of nasopharyngeal cancer. In: R.E. Wittes (ed.), Head and Neck Cancer, pp. 13-34, John Wiley and Sons, Ltd., Sussex, England (1985).

LEVINE, P.H., CONNELLY, R.R., and MCKAY, W.F., The influence of residence, race, and place of birth on the incidence of nasopharyngeal carcinoma. In: U. Prasad, D.V. Ablashi, P.H., Levine, and G.R. Pearson (eds.), Nasopharyngeal Carcinoma: Current Concepts, pp. 143-156, University of Malaya Press, Kuala Lumpur (1983).

LEVINE, P.H., CONNELLY, R.R., and MCKAY, W.F., Burkitt's lymphoma in the United States: cases reported to the American Burkitt Lymphoma Registry compared with population-based incidence and mortality data. In: G.T. O'Conor (ed.), Burkitt's Lymphoma: A Human Cancer Model, International Agency for Research on Cancer, Lyon (in press).

LEVINE, P.H., CONNELLY, R.R., and MCKAY, F.W., Recent findings in nasopharyngeal cancer in the United States. (submitted).

LEVINE, P.H., KAMARAJU, L.S., CONNELLY, R.R., BERARD, C.W., DORFMAN, R.F., MAGRATH, I., and EASTON, J.M., The American Burkitt's lymphoma registry: Eight years experience. Cancer, 49, 1016-1022 (1982).

LEVINE, P.H., O'CONOR, G.T., and BERARD, C.W., Antibodies to Epstein-Barr virus (EBV) in American patients with Burkitt's lymphoma. Cancer, 30, 610-615 (1972).

LEVINE, P.H., SANDLER, S.G., KOMP, D.M., O'CONOR, G.T., and O'CONNOR, D.M., Simultaneous occurrence of "American Burkitt's Lymphoma" in neighbors. N. Engl. J. Med., 288, 562-563 (1973).

MAGRATH, I., Clinical and pathobiological features of Burkitt's lymphoma and their relevance to treatment (this volume).

89

MILLER, R.W., Genetic and familial factors. In: P.
Calabresi, P.S. Schein, and S.A. Rosenberg (eds.), Basic
Principles and Clinical Management of Cancer, Macmillan
Publishing Company, New York (in press).

PIKE, M.C., WILLIAMS, E.H., and WRIGHT, B.H., Burkitt's
tumour in the West Nile District of Uganda, 1961-5. Br.
J. Med., 2, 395-399 (1967).

PIKE, M.C., and MORROW, R.H., Some epidemiological
problems with "EBV + Malaria gives BL"--A review. In:
P.M. Biggs, G. de-The, and L.N. Payne (eds.).
Oncogenesis and Herpesviruses, pp. 349-350, International
Agency for Research on Cancer, Lyon (1972).

RAAB-TRAUB, N., HUANG, D., YANG, C.S., and PEARSON, G.,
EBV DNA content and expression in nasopharyngeal
carcinoma (this volume).

SAXINGER, W., BLATTNER, W.A., LEVINE, P.H., CLARK, J.,
BIGGAR, R., HOH, M.W., MOGHISSI, J., MOURALI, N., NKRUMAH,
F.K., WILSON, P.J.L., JACOBSON, R., CROOKES, R., STRONG,
M., ANSARI, A.A., and GALLO, R.C., Human T-cell leukemia
virus (HTLV-1) antibodies in Africa. Science, 225,
1473-1476 (1984).

SAXINGER, W.C., LEVINE, P.H., DEAN, A.G., DE-THE, G.,
SARNGADHARAN, M.G., and GALLO, R.C., Evidence for exposure
to the HTLV-III in Uganda prior to 1973 (submitted).

SHANMUGARATNAM, K., Nasopharynx. In: D. Schottenfeld,
and J.F. Fraumeni, Jr. (eds.), Cancer Epidemiology and
Prevention, pp. 536-553, W.B. Saunders Co., Philadelphia
(1982).

SIMONS, M.J., Genetic aspects of EBV-associated
malignancies (this volume).

SIMONS, M.J., and SHANMUGARATNAM, K. (eds.), The Biology
of Nasopharyngeal Carcinoma, UICC Technical Report Series,
Hans Huber Publishers, Berne (1982).

9

GENETIC ASPECTS OF EBV–ASSOCIATED MALIGNANCIES

MALCOLM J. SIMONS

Immunogene Typing Laboratory,
Immunodiagnostic Centre,
Toorak, Victoria, Australia.

EBV is strongly associated with two malignancies, Naso-pharyngeal carcinoma (NPC) and Burkitt's lymphoma (BL). Of the two main aspects of genetic consideration, more information is emerging concerning genetic mechanisms of carcinogenesis in BL. Genetic factors responsible for excess cancer occurence are better understood in NPC than in any other common human cancer.

The two malignancies in which there is strong suspicion of EBV involvement are Nasopharyngeal carcinoma (NPC) and Burkitt's lymphoma (BL). EBV is also associated with a variety of lymphoproliferative diseases including malignant lymphoma but apparently not including Hodgkin's disease (Purtilo, 1983). The program for this symposium indicates that mechanisms and models of BL aetiology are to receive close attention, as will recent developments in oncogene molecular biology. These presentations can be expected to deal with one of the two main aspects of genetic consideration, namely that of genetic mechanisms of background incidence cancer development. Since my overview is the only one scheduled to address NPC, I will focus on the second main aspect of genetic factors responsible for excess NPC occurance in high risk groups.

Chromosomal aberrations are thought to play a central role in neoplasia, allowing the generalisation that all cancer is in one sense a genetic disease. For many cancers there are no distinguishing cytogenetic features detectable at the microscopic level. In BL there are exciting developments at the chromosomal and DNA molecular levels using cytogenetic banding and oncogene probe technologies. The emerging molecular description of genetic phenomena involving oncogene activation epitomises the former aspect of genetic consideration concerning genetic processes underlying background cancer occurence (Klein, 1981, 1983). To date, there is no information in NPC similar to the chromosomal rearrangements identified in BL. There is a pressing requirement to apply cytogenetic and molecular biological techniques to assess whether there are chromosomal alterations accompanying NPC and, if so, whether they occur in either background or in excess incidence patient populations, or both.

The epidemiological data does not suggest a major role for genetic elements in the excess risk for BL. Rather it seems that the excess incidence reflects interactive environmental exposure events. In complete contrast, the maintenance of high NPC incidence among Chinese is perhaps the best example of an epidemiological pattern suggesting an important genetic component (Simons and Shanmugaratnam, 1982). Furthermore, NPC stands alone as the only common human malignancy in which the strong genetic effect suspected of underlying excess incidence has been assigned to a polymorphic gene system for which the chromosomal localization is known.

Some introductory comments about genetics and neoplasia may be useful in order to better comprehend concepts of genetic-environment interaction, so that the evidence for genetic effects in NPC can be better appreciated. Comments which suggest some confusion in conceptual comprehension of the role of genetic and environmental factors in cancer causation include the following:
(i)   the HLA gene complex comprises only a minute fraction of the human genome so HLA genes can not be that important;
(ii) the HLA BW46 gene is only present in Chinese, so the genetic findings are not relevant to NPC patients of other ethnic groups;
(iii) incidence estimates among Hawaiian and mainland USA Chinese patients indicate a decline in US-born Chinese NPC incidence, so environmental effects are predominant, although genetic factors cannot be excluded;
(iv) the high NPC incidence maintained among overseas

Asian Chinese could be due to persistence of life-style habits rather than to genetics;

Simplistically stated, the genetics-versus-environment notion that is so widespread is a misleading polarisation of the two elements, implying that the roles of genetic factors and environmental agents are, in the extreme, mutually exclusive alternates.

There are examples that approach such extreme situations where genetic predisposition confers risks of 1000-fold or more on the occurence of malignancies. Retinoblastoma, squamous cell carcinoma in xeroderma pigmentosum patients, and colon cancer in patients with polyposis coli are disorders in which the importance of genetic factors in the aetiology of the cancer has been established. These and other disorders, including chromo-somal instability syndromes which serve as models of single disorders that predispose to malignancy, hereditary cancer syndromes in which the occurence of neoplasia is attribut-able to single gene effects, and primary immunodeficiencies associated with neoplasms, fall into the very high genetic risk category (see Mulvihill et al., 1977). Although no more than a very small fraction of cancers appear to be the result of such predominant genetic load, intense study of these rarities is providing clues on the control of gene expression that is expected to be of central importance in an understanding of cancer (Lancet, 1984b).

The question which is central to the 'genetics and cancer' issue is whether genetic susceptibility has a role in the pathogenesis of commonly occurring cancers. Since multiple case families are uncommon, and blood group marker associations are of relatively low strength, the casual observer could be excused for regarding genetic suscept-ibility as relatively unimportant in cancer occurence. The excess incidences of gallbladder carcinoma in American Indians, hepatocellular carcinoma in African blacks, and NPC among Southern Chinese are examples of common malignancies in which prominent genetic elements are suspected (Doll and Peto, 1981). Although these examples are the exception rather than the rule, the possibility exists that among common cancers there may be individuals in whom genetic predisposition confers a risk of 5, 10 or even several tens-fold higher than that in the general population. However, detection of such genetic effects among individuals at increased risk is difficult. The wide variability in cancer occurence even among genetically identical laboratory animals maintained under common environmental conditions, whereby some animals will die of cancer in middle and later age, while others will live to old age and succumb without evidence of malignancy, illustrates the difficulty of discerning genetic effects even in inbred populations living under conditions ideal for study. Even greater difficulty can be expected in outbred humans. Nonetheless, a relatively modest familial risk of the order of 3-fold is consistent with the existence of a major, single gene, genetically mediated susceptibility (Day and Simons, 1976). Peto (1980) has also explained that unimpressive relative risks may reflect substantial genetic proneness to disease occurence because the relative risk in genetic susceptibles compared with non-susceptibles will commonly be manyfold greater than the relative risk in relatives compared with the general population. The reasons are that:

1. since cancer patient populations include those with sporadic disease as well as those with familial risk, not all cancer patients are genetic susceptibles.

2. the relatives of sporadic disease patients are also not susceptible, so for every one non-familial risk patient there will be more than one relative falsely assigned as a relative at risk.

3. even among the cancer patients with familial risk there may be more than one genetic mechanism, each of which is likely to be associated with a separate relative risk.

4. the general population includes susceptibles as well as non-susceptibles.

The observed relative risk is thus diminished by these four factors from the 'true' relative risk conferred by any single genetic process. Assuming a unigenic familial risk, in the situation where susceptibility is due to a recess-ively inherited gene, the required information is that of the risk among homozygotes relative to that among hetero-zygotes and non-carriers. For a dominant gene the true incidence is given by the combined risk of homozygotes and heterozygotes divided by the risk in non-carriers. Considering a range of frequencies of the putative disease susceptibility 'gene', and accepting certain assumptions concerning susceptibility, a constant increase in disease incidence with age due to susceptibility, and a low risk in susceptibles, the lowest incidence ratios for dominant and recessive models corresponding to low sibling relative risk have been calculated for both the dominant and recessive models (Peto, 1980). Relative risks of the order of 3-fold can be associated with incidence ratios as high as 10:1 to 50:1. First degree relatives of patients with several types of cancer (lung, stomach, large intestine, uterus and breast) have relative risks of approximately 3. In the case of breast cancer, such relative risks have been interpreted as "not indicative of a strong hereditary effect, (rather) it is more in keeping with a polygenic mechanism, involving the action of several genes, presumably with small effects and accounting for only a relatively small portion of the total variation in breast cancer" (Anderson, 1976). Rather than revealing only small genetic variation, such low number risks may reflect major genetic effects (Day and Simons, 1976; Peto, 1980; Tulinius et al., 1982).

Relative risks in the range of 1.5 - 2.0, which can be associated with incidence ratios of around 10:1 and thereby reflect substantial genetic variation, are difficult to detect by traditional family studies. However, low number (approx. 2-fold) relative risks can be revealed if biomarkers exist which are disease associated and which can therefore be used to identify individuals with increased risk. An example which illustrates the limitations of classical family analysis, and the usefulness of identifying disease-associated biomarkers, is Ankylosing Spondylitis (AS). Only when the association with HLA-B27 was established was it possible, firstly, to demonstrate that the major inheritence involved an autosomal dominant trait, and secondly, to estimate the now-known penetrance of 8% in males and 1% in females with any precision, yet the relative risk associated with HLA-B27 in AS is very high at around 100 (Peto, 1980).

With this background information, the position in NPC can now be evaluated. "NPC remains the most convincing example of a human tumor associated with distinctive HLA patterns" (Lancet, 1979). The association of the major histocompatibility complex (MHC) HLA B locus gene, HLA-BW46, with susceptibility to NPC confers a relative risk of approximately 2 (Simons et al., 1978). The higher frequency of HLA-BW46 in newly diagnosed NPC patients has been repeatedly established in Singapore Chinese (Chan et al., 1983a), and has been similarly shown in overseas Chinese in Malaysia and Hong Kong (Simons et al., 1977) as well as in

NPC patients in the Peoples Republic of China (Simons, 1981a). In the last NPC symposium it was reported that: "Studies of HLA type....., have not yet revealed a pattern useful in predicting susceptibility or response to therapy" (Ablashi et al., 1983). However, the data in that paper reveals the frequency of HLA-BW46 in Cantonese to be not less than 44%, an increased frequency similar to that observed in previous studies of Singapore, Malaysian and other Chinese patients, and representing a relative risk of approx. 2. Recent HLA genetic findings in NPC (Simons, 1981a; Chan et al., 1983a) indicate associations additional to that of HLA-BW46, including a second B locus marker of susceptibility (BW58). Thus there are at least two distinct HLA patterns associated with susceptibility. There appears to be a particularly strong association of BW58 with early age ($\langle$30 years) of NPC onset (RR 3.8)(Chan et al., 1983a). The patterns associated both with susceptibility to NPC development and with survival from the disease are detailed elsewhere (Simons, 1981a; Simons and Shanmugaratnam, 1982; Chan et al., 1983a, 1983b).

The associations of HLA-BW46 and HLA-BW58 with NPC have been revealed by HLA phenotyping for the allelic products of separate loci. It is important to remember that initial interest in searching for HLA-associated risk factors in human cancer was prompted by the well-documented role of H2 genetics in murine leukemogenesis. The H2 associations involved haplotypes (combinations of genes on each of the pair of homologous chromosomes), not phenotypes, yet most human studies have been limited to phenotypes, and most discussion still revolves around the significance of single gene frequency differences. The unit of inheritance is the chromosome, not single genes. Thus, among unrelated cancer patients, HLA-associated genetic risk can best be sought in cancer patients by determining the frequencies of combinations of antigens as haplotypes and as co-occuring phenotypes. Where the risk can be expected to be low, as in the background incidence populations (Caucasians, Indians), haplotyping may well be essential for the detection of HLA or other biomarker-associated risk. Phenotyping a few tens of unrelated, heterogeneous patients cannot be regarded as much more than a fishing trip, simply because the order of magnitude of the expected risk is lower than can be detected by the test system using the accepted, conservative, corrective statistics.

The two HLA B locus genes associated with a risk for NPC are the same genes which contribute to the two haplotypes showing the strongest linkage disequilibrium in Chinese. It is therefore likely that the co-occuring phenotypes revealed by HLA typing provide indirect information on haplotype occurence. For example, virtually all HLA-AW33 subjects have HLA-BW58. However, not all HLA-BW58 subjects have HLA-AW33, not all HLA-BW46 individuals are HLA-A2 positive, and even fewer HLA-A2 positive subjects also have HLA-BW46. Furthermore, individuals who are phenotypically identical at the HLA A and B loci will differ at other loci in the D region (HLA-DR, DQ, DP) because of recombination, so HLA phenotyping is an unsatisfactory way of predicting haplotypes in unrelated subjects.

Family studies, which are the only satisfactory way of determining haplotypes, have been undertaken in Singapore Chinese NPC patients. The Singapore data indicates that the sequence of HLA genes on the A2, BW46- bearing haplotype in NPC may differ from that in normals (Chan and Wee, 1981; Chan et al., 1983b). In normal subjects BW46 is in linkage with DRW9. In NPC patients BW46 is increased in frequency, not in conjunction with DRW9, but with an as yet undefined DRW gene. A similar loss of linkage between the HLA genes of a commonly occuring haplotype has also been observed in Caucasian Cervical cancer (CaCx) patients among whom the BW44/ DRW7 linkage appears to be lost (Simons, 1981b). These observations suggest that a phenomenon involving loss of linkage is not unique to NPC, and that there will be a sub-population of only one or a few haplotypes among the common gene marker-bearing haplotypes that will prove to be strongly associated with NPC.

When family studies are extended to include multiple NPC cases, in addition to providing HLA haplotype assignments, information is obtained on mode(s) of genetic inheritence. Multiple case family studies also provide direct estimates of the full relative risk associated with biomarkers such as HLA through haplotype similarity in sibling/first cousin pairs of patients, avoid control group selection problems since within family segregation provides its own control, and avoid situations where different HLA genes are associated with a single disease in different ethnic groups, or where a particular gene is related to a disease in one population but apparently not in another, by considering HLA haplotypes rather than single genes.

Information from 21 families investigated in three separate studies (Hong Kong: 5 families - Ho, 1976; Guangdong Province: 5 families - Ou, 1983; Kwangsi Province: 11 families - Degos et al., 1984) indicates no requirement for 2-haplotype concordance, and hence weighs against a simple recessive model. Since there is also no requirement for the co-occurence of the two haplotypes bearing the HLA genes associated with high risk in the patient population, it is likely that each has a dominant mode of inheritance. The occurence of haplotypes other than those showing an association at the population level indicates that the disease-genetic association involves linkage with HLA genes present on chromosomes which are not discernible at the NPC patient population level.

At the recent Malaysian NPC symposium the view was repeatedly expressed that "a proper genetic study" was urgently needed in order to determine whether genetic risk had a role in NPC occurence (Klein, 1983). The study of familial risk for breast cancer in Iceland (Tulinius et al., 1982), which is possibly the most complete application of traditional genealogical approaches to the study of any human cancer, was referred to as an example for human geneticists interested in NPC to follow. As previously stated "it is doubtful if the informational infrastructure exists in any country where NPC is prevalent for an investigation as comprehensive as that of familial breast cancer in Iceland" (Simons and Shanmugaratnam, 1982). Furthermore, it is not obvious that such a study would contribute either to evidence for, or to an understanding of, the genetics of excess risk NPC more than is presently known from biomarker polymorphic genetics. In breast cancer, family studies have provided only the information that the highest risk marker is a positive family history. No information on the chromosomal localisation of any cancer-associated gene(s) can be revealed by such family studies. In NPC, by contrast, evidence for the involvement of chromosome 6-located HLA genes in multiple aspects of NPC occurence has been established beyond reasonable doubt. Also, there is preliminary evidence for the involvement of non-HLA genes (Kirk et al., 1978). Thus, the strong genetic element that is suspected in NPC is likely to have multiple components, and not be simply a single genetic system. A traditional genetic study of Chinese NPC families would be expected to reveal a genetic risk not less than that already shown to be associated with the HLA and non-HLA systems, and to provide a single, combined risk that fails to distinguish the separate genetic components. The genetic risk attributable to the HLA and non-HLA systems is already of a similar order of magnitude to that revealed in the highest risk subgroup (first degree relatives of premenopausal, bilateral cancer patients). Therefore, it is more appropriate to fully utilise existing information and to characterise the known and suspected polymorphic biomarkers, rather than to proceed with classical family studies as if no genetic information was available and to rediscover the already known genetic risk.

Carcinogenesis is a multi-stage process, so several levels of genetic action can be expected in the total risk attributable to genetically-determined phenomena. Animal models of genetic susceptibility and resistance provide some idea of the order of genetic complexity that can be anticipated in the human situation. For example, at least two levels of genetic resistance are recognised in Marek's disease, one at the level of target lymphoid cells for virus infection and transformation, and the second at the level of host immunological responses against virus and tumour antigens (Powell et al., 1982). While the latter level appears to be associated with chicken MHC genes equivalent to the human HLA gene system, the former is independant of MHC genes. The non-MHC genetic effect acts to influence susceptibility of T-lymphocytes to infection and transformation by Marek's disease virus. There may be some similarity with human BL in which B lymphocytes are both integral to immune function and themselves the target of EB viral infection. The relevant genetics is likely to involve that of the C3d receptor, or a closely similar structure, which serves as the EBV receptor.

One possibility for a non-MHC gene system in BL is that of the immunoglobulin heavy and light chain genes as revealed by allotypic markers. An interactive effect between Gm and Km homozygosity and elevated antibody titres to EBV antigens has been claimed (Biggar et al., 1984). However, insufficient allotypes were investigated to allow discrimination between homozygotes and heterozygotes for Gm and/or Km since, if an African population is not typed for G3m (b4,b5,s and c5), it is not possible to conclude that subjects with Gm (1,17;5,13) are homozygous, and if 50% of the subjects are Km(1) positive, then about 10% have to be Km(1+3−). Further studies are required which utilise reagents suitable for sufficient characterisation of the genes present in the different ethnic populations.

There is no substantive data on immunoglobulin allotypes in NPC. A2m allotyping may be of particular interest in view of the regular occurence of antibody to EB VCA of the IgA class. The IgA subclass distribution of the anti-EBVCA antibody is not known but, if the IgA2 subclass is involved , any distortion of the A2m(1) and A2m(2) gene frequencies will be readily detectable in Chinese among whom the gene frequencies are approximately 0.45 and 0.55 respectively, with corresponding phenotype frequencies for 1-1, 1-2 and 2-2 of 20-25%, 45-50% and 35-25% respectively. Investigations of this type can be expected to identify genetic variability additional to that associated with the HLA complex, and to clarify the contribution of immuno-globulin genes to the non-HLA component of inherited susceptibility to NPC.

The knowledge of HLA and other genetic associations with NPC can be applied to testing the role of environmental agents suspected of inducing/precipitating the disease process, and of testing hypotheses concerning causal mechanisms. EBV is a prime suspect as an aetiological agent in NPC, and in at least a proportion of BL cases. Questions that arise are why there is a failure of maintenance of the EBV latency state, whether a deficiency of immunoregulation is involved, whether the advent of immune responsiveness involving IgA antibody is an integral part of the EBV-related carcinogenic process, and whether immunogenetic mechanisms underlie the failure to maintain EBV genome suppression and/or the fate of antigen-expressing activated cells? An obvious first choice for study was the possible association between EBV infection/immunity and HLA gene type. Rickinson, Pope and their collegues (Rickinson et al., 1980; Pope et al., 1983) have identified HLA restriction of T-lymphocyte mediated cytotoxicity against EBV-specified antigens. There is a suggestion of haplotype preferential restriction of cytotoxicity which, if verified, may provide crucial insights into molecular mechanisms involving immunogene products. While most attention has been given to the role of haplotypes it should be remembered that, at the cell surface interface between gene products and environ-mental agents, it is the spatial arrangement of molecules comprising the phenotype which is the functionally relevant gene product combination. It follows that the trans as well as the cis configurations of MHC and other gene products as co-occuring phenotypes are the elements of functional significance. This fact is of fundamental importance to evolutionary considerations of polymorphism-based genetically inherited variation underlying genetic individuality. It is of major practical importance in confusing the analysis of family data seeking to detect disease inheritance by haplotype sharing. Assays of HLA-restricted, EBV-specific immunity exhibiting inter-allelic preference may substantiate the primacy of co-occuring phenotypes as phenogroups in genetic risk for disease.

NPC patients have diminished EBV-specific, T-cell mediated immunity as revealed by deficiency of regression of EBV transformation (Chan and Chew, 1981; Moss et al., 1983), but accompanying HLA studies have not yet been reported. There is a need to extend these assays of EBV-related T-cell function to T lymphocyte helper, suppressor and proliferative assays, and combine them with genetic typing.

It should not be forgotten that most of the class I MHC antigens in the mouse are not determined by H2 genes. An analogous situation in the human would predict the existence of class I molecules not encoded by the HLA A, B or C loci, but by the human homologue of the Qa/Tla complex which is likely to be relevant to oncogenesis and tumour immunity (Brickell et al., 1983; Lancet, 1984a).

The response to radiotherapy and general survival pattern of low incidence Caucasian and high incidence Chinese NPC patients is quite similar, suggesting that the difference between high and low incidence populations is confined to the occurence of NPC and not to disease course and prognosis. Once NPC occurs, the outcome seems to be independant of ethnic differences in risk. Any model of the aetiopathogenesis of excess risk for NPC susceptibility must account for the cumulative incidence of 1.5% in Cantonese males, and for a 2-fold lesser incidence in Cantonese females and in Hokkien males compared to Cantonese males. If EBV exposure is a requirement for NPC development and infectivity approaches 100%, and since the risk already attributable to MHC genes is of the order of 40-50% (Simons and Shanmugaratnam, 1982), a minimum model requires only a third component to account for the peak incidence in the highest incidence, Cantonese male population.

Application of the strategies considered here, with special attention to restriction of patient heterogeneity (Simons and Amiel, 1977; Simons, 1979; Simons and Shanmugaratnam, 1982) and consecutive family ascertainment, may enable the essential ecogenetic issues concerning heritable variation in response to environmental exposures to be clarified. Concurrent investigation of DNA restricted fragment length polymorphisms (RFLP) in the MHC region, which can be expected to assist in characterisation of sero-indistinguishable HLA haplotypes, and of RFLPs elsewhere in the genome, and search in tumour tissue for chromosomal rearrangements by cytogenetic examination and by use of oncogene probes , may reveal any NPC-related molecular genetic characteristics. These approaches can be expected to provide information on inherited or somatic events underlying differences between familial and non-familial disease. It is a realistic hope that there may be a convergence of ecogenetic and molecular genetic information towards a better understanding of the aetiopathogenesis of NPC by the 2nd conference in 1986.

ACKNOWLEDGEMENTS

The early genetic studies of NPC which were crucial in establishing the role of HLA-associated genes were supported by IARC research agreements under contract to NCI, NIH, and undertaken at the WHO Immunology Research and Training Centre, University of Singapore. I wish to acknowledge the stimulating and productive discussions with Prof. Chan Soh Ha, Dr. Nicholas Day, Prof. K. Shanmugaratnam and Dr. Brian Tait which have contributed to many of the ideas developed in this paper.

REFERENCES
ABLASHI, D.V., PRASAD, U., PEARSON, G.R., PRATHAP, K., ARMSTRONG, G.R., FAGGIONI, A., YADAV, M., EASTON, J.M., CHAN,S.H. and LEVINE, P.H. EBV-related studies with clinico-pathological correlation in Malaysian NPC. In: Prasad, U. et al.(eds), NPC: Current Concepts, pp. 163-171, University of Malaya, Kuala Lumpur (1983).

ANDERSON, D.E. Familial and genetic predisposition. In: B.A. Stoll (ed), Risk factors in breast cancer, pp. 3-24,Wm. Heinemann Medical Books (1976).

BIGGAR, R.J., PANDEY, J.P., HENLE, W., NKRUMAH, F.K. and LEVINE, P.H. Humoral immune response to Epstein-Barr virus antigens and immunoglobulin allotypes in African Burkitt's lymphoma patients. Int. J. Cancer, 33, 577-580 (1984).

BRICKELL, P.M., LATCHMAN, D.S., MURPHY, D., WILLISON, K. and RIGBY, P.J.W. Activation of a Qa/Tla class I major histocompatibility antigen gene is a general feature of oncogenesis in the mouse. Nature, 306, 756-759 (1983).

CHAN, S.H. and CHEW, T.S. Lack of regression in Epstein Barr Virus infected leucocyte cultures of nasopharyngeal carcinoma patients. Lancet, 2, 1353 (1981).

CHAN, S.H. and WEE, G.B. HLA and nasopharyngeal cancer in Singapore Chinese. In: M.J. Simons and B.D. Tait (eds), Proceedings of the second Asia and Oceania histocompatibility workshop conference, pp. 495-496, Immuno-publishing, Melbourne (1981).

CHAN, S.H., DAY, N.E., KUNARATNAM, N., CHIA, K.B. and SIMONS, M.J. HLA and Nasopharyngeal carcinoma in Chinese - a further study. Int. J. Cancer, 32, 171-176 (1983a).

CHAN, S.H., WEE, G.B., KUNARATNAM, N., CHIA, K.B. and DAY, N.E. HLA locus B and DR antigen associations in Chinese NPC patients and controls. In: U. Prasad et al.(eds), NPC: Current Concepts, pp. 307-312, University of Malaya, Kuala Lumpur (1983b).

DAY, N.E. and SIMONS, M.J. Disease susceptibility genes - their identification by multiple case family studies. Tissue Antigens, 8, 109-119 (1976).

DEGOS, L., LEPAGE, V., DE THE, G.B., BLANC, H., FEINGOLD, N., LU, S.T. and ZENG, Y. HLA genotypes, Gm genotypes, IgA anti-EBV antibodies in multiple family cases of naso-pharyngocarcinoma in China. In: 9th. International histocompatibility workshop and conference, Abstract 9W258, p.59, Vienna (1984).

DOLL, R. and PETO, R. The causes of cancer. Oxford University Press, Oxford (1981).

HO, J.H.C. Personal communication (1976).

KLEIN, G. The role of gene dosage and genetic transposition in carcinogenesis. Nature, 294, 313-318 (1981).

KLEIN, G. Summary of the symposium . In: U. Prasad et al. (eds), NPC: Current Concepts, pp.443-453, University of Malaya, Kuala Lumpur (1983).

KIRK, R.L., BLAKE, N.M., SERJEANTSON, S., SIMONS, M.J. and CHAN, S.H. Genetic components in susceptibility to nasopharyngeal carcinoma. In: G. de The and Y. Ito (eds), Nasopharyngeal carcinoma: aetiology and control, pp.283-297, IARC Scientific Publications No. 20, IARC, Lyon (1978).

LANCET. Some progress with nasopharyngeal carcinoma, 2, 959-960 (1979).

LANCET. Genes, cancer and the immune system, 1, 62 (1984a).

LANCET. Clues from familial cancer, 1, 1219-1220 (1984b).

MOSS, D,J., CHAN,S.H., BURROWS, S.R., CHEW, T.S., KANE, R.G., STAPLES, J.A. AND KUNARATNAM, N. Epstein Barr virus specific T-cell response in nasopharyngeal carcinoma patients. Int. J. Cancer, 32, 301-305 (1983).

MULVIHILL, J.J., MILLER, R.W. and FRAUMENI, J,F. (eds), Genetics of human cancer. Progress in cancer research and therapy, Vol. 3., pp.1-519 (1977).

OU, B.X. Personal communication (1983).

PETO, J. Genetic predisposition to cancer. In: J. Cairns (ed), Banbury report 4 - Cancer incidence in defined populations, pp.203-213, Cold Spring Harbor , New York (1980).

POPE, J.H., MISKO, I.S. and MOSS, D.J. Specificity of the Cytotoxic T Cell Response to EB Virus in Vitro. In: Prassad, U. et al.(eds), NPC: Current Concepts, pp.299-305, University of Malaya, Kuala Lumpur. (1983).

POWELL, P.C., LEE, L.F. MUSTILL, B.M. and RENNIE, M. The mechanism of genetic reistance to Marek's disease in chickens. Int. J. Cancer, 29, 169-174 (1982).

PURTILO, D.T. Immunopathology of X-linked lymphoproliferative syndrome. Immunology Today, 4, 291-297 (1983).

RICKINSON, A.B., WALLACE, L.E. and EPSTEIN, M.A. HLA-
restricted T-cell recognition of Epstein-Barr virus-infected
B cells. Nature (Lond) 283: 865-867. (1980).

SIMONS, M.J., WEE, G.B., SINGH, D., DHARMALINGHAM, S., YONG,
N.K., CHAU, J.C.W., HO, J.H.C., DAY, N.E. and DE THE, G.
Immunogenetic aspects of nasopharyngeal carcinoma. V.
Confirmation of a Chinese-related HLA profile (A2, Singapore
2) associated with an increased risk in Chinese for
nasopharyngeal carcinoma. Natl. Cancer Inst. Monograph No.
47, pp.147-152 (1977).

SIMONS, M.J. and AMIEL, J.L. HLA and malignant diseases. In:
J. Dausset and A. Svejgaard (eds), HLA and disease, pp.
212-232, Munskgaard, Copenhagen (1977).

SIMONS, M.J., CHAN, S.H., WEE, G.B., SHANMUGURATNAM, K.,
GOH, E.H., HO, J.H.C., CHAU, J.C.W., DHARMALINGHAM, S.,
PRASAD, U., BETUEL, H., DAY, N.E. and DE THE, G. Nasopharyngeal carcinoma and histocompatibility antigens. In: G.
de The and Y. Ito (eds), Nasopharyngeal carcinoma: aetiology
and control, pp. 271-282, IARC Scientific Publications No.
20, IARC, Lyon (1978).

SIMONS, M.J. Interaction between genetic and environmental
factors in human cancer. In: J.M. Birch (ed), Advances in
Medical Oncology, Research and Education, Vol. 3 -
Epidemiology, pp. 169-178, Pergamon Press, Oxford (1979).

SIMONS, M.J. Nasopharyngeal carcinoma (NPC), including
analysis of HLA gene patterns in Chinese patients with
cervical and hepatocellular carcinoma. In: M.J. Simons and
B.D. Tait (eds), Proceedings of the second Asia and Oceania
histocompatibility workshop conference, pp. 369-378, Immuno-
publishing, Melbourne (1981a).

SIMONS, M.J. Cervical carcinoma (CaCx). In: M.J. Simons and
B.D. Tait (eds), Proceedings of the second Asia and Oceania
histocompatibility workshop conference, pp. 363-366, Immuno-
publishing, Melbourne (1981b).

SIMONS, M.J. and SHANMUGARATNAM, K. (eds), The biology of nasopharyngeal carcinoma. UICC Technical Report Series, Vol. 71, UICC, Geneva (1982).

TULINIUS, H., DAY, N.E., BJARNASON, O., GEIRSSON, G., JOHANNESSON, G., DE GONZALEZ, M.A.L., SIGVALDASON, H., BJARNADOFFIR, G. and GRIMSDOTTIR, K. Familial breast cancer in Iceland. Int. J. Cancer, 29, 365–371 (1982).

10

PATHOLOGY OF EPSTEIN-BARR VIRUS (EBV)-ASSOCIATED

DISEASE   (THE LYMPHATIC SYSTEM)

Gerhard R.F.Krueger

Immunopathology Laboratories,
Pathology Institute, University of Cologne, 5000 Cologne 41, FRG

SUMMARY

Morphological changes in EBV infections are
determined by the specific activity of the virus
and by the host's immune response. EBV causes a
T-cell independent polyclonal B-cell proliferation, a T-suppressor cell activation, and is able
to transform infectable B-lymphocytes.
Healthy individuals exhibit upon EBV infection a
marked hyperplasia of lymphatic B-zones with immunoblastic transformation and plasmacytosis. Associated is a diffuse T-zone hyperplasia with increase in cytotoxic T-lymphocytes. The latter with
K- and NK-cells destroy transformed cells and limit the disease: infectious mononucleosis.
In T-cell deficiency, T-cell independent B-cell
proliferation progresses mimicking malignant lymph
oma. Such patterns are observed in persistent infectious mononucleosis, XLP syndrome, some transplant recipients and AIDS (partly complicated by
CMV and HTLV3 organisms).
Immune deficiency also allows atypical cells to
grow resulting in various malignant neoplasms such
as BURKITT-type tumors, KAPOSI's sarcoma etc. Also
in NPC EBV-carrying lymphocytes and immunologically responding cells probably assist in EBV transfection, transformation and growth of epithelial
cells.

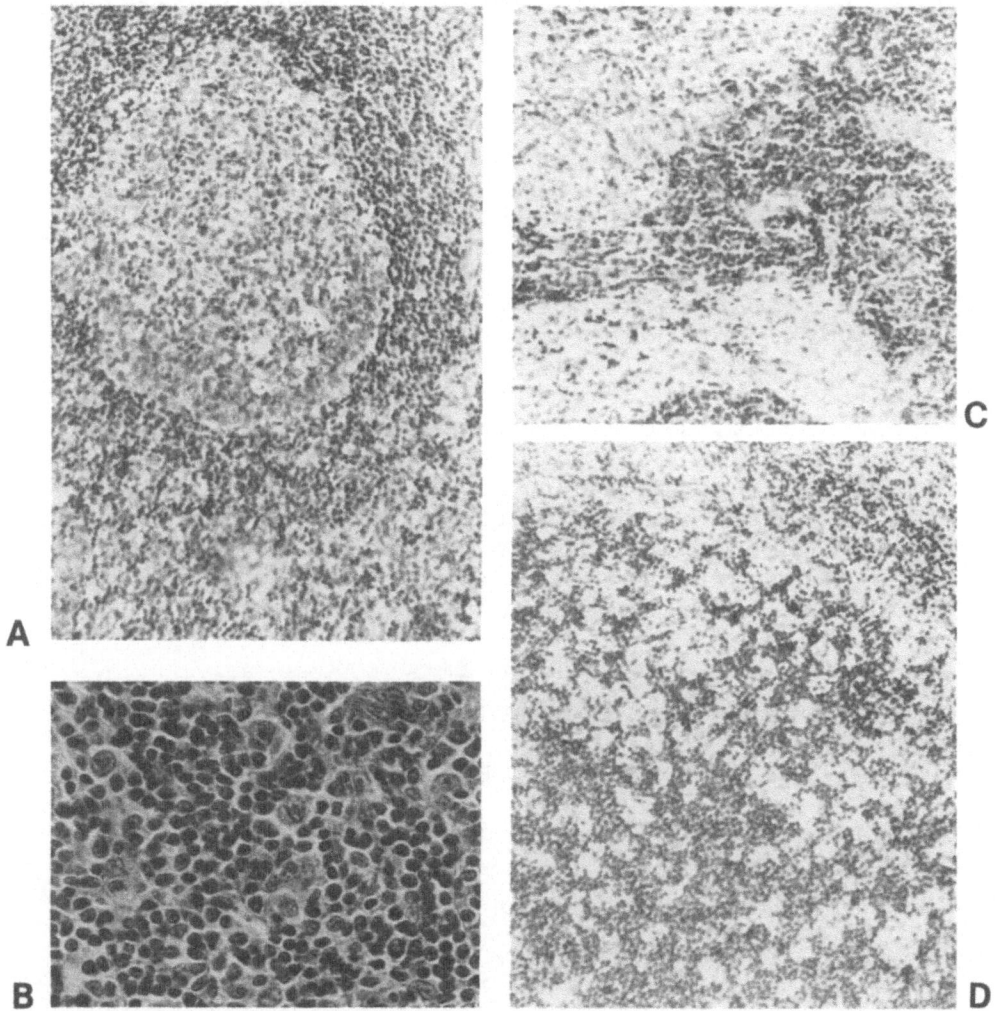

Figure 1. Reactive hyperplasia in an immunologically intact person (Lymph node). a) B-zone (follicular; H&E, 150x); b) T-zone (paracortical; H&E,
375x); c) sinus histiocytosis (phagocytic; H&E,
150x); d) reticulo-histiocytosis (phagocytic; H&E,
150x)

The lymphoid system allows in a unique way to correlate specific stimulation and function with cytological and structural changes (Cottier et al.,
1973; Syrjänen, 1982). Several functional units
are identified which respond with hypertrophy and
atrophy upon antigenic stimulation and toxic influences (Fig.1) The lymph node cortex hypertrophies in activation of the B-cell system leading
to antibody formation (follicular hyperplasia);
production of antibodies is accompanied by plasma
cell differentiation in the medullary cords.
The lymph node paracortex hypertrophies in activation of the T-cell system (diffuse hyperplasia)
leading to T-cell associated functions (T-cell
cytotoxicity, T-cell immune regulation). Sinus
endothelia and diffusely scattered macrophages hypertrophy in activation of the phagocytic system
(sinus histiocytosis, diffuse reticulo-histiocytosis). Cytologic changes can be monitored by various cell marker studies (Fig.2; Table 1; Warnke
and Levy, 1981; Sesterhenn et al., 1976).

TABLE I

CELL POPULATIONS IN REACTIVE LYMPH NODES[1]

T-cells	T_4/T_8 Cells	B-cells[2]	NK-cells[3]
26-40	2.8-1.5	30-40	7-10

[1] % of total mononuclear cells

[2] polyclonal (SIgG 8-12, SIgM 12-17, SIgA 6-12
SIgD 4-8, SIgE 0-4%)

[3] natural killer cells (Leu7 marker)

Infections will induce a balanced T- and B-cell
response (follicular and paracortical hyperplasia)
provided the functional units of the lymphatic system are intact, and there is no additional "toxic"
influence by the infectious organism besides it's
antigenic nature.

Epstein-Barr-Virus (EBV) is a lymphotropic virus
which replicates in cells of the B-lymphocyte lineage and thus exerts effects on the immune system

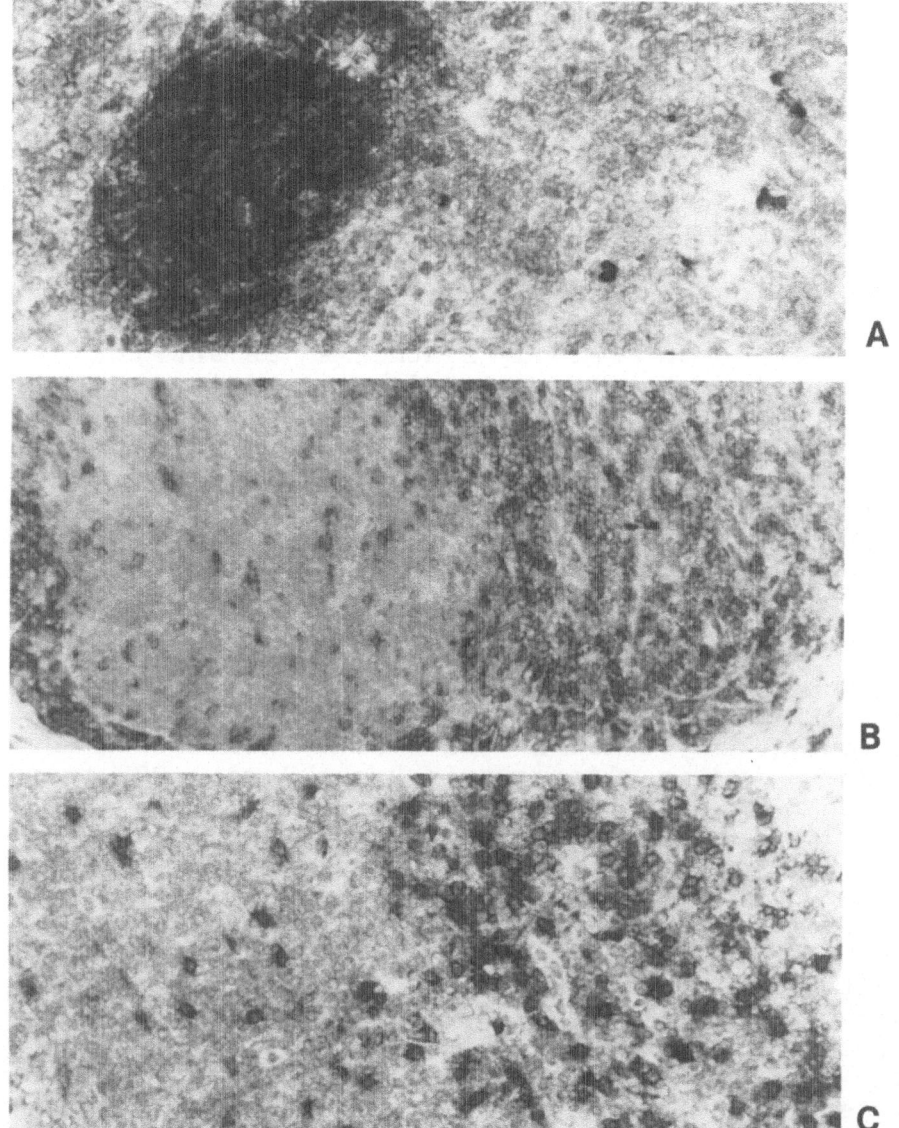

Figure 2. Immune histology in reactive lymph node hyperplasia (immune peroxidase method on frozen sections). a) Follicular B-lymphocytes (OKB7, 240x); b) T_4-helper cells in paracortex (OKT4, 240x);

c) T_8-suppressor cells in paracortex (OKT8, 375x)

in addition to it's antigenic stimulation. Some of
the effects are summarized in Table 2 (Rosen et al
1977; Fong et al., 1982; Purtilo and Sakamoto,
1981; Sonnabend et al., 1983; Epstein and Achong,
1979; Krueger, 1984).

TABLE II

EPSTEIN-BARR-VIRUS EFFECTS (DIRECT & INDIRECT)

+ Infection of B-lymphocytes - lytic, transform
                              ative, latent
+ Induction of antibody synthesis (as antigen)
+ Induction of cell-mediated cytotoxicity (against
                              EBV-infected cells)
+ T-cell independent polyclonal B-cell stimulat-
                              ion (as "mitogen")
+ T-suppressor cell activation
+ Induction of immune interferon
+ Infection or lympho/epithelial transfection
  of epithelial cells (nasopharynx, salivary
                              glands, thymus)

Consequently, hyperplastic changes in lymph nodes
after EBV infection are in excess of the usual
response: polyclonal B-cell stimulation causes an
advanced lympho-plasmacytoid hyperplasia; EBV-trans
formed B-lymphocytes will stimulate markedly the
T-cell defense (excessive paracortical hyperplasia).
T-cell factors may activate the phagocytic system
(reticulo-histiocytosis, epithelioid cell reaction
and eventual granuloma formation).
The balanced host response against EBV occurs in
healthy individuals with intact immune system.
Early childhood infection leads usually to silent
seroconversion without clinically or pathologic-
ally manifest disease (DeThé, 1980), but with persistent latent virus. Adolescent infection causes
a self-limited lymphoproliferative syndrome and
virus persistence: Infectious mononucleosis (Fig.
3).
Lymphoproliferation of the above described pattern
can be rather extensive simulating malignant lymph
oma. Between 6-20% of blood lymphocytes in the
early phase of the disease are EBNA-positive trans-

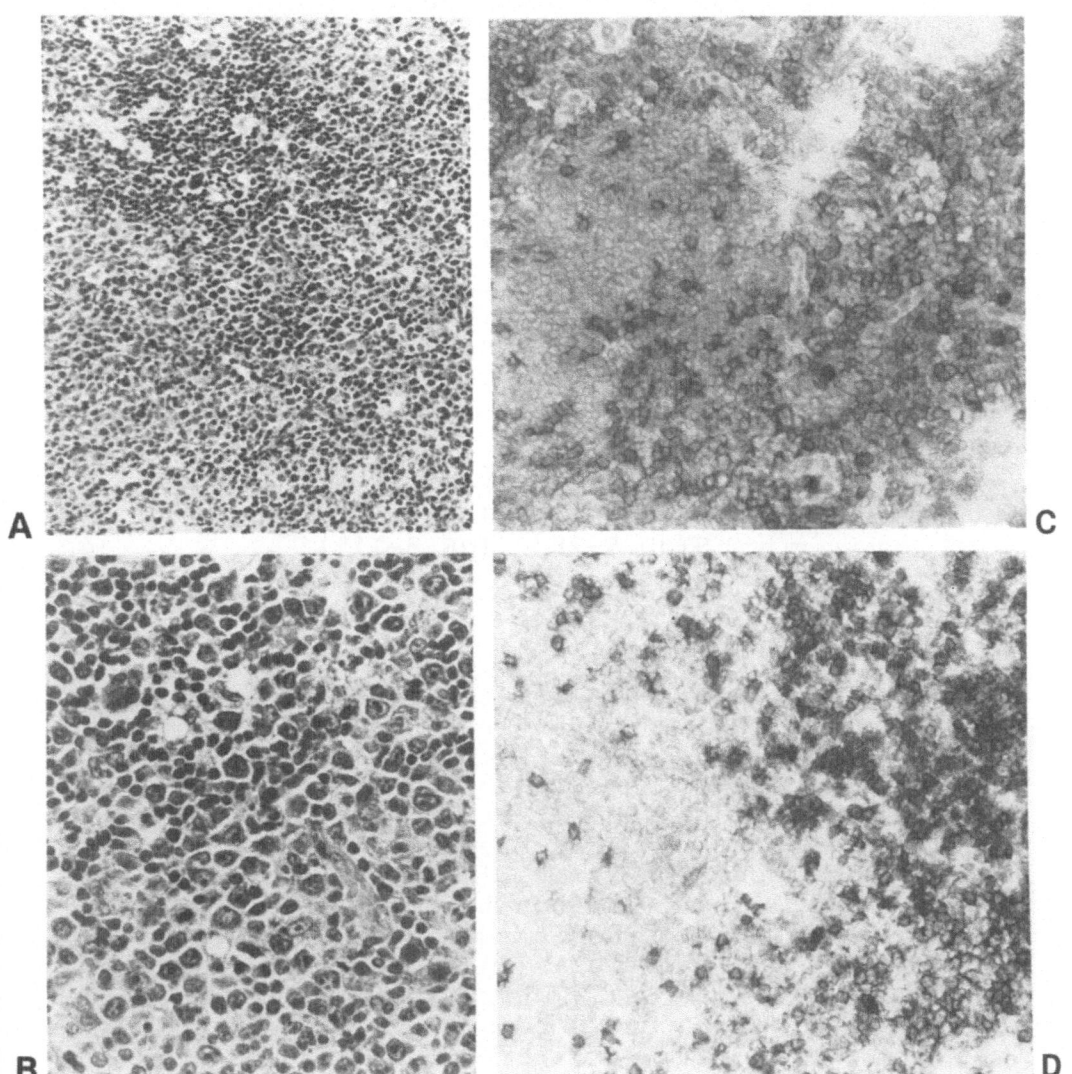

Figure 3. Marked paracortical hyperplasia in infectious mononucleosis of an immunologically intact patient. a) Lower magnification (H&E, 150x); b) higher magnification showing many blast cells (H&E, 375x). c) T_4-helper cells in paracortex (OKT4, 240x); d) increased numbers of T_8-suppressor/cytotoxic cells in paracortex (OKT8, 240x)

formed cells (Klein et al., 1981; Robinson et al., 1981; Lennert et al., 1981). Immune cytologic investigations always demonstrate a polyclonal lymphocyte proliferation (Table 3).

TABLE III

CELL POPULATIONS IN UNCOMPLICATED IM LYMPH NODES[1]

T-cells	T_4/T_8 cells	B-cells[2]	NK-cells[3]
35-45	1.1-0.9 init.	25-35	5-20
	1.5-3.1 later[4]		

[1] % of total mononuclear cells (5 patients)
[2] polyclonal (SIgG 8-12, SIgM 9-38, SIgA 5-21, SIgD 5-28, SIgE 3-11%)
[3] natural killer cells (Leu7 marker)
[4] in clinically convalescents

Recovery from the self-limited lymphoproliferative syndrome of IM is apparently dependent upon effective host defense by K-cells, NK-cells, cytotoxic T-cells and macrophages (Henle and Henle, 1979). Deficiency in one segment of the host's defense will imbalance the entire reaction to EBV infection and in addition will allow superimposed infections and malignant lesions to occur. The morphologic reaction patterns in such cases can be deduced from Table 1: regular follicular hyperplasia, which is T-helper cell controlled, will disappear. Instead, there occurs a progressive T-cell independent polyclonal B-cell proliferation and EBV-transformed lymphocytes persist and proliferate. B-lymphocytes may be in 50% or more EBNA positiv The T-zone becomes depopulated and atrophic. Total T-cell counts are decreased and T-helper cells in favor of T-suppressor cells are reduced. Superimposed infections may add inflammatory and necrotic lesions. Eventually neoplasms develop: B-immunoblastic lymphomas, Kaposi's sarcoma or other. Some of the resulting clinico-pathologic entities are summarized in Table 4.

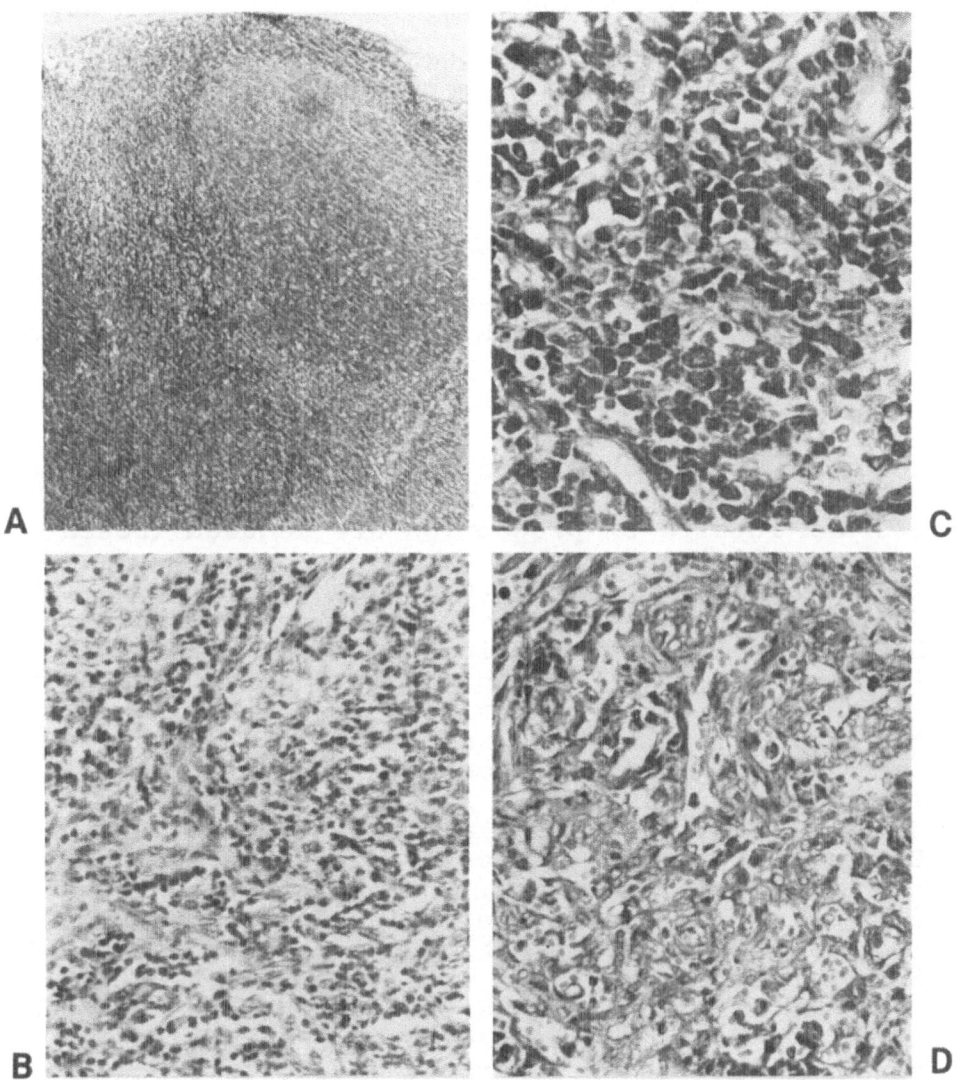

Figure 4. Lymph node changes in acquired immune
deficiency syndrome (AIDS). a) Lymphadenopathy
syndrome (stage I) with marked cortical and paracortical hyperplasia (H&E, 68x); b) AIDS with paracortical atrophy (stage II)(H&E, 240x); c) diffuse
atypical polyclonal lymphoproliferation in entire
lymph node (stage III)(Giemsa, 375x); d) Kaposi's
sarcoma with mixed atypical fibro-vascular proliferation (stage III) (H&E, 240x)

TABLE IV

EBV-RELATED CLINICO-PATHOLOGIC CONDITIONS IN
IMPAIRED HOST RESPONSE

+ Persistent infectious mononucleosis
+ Acute lethal infectious mononucleosis
+ Hypogammaglobulinemia
+ Aplastic anemia and other hematopoietic def-
  iciencies
+ Atypical polyclonal B-cell lymphoproliferation
+ Malignant lymphomas (Burkitt's lymphoma, B-cell
                      immunoblastic lymphoma)
+ Nasopharyngeal carcinoma

Accordingly, T-suppressor cell activity is increased in chronic persistent IM, NK-cell functions,
IL2 and interferon production are reduced (Jones
et al., 1984).
The development of structural changes in immune
deficiency and associated viral disease (including
EBV) is well documented in acquired immune deficiency syndrome (AIDS). Antigenic overloading, auto-
immunization by sperm antigen cross-reacting with
the host's lymphocytes, and T-helper cell lysis
by HTLV3 may initiate T-cell immune deficiency
(Sonnabend et al., 1983; Klatzman et al., 1984;
Gallo et al., 1984). Various organisms including
EBV and CMV stimulate T-cell independent polyclonal B-cell proliferation. Common infectious organisms in such patients indicating antigen overloading are summarized in Table 5, some of their effect
in Table 6.
Lymph node changes during the disease develop from
extensive follicular and diffuse hyperplasia to
paracortical (T-zone) atrophy and diffuse polyclonal lymphoplasmacytic proliferation mimicking malignant lymphoma (Figs.4 and 5)(Krueger et al.,
1983; Reichert et al., 1983; Guarda et al., 1983).
Immune cytologic investigations demonstrate more
clearly the quantitative T-cell defect with increasing B-lymphocytes (Table 7). The disturbed
T-cell immune regulation becomes clinically overt
in polyclonal hypergamma globulinemia and autoimmune reactions, circulating immune complexes,

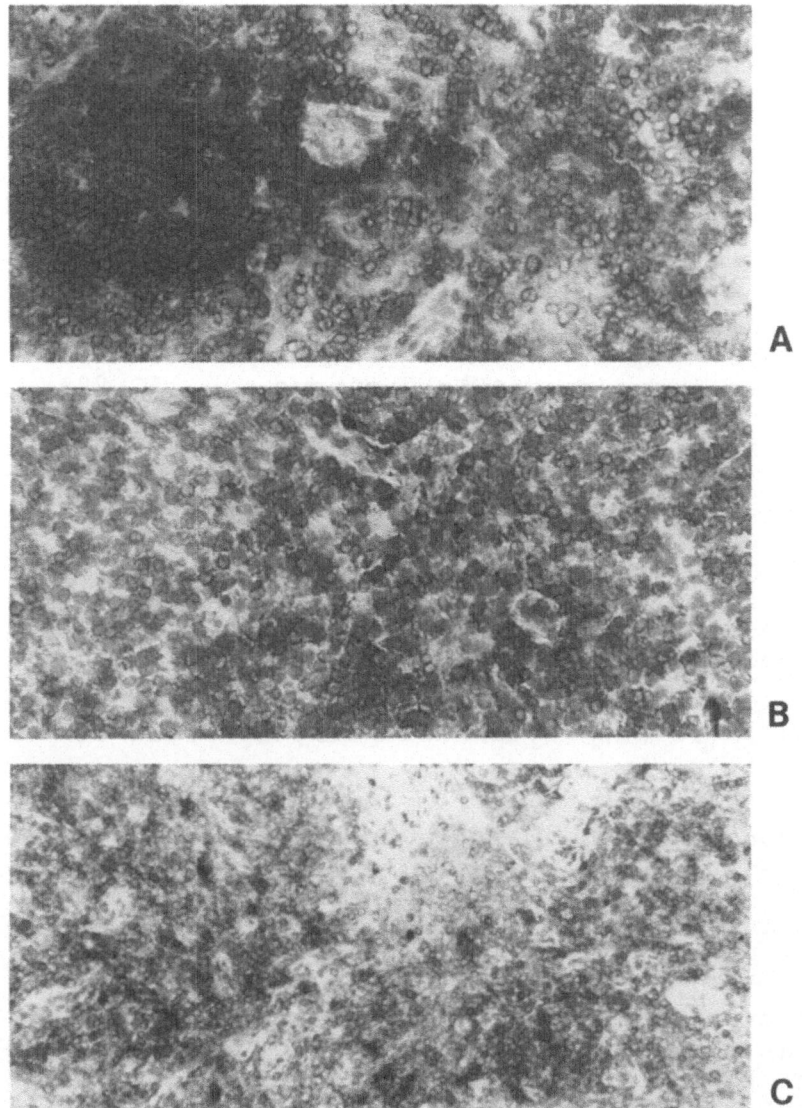

Figure 5. Cell populations in lymph node of AIDS
patient (stage II and III). a) B-lymphocytes are
diffusely increased in follicular remnant and in
paracortex (OKB7, 240x); b) diffusely increased
T_8-suppressor cells (OKT8, 240x); c) T_4-helper
lymphocytes are markedly reduced in all regions
(OKT4, 150x)

skin test anergy, depressed NK-cell activity, as
well as defective chemotaxis and phagocytosis
(Lane et al., 1983; Sonnabend et al., 1984; Groopman, 1984).

TABLE V

PERSISTENT OR ACUTE INFECTIONS IN PATIENTS WITH
LYMPHADENOPATHY SYNDROME AND AIDS

+ Epstein-Barr-Virus (EBV)   86-100%
+ Cytomegalovirus (CMV)   94-100%
+ HTLV3   62-97%
   (in lymphadenopathy syndrome   50-89%)
+ Hepatitis B & A   29-52%
+ Chlamydia organisms   -23%
+ Entameba coli & histolytica   -18%
+ Neisseria gonorrhea   -17%
+ Campylobacter-like organisms   -16%
+ Treponema pallidum   -15%
   in addition:
   Pneumocystis carinii, Herpes simplex, Candida
   species, Escherichia coli

(from 268 patients at 3 centers; 86 with AIDS for
HTLV3, 190 with lymphadebopathy syndrome for
HTLV3)

A second syndrome which even more clearly demonstrates the EBV-associated pathology in immune
deficient individuals was described by David Purtilo and his group (Purtilo et al., 1977; Purtilo,
1981; Purtilo, 1984): X-linked lymphoproliferative
syndrome (XLP). This immunodeficiency syndrome is
characterized by a defective T-lymphocyte function as summarized in Table 8 predisposing the patient to enhanced uncounteracted EBV effects.
Pathologic changes in XLP depend upon the stage
of disease and upon the degree of immune disturbance including both aplastic and necrotic lesions
in lymphatic and hematopoietic tissues as well as
polyclonal lymphoproliferation as described above ("aproliferative and proliferative syndromes").
True malignant lymphomas of B-immunoblastic type
arise late in the course of the disease (Fig.6).

117

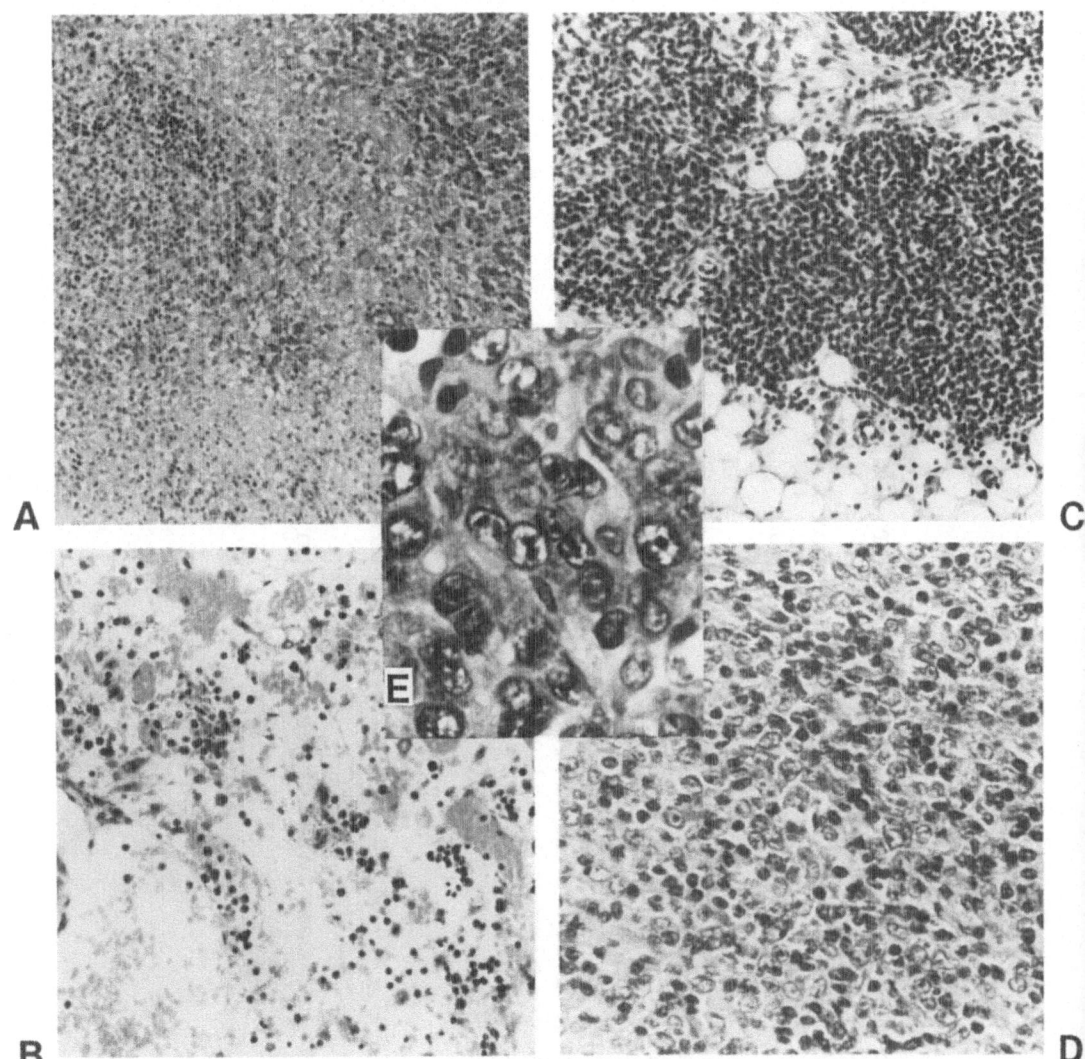

Figure 6. Lymphatic and hematopoietic tissue lesions in EBV-infected immunodeficient patient with
XLP syndrome. a) extensive necrosis of lymph node
(H&E, 150x); b) hematopoietic hypoplasia in bone
marrow (H&E, 150x); c) thymic atrophy (or hypoplasia ?) (H&E, 150x); d) diffuse polyclonal B-
lymphocyte proliferation (note a certain similarity of cell composion in infectious mononucleosis)
(H&E, 240x); e) occasional B-cell immunoblastic
lymphoma (H&E, 600x)

TABLE VI

DISTURBED IMMUNE REGULATION IN AIDS BY INFECTIOUS AND OTHER ANTIGENS

Disturbance	Agents				
	HTLV3	EBV	CMV	E.coli	Semen
T-helper cell lysis	+				
Thymic epithelial cell lysis		+			
T-suppressor cell activation		+	+		
Polyclonal B-cell stimulation		+	+		
Suppressor-monocyte activation			+		
Induction of autoantibody formation[1]			+		+
Facil. of immune complex formation[2]		+	+	+	+
(Malignant) transformation of B-lymphoc.		+			
Induction of angiogenesis (& fibrosis)[3]			+		

[1] including autoantibodies against lymphocytes suggestively induced by sperma

[2] circulating immune complexes interfere with effective cytotoxic T-cell and macrophage function

[3] "Co-pathogen" for the development of Kaposi's sarcoma ?

TABLE VII

CELL POPULATIONS IN AIDS LYMPH NODES[1]

Stage[2]	T-cells	T_4/T_8 cells	B-cells[3]
I	18-35	4.0-2.0	28-49
II	9-24	4.0-0.5	35-52
III	6-15	0.5-0.12	48-75

[1] % of total mononuclear cells

[2] stage I: lymphadenopathy syndrome (15 patients)
stage II: early AIDS with opportunistic infect.)
stage III: late irreversible AIDS (10 patients
          in stage II & III)

[3] polyclonal with extensive intracytoplasmic Ig

TABLE VIII

DEFECTS OF HOST DEFENSE IN XLP (DUNCAN-SYNDR.)

+ Defective antibody production against EBV (EBNA)
+ Defective memory T-cell production
+ Deficient secondary immune response
  (no IgM/IgG switch)
+ Deficient polyclonal Ig production by B-cells
  in vitro (upon antigenic stimulation)
+ Defective leukocyte migration inhibition by
  specific antigen
+ Inversion of T_4/T_8 ration

+ Inconstant defect of NK-cell function
+ Thymic epitheliolysis (with secondary defect in
  T-cell maturation?)

The thymus gland becomes progressively infiltrated
by plasma cells and atrophies. Of special interest
with regard to the apparent potential of EBV to
also infect epithelial cells is the observation of
Purtilo of thymic epitheliolysis in XLP (Purtilo,
1984). This concurs with a report of others (Seemayer et al., 1984) that thymic involution with

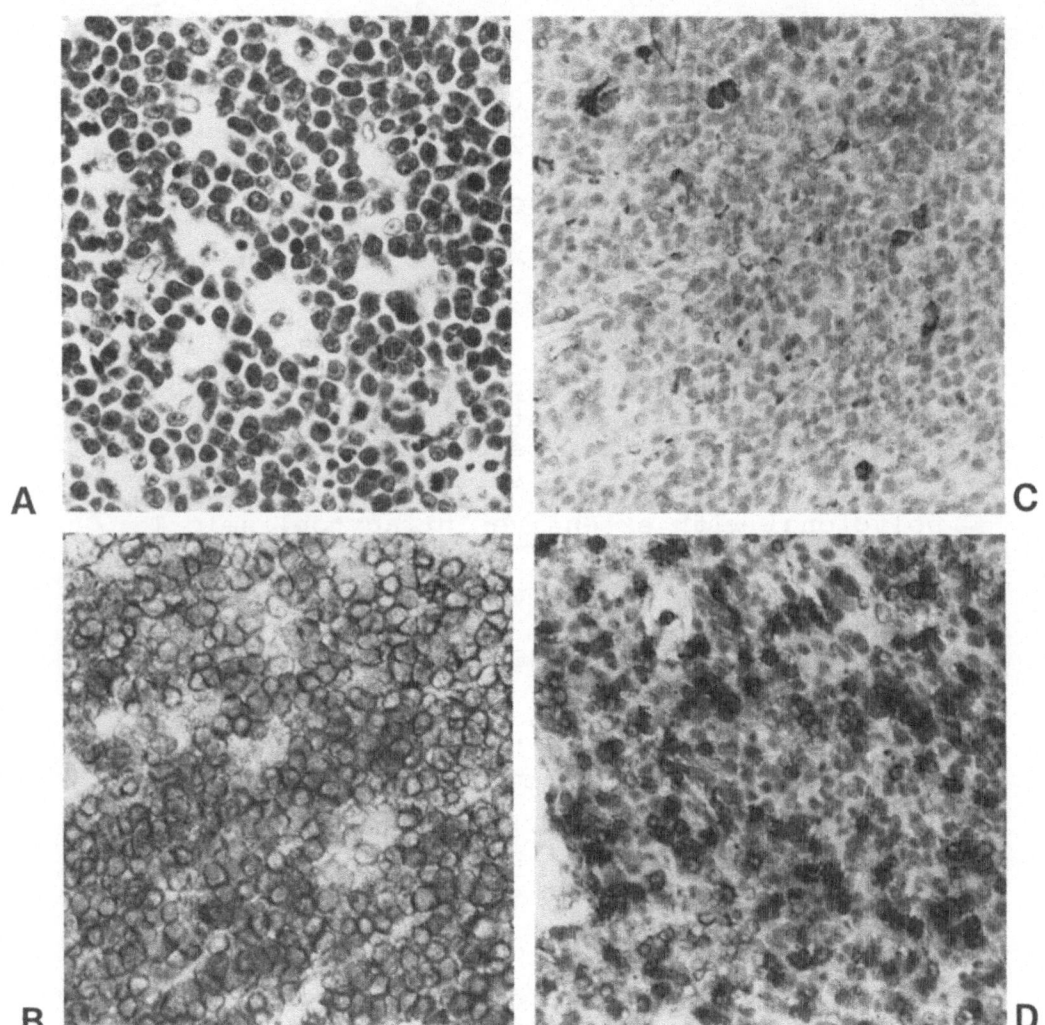

Figure 7. Malignant lymphoma of Burkitt's type
(EBV-positive). a) diffuse lymphoblastic proliferation with nuclear-debris-macrophages ("starry
sky phenomenon")(H&E, 375x); b) monoclonal tumor-
B-cell proliferation (OKB7, 375x); c) few scattered T_8-suppressor cells in lymphoma tissue(OKT8;
240x); d) loosely arranged T_4-helper cells in the
tumor (OKT4; 240x)

epithelial injury may also be seen in AIDS patients.
XLP is an extremely useful model for all pathologic conditions that may develop in EBV-infected
immunodeficient patients. The resulting clinico-
pathologic diseases essentially are those shown
in Table 4 (except for nasopharyngeal carcinoma
which has not yet been observed).

Infectious mononucleosis in the elderly and lethal infectious mononucleosis in defective host response may develop similar pathologic changes including also patients with allotransplants on immunosuppressive medication and patients with inherited immune deficiency syndromes other than
XLP (Linder and Purtilo, 1984; Reece et al., 1981;
Hanto et al., 1981). In acquired immune deficiency
syndromes, EBV infection must not be necessarily
a primary exogenous infection but rather an "endogenous re-infection" or virus reactivation such
as by CMV in AIDS. Thus, in the middle European
and North American population with life-long latent EBV in 85-95% essentially anybody may succumb from EBV-associated disease provided his normal T-cell controlled defense system will become
defective.

Two additional EBV-associated diseases with a different course must be mentioned: Burkitt's lymphoma (BL) and nasopharyngeal carcinoma (NPC).
Burkitt's lymphoma as compared to IM results apparently from unrestricted growth of EBV-transformed B-lymphocytes frozen in a certain narrow range
of differentiation. The histologic picture resembles a monoclonal proliferation of such cells (Fig.
7), more than 95% of which are carrying multiple
copies of EBV-DNA in classical (African) BL (Olweny et al., 1977; Geser et al., 1983). Instead of
the multicellular response to EBV in the diseases
mentioned so far, we are facing in BL a predominantly monocellular response suggesting a massive
infection and the presence of large numbers of
cells susceptible to infection and transformation
(DeThé, 1980). In classical BL, this situation apparently arises from the infection at young age

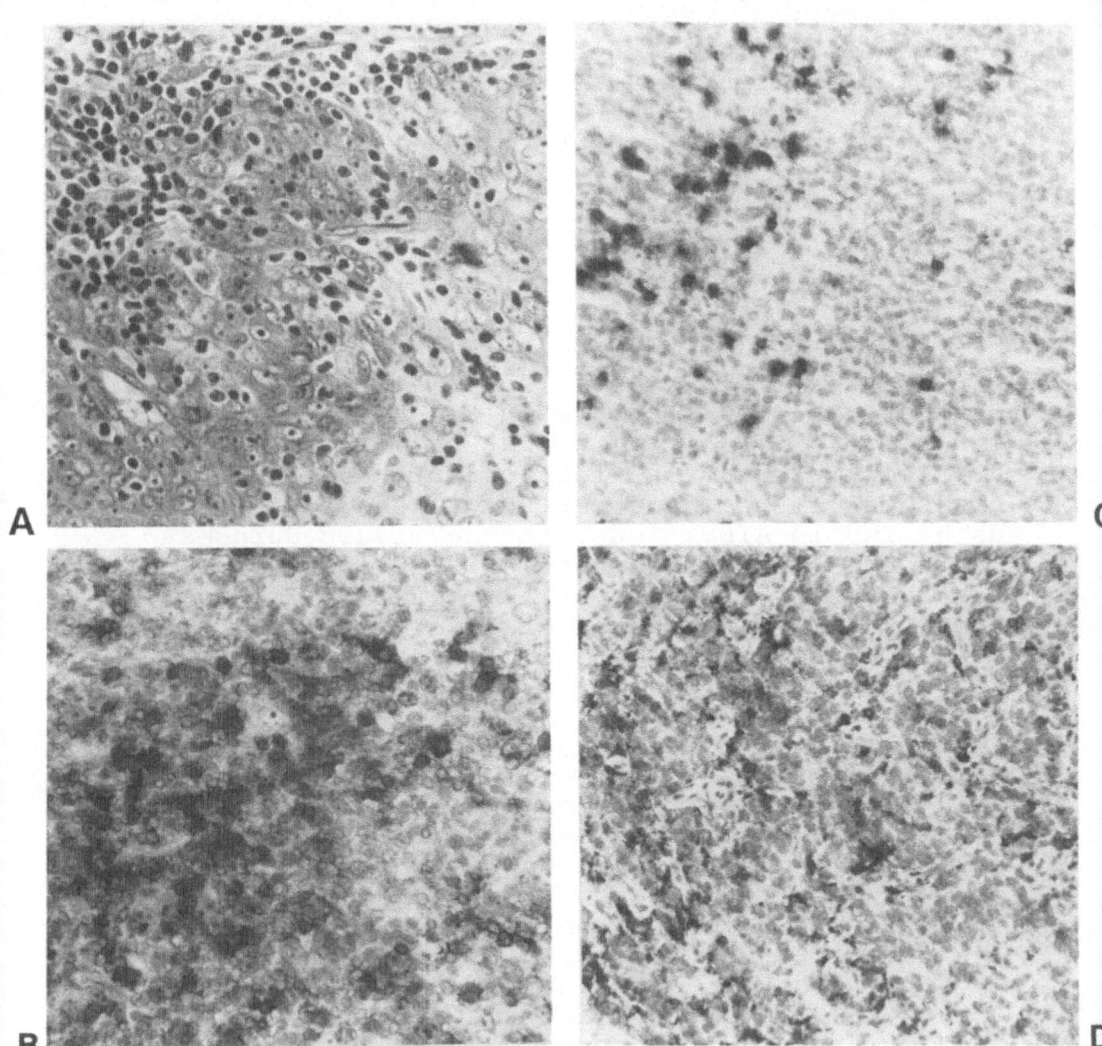

Figure 8. Lymphocyte populations in biopsy of
nasopharyngeal carcinoma (NPC). a) NPC of undifferentiated type with lymphoid stroma (H&E; 240x);
b) B-lymphocytes in lymphoid stroma bordering the
tumor cell nests (OKB7, 240x); c) few scattered
T_8-suppressor cells in the tumor tissue (OKT8,
240x); d) diffusely scattered T_4-helper cells in
the tumor (OKT4, 150x)

of patients with restricted immune responsiveness
and an essentially synchronized proliferation of
infectable B-lymphocytes due to persistent antigenic stimulation such as by malaria parasites
(Krueger and O'Conor, 1972). Evidence for massive
infection comes from significantly elevated anti-
EBV antibody titers in young children at risk for
BL and from the multiplicity of EBV genomic information in BL cells (Henle et al., 1979; Epstein and
Achong, 1979), evidence for the unusual persistent
stimulation of the B-lymphocyte system from the
marked hypergamma-globulinemia in malaria-infested
mice and man (Sizaret et al., 1972; Michaux, 1966).
A similar hyperimmunization of a rabbit with a
defined protein antigen is followed by a cyclic
response of massive induction of antibody synthesis with the respective cell proliferation and subsequent phases of unresponsiveness.
Massive infection by EBV of synchronously proliferating B-lymphocytes thus appears an essential
precondition in the pathogenesis of BL. Instability of the host's chromosomes with chromosomal aberrations (8-14q+ translocation) another decisive
factor in BL development (Klein, 1981).

Nasopharyngeal carcinoma (NPC), finally, arises
from EBV-transformed epithelial cells in lymphoepithelial tissues (Krueger, 1984). Squamous epithelia of a certain low degree of differentiation
carry EBV-DNA and proliferate to give rise to
squamous cell carcinomas, non-keratinizing or undifferentiated type. Characteristic of this tumor
is the close association of lymphoid cells and proliferating carcinoma indicated by the traditional
term "lymphoepithelial carcinoma" (Fig.8). The tumor originates in lymphoepithelial tissues of Waldeyer's ring, not in other similar sites like Pey-
er's patches, appendix etc., i.e. it develops apparently where epithelia of the ectoderm are in
close contact with primary EBV-infected B-lymphocytes. The mechanism of epithelial transfection
by EBV is still hypothetical; lympho-epithelial
cell fusion or EBV receptor transfer is discussed
(Wolf et al., 1981; Volsky et al., 1980). The nasopharynx naturally harbors various infectious org-

anisms including paramyxoviruses which may possibly support transfection mechanisms (Krueger, 1984).
Whatever the pathogenesis of NPC, it appears that
the lymphoid system plays an important role in the
initiation and maintenance of tumor cell growth
besides possible defense mechanisms. In this regard it is of special interest to note that NPC
differs to aforementioned lesions in that T-helper cells are not diminished in tumor biopsies
but relatively or absolutely increased (Table 9;
Fig.8); B-lymphocytes contain to a large part IgA;
the number of NK-cells, representatives of the
host defense, varies markedly and is apparently
stage-dependent (Wustrow et al., 1981; Jondal and
Klein, 1975).

TABLE IX

CELL POPULATIONS WITHIN NPC TISSUE[1]

T-cells	T_4/T_8 cells[2]	B-cells[3]	NK-cells[4]
35-62	1.8-1.0	16-31	2-15

[1] % of non-tumorous mononuclear cells (32 patients
with undifferentiated carcinoma & lymphoid
stroma only)
[2] 6 patients
[3] polyclonal (CIgG 5-10, CIgM 5-10, CIgA 54-78%)
[4] quite variable, apparently stage-dependent

Sundar, Kamaraju and collaborators (Sundar et al.,
1983; Kamaraju et al., 1983) were able to show
that specific IgA-anti-EBV antibodies may block
the tumor-associated K-cell activity. Thus, the
typical lymphoid component of NPC may as well support tumor growth and not necessarily indicate
effective host defense.

In essence, EBV infection in man causes quite variable clinico-pathologic responses reaching from
lymphopoietic and hematopoietic aplasia/hypoplasia
with immune deficiency to limited lymphoprolifer-

ation with autoimmune phenomena as well as to
malignant cell proliferation.
Prime target for malignant transformation is the
B-lymphocyte, yet under certain conditions apparently also specific epithelial cells become infected and transformed. The type of clinico-pathologic response developing after EBV infection
in the individual case is widely determined by
the infectious dose, by the mass of available
infectable and transformable cells, by the status
of the host's immune response, and apparently also
by certain genetic conditions. Detailed individual investigations of these parameters may provide a valuable tool to determine the outcome and
prognosis of EBV infection in a given patient.

REFERENCES

Cottier, H., Turk, J. and Sobin, L., A proposal
   for a standardized system of reporting human
   lymph node morphology in relation to immunological functions. J.Clin.Path. 26, 317-331 (1973)

DeThé, G., Role of Epstein-Barr virus in human disease: infectious mononucleosis, Burkitt's lymphoma, and nasopharyngeal carcinoma. In: G.Klein
   (ed.), Viral oncology, Raven Press, New York
   (1980)

Epstein, M.A. and Achong, B.G. (eds.), The Epstein-
   Barr virus, Springer Verlag, Berlin (1979)

Fong, S., Vaughan, J.H., Tsoukas, C.D., and Carson,
D.A., Selective induction of autoantibody secretion
   in human bone marrow by Epstein-Barr virus.
   J.Immunol. 129, 1941-1943 (1982)

Gallo, R.C., Salahuddin, S.Z., Popovic, G.M.,
Shearer, G.M., Kaplan, M., Haynes, B.F., Palker,
T.J., Redfield, R., Oleske, J., Safai,B., White,G.,
Foster, P., and Markham, P.D., Frequent detection
   and isolation of cytopathic retrovirus (HTLV III)
   from patients with AIDS and at risk for AIDS.
   Science 224, 500-503 (1984)

126

Geser, A., Lenoir, G.M., Anvret, M., Bornkamm, G.,
Klein, G., Williams, E.H., Wright, D.H., and
DeThé, G., Epstein-Barr virus markers in a series
of Bu-kitt's lymphomas from the West Nile district, Uganda. Eur.J.Cancer Clin.Oncol. 19,
1393-1404 (1983)

Groopman, J.E. (ed.), AIDS. In: J.W.Yarbro (ed.),
Seminars in oncology. Grune & Stratton, New York
(1984)

Guarda, L.A., Buttler, J.J., Mansell, P., Hersh,
E.M., Reuben, J., and Newell, G.R., Lymphadenopathy in homosexual men: morbid anatomy with
clinical and immunological correlations.
Am.J.Clin.Pathol. 79, 559-568 (1983)

Hanto, D.W., Frizzera, G., Purtilo, D.T., Sakamoto,
K., Sullivan, J.L., Saemundsen, A.K., Klein, G.,
Simmons, R.L., and Najarian, J.S., Clinical spectrum of lymphoproliferative disorders in renal
transplant recipients and evidence for the role
of Epstein-Barr virus. Cancer Res. 41, 4253-
4261 (1981)

Henle, G. and Henle, W., The virus as etiologic
agent of infectious mononucleosis. In: M.A. Epstein and B.G.Achong (eds.), The Epstein-Barr
virus. Springer Verlag, Berlin (1979)

Henle, W., Henle, G., and Lenette, E.T., The Epstein-Barr virus. Sci.American 241, 48-53 (1979)

Jondal, M. and Klein, G., Classification of lymphocytes in nasopharyngeal carcinoma (NPC) biopsies. Biomed. 23, 163-165 (1975)

Jones, J.F., Straus, S.E., Tosato, G., Kibler, R.A.,
Hicks, M., Lucas, D., Preble, O., Lawley, T.,
Armstrong, G., and Blaese, R.M., Immune assessment
of patients with chronic active EBV (CAEBV) inf-
ection. Proc.Ist.Internal.Sympos. on EBV and
Assoc.Malignant Dis., Loutraki, Sep.24-28 (1984)

Kamaraju, L., Levine, P.H., Sundar, S.K., Ablashi,
D-V., Faggioni, A., Armstrong, G.R., Bertram, G.,

127

and Krueger, G.R.F., Epstein-Barr virus related
lymphocyte stimulation inhibitor: a possible
prognostic tool for undifferentiated nasopharyngeal carcinoma. J.Nat.Cancer Inst. 70, 643-647,
(1983)

Klatzman, D., Barré-Sinoussi, F., and Gluckman,
J.C., Selective tropism for the helper-inducer
lymphocyte subset of a new human retrovirus
(LAV) associated with the acquired immune deficiency syndrome. Science 225, 59-62 (1984)

Klein, G., The role of gene dosage and genetic
transpositions in carcinogenesis. Nature 294,
313-318 (1981)

Klein, E., Ernberg, I., Masucci, M.G., Szigeti, R.,
Wu, Y.T., Masucci, G., and Svedmyr, E., T-cell
response to B-cells and Epstein-Barr virus antigens in infectious mononucleosis. Cancer Res.
41, 4210-4215 (1981)

Krueger, G.R.F., Nasopharyngeal carcinoma. In:
D.T.Purtilo (ed.), Immune deficiency and cancer.
Plenum Med.Book Co., New York (1984)

Krueger, G. and O'Conor, G.T., Epidemiologic and
immunologic considerations on the pathogenesis
of Burkitt's tumor. Rec.Res.Cancer Res. 39, 211-
224 (1972)

Krueger, G.R.F., Papadakis, T., and Michel, R.,
Spezielle pathologisch-anatomische Aspekte des
AIDS, acquired immune deficiency syndrome.
Z.Hautkr. 59, 507-522 (1983)

Lane, H.C., Masur, H., Edgar, L.C., Whalen, G.,
Rook, A.H., and Fauci, A.S., Abnormalities of B-
cell actication in patients with acquired immune
deficiency syndrome. N.Engl.J.Med. 309, 453-458
(1983)

Lennert, K., Schwarze, E.W. and Krueger, G., Lymph-
knotenveränderungen durch Virusinfektionen.
Verh.dtsch.Ges.Path. 65, 151-171 (1981)

Linder, J. and Purtilo, D.T., Infectious mononu-
cleosis and complications. In: D.T.Purtilo (ed.)
Immune deficiency and cancer. Plenum Med.Book
Co., New York (1984)

Michaux, J.L., Les immunoglobulins des Bantous a
l'etat normale et pathologique. Ann.Soc.Belge
Med.Trop. 46, 483-674 (1966)

Olweny, C.L.M., Atine, I., Kaddu-Mukasa, A.,
Owor, T., Anderson, M., Klein, G., Henle, W.,
and DeThé, G., Epstein-Barr virus genome studies
in Burkitt and Non-Burkitt lymphomas in Uganda.
J.Nat.Cancer Inst. 58, 1191-1196 (1977)

Purtilo, D.T., Immune deficiency predisposing to
Epstein-Barr virus-induced lymphoproliferative
diseases: the X-linked lymphoproliferative syndrome as a model. Adv.Cancer Res. 34, 279-312,
(1981)

Purtilo, D.T., Hematopathology of X-linked lymph-
oproliferative syndrome. In: D.T.Purtilo (ed.),
Immune deficiency and cancer. Plenum Med.Book
Co., New York (1984)

Purtilo, D.T. and Sakamoto, K., Epstein-Barr virus
and human disease: immune responses determine
the clinical and pathological expression. Human
Pathol. 12, 677-679 (1981)

Purtilo, D.T., Yang, J.P.S., Allegra, S., DeFlorio,
D., Hutt, L.M., Soltani, M., and Vawter, G., Hematopathology and pathogenesis of the X-linked
recessive lymphoproliferative syndrome. Am.J.
Med. 62, 225-233 (1977)

Reece, E.R., Gartner, J.G., Seemayer, T.A., Joncas,
J.H., and Pagano, J.S., Epstein-Barr virus in a
malignant lymphoproliferative disorder of B-
cells occurring after thymic epithelial transplantation for combined immunodeficiency. Cancer
Res. 41, 4243-4247 (1981)

Reichert, C.M., O'Leary, T.J., Levens, D.L.,
Simmrell, C.R., and Macher, A.M., Autopsy pathol-

ogy in the acquired immune deficiency syndrome.
Am. J. Pathol. 112, 357-382 (1983)

Robinson, J.E., Smith, D., and Niederman, J.,
Plasmacytic differentiation of circulating Epstein-Barr virus-infected B-lymphocytes during
acute infectious mononucleosis. J.Exp.Med. 153,
235-244 (1981)

Rosen, A., Gergely, P., Jondal, M., and Klein,G.,
Polyclonal Ig production after Epstein-Barr
virus infection of human leukocytes in vitro.
Nature 267, 52-54 (1977)

Seemayer, T.A., Laroche, A.C., Russo, P., Male-
branche, R., Arnoux, E., Guerin, J.M., Pierre, G.,
Dupuy, J.M., Gartner, J.G., Lapp, W.J., Spira,T.
J., and Elie, R., Precocious thymic involution
manifest by epithelial injury in the acquired
immune deficiency syndrome. Hum. Pathol. 15,
469-474 (1984)

Sesterhenn, K., Krueger, G.R.F., Uhlmann, Ch.,
Ablashi, D.V., Samii, H., Wustrow, F., and Fischer,
R., Klassifikation maligner Lymphome des Halsbe-
reiches: Kombinierte morphologische, immunzyto-
logische, serologische und zellkulturuntersu-
chungen. Laryngol.Rhinol.Otol. 55, 823-832 (1976)

Sizaret, P., O'Conor, G.T., Kretschmar, W., Bau-
mont, R., and Laval, M., Serum protein patterns
in mice following primary and challange infection
with plasmodium berghei. (ref.to Krueger &
O'Conor, 1972)

Sonnabend, J., Witkin, S.S., and Purtilo, D.T.,
Acquired immunodeficiency syndrome, opportunistic infections, and malignancies in male homosexuals. A hypothesis of etiologic factors
in pathogenesis. J.Amer.Med.Ass. 249, 2370-
2374 (1983)

Sonnabend, J.A., Witkin, S.S., Purtilo, R.B., and
Purtilo, D.T., The syndrome of acquired immune
deficiency among a subset of homosexual men. In:

Purtilo, D.T., Immune deficiency and cancer.
Plenum Med.Book Co., New York (1984)

Sundar, S.K., Ablashi, D.V., Kamaraju, L.S., Levine, P.H., Faggioni, A., Armstrong, G.R., Pearson, G.R., Krueger, G.R.F., Hewetson, J.F., Bertram, G., Sesterhenn, K., and Menezes, J., Sera
from patients with undifferentiated nasopharyngeal carcinoma contain a factor which abrogates specific Epstein-Barr virus antigen-induced lymphocyte response. Int.J.Cancer 29, 407-
412 (1982)

Syrjänen, K.J., The lymph nodes. Reaction to experimental and human tumors. Exp.Pathol. (Suppl.)
8, 1-123 (1982)

Volsky, D.J., Shapiro, I.M., and Klein,G., Transfer of Epstein-Barr virus receptors to receptor-
negative cells permits virus penetration and
antigen expression. Proc.Natl.Acad.Sci. USA 77,
5453 (1980)

Warnke, R. and Levy, R., Tissue section immunological methods in lymphomas. In: R.A.DeLellis
(ed.), Diagnostic immunohistochemistry. Masson
Publ.Inc. USA, New York (1981)

Wolf, H., Bayliss, G.J., and Wilmes, E., Biolog-
ical properties of Epstein-Barr virus. In:
E.Grundmann, G.R.F.Krueger, and D.V.Ablashi(ed.)
Nasopharyngeal carcinoma. G.Fischer Verlag,
Stuttgart (1981)

Wustrow, J., Karpinski, A., Haas, W., Krueger, G.
R.F., Bertram, G., and Sesterhenn, K., Correlation of hi-tological NPC tumor types with local
and perpheral T- and B-cell values. In: E.Grund-
mann, G.R.F.Krueger, and D.V.Ablashi (eds.),
Nasopharyngeal carcinoma. G.Fischer Verlag,
Stuttgart (1981)

11

The "Family Study" Approach to Investigating the Role of
Genetic Factors in Nasopharyngeal Carcinoma

Sherri J. Bale, Ph.D.,[1] Allen E. Bale, M.D.,[2]
Paul H. Levine, M.D.[2]

[1]Environmental Epidemiology Branch
[2]Clinical Epidemiology Branch
National Cancer Institute
Bethesda, Maryland 20205

SUMMARY

A number of reports strongly indicate the importance of
genetics in determining the outcome of EBV infection, the
most apparent example being the X-linked recessive
lymphoproliferative syndrome. In a cancer such as
nasopharyngeal carcinoma (NPC), where genetic factors may
modify the effect of environmental influences, emphasis to
date has been placed on laboratory studies in defining the
genetic component. Since these cancers frequently occur in
areas where logistic considerations prevent application of
many laboratory assays, particular attention should be given
to utilizing epidemiologic techniques which have proven
satisfactory for other neoplasms. In this report, we
describe the "family study" approach, including the
collection, analysis and interpretation of the data. These
methods could be utilized in a hospital setting where such
patients are seen, thereby improving our knowledge of the
relative contributions of genetics and environment in the
etiology of these diseases.

INTRODUCTION

Several lines of evidence suggest that there is a
genetic component to nasopharyngeal carcinoma (NPC)
susceptibility. Demographic studies of the Southern Chinese
indicate a factor apparently determined by ethnicity (Menck

and Henderson, 1979). HLA typing of affected and unaffected
Singapore Chinese has shown that individuals with the HLA-B
type Bw46 have an increased risk for the disease (Simons, et
al., 1978), and familial clustering has been reported (Nevo,
et al., 1971; Ho, 1972; Williams and de Thé, 1974;
Lanier, et al., 1979). However a simple dominant,
recessive, or X-linked model for inheritance is probably not
consistent with observed data. Some proportion of cases may
be attributable entirely to the effect of a single gene, but
probably most cases result from the interaction of a single
gene or multiple genes with environmental factors,
especially the Epstein-Barr virus, which is thought to be
etiologically important in some forms of this malignancy (de
Thé, 1980).

Carefully planned family studies are one approach to
clarifying the genetics of NPC and may be the most
practical. In this report we suggest a plan that can be
carried out in a clinic or hospital setting where NPC is
frequently seen. This approach can be used to provide
information on the relative role of genetic and
environmental factors in the etiology of this malignancy.
This "how to" proposal is divided into several sections
according to the following outline:

I. Data collection

A. Defining the disease
B. Collection of pedigrees

II. Data Analysis

A. Evaluating the contribution of genetic and
environmental factors and determining an
inheritance pattern (segregation analysis)
B. Gene mapping/identification of family members at
high risk (linkage analysis).

I. DATA COLLECTION

A. Defining the Disease

The first step in family studies must be establishment
of diagnostic criteria likely to identify all carriers of
the NPC gene (or genes). Those individuals with biopsy
proven NPC would clearly be considered to carry the gene;

but from a geneticist's point of view it is important to
identify individuals who are genetically susceptible but who
have not developed overt disease. Such individuals may be
identified through detection of precursor lesions, relevant
laboratory markers, or other diseases and/or abnormalities
consistently associated with NPC. A constellation of any of
the above with NPC could constitute an "NPC syndrome," which
might emerge through consistent collection of data.

For NPC, one precursor lesion that has been reported is
nasopharyngeal hyperplasia which has been noted to precede
NPC on several occasions and has been associated with
elevated IgA antibodies to EBV viral capsid antigen (VCA)
(Li, et al., 1983). Elevated IgA anti-EBV VCA antibodies
also may be part of an NPC syndrome since they have been
reported to precede the development of symptomatic NPC and
are now routinely used in studies to screen for NPC (Zeng, et
al., 1984).

Physical findings may also be markers for the NPC gene.
Most known cancer susceptibility genes have pleiotropic
effects -- the gene manifests itself by a variety of signs
and symptoms rather than just one. A striking example is
neurofibromatosis (NF), a disease which imparts only a
moderate cancer risk to affected individuals (Riccardi,
1981). If investigators studying this disease were to
concentrate on a single associated malignancy, such as
neurofibrosarcoma, the detection of a genetic pattern would
be very difficult because relatively few individuals with NF
develop neurofibrosarcoma. The recognition that individuals
with multiple café-au-lait spots and a variety of benign
tumors also carry the "neurofibrosarcoma susceptibility gene"
allows for elucidation of a clear autosomal dominant pattern.
Thus far no tumor has been consistently associated with NPC
but this may in part be due to the absence of systematic
searching for such an association. In addition to recording
other malignancies and non-malignant diseases in NPC
families, dysmorphic features should also be noted. The
frequency of these features must be established to be rare in
an ethnically similar control group. In a manner analogous
to neurofibromatosis, any demonstration of consistent
anomalies in affected individuals would allow for detection
of "NPC gene" carriers who are not affected with NPC.
Radiographic and laboratory studies may also be appropriate
modalities to search for non-tumor findings.

Statistical analysis to detect an excess of second malignancies in NPC patients may help expand the definition of the disease. If it is known that NPC is associated with other malignancies, it may be reasonable to include these malignancies in the definition of the syndrome.

B. Collection of Pedigrees

A "proband" is an individual of extreme phenotype who brings a family to the attention of an investigator. In studies of NPC, a proband is a person who has the NPC syndrome as defined by the methods in the previous section. For studies of traits with a suspected genetic component, it is imperative that probands are selected for participation based on a well-defined rule of ascertainment.

A reasonable method of ascertainment for a disease with as severe a presentation as NPC is through hospital admissions. For example, such a rule might be stated as, "Probands are defined as all individuals with the discharge diagnosis of NPC who were admitted to Hospitals A, B and C between the years X and Y." Alternatively, if the disease is a reportable one and a reliable population based registry exists, such as the Connecticut, Singapore or Hong Kong Tumor Registries, probands could be identified from all registered cases of NPC. The important aspect to note in both these examples is that ascertainment of cases is defined within a known period of time and over a known geographic space, without respect to family history of disease, known environmental exposures, or co-incident physical/medical findings in the probands.

Non-adherence to an ascertainment rule may lead to the inference of an incorrect etiologic mechanism of the disease trait. In contrast to the method just proposed, consider the "referral method." In this situation, the clinician sees an individual with NPC and is impressed, for example, that the patient's father died of the disease several years earlier. He refers this interesting case to a colleague at the University Medical Center whom he knows has an interest in the familial aspects of NPC. If this colleague collects all of his study subjects (i.e., probands) in this manner, a strong referral bias has been introduced into the study. The investigator now has a series of probands who have "interesting" family histories which are completely nonquantifiable. There is no definition of space and time from which the probands were selected and no idea what proportion of all NPC cases seen by the clinicians are being included in the study. Such a referral bias can lead to a faulty conclusion with respect to the genetic component of phenotypic variance, and even of the inheritance pattern itself.

Once probands for a study of familial disease have been ascertained, the study extends to the proband's relatives. We recommend a general method of pedigree collection such that the resulting data can be restructured in a variety of ways depending upon the specific question (segregation pattern or evidence for genetic linkage) being addressed. This is a modified version of the rules followed by the Edinburgh Cytogenetics Registry.

The first step is to collect information on all first-degree relatives (parents, offspring and siblings) of the proband. Dates of birth and death of each person, as well as vital status and relevant clinical information will be needed. For any relative who is found to be affected or have been affected by NPC or another tumor shown to be part of the "NPC phenotype," both the dates and methods of diagnosis should be recorded. Information on second-degree relatives should then be collected. These include the aunts, uncles, nieces, nephews and grandparents on both the maternal and paternal side of the proband. Finally, information on first cousins should be obtained. Figure 1 diagramatically depicts the scheme for collection of data on relatives of ascertained probands.

The critical step in constructing the pedigree for each proband is that whenever data on any sibship is collected, data on the whole sibship must be collected. In other words, information on affected family members should not be collected preferentially while the corresponding information on the unaffected siblings is ignored. This helps determine the "stopping point" in collecting family data - the point at which full information for any sibship is not available. Errors at this level of the data collection will strongly bias the results of any further genetic analyses.

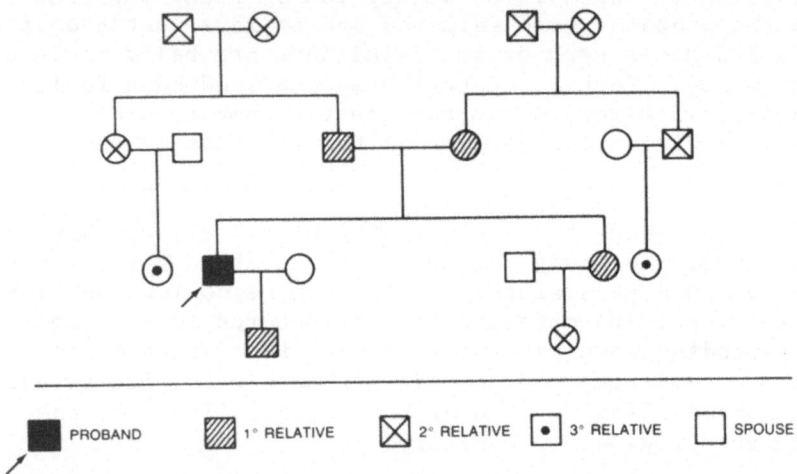

Figure 1. The scheme for collection of pedigree information. First-degree relatives include the parents, offspring and siblings of the proband. Second-degree relatives include the aunts, uncles, nieces, nephews and grandparents of the proband. Third-degree relatives include first cousins. Information about spouses (relatives of the proband by marriage) should also be collected whenever these individuals have produced offspring with the blood relative of the proband.

II. DATA ANALYSIS

A. Segregation Analysis

Segregation analysis is the technique used by geneticists to determine the inheritance pattern of a trait in families. In a practical sense, the resulting information provides a useful tool for several purposes, including genetic counseling, planning preventive and therapeutic measures and designing further investigations (Lalouel and Morton, 1981). Since the complex methodology requires expertise in statistical genetics as well as a major computer facility, the use of this important technique by clinical investigators requires a collaborative effort with a human population geneticist or genetic epidemiologist. Such collaboration can be arranged through correspondence, an important point when attempting research in developing regions where this type of expertise may not be available on site.

In order to interpret the results of any segregation analyses it is first necessary to understand the subtle differences between the terms "familial" and "genetic" and how they relate to each other. The word "familial" refers to the clustering of a trait (also called a phenotype) in a family. For example, obesity might be a characteristic of one family, while tall stature is a characteristic of another. However, the observation of familiality does not necessarily mean that the trait is caused by the effect of genes. To be "genetic" it must be shown that the trait is passed on from parent to child in a defined manner, following Mendel's laws of segregation and independent assortment (Table 1).

Table 1

MENDEL'S LAWS

<u>Mendel's First Law</u> – The Principle of Segregation: When a person is formed by the union of egg and sperm he receives corresponding genetic materials from both parents. These parental contributions separate when he, in turn, produces gametes, so that each gamete he produces contains either the maternal or paternal contribution of any specific piece of genetic material.

<u>Mendel's Second Law</u> –The Principle of Independent Assortment: When gametes are formed the genetic material contributed by one parent for several traits need not remain together. These materials have as much likelihood of passing into different gametes as of passing into the same one. (Levitan and Montague, 1977).

Therefore, genetic traits are a subset of all familial traits. Another subset of familial traits are those produced solely by shared environmental exposures, such as skin cancer in a group of family members exposed to arsenic. A third subset is composed of those traits which result from the interaction of genes and environment.

Geneticists often speak of single gene effects. A single gene for our purposes is one which acts in a purely Mendelian manner. In other words, the phenotype controlled by such a gene is virtually immune to the effects of other genes or of the environment. Single genes follow one of several patterns of inheritance, including autosomal dominant, autosomal recessive, X-linked dominant and X-linked recessive. Achondroplastic dwarfism is an example of a dominant disease; phenylketonuria is autosomal recessive. The X-linked lymphoproliferative syndrome, as described by Purtilo, et al. (1975), follows an X-linked recessive pattern while incontinentia pigmenti is thought to be X-linked dominant.

Several non-Mendelian patterns of inheritance can also be recognized. Some traits are controlled by several genes, each having a small additive effect. An example of such a polygenic trait is height. Multifactorial traits owe their expression to the effects of both polygenes and environmental influences. Intelligence is believed to be an example of multifactorial inheritance. Recently, a "mixed" model (Morton, 1974; Morton and MacLean, 1974; Morton et al., 1983), which involves the effects of a single major gene and contributions from both polygenes and environmental factors, has been hypothesized as the etiology of some human traits. The familiality of blood pressure and cholesterol levels may be explained by the mixed model of inheritance.

Traits whose familiality are due only to environmental exposures follow a sporadic pattern. This means only that there is no genetic influence on the resulting phenotype, and that the observed clustering of the trait in a family would disappear if the responsible environmental pressure were removed.

With this background, it is now possible to discuss segregation analysis in NPC, with the aim being to determine if the observation of familiality is the result of genetic factors alone, environmental factors alone, or a combination

139

of the two.

NPC does not appear to be due to a fully penetrant
Mendelian-acting single gene. (A fully penetrant gene is
one which always causes a visible effect when present in an
individual.) However, as we have seen, this in no way
precludes further investigation into the familial aspect of
the disease. For example, one may consider that the NPC
phenotype in an individual is due to a "susceptibility" gene
or genes. If this is the case, a person would inherit a
liability to develop NPC, an idea somewhat distinct from the
idea of an "NPC gene." This liability could be measured as
the amount of susceptibility to the disease, which, in turn,
could be related either to the number of polygenes he has
inherited, or to the strength of the particular single gene.
One can evaluate the relative effects of the genetic
component (the liability) and the environmental component
with respect to the expression of NPC in an individual.

The results of segregation analysis can allow one to
determine the most probable pattern of NPC inheritance;
either one of the Mendelian models, one of the non-Mendelian
models or the sporadic model. The analysis can be carried
out on nuclear families (first degree relatives of the
proband) or on extended pedigrees.

Another method, known as path analysis (Wright, 1920;
Li, 1977; Rao, et al., 1984), can evaluate the relationship
between a particular exposure variable (such as EBV
infection) and a genetic component to the outcome of
disease. This procedure is only applied to two-generation
nuclear families (the affected individual, his siblings and
parents). Each of these results can help investigators
provide appropriate counseling and focus preventive and
treatment measures correctly.

B. Linkage Analysis

Linkage analysis is a method of gene mapping based on
observations of the disease status of family members along
with knowledge of each individual's type with respect to
many polymorphic marker loci (i.e., red cell types, plasma
protein types, red cell enzyme types) (Conneally and Rivas,
1980). More specifically, the relative position of two
genes, one perhaps with a known chromosomal location, is
determined by observing how often they are transmitted

a.

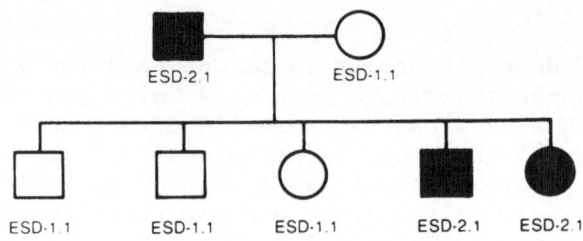

b.

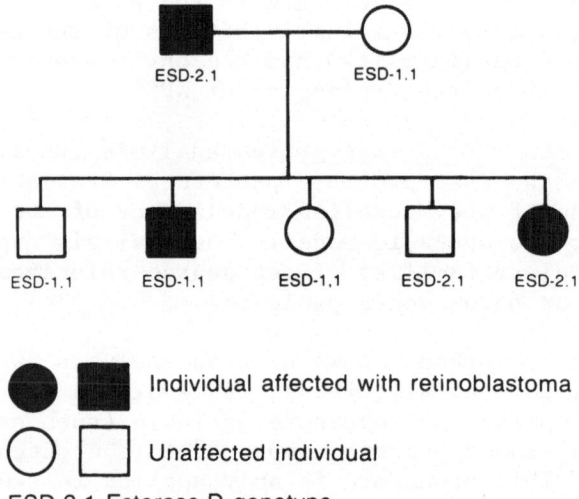

● ■ Individual affected with retinoblastoma

○ □ Unaffected individual

ESD-2,1 Esterase-D genotype

Figure 2a shows an actual family in which all children affected with retinoblastoma inherited the disease gene and the esterase D type 2 (ESD 2) gene from the affected parent. All unaffected children inherited the non-disease gene and the ESD 1 gene from the affected parent. In both cases the unaffected parent contributed a non-disease gene and an ESD 1 gene.

Figure 2b shows the expected outcome if there were no linkage. The ESD and retinoblastoma genes assort independently in this fictitious family.

together from parent to child. If two genes are linked then
the segregation ratios predicted by Mendel's laws are
distorted. In other words, the principle of independent
assortment will not hold true for linked genes. One example
of such distortion is seen in families where retinoblastoma
occurs. All affected children receive the same esterase-D
allele from their affected parent and all unaffected children
receive a different esterase-D allele (Figure 2a) (Sparkes, et
al., 1983). In the absence of linkage, these genes would
assort independently and no correlation between esterase-D
type and retinoblastoma affection would be expected (Figure
2b). Results of linkage analysis are reported as lod scores,
which indicate the strength of the evidence for linkage, and
"theta" values, which represent a measure of the distance
between the two genes.

The concepts of linkage and association are often
confused. Association always refers to a relationship in
the general population between specific observable traits
which may or may not be genetic. For example, there is an
association between nasopharyngeal carcinoma and the HLA-B
type Bw46 (Simons, et al., 1978). This does not necessarily
mean that there is an NPC gene located on the same
chromosome as the HLA gene, but it may indicate that for
some physiologic reason, perhaps an unusual response to the
ubiquitous EBV, those individuals who have HLA type Bw46 are
more prone to develop nasopharyngeal carcinoma. Another
example of association is HLA-B27 and ankylosing
spondylitis. Relative risk is often used as a measure of
association (Rosenberg and Kidd, 1977).

Knowledge of linkage between the NPC gene and a known
genetic marker would be useful practically in allowing for
detection of those family members who are at high risk but
are unaffected at the time of examination. Detection of
linkage could also confirm an inheritance pattern suggested
by segregation analysis and set the stage for isolation of
the susceptibility gene itself. Unlike segregation
analysis, this type of study can be performed with families
not collected according to stringent rules. Prior knowledge
of the inheritance pattern is required, however, and this
must derive from the segregation analysis of appropriately
collected families. Data collection involves determining
which family members are affected and which are not and then
defining the phenotypes of each individual with respect to a
panel of known markers (including blood types, biochemical

markers, and DNA polymorphisms). Analysis is virtually
always performed by a computer program which tests for
cosegregation of the disease gene with each genetic marker.
Since linkage analysis is a statistical test, larger and
multigenerational families tend to give significant results
more often.

Problems with this methodology include the requirement
for access to appropriate laboratory facilities, the high
cost of marker studies, and the need for a high speed
computer and some expertise in statistical genetics. The
fact that HLA data has been collected on NPC patients in the
past suggests that the first two obstacles can be
surmounted. Collaboration with a statistical geneticist
would obviate the latter problem.

CONCLUSION

Genetic studies can and should be designed and carried
out by clinicians doing research in disease endemic
geographic areas, although it is essential that
investigators understand the biases introduced into the data
by following irregular schemes of ascertainment. The actual
procedures of segregation, path and linkage analysis will,
however, require collaboration with an individual trained in
the methods of statistical genetics as well as sufficient
computer support. Well designed and executed studies of
this type will contribute much toward the understanding of
the genetic component of EBV-related malignancies.

ACKNOWLEDGEMENTS

We thank Ms. Millie Jacobus for her excellent typing
and Robert J. Biggar, M.D. for helpful comments.

REFERENCES

Conneally, P.M. and Rivas, M.L., Linkage analysis in man.
In: H. Harris and K. Hirschhorn (eds.). Advances in Human
Genetics, Vol. 10, pp. 209-266, Plenum Publishing Corp.
(1980).

de Thé, G., Role of Epstein-Barr virus in human diseases:
Infectious mononucleosis, Burkitt's lymphoma, and
nasopharyngeal carcinoma. In: Viral Oncology (Ed. Klein,
G.), Raven Press, New York (1980), pp. 769-797.

Ho, H.C., Current knowledge of the epidemiology of nasopharyngeal carcinoma - a review. In: Oncogenesis and Herpes viruses (Eds. Biggs, P. M., de Thé, G. and Payne, L.N., eds.), IARC Sci Pub, No. 2, IARC, Lyon (1972), pp. 357-366.

Lalouel, J.M. and Morton, N.E., Complex segregation analysis with pointers. Hum Hered 31, 312-321 (1981).

Lanier, A.P., Bender, T.R., Tschopp, C.F. and Dohan, P., Nasopharyngeal carcinoma in an Alaskan Eskimo family: Report of three cases. J Natl Cancer Inst 62, 1121-1124 (1979).

Levitan, M. and Montagu, A., Textbook of Human Genetics. Oxford University Press, New York, pp. 155-191 (1977).

Li, C.C., Path Analysis-A Primer. Boxwood Press, Pacific Grove, California, (1977).

Li, Z.Q., Chen, J.J. and Li, W.J., Early detection of nasopharyngeal carcinoma (NPC) and nasopharyngeal hyperplastic lesion (NPHL) with its relationship to carinomatous change. In Prasad, U., Ablashi, D.V., Levine, P.H. and Pearson, G. R.: Nasopharyngeal Carcinoma: Current Concepts, Kuala Lumpur, University Malaya Press, 1983.

Menck, H.R. and Henderson, B.E., Cancer incidence rates in the Pacific Basin. National Cancer Inst Monogr, 53, 119-124 (1979).

Morton, N.E., Analysis of family resemblence. I. Introduction. Am J Hum Genet 26, 318-330 (1974).

Morton, N.E. and MacLean, C.J., Analysis of family resemblance. III. Complex segregation of quantitative traits. Am J Hum Genet 26, 489-503 (1974).

Morton, N.E., Rao, D.C. and Lalouel, J.M., Methods in Genetic Epidemiology. S. Karger, Basel, Switzerland (1983).

Nevo, S., Meyer, W. and Altman, M. Carcinoma of nasopharynx in twins. Cancer 28, 807-809 (1971).

Purtilo, D.T., Yang, J.P.S., Cassel, C.K., Harper, P., Stephenson, S.R., Landing, B.H. and Vawter, G.F., X-linked recessive progressive combined variable immunodeficiency (Duncan's Disease). Lancet 1, 935-950 (1975).

Rao, D.C., McGue, M., Wette, R. and Glueck, C.J., Path analysis in genetic epidemiology. In: A. Chakravarti (ed.). Human Population Genetics-The Pittsburgh Symposium, pp. 35-81, Van Nostrand Reinhold Co., New York (1984).

Riccardi, V.M., Von Recklinghausen neurofibromatosis. New Engl. J. Med. 305, 1617-1626 (1981).

Rosenberg, L.E. and Kidd, K.K., HLA and disease susceptibility: A primer. New Engl. J. Med. 297, 1060-1062 (1977).

Simons, M.J., Chan, S.H., Wee, G.B., Shanmugaratnam, K., Goh, E.H., Ho, J.H.C., Chau, J.C.W., Darmalingam, S., Prasad, U., Betuel, H., Day, N.E. and de Thé, G., Nasopharyngeal carcinoma and histocompatibility antigens. In: Nasopharyngeal Carcinoma: Etiology and Control (Eds. de Thé, G. and Ito, Y.), IARC Sci Publ, No. 20, IARC, Lyon (1978), pp. 271-282.

Sparkes, R.S., Murphree, A.L., Lingua, R.W., Sparkes, M.C., Field, L.L., Funderburk, S.J., Benedict, W.F., Gene for hereditary retinoblastoma assigned to human chromosome 13 by linkage to esterase D. Science 219, 971-973 (1983).

Williams, E.H. and de Thé, G., Familial aggregation in nasopharyngeal carcinoma. Lancet 2, 295-296 (1974).

Wright, S., The relative importance of heredity and environment in determining the piebald pattern of guinea pigs. Proc Natl Acad Sci USA 6, 320-322 (1920).

Zeng, Y., Shen, S-J., Deng, H., Ma, J-L., Zhang, Q., Zhu, J-S. and Cheng, J-R., Early nasophargngeal carcinoma among IgA/VCA antibody positive individuals detected by anticomplement immunoenzymatic method. Chinese Medical Journal 97, 155-157 (1984).

12

AN EBV-ASSOCIATED SALIVARY GLAND CANCER

Anne P. Lanier, M.D., M.P.H.
Sarala Krishnamurthy, M.D., M.P.H.
Susan E. Clift, M.D.
Kathy T. Kline, M.D.
Centers for Disease Control and
  Alaska Native Medical Center
Anchorage, Alaska 99501

Georg W. Bornkamm, M.D.
Institut for Virologie im Zentrum fur Hygiene
Freiburg im Breisgau, West Germany

Werner Henle, M.D.
The Children's Hospital of Philadelphia
Philadelphia, Pennsylvania 19104

Allen Gown, M.D.
Department of Pathology
University of Washington School of Medicine
Seattle, Washington 98195

David Thorning, M.D.
Department of Pathology
Veterans Administration Medical Center
Seattle, Washington 98108

SUMMARY

This is a case report of a 57-year-old Alaskan Native
woman diagnosed with malignant lymphoepithelial lesion
of the parotid gland. Extensive morphologic studies
documented that the lesion was epithelial with a
predominantly T-cell lymphocytic infiltrate. Tests for
EBV demonstrated the tumor was positive for EBNA and EBV
DNA, while adjacent non-malignant tissue was negative.

Eskimos of Alaska, Canada, and Greenland are at increased risk for cancers of both the nasopharynx and salivary gland (Lanier et al, 1980; Nielsen et al, 1977, 1978; Wallace et al, 1963; Mallen and Shandro, 1974). The excess risk of salivary gland cancer has been found to be largely due to the occurrence of unusual tumors classified as malignant lymphoepithelial lesions (Wallace et al, 1963; Arthaud, 1972; Nielsen et al, 1977). These malignant lymphoepithelial lesions (MLEL) are characterized by islands of anaplastic cells in a dense lymphocytic background.

Although the majority of tumors of this type have been reported to date among Eskimos, there have also been reports of malignant lymphoepithelial lesions in Caucasians and Blacks of Europe and America, and most recently, Japanese and Chinese (Ferlito and Donati, 1977; Gravanis and Giansanti, 1970; Nagao et al 1983; Redondo et al, 1981; Dong and Lo, 1983).

In Alaska during the time period 1966 through 1980, 16 Alaskan Natives (Eskimos, Indians and Aleuts) developed malignant tumors of the major salivary glands. The observed to expected ratio (based on U.S. white rates) was significantly high (4.7) in females and high (1.7), but not significantly high, in males. Approximately twice as many females as males developed the cancers. The age range was 17 to 70, however, all but one salivary gland cancer patient was diagnosed under age 60. Twelve of the 16 salivary gland cancers were MLEL. To date, tumor tissues from Eskimo patients with MLEL, (two from Alaska and one from Greenland) have been reported to be positive for EBV by DNA hybridization techniques (Lanier et. al, 1981; Saemundsen et. al., 1982).

The presentation in January, 1983, of a 57-year-old Alaskan Native woman with a salivary gland tumor which was classified as a malignant lymphoepithelial lesion, provided an opportunity to study the tumor for EBV and morphologic characteristics, and to evaluate the patient for evidence of autoimmune disease, such as primary and secondary Sjogren's syndrome.

Histologically the 1.5 cm tumor included islands of neoplastic cells within a dense lymphocytic background diagnostic of MLEL. The key features of the tumor islands include the increased nucleo-cytoplasmic ratio, the increased mitotic rate, and a relative lack of differentiation. On the periphery of the tumor were areas that morphologically resembled benign lymphoepithelial lesions.

Multiple nasopharyngeal biopsies were negative. On immunocytochemistry, the neoplastic cells reacted with anticytokeratin antibodies; and the lymphocytic infiltrate reacted to antibodies of both pan-T-cell and pan B-cell specificities, but with a T-cell predominance. Additional staining indicated that suppressor T-cells predominated over helper T-cells. On electron microscopy the neoplastic cells showed evidence of epidermoid differentiation, including basement membranes and desmosomes.

EBV studies included touch prints for EBNA by immunofluorescence, as well as for EBV DNA by hybridization of extracted DNA. The patients tumor tissue was positive for both EBNA and EBV DNA, while tissue from the same patient, namely adjacent non-neoplastic parotid gland and ipsilateral normal submandibular gland were negative for EBNA and EBV DNA. The patient's cervical node was also negative for EBNA. In addition, 6 pleomorphic adenomas, a benign lymphoepithelial lesion of the parotid gland, and normal salivary gland tissues from 7 other patients tested were negative for EBNA and EBV DNA. All EBV DNA hybridizations described above were tested simultaneously and only the patient's tumor tissue was positive.

The patient's serum was tested for EBV antibodies on 3 separate samples (Table 1) Except for the presence of IgA anti-VCA antibodies in 2 of the 3 samples, the results show evidence of past primary infection and are not typical of the serological pattern of NPC or Burkitt's lymphoma.

As indicated, the patient did have a nonspecific anti-nuclear antibody when serum was tested on an EBV-negative cell line. Serum was also positive (2+) for ANA when tested on a Hep-2 cell line, but did not react with calf thymus or human spleen extracts. Antibodies to Sjogren's syndrome antigens (SS-A and SS-B) were not

Table 1: SERUM ANTIBODIES TO EPSTEIN-BARR VIRAL ANTIGENS
(Reciprocal titres)

Serum	Viral Capsid Antigen		Early Antigen-Diffuse and Restricted				Nuclear Antigen
	IgA VCA	IgG VCA	IgA EA-D	IgG EA-D	IgA EA-R	IgG EA-R	EBNA
Pre-op	10	160	<10	<10	<10	<10	$\geq$320 (ANA*)
At dx	10	160	<10	<10	<10	<10	$\geq$320 (ANA*)
Post-op	<10	160	<10	<10	<10	<10	$\geq$320 (ANA*)

*Non specific anti-nuclear antibody present

detected. Clinically, the patient described a 1-year history of episodes of arthritis, but was rheumatoid factor negative and had no objective signs of arthritis at the time of her tumor.

The patient was treated with excision of the tumor followed by irradiation and was well without evidence of tumor 9 months following excision.

Evaluation of this patient and tumor confirms that this neoplasm was a primary salivary gland lesion and not metastatic from the nasopharynx. Histologically, it fits the classification of MLEL most frequently described in Eskimos. The neoplastic cells were poorly differentiated epidermoid. The lymphocytic infiltrate was polyclonal and not monoclonal and predominantly T-cells not B-cells. EBV studies of the tumor were positive, although the EBV picture serologically was not remarkable.

This report, plus the 3 previous EBV-positive MLEL case reports in the literature, suggest that salivary gland tumors that fit the classification of MLEL, may deserve inclusion in the list of EBV-associated malignancies.

REFERENCES

ARTHAUD, J.B., Anaplastic parotid carcinoma ("Malignant lymphoepithelial lesion") in seven Alaskan Natives. Am. J. Clin. Pathol., 57, 275-86 (1972).

DONG, H., LO, G., Malignant lymphoepithelial lesions of the salivary glands with anaplastic carcinomatous change. Cancer 52, 2245-2252 (1983).

FERLITO, A., DONATI, L.F., Malignant lymphoepithelial lesions. J. Laryngol. Otol., 91, 869-85 (1977).

GRAVANIS, M.B., GIANSANTI, J.S., Malignant histopathologic counterpart of the benign lymphoepithelial lesion. Cancer, 26, 1332-42 (1970).

LANIER, A.P., BENDER, T.R., BLOT, W.J., FRAUMENI, J.F., Jr., Cancer in Alaskan Indians, Eskimos and Aleuts. J. Natl. Cancer Inst., 65, 1157-1159 (1980).

LANIER, A.P., BORNKAMM, G.W., HENLE, W., et al., Association of Epstein-Barr virus with nasopharyngeal carcinoma in Alaskan Native patients: serum antibodies and tissue EBNA and DNA. Int. J. Cancer, 28 301-305 (1981).

MALLEN, R. W., SHANDRO, W. G., Nasopharyngeal carcinoma in Eskimos. Can. J. Otolaryngol., 3, 175-179 (1974).

NAGAO, K., MATSUZAKI, O., SAIGA, H., et al, A histopathologic study of benign and malignant lymphoepithelial lesions of the parotid gland. Cancer, 52, 1044-52 (1983).

NIELSEN, N.H., MIKKELSEN, F., HANSEN, J.P., Incidence of salivary gland neoplasms in Greenland with special reference to an anaplastic carcinoma. Acta. Pathol. Microbiol. Scand. (Sect. A), 86, 185-193 (1978).

NEILSEN, N. H., MIKKELSEN, F., and HART HANSEN, J. P., Nasopharyngeal cancer in Greenland. Acta Pathol. Microbiol. Scand., 85, 850-858 (1977).

REDONDO, C., GARCIA, A., VASQUEZ, F., Malignant lymphoepithelial lesion of the parotid gland: poorly differentiated squamous cell carcinoma with lymphoid stroma. Cancer, 48, 289-292 (1981).

SAEMUNDSEN, A.K., ALBECK, H., HANSEN, J.P.H, et al., Epstein-Barr virus in nasopharyngeal and salivary gland carcinomas of Greenland Eskimos. Br. J. Cancer, 46, 721-28, (1982).

WALLACE, A. C., MacDOUGALL, J. T., HILDES, J. A., LEDERMAN, J. M., Salivary gland tumors in Canadian Eskimos. Cancer, 16, 1338-52 (1963).

13

NASOPHARYNGEAL CARCINOMA : EARLY DETECTION AND IGA-RELATED

PRE-NPC CONDITION. ACHIEVEMENTS AND PROSPECTIVES

ZENG, Yi and de THE, Guy

INSTITUTE OF VIROLOGY-BEIJING- PRC and
CNRS LABORATORY, FAC. OF MED. A. CARREL, LYON
FRANCE

INTRODUCTION

Undifferentiated carcinomas of the nasopharynx (or NPC) represent a major cancer killer for more than 200 million people in South China, as well as in large areas of the South East Asia, North and East Africa and in Eskimo populations. This cancer is closely associated with the ubiquitous Epstein Barr herpes virus. In contrast with the situation observed in Burkitt's Lymphoma, the association between EBV and NPC is constant in every part of the world where NPC is observed, and unrelated to its level of incidence. Such an association, most probably causative in nature, has recently been reviewed (de-Thé, 1982,1984).

Following the observation of Wara, W.M. et al in 1975, that NPC patients had high level of IgA antibodies, Henle and Henle (1976) and Desgranges and de-Thé (1978) showed that such IgA were directed against VCA and EA and were regularly observed in NPC patients from Chinese, Arabic and Caucasian origins, but absent in patients with other ENT tumor. These data urged us to use the IgA/VCA test for early detection of this tumor in the endemic areas of South China (Zeng et al, 1979, 1980, 1982). We shall review first these population surveys, then discuss the pre-NPC conditions associated with rising titers of IgA.

The interplay between an ubiquitous EB Virus and
nasopharyngeal carcinoma stresses the need for other
environmental factors, possibly related to life-style, and
to the reactivation of EBV.

I - EARLY DETECTION OF NPC IN HIGH RISK POPULATIONS
OF SOUTH CHINA

A. 1978-1980 : Survey in Zang-Wu County

A major survey was implemented in 1978, in a rural
area of the Eastern part of the Guang-Xi autonomous region
(Zeng et al, 1979, 1980). The County of Zang-Wu comprises
15 communes with a total population of 450.000. Starting in
1978, individuals aged 30 and above, were registered and a
small amount of blood was collected. The sera were tested
for IgA/VCA antibodies by the immunoenzymatic test (see
Zeng et al, 1980). As seen in table 1, a total of 148 029
persons were thus screened and 3 533 were found to have
IgA/VCA antibodies. The positive sera were then titered and
13 % showed titer superior or equal to 80, representing
460 individuals. All the 3 533 IgA/VCA positive persons,
were then clinically examined, and 55 NPC patients
uncovered. The clinical stages at which the patients were
recognized are given in table 1, where it can be seen that
the majority were at stage 2 and 3 of the disease.

The clinical follow-up of the IgA positive individuals
for 1 to 3 years led to the diagnosis of another 32 cases.
The distribution of these cases according to stages, was
not dramatically different from that of the main survey
although there was a shift to early ages of detection (see
table 1).

B. Wu-Zhou City

The City of Wu-Zhou, located in the centre of the
Zang-Wu County, with a population of 170.000, was an ideal
place to try and implement an early detection of NPC with
a systematic survey and follow-up of the town dwellers.
Previous cancer registration in the town of Wu-Zhou, showed
that the mean annual incidence of NPC was around 17 to 20
per 100 000. As seen in table 2, survey of nearly 21 000
individuals aged 40 and above in Wu-Zhou, showed that

Table 1 : Detection of NPC Cases in Zang-Wu County during 1978-1980 by IgA/VCA Test and during 1 to 3 years follow-up.

	N° of surveyed person	Ca in situ	Stages of NPC detected								TOTAL NPC detected
			I	%	II	%	III	%	IV	%	
1978-1980 Serological mass survey	148 029	1	12	22%	19	34%	17	31%	6	11%	55
Follow-up for 1-3 years of IgA/VCA+	3 478	0	10	31%	9	28%	11	34%	2	6%	32
TOTAL		1	22		28		28		8		87

Table 2 : Survey in Wu-Zhou City

	N° persons examined	N° IgA/VCA positive	%	N° NPC	Stages of NPC							NPC prev. rate in survey	NPC prev. in IgA/VCA + indiv.
					I	%	II	%	III	%	IV		
General Populat.	20 726	1 136	5.5%	35	15	43%	17	48.5%	3	8.5%	0	1.5 %	27 %
Chemical Factory	216	22	10%	3								14 %	136 %

1 136 (5.5%) persons had IgA/VCA antibodies. Clinical examination of this later group allowed the detection of 31 cases of NPC, the clinical stages being given in table 2. It is remarkable to see that in this survey of the Wu-Zhou City, the proportion of stage 1 and 2 represented 90% of the tumor detected. This was due to the fact that clinical detection of the tumor had been efficient since a few years in this town. The prevalence rate of NPC in the surveyed population thus reached 150/100 000, a very high figure if one considers that patients at stage 3 and 4 of the disease present in hospitals were not included.

A serological survey, carried out in a chemical factory, detected more NPC than expected. As seen in table 2, 216 individuals were tested in this factory and 22 or (10%) were found IgA/VCA positive. Among those, 3 were discovered having NPC. Although this may be due to chance, an experimental investigation of the chemicals handled in this factory has been implemented.

C. Laucheng County

Situated in a Northern part of the Guang-Xi autonomous region, Laucheng County is inhabited by two sub-language groups, namely the Molaos and the Hans. As seen in table 3, the survey of 15.324 individuals from the Molaos minority gave a prevalence of 1% IgA/VCA positive (151 persons) and 7 cases of NPC were detected. In the Han majority, 0.6% of the surveyed population had IgA/VCA antibodies and 6 cases of NPC were detected (Tao et al, in press).

D. IgA/EA Antibodies represent a better test for early detection of NPC

Whereas immunofluorescent (Desgranges and de Thé, 1978) and immunoenzymatic tests (Laboratory of Cancer Institute, 1978) detected IgA/EA antibodies in about 70% to 75% of NPC patients, the immuno-autoradiographic test developed by Zeng et al, 1983 (using 125 I-labelled antihuman IgA antibodies) detected IgA/EA antibodies in 96% of NPC patients with a GMT titer of 1:97. Using this later test, patients with chronic inflammation of the nasopharynx,

Table 3 : Survey of Laucheng County

Sub-language groups	N° persons examined	N° IgA/VCA positive	%	N° NPC detected	NPC prev. rate in survey	NPC prev. in IgA/VCA + indiv.
Molaos	15 324	151	1%	7	45/100 000	4.6%
Han	11 117	76	0.6%	6	54/100 000	7.9%

Table 4 : Comparison of IgA/VCA and IgA/EA in detecting NPC*

	N° persons surveyed	IgA VCA+	IgA EA+	NPC detected	% of NPC among IgA+
Wu-Zhou City	12 930	680.........		13	2%
			30.......	9	30%
Laucheng County	26 441	227.........		13	5,7%
			14.....	6	43%
Total NPC detected				28	

* by immunoenzymatic test

had IgA/EA antibodies in 22% of the cases with a GMT titer
of 1:9. Patients with malignant tumors other than NPC were
positive in 4% of the cases with a GMT titer of 1:5. All
normal individuals were found negative.

Table 4 gives the comparative results of both the
IgA/VCA and IgA/EA tests in detecting NPC in Wu-Zhou City
and Launcheng County. It can be seen that 30 to 43% of
individuals with IgA EA antibodies have a detectable NPC.

Such test, sensitive and specific, could replace the
IgA/VCA test for the early detection of NPC, since the
background noise of IgA/EA in non NPC individuals is very
low. Furthermore, IgA/EA test will be most instrumental
for detecting and investigating pre-NPC conditions (see
below).

E. AN ELISA TEST USING MONOCLONAL ANTIBODIES HAS BEEN
RECENTLY DEVELOPED (Pi et al, submitted for publication)

Such a test, which has nearly the same sensibility and
a   better specificity for IgA/VCA and IgA/EA than the
immunoenzymatic and immunoautoradiographical tests and
which is best adapted to field conditions, should ease the
implementation of population seroepidemiological surveys,
in large areas of South-East Asia, where this cancer is a
main killer.

II- IGA RELATED PRE NPC CONDITIONS

It appears as if the presence of IgA antibody to VCA
and to EA represents pre NPC conditions (de-Thé, Zeng et
al, 1983). In order to see whether the presence of IgA
antibodies corresponded to a specific viral activity in
nasopharyngeal mucosa, 56 individuals with IgA/VCA antibodies for more than 18 months, were clinically examined
and biopsied. Four NPC cases were found (two at early stages
of the disease) and further 14 individuals had detectable
EBV/DNA sequences and/or EBNA antigen in their nasopharyngeal mucosa without histopathological nor clinical evidence
of NPC (Desgranges et al, 1982). As it was not possible to
take nasopharyngeal biopsies from normal individuals
lacking IgA/VCA antibodies, exfoliated cells collected

from the nasopharynx (using a negative pressure apparatus
developped by Zhangjiang Medical College in 1976) in 62
IgA/VCA antibody positive and 39 IgA/VCA antibody negative
individuals were tested for the presence of EBV/DNA
sequences by spot followed by blot-hybridization
(Desgranges et al, 1983). As seen in table 5, 13 of the

62 IgA/VCA positive specimen (21%), and 6 out of the 39
IgA negative specimen (15.4%), were found to contain
EBV/DNA sequences. Among those, 20 IgA/VCA antibody
positive and 26 IgA/VCA antibody negative individuals
were followed a year later. Their exfoliated cells from
the nasopharynx were again collected, and tested for the
presence of EBV/DNA sequences. Three out of seven
individuals who showed a year previously EBV/DNA sequence
in exfoliated nasopharyngeal cells, failed to do so a year
later. In parallel, 2 out of 15 EBV/DNA negative exfoliated
cells became EBV/DNA positive a year later. Such results
suggest that the presence of EBV/DNA sequences in the
nasopharynx, and the presence of IgA/VCA or IgA/EA antibodies in the serum, are not directly related. Unfortunately, the cell type harbouring the EBV/DNA could not be
characterized in these studies. In situ, hybridizations
made by A.Wolf et al (unplublished) on a few samples from
IgA/VCA positive individuals suggested that the EBV/DNA
positive cells were of epithelial nature.

Using an anticomplement immunoenzymatic method (ACIE),
for the detection of EBNA (Shen et al, 1983), exfoliated
cells from the nasopharynx obtained from positive and
negative IgA/VCA individuals were tested by this anticomplement immunoenzymatic test. As seen in table 5, 34% of
IgA/VCA positive and 20% of IgA/VCA negative individuals
has detectable EBNA in cells which were considered as
epithelial. Thus the virus seems to be present in normal
conditions in nasopharyngeal mucosa. The development of
IgA/VCA antibodies must reflect a critical difference in
the local immune response against the EB viral infection.
In fact, IgG/EA antibodies are usually present in IgA/VCA
positive individuals, thus reflecting a reactivation of the
virus. That such reactivation takes place in the nasopharynx
is highly probable, but not yet established, nor the fact
that it precedes and not succeeds subclinical development
of NPC.

Table 5 : Comparative detection of EBV/DNA and EBNA in IgA/VCA positive and negative individuals.

	N° tested	EBV/DNA* positive	N° tested	EBNA** positive
IgA/VCA positive individuals	62	13 21%	26	9 34%
IgA/VCA negative individuals	39	6 15%	46	9 20%

* Desgranges et al, 1983
** Shen et al, 1983

Table 6 : Stability and fluctuations of IgA/VCA antibody over 3 years in relation to risk of NPC

	Persons IgA/VCA+	stability (no change in IgA/VCA Ab.)	Loss of IgA/VCA Ab. (retroversion)	Fluctuation in IgA/VCA Antibodies		
				Increase◀	Decline▼	Variations of IgA/VCA Ab.◀▼
N°	1138	455	398	81	162	42
%	100%	40%	35%	7.1%	14.2%	3.9%
NPC patients detected		6		15		

◀▼ 4 fold increase or decrease
◀▼◀ fluctuations

Stability and fluctuation in IgA/VCA antibody titers and relation to risk of NPC

In the County of Zang Wu 1 138 individuals with IgA/
VCA antibodies were followed for   3 years from both
the serological and clinical view-points. Table 6 shows
that 40% of them (455 individuals) exhibited stable
IgA/VCA titers. Among those, 6 developed NPC within 3
years of follow-up (1.3%). 398 individuals (35%) lost their
IgA/VCA antibodies within this period and no NPC was
discovered among them. IgA/VCA antibodies increased by 4
dilutions or more in 81 individuals and 15 of them developed NPC (18.5%). These results (Zeng et al, in preparation)
strongly support the hypothesis that EBV reactivation,
reflected by a specific serological profile (increasing
titers of IgG EA, IgA VCA, IgA EA) represents a pre-NPC
condition. Whether or not such conditions reflect our
inability to detect sub-clinical tumorous growth in the
submucosa of the nasopharynx remains to be determined.

III - PERSPECTIVE AND PRIORITIES

Early detection of NPC by the IgA/VCA or probably
better by the IgA/EA test is feasible today, and should
therefore be applied for the benefit of large populations
at risk for this tumour which represent approximately 230
millions persons around the world. Table 7 gives the
difference in the clinical stages of NPC patients diagnosed
in out-patients clinics, and of the patients detected during
the above described early detection schemes. The shift
towards early stages is obvious (43% versus 1.7% detected
at stage I).·Such a shift should have a critical impact on
mortality by NPC, if one considers the 5 year survival
rates after radiotherapy according to clinical stages. In
Shanghai, for example (Zeng personal communication), more
than 90% of NPC patients treated at stage I of the disease
exhibited a 5 year disease free survival. In contrast, NPC
patients diagnosed at stage IV and V, which represent the
majority in out-patients clinics of endemic areas, have less
than one year survival in 70% of the cases.

It is therefore of great interest to see that EBV serology has such a critical and practical impact for patients' care before the nature of the relationship between the virus and this cancer was uncovered. It is a clear example where important applications for public health can be implemented prior to the understanding of the mechanism involved. If the final proof that EBV is causally related to NPC is not yet at hand, the results shown in table 1, 4, 6 and 7 strongly favour an etiological role of the virus in the development of undifferentiated carcinomas of the nasopharynx.

The priorities in Prevention Research concerning NPC should focus on the understanding of the virological, molecular and immunological events taking place in the nasopharyngeal mucosa during the pre-NPC events. Such an understanding will in turn permit the implementation of primary prevention, either by anti EBV interventions or by eliminating co-factors. Such co-factors may be present in the immediate environment of individuals at risk (Ito et al, 1983, Zeng et al, 1983)   or possibly associated with bacterial flora in the nasopharynx (Zeng et al, in press).

Table 7 : Comparison of NPC stage from outpatient clinic and from   serological screening

		N° of cases	I	II	III	IV
				Stages		
Outpatient Clinic	N	1 066	18	312	556	180
	%	100%	1.7%	51.3%	51.3%	17.2%
Serological+ screening	N	35	15	17	3	
	%	100%	43%	48.5%	8.5%	

Acknowledgments : Some aspects of these studies were supported by the CNRS, and the Fondation pour la Recherche Médicale, Paris.

REFERENCES

Desgranges, C., and de Thé, G., IgA and nasopharyngeal carcinoma, in: Oncogenesis and Herpesviruses III (G. de Thé, Y. Ito, and F. Rapp, eds.), pp.883-891, IARC Scientific Publications N° 25, Lyon, (1978).

Desgranges, C., Bornkamm, G.W., Zeng, Y., Wang, P.C., Zhu, J.S., Shang, H., and de Thé, G., Detection of Epstein-Barr viral DNA in the nasopharyngeal mucosa of Chinese with IgA/EBV-specific antibodies, Int. J. Cancer 29:187-191 (1982).

Desgranges, C., Pi, G.H., Bornkamm, G.H., Legrand, C., Zeng, Y and de Thé, G., Presence of EBV-DNA sequences in naso-pharyngeal cells of individuals without IgA/VCA antibodies. Int. J. Cancer, 32: 543-545, (1983).

De Thé, G., Epidemiology of Epstein-Barr Virus and Associated Diseases in Man, in: The Herpesviruses, vol. 1, Roizman, B. (ed.), pp 25-103, Plenum Publishing Corporation, New York (1982).

De Thé, G., The role of the Epstein-Barr Virus (EBV) in the etiology and control of nasopharyngeal carcinoma (NPC). In: Cancer of the Head and Neck, Williams and Wilkins Publishers, (1984).

De Thé, G., Zeng, Y., Desgranges, C., and Pi, G.H. The Existence of Pre-Nasopharyngeal Carcinoma Conditions Should Allow Preventive Interventions. In: Nasopharyngeal Carcinoma: Current Concepts, Prasad et al,(eds.), University of Malaya, Kuala Lumpur, pp. 365-374, (1983).

Ito, Y., Ohigashi, H., Koshimizu, K., Zeng, Y., Epstein-Barr Virus-activating principle in the ether extracts of soils collected from under plants which contain active diterpene esters. Cancer Letters, 19,pp. 113-117, Elsevier Scientific Publishers Ireland Ltd, (1983).

Henle, G., and Henle, W., Epstein-Barr virus-specific IgA serum antibodies as an outstanding feature of nasopharyngeal carcinoma, Int. J. Cancer 17:1-7, (1976)

Laboratory of Tumor Viruses of Cancer Institute, Laboratory of Tumor Viruses of Institute of Epidemiology, Department of Radiotherapy of Cancer Institute, Department of Otolaryngology of Beijing Worker-Peasant-Soldier Hospital, Detection of EB virus-specific serum IgG and IgA antibodies

from patients with nasopharyngeal carcinoma, (1978). Acta
Microbiol, Sin. 18, 253-258

Pi, G.H., Zeng, Y., de Thé, G., Enzyme-linked immunosorbent
Assay for the detection of Epstein-Barr Virus IgA/EA antibody (submitted for publication).

Shen, S.J., Chen, C.P., Ma, J.L., Zhong, W., Zeng, Y..
Further study on detection of EBNA from nasopharyngeal
exfoliated cells of patients with nasopharyngeal carcinoma.
Acta Zhanjiang Medical College, 1, 34-37 (1983)

Tao, E.C., Wang, P.C., Wei, J.N., Li, E.J., Wei, R.F., Too,
C.M., Gu, S.T., Tan, S.M., Tang, H., Zeng, Y., Pi, G.H.
Serological Mass Survey of Nasopharyngeal Carcinoma in
Laucheng County, (submitted for publication).

Wara, W.M., Wara, D.W., Phillips, T.L., and Ammahh, A.
Elevated IgA in Carcinoma of the Nasopharynx, Cancer 35:
1313-1315 (1975)

Zeng, Y., Liu, Y.X., Wei, J.N., Zhu, J.S., Cai, S.L., Wang,
P.Z., Zhong, J.M., Li, R.C., Pan, W.J., Li, E.J., and Tan,
B.F. Serological mass survey of nasopharyngeal carcinoma.
Acta Acad. Med. Sin. 1, 123-126, (1979)

Zeng, Y., Liu, Y.X., Liu, Z.R., Zhen, S.W., Wei, J.N., Zhu,
J.S. and Zei, H.S. Application of an immunoenzymatic method
and an immunoautoradiographic method for a mass survey of
nasopharyngeal carcinoma, Intervirology, 13, 162-168,(1980)

Zeng, Y., Zhang, L.G., Li, H.Y., Jan, M.C., Zhang, Q., Wu,
Y.C., Wang, Y.S., and Su, G.R. Serological mass survey for
early detection of nasopharyngeal carcinoma in Wuzhou
City, China. Int. J. Cancer 29, 139-141, (1982)

Zeng, Y., Zhong, J.M., Li, L.Y., Wang, P.Z., Tang, H., Ma,
Y.R., Zhu, T.S., Pan, W.J., Liu, Y.X., Wei, J.N., Chen, J.Y.,
M.Y.K., Li, E.J., Tan, B.F. Follow-up studies on Epstein-
Barr, virus IgA/VCA Antibody Positive Persons in Zangwu
County, China. Intervirology. 20, 190-194, (1983)

Zeng, Y., Gi, Z.W., Wang, P.C., Tan, H.Z., Ito, Y., Induction
of Epstein-Barr VIrus Antigen in Raji cells and P3HR-1 cells
by culture fluids of anaerobes from nasopharynx of patients
with nasopharyngeal carcinoma and with other ear-nose-throat

diseases. (submitted for publication)

Zeng, Y., Miao, X.C., Jaio, B., Li, H.Y., Ni, H.Y., and Ito, Y., Epstein-Barr Virus activation in Raji cells with ether extracts of soil from different areas in China. Cancer Letters, in press.

Zeng, Y., Gong, M.G., Jan, M.G., Fun, Zeng, L.G. and Li, H.Y. Detection of Epstein-Barr Virus IgA/EA antibody for diagnosis of nasopharyngeal carcinoma by immunoautoradiography. Int. J. Cancer, 31, 599-601, (1983)

Zeng, Y., Zhong, J.M., Mo, Y.K., Miao, X.C., Epstein-Barr virus early antigen induction in Raji cells by Chinese medical herbs. Intervirology, 19, 201-204, (1983)

Zhangjiang Medical College, Diagnosis of nasopharyngeal carcinoma by cytological examination of exfoliated cells taken by negative pressure suction. Chin. med. J., 1, 45-47 (1976)

14

EVALUATION OF EPSTEIN-BARR VIRUS SEROLOGIC ANALYSIS IN

NORTH AMERICAN PATIENTS WITH NASOPHARYNGEAL CARCINOMA AND

IN COMPARISON GROUPS

H. Bryan Neel III, M.D., Ph.D., Gary R.

Pearson, Ph.D.,* and William F. Taylor, Ph.D.,

Mayo Clinic and Mayo Foundation, Rochester, MN

55905, U.S.A.

SUMMARY

This prospective cooperative study was initiated in
1978 to determine the value of serologic tests related to
Epstein-Barr virus (EBV), including the antibody response
to the EBV membrane antigen as measured by the
antibody-dependent cellular cytotoxicity (ADCC) assay, in
North American patients with different histopathologic
types of nasopharyngeal carcinoma (NPC). Serologic
testing is a useful diagnostic aid for patients with NPC,
particularly those in whom the tumors are small and
submucosal (difficult to see or occult). A large body of
clinical evidence, histopathologic data, and more recently
immunologic studies supports the concept that carcinomas
of the nasopharynx constitute two distinct diseases.
Today, these are classified as WHO type 1 tumors and
combined WHO types 2 and 3 tumors. The ADCC titers
obtained at diagnosis often predict the clinical course of
patients with WHO types 2 and 3 NPC regardless of the
stage of the disease. A low ADCC titer at diagnosis
portends a poor prognosis, and the determination of
antibody titers identifies patients in whom recurrent
disease is likely to develop after conventional
irradiation therapy.

*Present address: Georgetown University, Washington, DC.

This prospective cooperative study was initiated in 1978 to determine the value of serologic tests related to Epstein-Barr virus (EBV), including the antibody response to the EBV membrane antigen as measured by the antibody-dependent cellular cytotoxicity (ADCC) assay, in North American patients with different histopathologic types of nasopharyngeal carcinoma (NPC). It involves several institutions in the United States to provide a sufficient number of patients--mostly Caucasians--from different geographic locations and diverse ethnic backgrounds (Neel et al., 1980, 1981A, 1983, 1984, in press; Pearson et al., 1983). Other studies, primarily in Chinese patients, have found potential value for most of the antibody responses (Henle and Henle, 1966, 1976; Henle et al., 1970, 1971, 1977; Pearson et al., 1971; Ho et al., 1978; Naegele et al., 1982; Sundar et al., 1982; Ringborg et al., 1983).

This report explores the data on the NPC patients and on controls consisting of normal blood donors, patients with squamous cell carcinoma elsewhere in the head and neck, and patients with benign diseases commonly seen in an otorhinolaryngologic practice.

MATERIALS AND METHODS

Serum Donors

Serum samples were collected during the same time period from three groups of donors.

Group 1. One hundred fifty-one patients with NPC. Serum samples were collected at diagnosis, but before treatment. Tissue removed to establish the diagnosis of NPC was evaluated by a panel of four pathologists and classified according to the standards established by the World Health Organization (Shanmugaratnam and Sobin, 1978) as squamous cell carcinoma (WHO type 1), nonkeratinizing carcinoma (WHO type 2), or undifferentiated carcinoma (WHO type 3).

Group 2. Patients with various related diseases: 147 with squamous cell carcinoma of the nose or sinuses (excluding the nasopharynx), mouth or pharynx, ear, and other miscellaneous sites; 71 with other types of malignant

tumors (excluding squamous cell carcinomas) of the
nasopharynx, nose or sinuses, mouth or pharynx, and neck or
parotid gland; and 407 with various benign head and neck
diseases located at sites throughout the upper
aerodigestive tract and typical of those seen in
otorhinolaryngologic practice.

Group 3. Two hundred seventy-eight healthy persons
(blood donors).

Tests

All serum specimens were titrated for IgG antibodies
to EA(D) and IgA antibodies to VCA by indirect
immunofluorescence procedures (Henle et al., 1970, 1977;
Henle and Henle, 1976). All tests were performed at the
Mayo Clinic and at the Children's Hospital of
Philadelphia. There was good concordance between results
from the two laboratories.

ADCC data from 134 of the 151 patients with NPC were
analyzed. The ADCC radioimmunoassay measures antibodies
to an EBV membrane antigen component; the procedure has
been described in detail (Pearson and Orr, 1976; Pearson
et al., 1978; Mathew et al., 1981). The ADCC titer is the
highest dilution serum that caused significant ADCC
(Student's t test was used as a test of significance).
All sera were assayed at least three times. When titer
varied significantly in sequential serum samples from the
same patient, all sera were retested together in the same
assay to verify the changes in titers.

Statistical Methods

The IgG antibodies to EA(D) and IgA antibodies to VCA
discussed here have a potential for use as diagnostic
markers for NPC. We investigated this application by
estimating the sensitivity and specificity of the two
titers considered herein. The usual definitions apply:
sensitivity is the proportion of true NPC patients who
have positive markers; specificity is the proportion of
true non-NPC patients who have negative markers. The data
are analyzed in terms of positive and negative responses,
positivity being defined as a titer $\geq 1:10$. The percentage

of positive titers was considered in detail by using
fairly fine breakdowns of the various classes of patients.
Significance tests were based on appropriate contingency
table analysis. On occasion, the breakdown into subgroups
was such that very small numbers were involved, and so a
certain amount of pooling was done in conjunction with the
χ^2 test used with contingency tables. In every case the
groups were defined on the basis of sample size and
anatomy without reference to outcome.

Assessment of ADCC titer as it relates to prognosis
was based on segregation of the patients into groups with
high or low titers at the time of diagnosis. ADCC titers
>1:7,680 were classified as high, and titers <1:7,680 were
classified as low (Student's t test was used as a test of
significance). Actuarial survival was calculated by the
method of Kaplan and Meier (1958).

RESULTS

Sensitivity and Specificity of EA(D) and VCA
(IgA) in Diagnosis

The EA(D) (IgG) response is a sensitive marker for WHO
type 2 and type 3 patients; 94% and 83%, respectively, of
such patients had positive responses (Table 1). The test
is not sensitive for patients with WHO type 1 tumors; only
35% had positive responses. The test is not very specific
because only 62% to 71% of the responses were negative for
non-NPC patients and normal donors.

The VCA (IgA) response is equally as sensitive for WHO
type 2 and type 3 patients: 89% and 84%, respectively,
had positive responses. It is insensitive for WHO type 1
patients (16%). In contrast to the EA(D) response, it is
quite specific because 82% to 91% of the non-NPC patients
and healthy donors had negative responses. WHO types 2
and 3 are similar to each other. The low percentage of
both responses for WHO type 1 patients is similar to that
for the comparison groups, and normal donors had even
fewer positive responses (9%) than did non-NPC patients;
this difference is statistically significant for VCA
(IgA). As a marker for WHO type 2 or 3 tumors, the VCA
(IgA) response is sensitive (positive in 85%) and quite

Table 1. Antibody Responses by Group Studied and Type of Antibody

Group	No.	VCA (IgA)			EA(D) (IgG)		
		% pos.	% neg.	% right*	% pos.	% neg.	% right
NPC							
WHO type 2	18	89 } 85%	11	89	94 } 85%	6	94
WHO type 3	96	84	16	84	83	17	83
WHO type 1	37	16	84	16	35	65	35
Comparison groups†							
SC CA, head and neck	147	18	82	82	31	69	69
Other CA, head and neck	71	13	87	87	38	62	62
Benign disease, head and neck	407	14	86	86	37	63	63
Healthy donors	278	9‡	91	91	29	71	71

*Correct for NPC and thus useful response as marker for NPC.

†SC, squamous cell; CA, carcinoma.

‡Significantly lower than for the 625 patients in the patient comparison groups.

(From Neel, H.B., III, Pearson, G.R., and Taylor, W.F., Antibodies to Epstein-Barr virus in patients with nasopharyngeal carcinoma and in comparison groups. Ann. Otol. Rhinol. Laryngol., 93, 477-482 [1984]. By permission of the Annals Publishing Company.)

specific (negative in 85% of comparison patients and 91% of normal donors).

In patients with squamous cell carcinoma (excluding nasopharynx) with involvement of the tongue, 50% were VCA (IgA) positive and 62% were EA(D) positive, compared to 18% and 31%, respectively, of the patient comparison group as a whole. Both differences are significant for VCA (IgA) ($\underline{P}$ = 0.007) and for EA(D) ($\underline{P}$ = 0.02). Proximity of the tumor to Waldeyer's ring at the base of the tongue did not influence the incidence of positivity because the number of positive responses was equally divided between the anterior and the base of the tongue. With squamous cell carcinoma of the larynx, 23% of the patients had positive VCA (IgA) responses and 38% had positive EA(D) responses; although slightly higher than for the group as a whole, these are not significantly different. There are no significant differences between the glottic, supraglottic, and subglottic regions of the larynx.

Among patients with other types of malignant tumors (excluding squamous cell carcinomas) in the head and neck, 13% were VCA (IgA) positive and 38% were EA(D) positive. No comparisons were significant. Percentages of positive VCA (IgA) and EA(D) responses were high in patients with adenocarcinoma of the parotid, but the number of patients was small. The percentage of positive responses was higher in patients with leukemias than in the group as a whole; the two patients with positive responses had chronic lymphocytic leukemia. One patient had had hyperplastic nasopharyngeal lymphoid tissue initially and eventually chronic lymphocytic leukemia developed, but the serologic profile was typical of patients with active NPC. Patients with lymphoma of the nasopharynx did not have the same serologic profile as patients with NPC.

In the comparison patients with benign head and neck diseases, 14% had positive VCA (IgA) responses and 37% had positive EA(D) responses. Only the VCA (IgA) responses showed a significant difference among the subgroups ($\underline{P}$ < 0.05). Percentage of positive VCA (IgA) responses was significantly higher (31%) in patients with inflammatory nasal polyps or nasal papillomas than in the whole group. Twenty-two percent of the patients with benign epithelial diseases of the vocal cords (myxomatous polyps, leukoplakia/hyperkeratosis, nodules, granulomas,

etc.) had positive VCA (IgA) responses and 43% had positive
EA(D) responses (differences from group as a whole not
significant). In the large subset of patients with benign
adenoid hypertrophy, 20% had positive VCA (IgA) responses
(not significantly different from the percentage for the
group as a whole), and the percentage of the positive
EA(D) responses was the same as that for the group as a
whole (37%).

Relationships Among ADCC, WHO Tumor Type, and Stage of Disease

Patients with WHO 2 and 3 nasopharyngeal carcinomas
were grouped together because, as shown above and reported
previously (Neel et al., in press), their anti-EBV
serologic profiles were similar. To determine if ADCC
titers at diagnosis were related to the clinical stage of
disease, we segregated patients by the level of ADCC and
by histopathologic tumor types into the groups of disease
in the AJC and Ho staging systems (AJC, 1977; Ho, 1978;
Neel et al., 1981B). Most of the patients with WHO 1
carcinomas had stage IV disease (AJC system) or stage III
disease (Ho system) (Fig. 1 upper left and upper right).
The pattern was similar in patients with WHO 2 and 3
carcinomas when classified by the AJC system but the
distribution was more even among stages I, II, and III
when these patients were classified by the Ho system (Fig.
1 lower left and lower right). High and low ADCC titers
were seen in all stages (except in Ho stage V), and the
distribution of patients by high and low ADCC titers was
similar in each of the stage groupings. Therefore, no
relationship was apparent between the ADCC titer at
diagnosis and the clinical stage of the disease in the two
commonly applied staging systems.

Relationship Between ADCC Titers at Diagnosis and the Course of Disease

With WHO 1 carcinoma disease, progression (defined as
the appearance of clinically detectable recurrence,
usually metastatic disease) differed little among patients
who had high or low ADCC titers. The samples were small,
and no significant difference was found ($\underline{P}$ = 0.2). In the
combined WHO 2 and 3 group, disease progression occurred

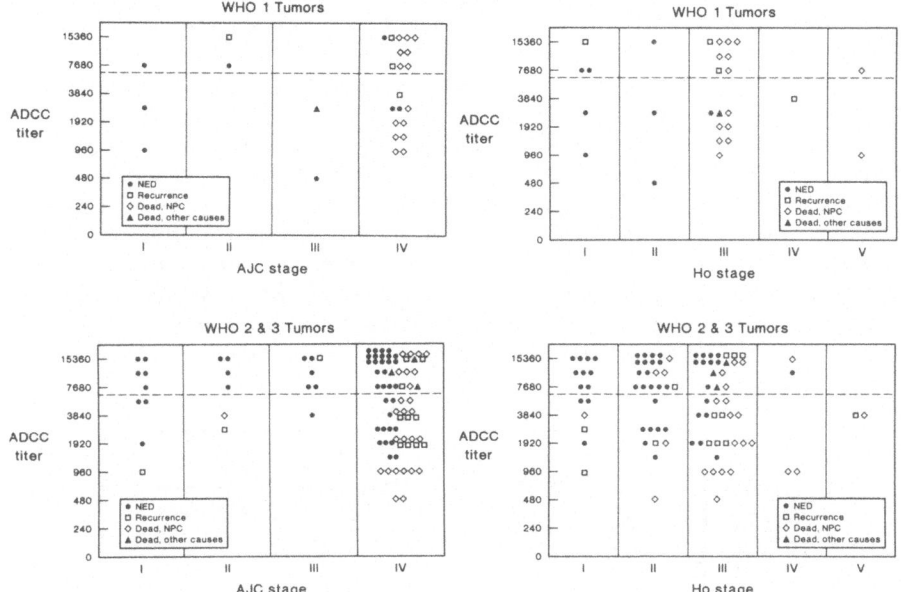

Fig. 1. Relationship of ADCC titer to clinical stage of disease. Patients with WHO 1 carcinomas (Upper) and WHO 2 or 3 carcinomas (Lower) were classified according to AJC (Left) and Ho (Right) systems. NED, no evidence of disease; NPC, nasopharyngeal carcinoma. (From Neel, H.B., III, Pearson, G.R., and Taylor, W.F., Antibody-dependent cellular cytotoxicity: relation to stage and disease course in North American patients with nasopharyngeal carcinoma. Arch. Otolaryngol. [in press].)

much more frequently in those with low titers ($\underline{P}$ = 0.0001). Approximately 75% of the patients with high titers remained disease-free for 3 years or longer, but only 34% of the patients with low titers remained disease-free at 3 years and only 20% were disease-free at 5 years. When disease did progress in patients with high ADCC titers, it did so within the first 3 years after diagnosis. After that, the number of patients who remained disease-free was relatively stable. In contrast, the number of patients with low titers who remained disease-free continued to decline steadily for 5 years or more.

A clear association is apparent between survival after treatment and a high or low ADCC titer at diagnosis. Of the patients who had high ADCC titers at diagnosis, 80% survived for 3 years or longer, whereas only 50% of the patients with low titers were alive at 3 years and only 35% at 5 years ($\underline{P}$ = 0.008) (Fig. 2). All of the deaths in the patients with high titers occurred during the first 3 years of observation, and deaths in the patients with low titers continued beyond 4 years. Actual death rates were about 2 times higher in the patients with low titers than in those with high titers during the first 3 years after diagnosis. Several of the patients who were studied are still alive with clinically apparent metastatic disease,

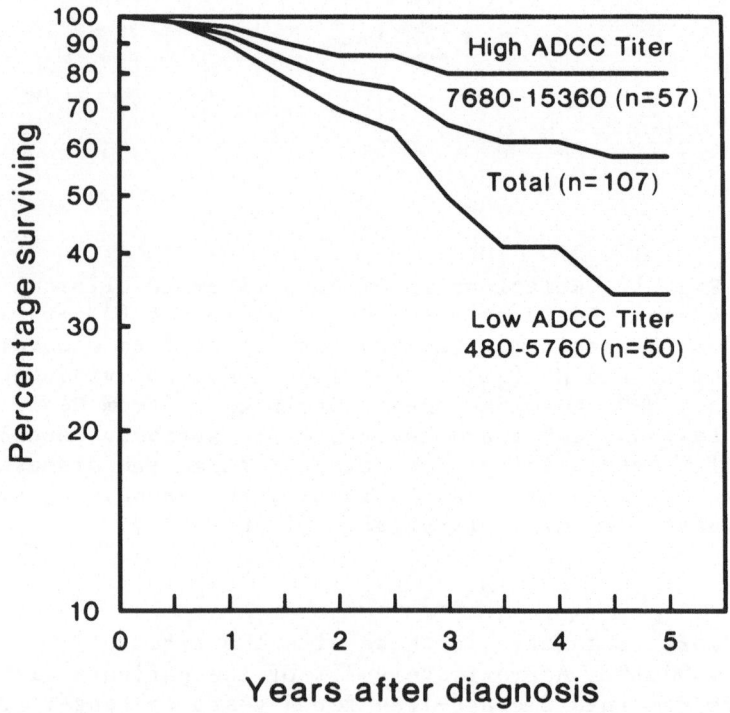

Fig. 2. Actuarial survival of patients with WHO 2 or 3 carcinomas on basis of ADCC titers at diagnosis. (From Neel, H.B., III, Pearson, G.R., and Taylor, W.F., Antibody-dependent cellular cytotoxicity: relation to stage and disease course in North American patients with nasopharyngeal carcinoma. Arch. Otolaryngol. [in press].)

and virtually all will eventually die as a consequence of
their disease.

Histopathology and Survival

In the most recent actuarial survival calculation for
all WHO histopathologic types of NPC in this study, at 5
years, 50% of the 151 patients were alive. When the
patients were divided into two groups based on
histopathologic type, however, there was a definite trend
toward more deaths from NPC in the WHO type 1 group than
in the type 2 and type 3 group (Fig. 3). At 5 years, 59%
of the 114 patients with WHO 2 and 3 tumors were alive
whereas less than 25% of the 37 patients with WHO 1 tumors
were alive. The difference is significant (P̲ < 0.0001).

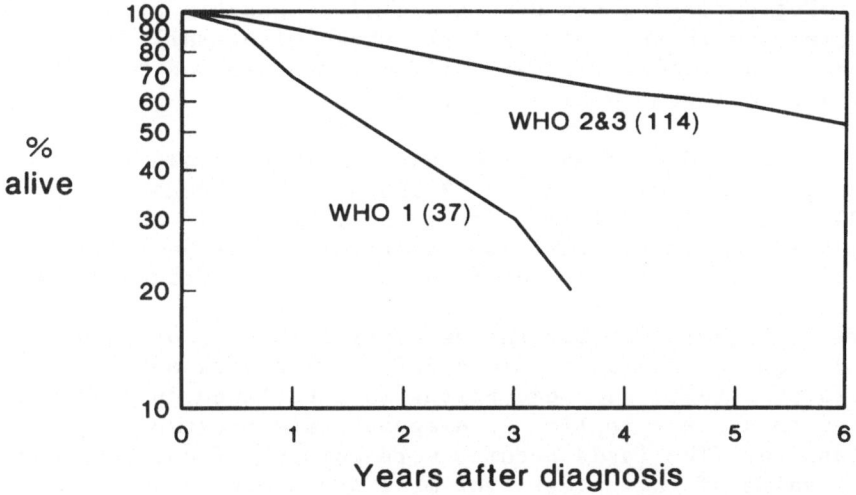

Fig. 3. Actuarial survival of 114 patients with WHO
type 2 or type 3 NPC and 37 patients with WHO type 1 NPC.

DISCUSSION

This study was initiated to determine, prospectively,
the clinical value, as it relates to diagnosis and
prognosis, of EBV-related serologic testing in patients
with NPC. It reaffirms that, in North American

patients--mostly Caucasians--antibodies to two EBV
antigens are present more frequently and at higher titers
in patients with WHO type 2 or 3 NPC than in patients with
WHO type 1 NPC or in comparison populations (Neel et al.,
1980, 1983; Pearson et al., 1983). The differences
between the various histopathologic groups of NPC were
particularly evident with both EA(D) (IgG) and VCA (IgA)
responses, and these tests clearly separate NPC into two
major categories. The more sensitive of the two tests
studied was the IgA antibody to VCA. The specificity of
the test can be improved by segregating the patients with
high (>1:10) VCA (IgA) titers from the patients with low
titers (<1:10), as shown in a previous study (Neel et al.,
1980), but then some degree of sensitivity is lost. Also,
it is important to recognize that, in patients with small
NPCs, 90% with stage I NPC (WHO type 2 or 3) have positive
VCA (IgA) responses (Neel et al., 1983). Therefore,
serologic testing has been a useful diagnostic aid in many
patients with NPC, particularly those with small
submucosal tumors that are difficult to see and those
with occult NPC (Neel et al., 1981A).

It should be emphasized that sera from about 15% of
patients with WHO type 2 or 3 NPC have been negative for
VCA (IgA). In addition, the sera of 9% to 18% of
individuals in the various comparison groups have been
positive and are considered to be "false-positives." We
believe that it is possible that some of these patients
are at higher risk for the development of NPC or have some
unrecognized cellular immune defect. As with any
serologic test, such exceptions must be taken into account
when assays are applied in diagnosis and treatment
planning. The issue becomes more complex if one considers
the value of these tests for mass screening in
high-incidence areas. The predictability of the test is
acceptable in the context of otorhinolaryngologic practice
because the VCA (IgA) test is quite sensitive and
specific.

It was interesting to find subgroups of patients--some
of them small--with a higher percentage of positive VCA
(IgA) responses within the comparison groups. These
subgroups were patients with squamous cell carcinoma of
the tongue, adenocarcinoma of the parotid, leukemia, or
benign nasal polyps. In clinical practice, all of these
disorders are easily differentiated from NPC by physical

examination, but the nasopharynx should be examined carefully. In none was the percentage of positive responses even close to that in the WHO types 2 and 3 group.

The ADCC titer at the time of diagnosis is often predictive of the prognosis. Certainly, patients with low titers have a greater risk for recurrence. The ADCC titer is apparently another factor, in addition to staging, that can be used to predict prognosis. Clinical staging is the traditional approach for predicting prognosis, but the ADCC titer can be used to segregate patients within the stage groups into those with "good" and with "poor" prognoses. Serologic testing eventually may become one of the methods for staging patients with WHO types 2 and 3 forms of NPC. Both ADCC and WHO tumor type would be important elements in a prognostic scoring system. The roles of age, sex, and duration of symptoms in such a system have not been determined yet.

It is not clear why serum ADCC titers are high in some patients and low in others. It is clear that low ADCC titers are more likely to portend progression of the disease and death from the disease than are high titers and that these titers do not seem to be related to tumor burden in the two clinical staging systems reported here or in the UICC system. However, in all the stage groups, the propensity for recurrences was greater among patients with WHO types 2 and 3 tumors with low titers. Possibly, the low titers reflect a deficiency of antibodies to the major ADCC epitope that is expressed on the major EBV-induced antigen (Qualtiere et al., 1982). However, a more likely explanation relates to studies from our laboratory in which low ADCC values were shown possibly to be caused by "blocking" of IgG-mediated ADCC by IgA antibodies (Pearson et al., 1978; Mathew et al., 1980, 1981; Bertram et al., 1983). Most patients with the poorly differentiated types of NPC have IgA antibodies to EBV antigens in their sera. IgA antibodies purified from the sera of patients with NPC block the ADCC reaction mediated by IgG anti-EBV antibodies.

If one assumes that ADCC functions against the tumor in vivo, then this antibody to membrane antigen might be active in specific immunity against NPC. Therefore, approaches to enhance the level of this antibody in the circulation or approaches to remove IgA from the circulation might have a therapeutic effect (Neel et al., 1983). Possibly, the blocking activity of IgA antibodies could be abrogated with high-titer serum, through active immunization with a vaccine against membrane antigen that expresses the ADCC epitope, or by plasmapheresis of IgA. It will be necessary to observe patients with NPC for a longer period and to continue to study more patients for several years before the prognostic value of any of the anti-EBV markers can be determined conclusively or before such therapeutic measures should be considered as routine adjuncts to conventional procedures for the treatment of patients with NPC.

ACKNOWLEDGMENT

Supported in part by Contract CP91006 from the Division of Cancer Cause and Prevention, National Cancer Institute, Bethesda, Maryland, and the St. Jude Fund for Research and Education in Otolaryngology.

REFERENCES

American Joint Committee for Cancer Staging and End-Results Reporting: Manual for Staging of Cancer 1977. Published by the American Joint Committee, 55 East Erie Street, Chicago, IL 60611 (1977).

Bertram, G., Pearson, G.R., Faggioni, A., Krueger, G.R.F., Sesterhenn, K., Ablashi, D.V., and Levine, P.H., A long-term study of EBV and non-EBV related tests and their correlation with the clinical course of nasopharyngeal carcinoma. In: U. Prasad, D.V. Ablashi, P.H. Levine, and G.R. Pearson (ed.), Proceedings of the Fourth International Symposium on Nasopharyngeal Carcinoma, pp. 115-124, University of Malaya Press, Kuala Lumpur (1983).

Henle, G. and Henle, W., Immunofluorescence in cells derived from Burkitt's lymphoma. J. Bacteriol., 91, 1248-1256 (1966).

Henle, G. and Henle, W., Epstein-Barr virus-specific IgA serum antibodies as an outstanding feature of nasopharyngeal carcinoma. Int. J. Cancer, 17, 1-7 (1976).

Henle, G., Henle, W., and Klein, G., Demonstration of two distinct components in the early antigen complex of Epstein-Barr virus-infected cells. Int. J. Cancer, 8, 272-282 (1971).

Henle, W., Henle, G., Zajac, B.A., Pearson, G., Waubke, R., and Scriba, M., Differential reactivity of human serums with early antigens induced by Epstein-Barr virus. Science, 169, 188-190 (1970).

Henle, W., Ho, J.H.C., Henle, G., Chau, J.C.W., and Kwan, H.C., Nasopharyngeal carcinoma: significance of changes in Epstein-Barr virus-related antibody patterns following therapy. Int. J. Cancer, 20, 663-672 (1977).

Ho, H.C., Ng, M.H., and Kwan, H.C., Factors affecting serum IgA antibody to Epstein-Barr viral capsid antigens in nasopharyngeal carcinoma. Br. J. Cancer, 37, 356-362 (1978).

Ho, J.H.C., Stage classification of nasopharyngeal carcinoma: a review. In: G. de-Thé and Y. Ito (ed.), Nasopharyngeal Carcinoma: Etiology and Control, pp. 99-113, International Agency for Research on Cancer, Lyon, France (1978).

Kaplan, E.L. and Meier, P., Nonparametric estimation from incomplete observations. J. Am. Stat. Assoc., 53, 457-481 (1958).

Mathew, G.D., Qualtiere, L.F., Neel, H.B., III, and Pearson, G.R., Immunoglobulin A antibody to Epstein-Barr viral antigens and prognosis in nasopharyngeal carcinoma. Otolaryngol. Head Neck Surg., 88, 52-57 (1980).

Mathew, G.D., Qualtiere, L.F., Neel, H.B., III, and Pearson, G.R., IgA antibody, antibody-dependent cellular cytotoxicity and prognosis in patients with nasopharyngeal carcinoma. Int. J. Cancer, 27, 175-180 (1981).

Naegele, R.F., Champion, J., Murphy, S., Henle, G., and Henle, W., Nasopharyngeal carcinoma in American children: Epstein-Barr virus-specific antibody titers and prognosis. Int. J. Cancer, 29, 209-212 (1982).

Neel, H.B., III, Pearson, G.R., and Taylor, W.F.,
Antibodies to Epstein-Barr virus in patients with
nasopharyngeal carcinoma and in comparison groups. Ann.
Otol. Rhinol. Laryngol., 93, 477-482 (1984).

Neel, H.B., III, Pearson, G.R., and Taylor, W.F., Antibody-
dependent cellular cytotoxicity: relation to stage and
disease course in North American patients with
nasopharyngeal carcinoma. Arch. Otolaryngol. (in press).

Neel, H.B., III, Pearson, G.R., Weiland, L.H., Taylor,
W.F., and Goepfert, H.H., Immunologic detection of occult
primary cancer of the head and neck. Otolaryngol. Head
Neck Surg., 89, 230-234 (1981A).

Neel, H.B., III, Pearson, G.R., Weiland, L.H., Taylor,
W.F., Goepfert, H.H., Pilch, B.Z., Goodman, M., Lanier,
A.P., Huang, A.T., Hyams, V.J., Levine, P.H., Henle, G.,
and Henle W., Application of Epstein-Barr virus serology
to the diagnosis and staging of North American patients
with nasopharyngeal carcinoma. Otolaryngol. Head Neck
Surg., 91, 255-262 (1983).

Neel, H.B., III, Pearson, G.R., Weiland, L.H., Taylor,
W.F., Goepfert, H.H., Pilch, B.Z., Lanier, A.P., Huang,
A.T., Hyams, V.J., Levine, P.H., Henle, G., and Henle,
W., Anti-EBV serologic tests for nasopharyngeal
carcinoma. Laryngoscope, 90, 1981-1990 (1980).

Neel, H.B., III, Taylor, W.F., Pearson, G.R., and Weiland,
L.H., Clinical staging of patients with nasopharyngeal
carcinoma. In: E. Grundmann, G. Krueger, and D. Ablashi
(ed.), Cancer Campaign, Vol. 5, pp. 73-79, Gustav Fischer
Verlag, Stuttgart (1981B).

Pearson, G.R., Henle, G., and Henle, W., Production of
antigens associated with Epstein-Barr virus in
experimentally infected lymphoblastoid cell lines. J.
Natl. Cancer Inst., 46, 1243-1250 (1971).

Pearson, G.R., Johansson, B., and Klein, G., Antibody-
dependent cellular cytotoxicity against Epstein-Barr
virus-associated antigens in African patients with
nasopharyngeal carcinoma. Int. J. Cancer, 22, 120-125
(1978).

Pearson, G.R. and Orr, T.W., Antibody-dependent lymphocyte cytotoxicity against cells expressing Epstein-Barr virus antigens. J. Natl. Cancer Inst., 56, 485-488 (1976).

Pearson, G.R., Weiland, L.H., Neel, H.B., III, Taylor, W., Earle, J., Mulroney, S.E., Goepfert, H., Lanier, A., Talvot, M.L., Pilch, B., Goodman, M., Huang, A., Levine, P.H., Hyams, V., Moran, E., Henle, G., and Henle, W., Application of Epstein-Barr virus (EBV) serology to the diagnosis of North American nasopharyngeal carcinoma. Cancer, 51, 260-268 (1983).

Qualtiere, L.F., Chase, R., and Pearson, G.R., Purification and biologic characterization of a major Epstein Barr virus-induced membrane glycoprotein. J. Immunol., 129, 814-818 (1982).

Ringborg, U., Henle, W., Henle, G., Ingimarsson, S., Klein, G., Silfuersward, C., and Strander, H., Epstein-Barr virus-specific serodiagnostic tests in carcinomas of the head and neck. Cancer, 52, 1237-1243 (1983).

Shanmugaratnam, K. and Sobin, L., Histological Typing of Upper Respiratory Tract Tumours. No. 19, World Health Organization, Geneva, Switzerland, pp. 32-33 (1978).

Sundar, S.K., Ablashi, D.V., Kamaraju, L.S., Levine, P.H., Faggioni, A., Armstrong, G.R., Pearson, G.R., Krueger, G.R., Hewetson, J.F., Bertram, G., Sesterhenn, K., and Menezes, J., Sera from patients with undifferentiated nasopharyngeal carcinoma contain a factor which abrogates specific Epstein-Barr virus antigen-induced lymphocyte response. Int. J. Cancer, 29, 407-412 (1982).

15

USE OF EPSTEIN-BARR VIRUS SEROLOGY IN THE DIAGNOSIS

OF NASOPHARYNGEAL CARCINOMA IN MALAYSIA

M. Yadav[*], N. Malliha[*], A.W. Norhanom[*] and U. Prasad[†]

[*]Department of Genetics and Cellular Biology,

University of Malaya [†]Department of Otorhinolaryngology,

University Hospital

SUMMARY

Sera from 129 Malaysian patients with NPC and controls were assayed for antibodies to EBV-related antigens. Histopathologically there were 30 WHO type 1, 20 WHO type 2 and 79 WHO type 3 tumour cases. There was no significant difference between the geometric mean titre (GMT) of the anti-EBV antibodies for the three WHO type tumours. In the Chinese, Malays and Kadazan patients the titres of IgG anti-EA and IgA anti-VCA antibodies increased with the stage of the disease; in the Chinese NPC patients the GMT titres decreased at Stage IV but in the Malay and Kadazan NPC patients the titres continued to increase. Morever, the GMT for anti-EBV antibodies were higher in the younger NPC patients ($\leq$39) compared to older patients at all stages of the disease. It appears that the IgA anti-VCA antibody response is specific to NPC and is a useful diagnostic aid when used alone or when used in combination with IgG anti-EA titres.

INTRODUCTION

In Malaysia, NPC is the most frequently diagnosed neoplasm of the upper respiratory tract with an annual average of 406 which represents an incidence of 2.96 per 100,000 population per year (Yadav et al., in press). The age adjusted incidence in Chinese, Malays and Indians per 100,000 is 16.5, 2.3 and 1.0 for males and 7.2, 0.7 and nil for females respectively (Armstrong et al., 1979). In the ethnic minority groups, like the Kadazans, a bimodal incidence with peaks during adolescence and late middle age have been noted (Rothwell, 1979; Yadav et al., in press). The current study was initiated to determine the value of EBV serology for the diagnosis of NPC in Malaysia. The patients were drawn from all geographic regions of the country and included the major racial/ethnic groups. Normal serum donors and those patients with other head and neck tumours were included for comparison.

MATERIALS AND METHODS

Patients: The study was initiated in 1982, and all patients histopathologically confirmed for NPC were enrolled. The age range was from 11 to 86 years; the median age was 43 years. By race the series consisted of 91 (70.5%) Chinese, 21 (16.2%) ethnic Malays, 4 (3.1%) Indian and 13 (10.9%) ethnic Kadazans.

The stage of the tumours was determined on the basis of the clinical data according to the Ho system (Ho, 1978a). Blood was collected at first examination and periodically at 4-6 months intervals. All serum samples were aliquoted and stored at -70°C (long periods) or -30°C (short periods) before use.

The biopsy taken from the postnasal space was evaluated by a panel of pathologists and classified according to the standards of the World Health Organization (WHO) into broad categories of squamous cell carcinoma (WHO 1; 37 patients = 28.7%); nonkeratinizing carcinoma (WHO 2; 20 patients = 15.5%), and undifferentiated carcinomas (WHO 3; 72 patients = 55.8%) (Shanmugaratnam and Sobin, 1978).

Immunovirologic assay: Coded serum samples from all patients and controls were tested for antibodies to EBV-induced viral capsid antigen (VCA) and early antigen (EA) by the fluorescence techniques previously described in detail (Henle and Henle, 1966; Henle et al., 1970). Briefly, smears of P3HR-1 cells fixed in cold acetone served for detection of antibodies to VCA. Raji cells superinfected with P3HR-1 virus which expressed about 15-20 percent EA-positive cells were used for titration of anti-EA antibodies. The geometric mean titre (GMT) was calculated from antibody-positive (1:10 dilution) sera in each group.

RESULTS

Anti-EBV Titres in Malaysian Patients

The incidence of positive antibody titres in the sera from patients with NPC differed significantly (p <0.05) from the incidence of positive titres in the two comparison control groups for the three types of antibodies titred (Table 1). All 129 NPC sera were positive for IgG antibodies to VCA as opposed to 96-98 percent of the sera from the two different control groups. The GMT of the NPC sera was 268 which significantly differed from the GMT of controls (28 and 36).

Table 1. Anti-EBV Antibody Titres in Sera from Malaysian NPC Patients and Controls.

Serologic tests	Serum donors		
	NPC (n = 129)	Other head and neck tumours (n = 128)	Normals (n = 130)
IgG Anti-VCA			
% Positives	100	97	98
GMT of positives	268*	36	28
Range	80-1280	10-160	10-640
IgA Anti-VCA			
% Positives	90	4	2
GMT of positives	39*	7	28
Range	10-160	10-20	10-80
IgG Anti-EA			
% Positives	96	25	1
GMT of positives	52*	9	1
Range	10-160	10-40	10-80

GMT = Geometric mean titre
* = Significantly higher titre than other groups; p < 0.05.

Both the IgA anti-VCA and IgG anti-EA showed high specificity for NPC. For IgA anti-VCA 90 percent were positive compared to less than 4 percent in controls; the GMT was 39 for NPC patients and 7-28 in controls. Similarly the IgG anti-EA was positive in 96 percent compared to 1-25 percent in controls; the GMT was 52 for NPC patients and 1-9 in controls. When the IgA anti-VCA and IgG anti-EA are taken together only 4 of 129 NPC patients (2.1%) remained negative.

Table 2 shows that the peak GMT of the IgA anti-VCA and IgG anti-EA occurred at 30-39 age group. The maximum frequency of the disease is at 40-49 years, almost a decade later. In the case of IgG anti-VCA antibody, the titers increased with age.

Table 2. Anti-EBV antibody titres of NPC patients in relation to the age in years

Age Group	IgG anti-VCA GMT	IgA anti-VCA GMT	IgG anti-EA GMT
10 - 29	250(17)**	38(15)	54(16)
30 - 39	264(29)	40(26)	60(27)
40 - 49	257(38)	32(33)	47(37)
50 - 59	277(24)	32(22)	45(23)
> 60	341(21)	37(20)	40(21)

** Parentheses indicate number of patients in each group.

Relationship between histopathologic type and anti-EBV titres

The titres of the sera were analysed according to the WHO classification in order to seek an association between EBV-serology and histopathology. There was no statistical difference in the serological profile among the 3 WHO types

for the three antibodies assayed (Table 3). Moreover, a
positive association was not found between EBV-serology
and histopathology when race was taken into consideration.

Table 3. Anti-EBV antibody titres in relation to histopathology using WHO classification.

Serologic tests	WHO 1 (n = 30)	WHO 2 (n = 20)	WHO 3 (n = 79)
IgG anti-VCA			
% Positive (GMT)*	100 (216)	100 (251)	100 (319)
IgA anti-VCA			
% Positive (GMT)*	87 (32)	85 (35)	94 (39)
IgG anti-EA			
% Positive (GMT)*	90 (47)	100 (44)	99 (52)

* Geometric mean titre of positives.

Relationship between anti-EBV titres and stage of disease

Analysis of EBV titres by stage included consideration
of age. In general the anti-EBV titres were higher in
patients younger than 39 years old compared to those older
for all three anti-EBV antibodies assayed for all stages
(Table 4). Furthermore, the antibodies in Stage I and II
were lower than those in Stage III and IV.

Table 4. Relationship between anti-EBV antibody titres and stage of disease

Serologic tests	Stage of Disease			
	I	II	III	IV
IgG anti-VCA				
< 39 years	285(6)*	226(16)	340(21)	320(3)
> 39 years	219(11)	215(26)	309(24)	279(22)
IgA anti-VCA				
< 39 years	53(5)	33(14)	58(20)	80(2)
> 39 years	34(9)	23(24)	45(22)	44(20)
IgG anti-EA				
< 39 years	25(6)	59(15)	65(20)	80(3)
39 years	25(11)	36(35)	63(23)	55(22)

* Geometric mean titer; number of patients in parenthesis.

In the Chinese, the titres decreased from Stage III to Stage IV, but in the Malays and Kadazans there was a marked increase in the anti-EBV titres from Stage I to IV for all three types of antibodies assayed (Table 5).

Table 5. Relationship between anti-EBV antibody titers, race and stage of disease.

	Stage of Disease			
	I	II	III	IV
Chinese(91)				
IgG anti-VCA	254(15)*	313(34)	355(27)	335(15)
IgA anti-VCA	40(14)	26(31)	45(25)	40(14)
IgG anti-EA	38(15)	43(33)	68(26)	56(15)
Malays and Kadazans(34)				
IgG anti-VCA	113(2)	160(8)	276(14)	394(10)
IgA anti-VCA	-	28(8)	36(14)	54(9)
IgG anti-EA	-	31(8)	66(14)	79(10)

* Geometric mean titre; parentheses show number of patients.

DISCUSSION

The most specific anti-EBV antibody response in the Malaysian NPC patients was that of IgA anti-VCA antibody. Sera from 90 percent of the NPC cases contained the antibody compared to 2-4 percent of the controls, and the GMT was significantly higher in the NPC patients compared to the two groups of controls used. The high specificity of the IgA anti-VCA antibodies to NPC has been noted previously in the Chinese and the American patients (Henle and Henle, 1976; Ho et al., 1976; Desgranges et al., 1977; Pearson et al., 1983). The IgG anti-EA antibody also shows a high specificity to NPC compared to the controls and 96 percent NPC cases were positive using this test. However, the application of the IgA anti-VCA and IgG anti-EA tests in combination is of greater sensitivity than either one or these tests used alone.

It is interesting to note that 7 percent of the individuals with head and neck tumours other than NPC and 2 percent of normal individuals were positive for IgA anti-VCA, albeit at low titre levels. Clinical observations and in some cases histopathology of postnasal biopsy did

not confirm NPC in these groups but these tests do not
exclude occult NPC. The normal individuals with elevated
IgA anti-VCA titres are presently being followed at regular
intervals and over a period of 2 years have not shown any
adverse developments. Such follow-up studies of individuals with raised IgA anti-VCA titres but clinically
negative for NPC would be useful in establishing whether
these antibodies precede symptoms of NPC (Ho et al.,
1978b; de The et al., 1983).

Serologic screening of the patients is a valuable
diagnostic aid in most NPC patients, particularly those
with small submucosal tumours that are difficult to
visualise clinically and those with occult NPC. On rare
occasions an occult primary may first be detected by nodal
metastasis. In a small percentage of NPC cases the anti-
EBV serology remains negative when clinical and histopathological signs are positive. The absence of IgA anti-VCA
is not because of selective IgA deficiency since these
patients often have normal serum IgA levels. Moreover,
selective IgA deficiency is extremely rare in Malaysia
(Yadav and Iyngkaran, 1979). There are no features that
uniquely separate NPC patients with negative IgA anti-VCA
titres from other NPC cases. Furthermore, the IgA anti-VCA
antibody remains negative in patients in whom the disease
reappears after treatment. Further studies are required
to identify the basis for the negative serological response in these NPC patients.

DNA hybridization studies indicate that the EBV
genome is regularly associated with undifferentiated carcinoma and nonkeratinizing carcinoma but only with 20
percent of the squamous cell carcinoma (Nonoyama et al.,
1973; Klein et al., 1974; Zur Hausen et al., 1974; Pagano,
1974; Andersson-Anvret et al., 1977). Moreover, Pearson
et al. (1983) noted that sera from patients with WHO 2
and 3 type NPC frequently contained antibodies, often
in high titres, to EBV antigens. In contrast patients
with WHO 1 type tumours had an EBV serologic profile
similar to those found in patients with other head and
neck neoplasms. Our serological data, however, fails
to distinguish the 3 histopathological types either
with IgG anti-EA or IgA anti-VCA antibody. Further
studies are needed to resolve the anomaly between the

188

present data and that of earlier published accounts.
However, it is pertinent to note that Shanmugaratnam et
al. (1979) did not find any evidence to support a difference between WHO 1, 2 and 3 type NPC tumours and concluded that they were variants of a fairly homogenous group
of neoplasms in Singapore. Similar conclusions were
reached by Prathap et al. (1983) who found the coexistence of more than one histological type within the
same biopsy specimen and the presence of amyloid and EB
nuclear antigen in all three histological types in
Malaysia.

The present data confirms the increase in titres of
antibodies to EBV-related antigens in NPC with an increase
in tumour burden (Henle et al. 1973, 1977; de The et al.,
1975; Lynn et al., 1973; Neel et al., 1983). There is a
step-wise increase in GMT with each stage of the disease
for the Chinese, Malays and Kadazans. The step-wise
increase was more marked in the Malays and Kadazans and
thus in Stage IV in this group there was no decrease in
GMT as seen in the Chinese. This decrease in GMT in
Stage IV has been attributed to depressed immunologic responses due to secondary effects (Neel et al., 1983). In
the Malays and Kadazans, the IgA anti-VCA titres were
undetectable in Stage I and this observation casts doubts
on whether these antibodies precede symptoms of NPC in
these groups.

The GMT for IgA anti-VCA are highest in the 30-39
age group almost a decade earlier than the mean peak
incidence of NPC in Malaysia. Moreover, the GMT are highest in NPC patients younger than 39 years old compared
to those older at all stages of the disease. This may
reflect a more vigorous immune response in the younger
patients. The present data would indicate that IgA anti-
VCA along with IgG anti-EA is a useful marker of NPC but
racial variations limits its use for early diagnosis.

ACKNOWLEDGEMENTS

Supported by University of Malaya Research Grants
Committee through the Institute of Advanced Studies.

REFERENCES

Andersson-Anvret, M., Forsby, N., Klein, G., Henle, W.:
Relationship between the Epstein-Barr virus and undifferentiated nasopharyngeal carcinoma: Correlated nucleic
acid hybridization and histopathological examination.
Int. J. Cancer 20: 486-494 (1977).

Armstrong, R.W., Kannan-Kutty, M., Dharmalingam, S.K.,
Ponnudurai, J.R.: Incidence of nasopharyngeal carcinoma in
Malaysia, 1968-1977. Brit. J. Cancer 40: 557-567 (1979).

Desgranges, C., De The, G., Ho., J.H.C., Ellouz, R.:
Neutralizing EBV-specific IgA in throat washings
of nasopharyngeal carcinoma (NPC) patients. Int. J.
Cancer 19: 627-633 (1977).

De The, G. Zeng, Y., Desgranges, C., Pi, G.H.: The
existence of pre-nasopharyngeal carcinoma condition should
allow preventive interventions. In: U. Prasad et al. (ed.)
Nasopharyngeal Carcinoma: Current concepts, p.365-274,
University of Malaya, Kuala Lumpur (1983).

De The, G., Day, N.E., Geser, A., Lavoue, M.F., Ho, J.H.C.,
Simons, M.J., Sohier, R., Tukei, P., Vonka, V., Zavadova,
H.: Seroepidemiology of the Epstein-Barr virus: preliminary
analysis of an international study - a review. Part 2,
pp. 3-16 In: Oncogenesis and herpesviruses II. de The,
G., et al. (eds). International Agency for Research in
Cancer, Lyon (1975).

Henle, G., Henle, W.: Immunofluorescence in cells derived
from Burkitt's lymphoma. J. Bact. 91: 1248-1256 (1966).

Henle, G., Henle, W.: Epstein-Barr virus-specific IgA
serum antibodies as an outstanding feature of nasopharyngeal carcinoma. Int. J. Cancer 17: 1-17 (1976).

Henle, W., Henle, G., Zajac, B.A., Pearson, G., Waubke, R.,
Scriba, M.: Differential reactivity of human sera with
early antigens induced by Epstein-Barr virus. Science
169: 188-190 (1970).

Henle, W., Ho, H.C., Henle, G., Kwan, H.C.: Antibodies
to Epstein-Barr virus-related antigens in nasopharyngeal
carcinoma. Comparison of active cases and long term
survivors. J. Nat. Cancer Inst. 51: 361-369 (1973).

Henle, W., Ho, J.H.C., Henle, G., Chau, J.C.W., Kwan, H.C.:
Nasopharyngeal carcinoma: Significance of changes in
Epstein-Barr virus-related antibody patterns following
therapy. Int. J. Cancer 20: 663-672 (1977).

Ho, J.H.C.: Stage classification of nasopharyngeal carcinoma: A Review In: Nasopharyngeal Carcinoma: Etiology and
Control, pp. 99-113, de The, G. and Ito, Y., (eds.) International Agency for Research in Cancer, Lyon (1978a).

Ho, J.H.C., Kwan, H.C., NG, M.H., De The, G.: Serum IgA
antibodies to EBV capsid antigens preceeding symptoms of
nasopharyngeal carcinoma. Lancet 1: 436-437 (1978b).

Ho, J.H.C., NG, M.H., Kwan, H.C., Chau, J.C.M.: Epstein-
Barr virus-specific IgA and IgG serum antibodies in
nasopharyngeal carcinoma. Brit. J. Cancer 34: 655-660
(1976).

Klein, G., Giovanella, B.C., Lindahl, T., Fialkow, P.J.,
Singh, S., Stehlin, J.S.: Direct evidence for the presence
of Epstein-Barr virus DNa and nuclear antigen in malignant
epithelial cells from patients with poorly differentiated
carcinoma of the nasopharynx. Proc. Nat. Acad. Sci.
U.S.A. 71: 4737-4741 (1974).

Lynn, T.C., Tu, S.M., Hirayama, T., Kawamura, A. Jr.:
Nasopharyngeal carcinoma and Epstein-Barr virus II:
Clinical course and the anti-VCA antibody. Jap. J.
Exp. Med. 43: 135-144 (1973).

Neel, III, H.B., Pearson, G.R., Weiland, L.H.,
Taylor, W.F., Goepfert, H.H., Pilch, B.Z., Goodman, M.,
Lanier, A.P., Huang, A.T., Hyams, V.J., Levine, P.H.,
Henle, G., Henle, W.: Application of Epstein Barr virus
serology to the diagnosis and staging of North American
patients with nasopharyngeal carcinoma. Otolaryngl. Head
Neck Sur., 91: 255-262 (1983).

Nonoyama, M., Huang, C.H., Pagano, J.S., Klein, G., Singh, S.: DNA of Epstein-Barr virus detected in tissue of Burkitt's lymphoma and nasopharyngeal carcinoma. Proc. Nat. Acad. Sci. U.S.A. 70: 3265-3268 (1973).

Pagano, J.A.: The Epstein-Barr virus and malignancy: Molecular evidence. Cold Spring Harbor Symp. Quant. Bio. 39: 797-805 (1974).

Pearson, G.R., Weiland, L.H., Neel, III, H.B., Taylor, W., Earle, J., Mulroney, S.E., Goepfert, H., Lanier, A., Talvot, M.L., Pilch, B., Goodman, M., Huang, A., Levine, P.H., Hyams, V., Moran, E., Henle, G., Henle, W.: Application of Epstein-Barr virus (EBV) serology to the diagnosis of North American nasopharyngeal carcinoma. Cancer 51: 260-268 (1983).

Prathap, K., Prasad, U., Ablashi, D.V.: The pathology of nasopharyngeal carcinoma in Malaysians, p. 55-64. In: Nasopharyngeal carcinoma: Current Concepts, Prasad, U. et al. (eds.) University of Malaya, Kuala Lumpur (1983).

Rothwell, R.I.: Juvenile nasopharyngeal carcinoma in Sabah (Malaysia). Clinical Oncology 5: 353-358 (1979).

Shanmugaratnam, K., Sobin, L.H.: Histological typing of upper respiratory tract tumours: International Histological Classification of Tumours, No. 19. World Health Organization, Geneve (1978).

Shanmugaratnam, K., Chan, S.H., De The, G. GOH, J.E.H., Khor, T.H., Simons, M.J., Tye, C.Y.: Histopathology of nasopharyngeal carcinoma. Correlations with epidemiology, survival rates and other biological characteristics. Cancer 44: 1029-1044 (1979).

Yadav, M., Iyngkaran, N.: Low incidence of selective IgA deficiency in normal Malaysians. Med. J. Mal. 24: 145-148 (1979).

Yadav, M., Tan, M.K., Singh, P., Dharmalingam, S.K. Nasopharyngeal carcinoma in Malaysians under the age of twenty years. Clinical Oncology (in press).

zur Hausen, H., Schulte-Holthausen, H., Wolf, H., Dorries, K., Egger, H.: Attempts to detect virus-specific DNA in human tumours II Nucleic acid hybridization with complementary RNA of human herpes group viruses. Int. J. Cancer 13: 657-664 (1974).

16

AN ANALYSIS OF THE RELATIONSHIP BETWEEN CLINICAL PATHOLOGY

AND SEROLOGICAL LEVEL OF EB VIRUS VCA-IgA ANTIBODY IN

NASOPHARYNGEAL CARCINOMA

Huang, Pei Sheng, Chen, Yu Dong, Shen, Yuin Yin

Fujian Medical College, Fuzhou, Fujian,

People's Republic of China

SUMMARY

Nasopharyngeal carcinoma (NPC) is one of the common malignant tumors in Fujian Province, China. By using the indirect immunoperoxidase labelled antibody method, the antibody level of 363 cases 5 and the relationship to histopathological changes were studied. Among these cases, 127 were untreated NPC with a positive rate for IgA anti-VCA antibody of 91.3% (GMT 1:33). Forty two of the 363 cases had other tumors of head and neck with a positive rate of 7.1% (GMT 1:8) while none of 53 cases of chronic nasopharyngitis and 141 cases of normal individual had detectable antibody. Since this assay has a significant specificity, it is used as an effective serological diagnostic tool for NPC in our clinic.

The serological level of VCA-IgA antibody in untreated NPC cases correlated with the duration of disease, the size of the metastatic lymph node, the proportion of NPC parenchyma and the degree of plasma cell infiltration in tumor tissues. No significant relationship was found between the VCA-IgA antibody level and the sex, age, original site of the tumor, mode of tumor growth and the relationship between HLA-type and survival in this patient population.

INTRODUCTION

Nasopharyngeal carcinoma (NPC), is one of the more common malignant tumors in Fujian. The close relationship between EB virus and NPC has been discussed (Henle and Henle, 1976; Ho et al., 1976; Klein et al., 1974). Although the level of EB virus VCA-IgA serum antibody is high in NPC patients, its clinical and pathological significance remains unsettled. As an attempt to solve some of the problems, the following studies were undertaken.

MATERIALS AND METHODS

This report was based on 363 individuals including 127 NPC patients and 236 control cases collected within 1980-1981 from our NPC clinic. With application of the indirect immunoperoxidase technique, EB virus VCA-IgA serum antibody was examined. 222 nasopharyngeal biopsy specimens from all of the patients (141 normal subjects were excluded) were obtained. Paraffin sections were routinely processed and studied. The PAP technique was used for IgA, IgG and IgM in tissue plasma cells from 68 cases of NPC and 40 cases of chronic nasopharyngitis.

RESULTS AND DISCUSSIONS

I: EB virus VCA-IgA serum antibody in NPC patients before treatment and in control groups.

The positive rate of VCA-IgA serum antibody in the 127 NPC patients before treatment was 91.3% (GMT 1:33) as compared to only 7.1% (GMT 1:79) in the 42 patients with other head and neck tumors. None of the serum samples from 53 patients with chronic nasopharyngitis or 141 normal subjects had detectable IgA anti-VCA antibody. Thus, the test has been routinely used in our NPC clinic as a simple and reliable serodiagnostic method for NPC.

II. Relation between level of EB virus VCA-IgA serum antibody and clinical characteristics of NPC patients.

Levels of VCA-IgA serum antibody in 127 NPC patients were analyzed according to clinical stages by using National TNM staging system proposed in 1979. A correlation between VCA-IgA serum antibody and disease stage was apparent (Table 1). Patients with a short survival (less than 1 year) were less likely to have elevated titers (10.4%) and had a lower GMT (1:29) than those with a longer survival (29%, GMT=1:47) (Table 2).

Table 1. Clinical Stages and VCA-IgA Level in Sera

of 127 NPC Patients

Clinical Stages	No. of Cases	No of Cases with VCA-IgA $\geq 1:5$	%	GMT
I	12	8	66.7	1:13
II	48	43	89.6	1:25
III	59	57	96.6	1:46
IV	8	8	100.0	1:52
Total	127	116	93.3	1:33

A relationship between the serum antibody and dissemination of the tumor was also noticed, but no difference was found relevant to age, sex or size of the primary tumors. The difference between patients with extensive lymph node metastasis (N_2+N_3) and those without or with few metastatic lymph nodes (N_0+N_1) was significant ($0.05>P>0.01$).

III. Relation between EB virus VCA-IgA serum antibody and histopathological characteristics of NPC.

Among 127 cases of NPC, 108 were poorly-differentiated squamous cell carcinoma (PDSCC). The level of serum antibody in poorly-differentiated adenocarcinoma (PDA) was comparatively low and that of the only case of well-differentiated squamous cell carcinoma (WDSCC) was negative (Table 2).

Pattern of the growth, structure of the tumor and infiltration of lymphoid cells in 80 NPC cases were also studied (Table 3). No relation was discovered between level of antibody and pattern of the growth but in the parenchymatous type of NPC, the percentage of positive serum antibody cases was 96.7% while in the stromatous type it was 66.7% ($P=0.05$).

Table 2. Clinical Characteristics and VCA-IgA Serum Antibody Level in 127 NPC Patients

	No. of Cases	No. Positive Cases*	%	P	No. of Cases with High Titer*	%	P	GMT
Sex Male	96	87	90.6	P>.05	15	15.6	P>0.05	1:31
Female	31	29	93.5			12.9		1:38
Age $\leq$ 50	80	75	93.8	P>0.05	9	11.3	P>0.05	1:31
> 50	47	41	87.2		10	21.3		1:36
Survival $\leq$ 1 yr.	96	87	90.6	P>0.05	10	10.4	0.05>P>0.01	1:29
> 1 yr.	31	29	93.5		9	29.0		1:47
$T_0 + T_1$	70	64	91.4		8	11.4		1:30
T_2	35	31	88.6	P>0.05	7	20.0	P>0.05	1:32
$T_3 + T_4$	22	21	95.5		4	18.2		1:45
N_0	26	19	73.1		3	11.5		1:18
N_1	51	48	94.1	$N_0:N_1+N_2+N_3$ 0.05>P>0.01	4	7.8	$N_0+N_1:N_2+N_3$ 0.05>P>0.01	1:30
N_2	46	45	97.8		10	21.7		1:48
N_3	4	4	100.0		2	50.0		1:80
$N_1 + N_2 + N_3$	101	97	96.0		16	15.8		1:38

* VCA-IgA Serum Antibody Titer $\geq$ 1:5
** VCA-IgA Serum Antibody Titer $\geq$ 1:160

Table 3. Histopathological Characteristics and VCA-IgA Serum Antibody Level in 127 NPC Patients

		No. of Cases	No. (%) of Positive Cases	No. (%) of Cases with High Titer	GMT
Histological classification	Undiff. Carc.	1	1 (100)	0 (0)	1:10
	Poorly-diff. Sq. Cell Carc.	108	99 (92)	18 (17)	1:36
	Vesiculo-nuc. Cell Carc.	10	10 (100)	1 (10)	1:16
	Poorly-diff. Adeno Carc.	7	6 (86)	0 (0)	1:9
	Well-diff. Sq. Cell Carc.	1	0 (0)	0 (0)	0
Growth pattern:	Massive Form	34	32 (94)	4 (12)	1:28
	Alveolar or Trabecular Form	31	27 (87)	5 (16)	1:31
	Diffuse Form	15	13 (87)	2 (13)	1:32
Tumor structure:	Parenchymatous Type	30	29 (97)	4 (13)	1:32
	Intermediate Type	41	37 (90)	7 (17)	1:33
	Stromatous Type	9	6 (67)	0 (0)	1:16

The level of serum antibody rose correspondingly with an increase of plasma cells in the tumors. The percentage of high-titer cases in small-amount plasma cell group was 6.9% with GMT 1:26; and in moderate and large-amount group was 31.8% with GMT 1:53. The difference was significant (0.05> P>0.01). No similar relation was found in lymphocytes (Table 4).

Table 4. Lymphocytes, Plasma Cells and VCA-IgA Serum Antibody in 80 NPC Patients

		No. of Cases	No. (%) of positive Cases	No. (%) of Cases with High Titer	GMT
Lymphocyte in tumor Parenchyma	Small amount	60	54 (90)	7 (12)	1:29
	Mod. -large amount	20	18 (90)	4 (20)	1:31
Lymphocytes in tumor	Small amount	24	20 (83)	3 (12.5)	1:25
	Mod. -large	56	52 (93)	8 (14)	1:33
Plasma cells in tumor tissue	Small amount	58	53 (91)	4 (7)*	1:26
	Mod. -large amount	22	19 (86)	7 (32)*	1:53

*0.05 > P >0.01

IV. Relation between level of EB virus VCA-IgA serum antibody and cytoplasmic Ig in plasma cells of NPC.

Cytoplasmic IgA, IgG and IgM in tissue plasma cells of 68 NPC and 40 chronic nasopharyngitis biopsies were studied with PAP technique and EB virus VCA-IgA serum antibody was assayed. IgA$^+$ plasma cells were much more common in NPC than in chronic nasopharyngitis. No such difference was shown in IgG+ plasma cells (Table 5). IgM plasma cells were seldom seen in either disease. The level of VCA-IgA serum antibody in NPC patients correlated with the amount of IgA$^+$ plasma cells in the tumor tissue. When the amount of IgA$^+$ plasma cells was less than 5 per high power field, the GMT was 1:19, and when it was more than 26 per high power field, the GMT was 1:59 (Table 6). This suggests that certain specific antigen is present during tumor growth causing reactionary proliferation of plasma cells in the tumor. These cells may take part in the production of EB virus VCA-IgA antibody which may later get into the blood or saliva.

Table 5. Amount of Plasma Cells in 68 NPC and
40 Chronic Nasopharyngitis

	No. of Cases	IgA$^+$ Plasma Cells					
		No. of Cases $\leq$10/HPF	%	P	No. of Cases $\geq$10/HPF	%	P
NPC	68	38	55.9		30	44.1	
				P>0.05			0.05>0.01
Chronic Nasopharyngitis	40	30	75.0		10	25.0	

	No. of Cases	IgG$^+$ Plasma Cells					
		No. of Cases $\leq$10/HPF	%	P	No. of Cases $\geq$10/HPF	%	P
NPC	68	48	70.6		20	29.4	
				P>0.05			P>0.05
Chronic Nasopharyngitis	40	25	62.5		15	37.5	

Table 6. IgA$^+$ plasma cells and VCA-IgA Serum Antibody
Level in 68 NPC Patients

Amount of IgA+ Plasma Cells	No. of Cases	GMT
$\leq$ 5/HPF	32	1:19
6-10/HPF	6	1:36
11-25/HPF	23	1:52
$\geq$ 26/HPF	7	1:59

REFERENCES

1. Henle, G and Henle, W: Epstein-Barr virus-specific IgA
    serum antibodies as an outstanding feature of naso-
    pharyngeal carcinoma.   Int J Cancer 17:1-17 (1976).

2. Ho, J.H.C, Ng, M.H., Kwan, H.C., Chau, J.C.M.: Epstein-
    Barr-Virus-Specific IgA and IgG serum antibodies in
    nasopharyngeal carcinoma.   Brit J Cancer 34:655-660
    (1976).

3. Klein G, Giovanella, B.C., Lindahl, T., Fialkow, P.J.
    Singh, S., Stehlin, J.S.: Direct evidence for the pre-
    sence of Epstein-Barr virus DNA and nuclear antigen in
    malignant epithelial cells from patients with poorly
    differentiated carcinoma of the nasopharynx.   Proc
    Nat Acad Sci (USA) 71:4737-4741 (1974).

17

BRIEF COMMUNICATION

FOSSA OF ROSENMULLER: THE SITE FOR INITIAL
DEVELOPMENT OF CARCINOMA OF THE NASOPHARYNX

U. Prasad[1], J. Singh[2] and R. Patmanathan[3]

[1]Department of E.N.T., [2]Department of Radiology and
[3]Department of Pathology, University of Malaya,
Kuala Lumpur, Malaysia

SUMMARY

Along with the fossa of Rossenmuller (FOR),
other walls of the nasopharynx have been considered
to be the site of origin of the carcinoma of the
nasopharynx (NPC), a disease presumed to be caused by
the Epstein-Barr virus. In this study, a detailed
analysis of clinical, radiological and histopathological
findings in 150 confirmed cases of NPC has been made
which suggests that all the carcinomas of the
nasopharynx originate from cells lining the FOR. A
possible interaction between these epithelial cells
(columnar or squamous) which are more prone to
onslaught by the environmental agent, the subepithelial
lymphocytes and the lymphotropic virus EBV could
explain the pathogenesis of this cancer.

INTRODUCTION

The fossa of Rosenmuller has been reported in
the past to be one of the common sites of origin of
nasopharyngeal carcinoma with varying frequencies
(Dawes, 1969; Oreskovic et al., 1968; Shanmugaratnam
1971; Ho, 1978; Prasad, 1972 and 1979). However,
these observations were largely based on routine
clinical examination of the nasopharynx, a procedure
which is often rather inadequately performed even by
specialists in the field. As a result of closer
examination of the FOR in 40 cases of occult primary
in NPC using telescope and/or fibrescope, supplemented
by computerized tomography (C.T. scan), Prasad (1983)
reported that the earliest lesion in those cases
could have commenced in the depth of the FOR. The
purpose of this paper is to establish that all the
carcinomas of the nasopharynx originate from the
cells lining the FOR.

MATERIALS AND METHODS

An analysis was made of 150 histologically confirmed cases of nasopharyngeal carcinoma, seen at
the University Hospital, Kuala Lumpur, Malaysia. The
following data were available for study:

(a) Full clinical data: This included the assessment of site, side and extent of tumour as noticed
on routine examination, sometimes assisted by
the use of nasopharyngofibrescope or nasopharyngeal telescope.

(b) C.T. scan: Serial axial views of the
nasopharynx at 5 mm cut, using high resolution
Pfizer fast scanner with contrast enhancement
in suitable cases, were examined in great detail.

(c) Histopathological diagnosis: The patients
were subdivided into 3 groups according to WHO
classification.

In 19 cases where there was some difficulty in
identifying the suitable area for obtaining representative
tissue, biopsy specimens were taken separately from
three sites (two from the fossa of Rosenmuller and
from the postero-superior wall or the roof) in each
case. Histological details were studied in relation
to the site from where they were obtained.

RESULTS

Based on clinical examination in those 150 cases
of NPC, it was felt that in 127 cases (about 85%)
there was involvement of one of the two fossae of
Rosenmuller. In 65 cases (43.33%) the lesion was
limited to the fossa (Fig. 1), while in 34 (22.6%) it
had gone beyond it, usually along its anterior wall,
causing the increased bulkiness of the torus, and at
times tending to occlude the Eustachian tube (Fig.2).
There was spread of tumour along the postero-superior
wall in 28 cases (18.7%); however, the mass was seen
in continuity with that along the FOR. Both the
fossa seemed to be affected in 15 cases (10%), while
in 3 (2%) there was no evidence of definite tumour,
even on fibrescopic examination.

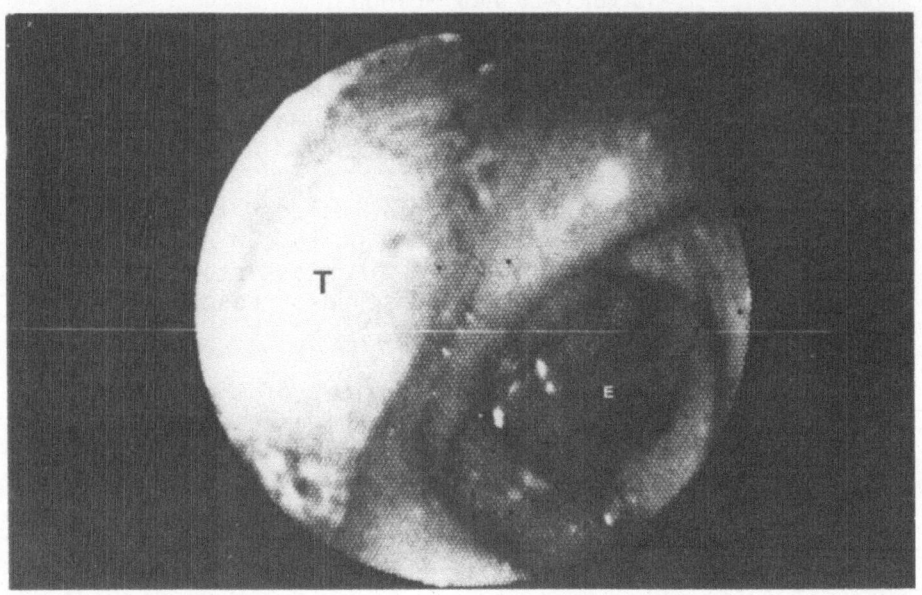

FIGURE 1 - Normal depth of the FOR is filled up by
submucosal tumour (T). The Eustachian
tube (E) is normal.

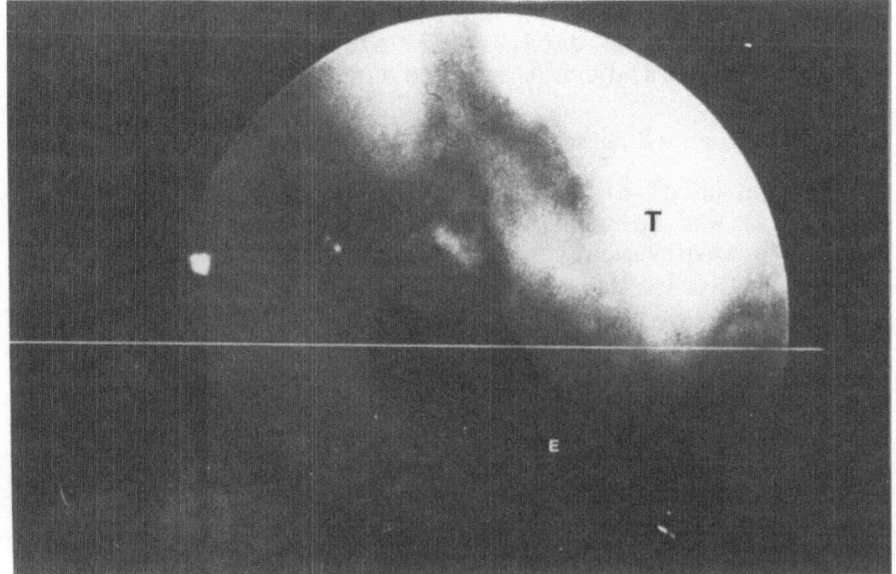

FIGURE 2 - Tumour (T) mass after filling the FOR is
overlapping the Eustachian tube orifice (E).

203

When the C.T. scan was taken into consideration, the picture was far more revealing. In 147 cases (98%) there was distortion of the FOR (Figs. 3 & 4) which included 37 cases (26.7%) with tumour limited to FOR, 80 cases (53.3%) with more soft tissue involvement, and 30 cases (20%) in which there was erosion of the skull bone (Fig. 5). The C.T. scan was non-contributory in only 3 cases (2%).

As for the histopathological diagnosis, 86% (126) belonged to WHO type III, while only 13 cases were grouped under WHO II. Although 11 patients (7.3%) were diagnosed as WHO I, all of them were either poorly differentiated or moderately differentiated squamous cell carcinoma. There was no case of well differentiated squamous cell carcinoma in this series. Also, there were often foci of more than one histological type of carcinoma in the same biopsy as reported earlier (Prathap et al., 1983).

Results of histological examination of biopsy specimens, obtained from three sites separately (2 FOR and the roof), in those 19 cases where macroscopically it was difficult to identify the tumour tissue, revealed certain rather interesting features. There were 12 cases (63.7%) in which the lesions involved the two fossa separately, with the intervening roof or the postero-superior wall free from tumour. In the other seven cases (36.3%), in addition to the roof, there was tumour mass in one (4 cases) or both (3 cases) fossae. In none of these cases was tissue from only the roof positive for NPC.

DISCUSSION

In the past, the site of origin of NPC was determined by simple clinical and macroscopic observation of the nasopharynx. Due to the difficult anatomical location, this area has been termed as a "blind spot" (Cantrill and Buschke, 1946), an "unknown region" (Hickley, 1951) and a "hidden cavity" (Davies, 1948). As such, the observations were not accurate. However, recently, due to the the use of fibrescopes and or telescopes, this concept is changing, and one can now inspect these areas in great detail. Even

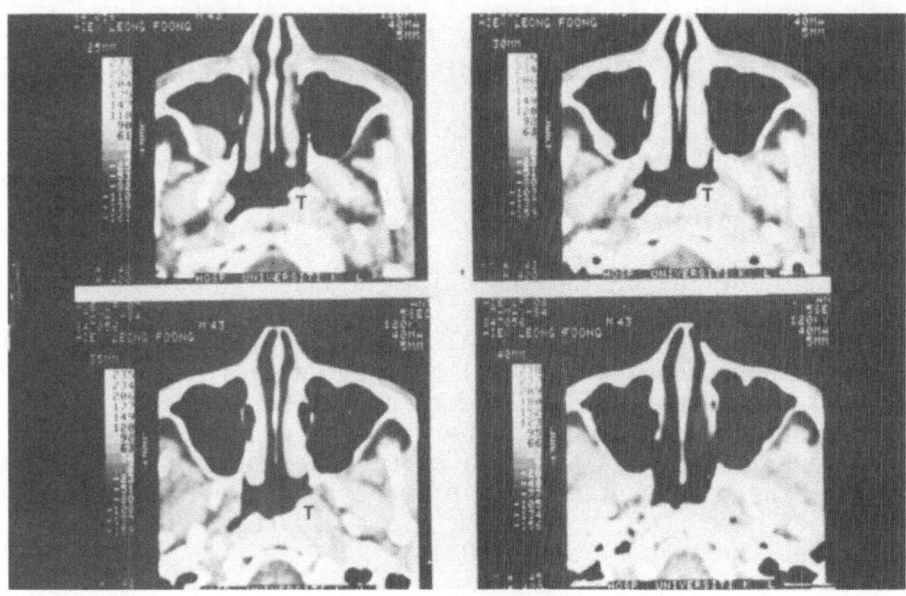

FIGURE 3 - Small tumour (T) confined to right FOR.
          Left E. tube, torus and FOR quite normal.

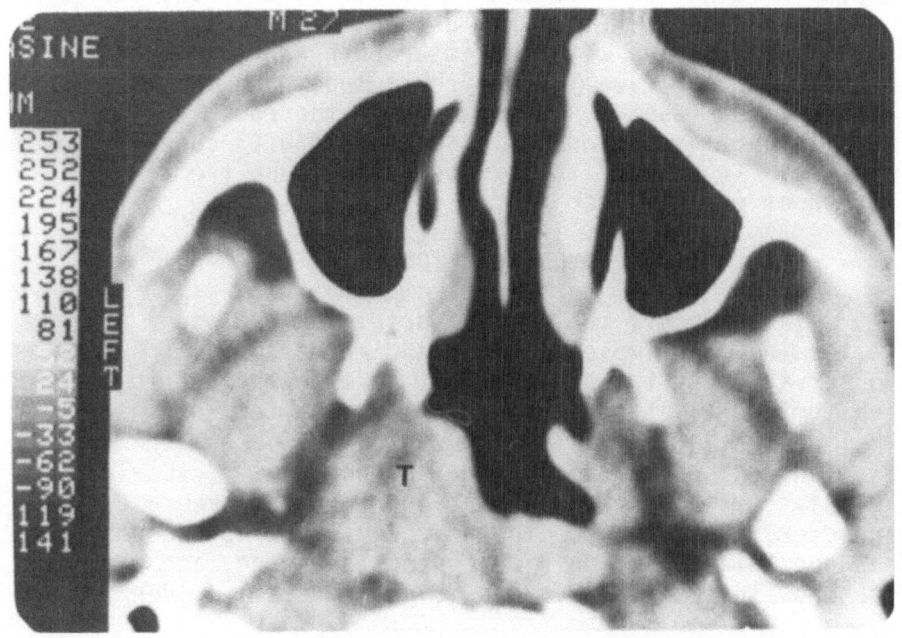

FIGURE 4 - Bigger tumour (T) from (L) FOR distorting the
          torus and blocking the tube. Opposite side normal

very small lesions can be identified and biopsy specimen can be taken for histopathological examination. As a result of these examinations, the number of so-called occult primary NPC has now dropped tremendously in our facilities. In fact, there were only 3 such cases in this series. More and more, early lesions involving only the depth of FOR (65 out of 150) are being diagnosed. Close study of these early lesions has helped to determine the site of origin of lesions in such cases. This was supported by the study of the C.T. scan. The C.T. imaging clearly demonstrated the details of soft tissue abnomality, even if they were submucosal and very minimal in extent.

In all those cases in which the clinicians observed the lesion to be along the postero-superior wall, C.T. scan revealed the distortion of FOR. In fact, when biopsies were obtained from multiple sites, in 12 out of 19 cases, both FOR were histologically positive for NPC. Thus with the application of new tools like C.T. scanning and fibrescopy, it was possible to establish that the FOR is the site of origin of all NPC.

Regarding its significance in the pathogenesis of NPC, while we know that these tumours originate from the epithelial cells (Prasad, 1974, 1978) and EBV is associated with them, we suggest that there is an interaction between the basal (totipotent) layer of these cells lining the FOR and the subepithelial (EBV-containing) lymphocytes, which are in close proximity at this site. Both are subjected to the effects of environmental agents more easily at this location (FOR being a recess).

REFERENCES

CANTRIL, S.T. and BUSCHKE, F., Malignant tumours of the nasopharynx. West. J. Surg., 54, 494-496 (1946).

DAVIES, E.D.D., Diagnosis and treatment of tumours of nasopharynx. J. Laryng. Otol., 62, 192-205 (1948).

206

DAWES, J.D.K., Malignant disease of the nasopharynx.
J. Laryng. Otol., 83, 21-28 (1969).

HICKLEY, H.L., Nasopharyngeal malignant tumour.
Arch. Otolaryng., 53, 53-67 (1951).

HO, J.H.C., Stage classification of nasopharyngeal
carcinoma: A review. In: G. de The and Y. Ito (ed.)
Nasopharyngeal carcinoma: Etiology and control.
International Agency for Research in Cancer, Lyon,
France, 99-113 (1978).

ORESKOVIC, M., PETRIC, K., AND PADOVAN, S., The
otoneuro-ophthalmologic diagnosis in tumours of the
nasopharynx. J. Laryng. Otol., 82, 575-601 (1968).

PRASAD, U., Cancer of the nasopharynx, a clinical
analysis with anatomico-pathological orientation.
J.R. Coll. Surg., Edinb., 17, 108-117 (1972).

PRASAD, U., Cells of origin of nasopharyngeal carcinoma: An electron microscopic study. J. Laryng. Otol.
88, 1087-94 (1974).

PRASAD, U., Nasopharyngeal carcinoma, etiological
aspects and electron microscopic observations., J.R.
Coll. Surg., Edinb., 23, 199-207 (1978).

PRASAD, U., Fossa of Rosenmuller and nasopharyngeal
carcinoma. Med. J., Malaysia, 33, 222-225 (1979).

PRASAD, U., Problem of occult primary nasopharyngeal
carcinoma. In: U. Prasad, D.V. Ablashi, P.H. Levine,
and G.R. Pearson (ed.) Nasopharyngeal carcinoma-
current concepts. Univ. of Malaya, Kuala Lumpur, 11-
15 (1983).

PRATHAP, K., PRASAD, U., and ABLASHI, D.V., The
pathology of nasopharyngeal carcinoma in Malaysians.
In: U. Prasad, D.V. Ablashi, P.H. Levine and G.R.
Pearson (ed.). Nasopharyngeal carcinoma- current
concepts. Univ. of Malaya, Kuala Lumpur, 55-63 (1983).

SHANMUGARATNAM, K., Studies on the etiology of
nasopharyngeal carcinoma. Int. Rev. of Exp. Path.,
361-413 (1971).

18

BRIEF COMMUNICATION

CARCINO-EMBRYONIC ANTIGEN (CEA) IN NASOPHARYNGEAL

CARCINOMA AND CHRONIC NASOPHARYNGITIS

Shen, Yuin Yin and Huang, Pei Sheng

Department of Anatomical Pathology, Fujian Medical College

Fuzhou, Fujian, People's Republic of China

When carcino-embryonic antigen (CEA) was discovered
in carcinoma of the colon and fetal colon mucosa (Gold and
Friedman, 1965), it was thought to be a specific antigen
and also a specific marker of human digestive system and
its carcinomas. Since then, additional case reports indicated that CEA was also present in carcinoma of various
organs such as lung, breast, ovary, uterine cervix, and
urinary bladder. Thus far no report has evaluated the
nasopharynx, where EBV appears to replicate, or nasopharyngeal carcinoma, which is suspected of being caused by EBV.
With the use of the immunoperoxidase method, the presence
and the distribution of CEA in nasopharyngeal epithelium
in relation to inflammation and carcinomatous change were
studied in our laboratory.

MATERIALS AND METHODS

Nasopharyngeal biopsy specimens from 52 cases of
chronic nasopharyngitis and 94 cases of untreated nasopharyngeal carcinoma were collected. Tissues were
routinely fixed in 10% neutral buffered formalin and
paraffin-embedded. 5u-thick sections were prepared for
H+E stain and also for immunohistochemical staining.

The peroxidase-anti-peroxidase (PAP) immunohistochemical staining technique of Sternberger was performed;
rabbit anti-human CEA as primary antibody, goat anti-rabbit
serum as linking antibody and rabbit peroxidase-antiperoxi-
dase complex as the final step. All reagents and antibodies were supplied by Immulok, CA, USA. Slides were

immersed in 3% hydrogen peroxidase for 5 minutes and also
in swine serum for 20 minutes prior to staining procedure.
After completion of the immunologic staining, slides were
counterstained with Mayer's hematoxylin.

Negative and positive controls were employed. The
negative control was performed by substituting nonspecific
rabbit serum for the primary antibody. The negative control was always negative. The positive control specimen
was an adenocarcinoma of the colon which was strongly positive for anti-CEA antibody with the immunoperoxidase stain.

RESULTS AND DISCUSSION

Distribution of CEA in nasopharyngeal epithelium

Among 52 cases of chronic nasopharyngitis, the nasopharyngeal mucosa of 42 cases was overlined by pseudo-
stratified ciliated columnal epithelium with 15 cases
(53.7%) showing a positive CEA reaction. It appeared to
be brownish granular or globular in shape and continuous
or scattered in distribution more along the distal portion
or free end of the epithelial cells. Its close association
with inflammation in the nasopharynx was obvious as no CEA
was demonstrated in ten biopsies of mild nasopharyngitis
but CEA was found more often when inflammation became
prominent (15/32 or 46.9%). Similar findings were
reported by Goldenberg (1978) on mucosal epithelium of the
colon and Jautzke (1982) on mucosal epithelium of the
urinary bladder.

Stratified squamous epithelium was found in 21 cases.
In the superficial layers of the squamous epithelium, CEA-
like material was present along the cell membrane showing
a linear or honey-comb appearance, and the cells in the
basal layer were spared. It was most likely the keratin
or keratin precursor in the squamous epithelial cells that
crossreacted with anti-CEA antibody.

Non-specific crossreacting antigen was also present in
the cytoplasm of granulocytes, monocytes and macrophages.

Distribution of CEA in NPC

Among 94 cases of NPC, 27 (28.7%) showed a positive
reaction. According to our National classification of NPC

proposed in 1979, there were 12 cases in this group classified as well-differentiated and moderately-differentiated
squamous cell carcinoma (so-called keratinizing squamous
cell carcinoma). The frequency of positive reaction in
these two types was 71.4% (5/7) and 40% (2/5) respectively.
As the distribution and the appearance of the brownish
products in the tumors were corresponding to that seen in
the overlying squamous epithelium, it was also considered
as a cross-reacting phenomenon of keratin and keratin
precursor with anti-CEA antibody. The incidence of CEA
in poorly-differentiated adenocarcinoma (53.8%, 7/13) was
much higher than that in poorly-differentiated squamous
cell carcinoma (17.6%, 9/51) ($p<0.05$). Either from the
histogenic and morphologic standpoint or antigenic standpoint, we believe that they are two different types of
tumors but tumors composed of mixture of various types of
tumor cells are not uncommon in this area.

In comparison with the undifferentiated type of NPC,
the incidence of CEA in vesiculo-nuclear cell carcinoma
(VNCC) was slightly higher (25%, 3/12) but lower than
the other types. As in our previous studies, the present
observation again suggests that VNCC is different from
undifferentiated type of nasopharyngeal carcinoma. It
is a poorly-differentiated carcinoma originating from the
cells at the basal layer of either types of nasopharyngeal
epithelium including transitional epithelium with the
potential for biphasic differentiation.

From the group of 77 non-keratinizing NPC, we found
that the incidence of CEA increased correspondingly with
extension and dissemination of the tumor mass (Table 1).
Furthermore, from 26 cases with follow-up data after radiotherapy, the 5-year survival rate in CEA+ ones was 21.4%
(3/14) and in CEA$^-$ it was 41.7% (5/12).

Table 1: Relationship between CEA and Clinical Stages of
Non-keratinizing Carcinoma of Nasopharynx

	No. of Cases	CEA^+ No. of Cases	%
T_0	28	6	21.4
T_1 -- T_4	50	17	34.0
N_0	16	3	18.8
N_1 -- N_3	62	20	32.3

CONCLUSION

The results of this study suggest that CEA is present in nasopharyngeal mucosa with inflammation and in some nasopharyngeal carcinomas. Until monoclonal anti-CEA antibodies for identification of the different antigenic determinants of CEA become available, CEA as a tumor marker of NPC is not specific but it may help in detection of tumour dissemination and predict the therapeutic effect and prognosis of NPC patients.

REFERENCES

Gold, P. and Friedman, S.O. Specific carcinoembryonic antigen of the human digestive system. J. Exp. Med. 122: 467-481 (1965).

Goldenberg, D.M. Immunocytochemical detection of carcino-embryonic antigen in conventional histopathology specimens. Cancer 42: 1546-1553 (1978).

Jautzke, G., et al. Immunohistochemical demonstration of carcinoembryonic antigen (CEA) and its correlation with grading and staging on tissue sections of urinary bladder carcinoma. Cancer 50: 2050-2056 (1982).

19

BRIEF COMMUNICATION

HLA-ANTIGENS IN NPC PATIENTS FROM A LOW - RISK

AREA (Cologne, FRG)

G. Bertram,* J. Kruger,** K. Sesterhenn***

*ENT - Clinic and **Department of Transfusion Medicine

University of Cologne; ***ENT - Clinic of the University

of Hamburg

SUMMARY

HLA typing of 76 low-risk patients with malignant
tumors of the nasopharynx (n = 65) and histomorphological-
ly identical carcinomas of the mesopharyngeal region
(n = 11) showed no significantly different frequency of
any HLA-antigen. A slight elevation of B-27 was shown for
so-called lympholepithelial carcinomas of the nasopharyngeal region. Neither HLA Bw46 (Sin-2) nor any other NPC-
linked antigen of high-risk groups has been found in our
patients. Apparently, susceptibility to NPC in the Caucasian ethnic group is neither determined nor codetermined
by class I antigens of the HLA system.

INTRODUCTION

Nasopharyngeal carcinoma (NPC) is the most important
malignant tumor in Southeast Asia. Although previously
reported incidence rates in Germany (Waterhouse et al.,
1976) are low (Table 1), data from our own prospective
study (1979-1983), show markedly elevated rates (Fig. 1).

Table 1: Incidence of NPC in Germany. Compared to the
data of Waterhouse et al., 1976*, Cologne data
(Bertram and Stutzer, 1984 - unpublished)** show
a significantly higher rate for males. n.i. = no
incidence for whole FRG known.

LOCATION	INCIDENCE	
	M	F
German Democratic Republic*		
Federal Republic of Germany (FRG)	n.i.	n.i.
Saarland, FRG*	0.4	0.2
Hamburg, FRG*	0.1	0.1
Cologne, FRG**	1.1	0.3

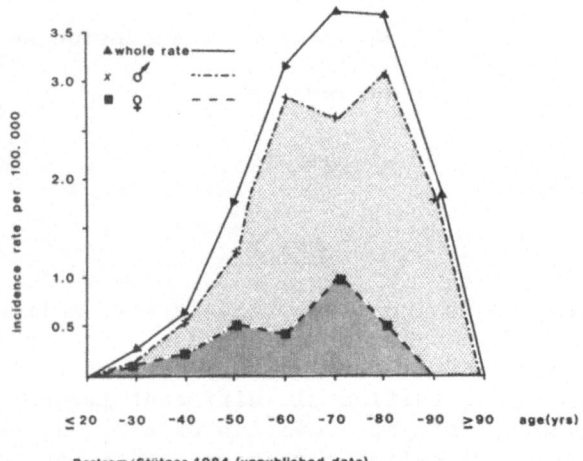

Bertram/Stützer,1984 (unpublished data)

Figure 1: Age adjusted incidence rates in German
Caucasians (Cologne). Incidence rates have been
standardized according to the population of North-
Rhine-Westfalia, FRG (December 1st, 1981).

In high-risk areas the prognosis of the tumor is
strongly associated with the incidence of HLA Bw46
and other antigens of the different HLA loci (Simons
et al., 1974 and 1975; Chan and Simons, 1977; Chan et
al., 1983). Comparable findings of HLA-association for
low-risk tumor patients have not been found up to now
(Kruger et al., 1981; Beigel et al., 1983).

MATERIALS AND METHODS

All NPC-patients (n = 57) and controls (n = 19 with tumors and n = 800 healthy persons) were of Caucasian ethnic group. Data of a high-risk control group are those of Simons et al., 1978. Histologic classification of all patients with malignant nasopharyngeal and mesopharyngeal tumors followed the criteria of Rappaport (1966), Lennert (1978), Shanmugaratnam and Sobin (1978), and Krueger and Wustrow (1981). The 57 NPC patients can be subdivided into groups of 39 patients with so-called lymphoepithelial carcinoma and 18 patients with squamous cell carcinoma and non-keratinizing carcinoma without lymphoid stroma. Tissue typing was done by the two-stage microlymphocytotoxicity test according to the NIH Tissue Typing Manual as described elsewhere (Kruger et al., 1981). Test sera with high antibody titers against Bw46 have been kindly provided by Dr. S.H. Chan. Phenotype frequencies were compared with those of a German control population (Lenhard, 1979) by Fisher's Exact Test.

RESULTS

The frequencies of locus A and B antigens among NPC and mesopharyngeal tumor patients are shown in table 2. The incidence of the tested antigens was rather similar in our groups as compared to Lenhard's controls. Antigens with significant differences among patients of low- and high-risk regions and their corresponding controls are given in table 3. Bw46, A-2, A-11, and B-17 antigens, which are changed significantly in patients in high-risk regions (Chan and Simons, 1977, Chan et al., 1983, Simons et al., 1974, 1975, and 1978), showed no aberrations in our low-risk area patients. An elevation of A-3 and B-5, apparently elevated in our first study (Kruger et al., 1981), was not confirmed in the present study.

DISCUSSION

The results of investigators of patients in high-risk regions demonstrate that HLA antigens play a role in determining both susceptibility for NPC and prognosis (Chan et al., 1983; Simons et al., 1975 and 1978). Former

ANTIGENS	COLOGNE			CHINESE *
	all NPC	lymphoepithelial tumors		
	(n=57)	NPC (n=39)	mesoph. r. (n=11)	NPC (n=141)
A1	16/57	8/39	3/11	1/141
A2	28	17	6	86 p
A3	25	18	4	1
A9	11	9	∅	47
A10	5	5	2	13
A11	3	3	2	57 p
Aw19	∅	∅	∅	22
A28	2	1	∅	∅
A29	2	2	2	1
Aw30/31	4	2	∅	∅
Aw32	7	6	∅	∅
blank	11	7	3	52
B5	17/57	10/39	2/11	19/141
B7	19	10	2	1
B8	12	5	4	1
B12	11	10	4	2
B13	1	1	1	19
B14	4	3	∅	∅
B15	5	5	3	26
Bw16	2	1	1	14
B17	5	4	1	40 p
B18	4	4	∅	1
Bw21	3	3	∅	∅
Bw22	2	2	∅	9
B27	3	∅ p	∅	4
Bw35	4	4	1	8
B37	3	3	∅	2
B40	7	4	2	53
Bw41	1	1	1	∅
Bw46	∅	∅	∅	48 p
blank	11	8	1	37

Table 2: Frequency of HLA antigens among low-risk and high-risk patients. * according to data from SIMONS et al., 1978; p marks significant changes in frequency compared to the frequency in healthy controls.

215

investigations (Kruger et al., 1981) of patients in a low-risk area showed that susceptibility to NPC might be associated with certain HLA antigens, too. These findings have not been confirmed in this study. In NPC patients outside of high-risk regions HLA-Bw46 has never been found. Other antigens associated with NPC-prognosis in Southeast Asia show normal frequencies in patients as well as in healthy controls within our low-risk region.

Table 3: Comparison of data of the significantly altered and other important antigens in NPC patients. % frequency of antigens; p significance; * significant only for patients with so-called lymphoepithelial carcinoma.

ANTIGENS	AREA			
	high-risk		low-risk	
	%	p	%	p
A2	61.0	0.064	49.1	
A3	1.0		43.8	
A11	40.4	0.004	5.3	
B5	13.5		29.8	
B13	13.5		1.8	
B17	28.4	0.02	8.8	
B27	2.8		(5.2)*	0.045*
Bw46	34.0	0.008	0.0	

REFERENCES

Beigel, A., Peulen, J.F., Westphal, E.: Verteilungsmuster von Histokompatibilitatsantigenen (HLA) bei Nasopoharynx-tumoren. Arch. Otorhinolaryngol. 237, 285-288 (1983).

Bertram, G., Stutzer, H.: Unpublished data.

Chan, S.H., Simons, M.J.: Immunogenetics of Naopharyngeal carcinoma. Ann. Acad. Med. Sing., 342-346 (1977).

Chan, S.H., Wee, G.B., Kunaratnam, N., Chia, K.B., Day, N.E.: HLA locus B and DR antigen associations in Chinese NPC patients and controls. In: U. Prasad et al. (eds.): Nasopharyngeal carcinoma: current concepts, 307-312, Univ. Malaya Press. Kuala Lumpur (1983).

Krueger, G.R.F., Wustrow, J.: Current histological classification of Nasopharyngeal carcinoma (NPC) at Cologne University. In: E. Grundmann et al. (eds.): Nasopharyngeal carcinoma. Cancer Campaign, 5, 11-15, Fischer Verlag. Stuttgart, New York (1981).

Kruger, J., Ieromnimon, V.. Dahr, W.: Frequencies of HLA antigens in patients with NPC. In: E. Grundmann et al. (eds.): Nasopharyngeal carcinoma. Cancer Campaign, 5, 201-203, Fischer Verlag. Stuttgart, New York (1981).

Lenhard, V.: Das HLA-System und seine Bedeutung fur die Klinische Transplantation. Lab. Med., 3, 12-20 (1979).

Lennert, K.: Malignant lymphomas other than Hodgkin's disease: histology, cytology, ultrastructure, immunology. Springer Verlag. Berlin, Heidelberg, New York (1978).

Rappaport, H.: Tumors of the hematopoietic system. Atlas of tumor pathology. Armed Forces Institute of Pathology. Washington, D.C. (1966).

Shanmugaratnam, K., Sobin, L.H.: Histological typing of upper respiratory tract tumors. WHO, Geneva (1978).

Simons, M.J., Wee, G.B., Day, N.E., Morris, P.J., Shanmugaratnam, K., De-The, G.: Immunogenetic aspects of Nasopharyngeal carcinoma. I. Differences in HL-A profiles between patients and comparison groups. Int. J. Cancer, 13, 122-134 (1974).

Simons, M.J., Wee, G.B., Day, N.E., Chan, S.H., Shanmugaratnam, K., De-The, G: Probable identification of an HL-A locus antigen associated with a high risk of Nasopharyngeal carcinoma. Lancet, i, 142-143 (1975).

Simons, M.J., Chan, S.H., Darmalingam, S., Wee, G.B.,
Shanmugaratnam, K., Prasad, U., Goh, E.H., Betuel, H.,
Ho, J.H.C., Chan, J.C.W., Day, N.E., De-The, G: Naso-
pharyngeal carcinoma and histocompatibility antigens.
In: G. De-The et al. (eds.): Nasopharyngeal carcinoma:
etiology and control, 271-282, IARC Scientif. Publ.,
No. 20, Lyon (1978).

Waterhouse, J., Muir, C., Correa, P., Powell, J. (eds.)/
Cancer incidence in five ontinents. Vol. 3, IARC Scientif.
Publ., No. 15, Lyon (1976).

ACKNOWLEDGEMENTS

The authors thank Mrs. A. Zielonki and Mrs. S.
Hillebrand for excellent technical assistance, Dr. S.H.
Chan for probes of Bw46 test sera, and Dr. W. Haase
for statistical assistance.

MOLECULAR BIOLOGY OF EBV

20

PERSISTENCE AND EXPRESSION OF THE EPSTEIN-BARR VIRUS GENOME

IN LATENT INFECTION AND GROWTH TRANSFORMATION OF LYMPHOCYTES

Elliott Kieff, Kevin Hennessy, Timothy Dambaugh,

Takumi Matsuo, Sue Fennewald, Mark Heller, Lisa

Petti and Mary Hummel

Kovler Viral Oncology Laboratories

The University of Chicago

910 E. 58th Street

Chicago, Illinois   60637

Epstein-Barr Virus (EBV) is believed to be an important etiologic agent of nasopharyngeal cancers (NPC) especially of the anaplastic type since (i) NPC cells invariably harbor EBV (Wolf et al., 1975; Huang et al., 1974; Klein et al., 1974); (ii) EBV is believed to be a cause of human B lymphocyte tumors (Epstein and Achong, 1978); (iii) the immune response to EBV infection is predictive of NPC development and of prognosis (Henle et al., 1970; Zeng et al., 1980; Pearson et al., 1984); and (iv) most NPC tumors originate from an unique anatomic site where there is closest proximity between lymphoid cells in which EBV is latent and epithelial cells in which tumor originates (Prasad 1981). Because of difficulties in infecting epithelial cells with EBV and in propagating NPC cells in vitro, current knowledge of EBV-induced cell proliferation comes mostly from the human B-lymphocyte tumor model. Not only is EBV almost always present in Burkitt lymphoma cells (Nonoyama et al., 1973; Lindahl et al., 1974); but also, virus infection of lymphocytes in

222

vitro leads to cell proliferation (Henle et al., 1967; Pope
et al., 1968), virus-infected cells from tumors in nonhuman
primates (Miller et al., 1977), and virus-transformed cells
of varying stages of oncogenic potential can be grown in
continuous culture (Nilsson 1971). Thus, the biochemistry
of EBV persistence and gene expression in lymphoid cells
has been amenable to investigation. Some biochemical
analyses have also been done of human NPC biopsy specimens
and of NPC cells passaged in nude mice (Klein et al., 1974;
Raab-Traub 1983). The purpose of this brief overview is to
describe current knowledge of virus persistence and virus
gene expression in growth transformed or malignant
lymphocyte or NPC cells and to indicate some important
areas for future research.

INTRACELLULAR EBV GENOMES

The virion EBV genome is a linear double-strand DNA
molecule with nonrandom single-strand breaks (Pritchett et
al., 1975). The entire genome has been cloned (Given and
Kieff 1978; Dambaugh et al., 1980) and sequenced (Baer et
al., 1984). The genome is 170 kbp. There are unique and
tandemly repeated DNA elements. Since the number of tandem
repeats varies among different EBV isolates and among
different molecules of a given isolate, it is useful to
consider the genome as being organized as TR-U1-IR1-U2-IR2-
U3-IR3-U4-IR4-U5-TR where TR are direct-tandem 550 bp
repeats (Given et al., 1979); IR1, 2 and 4 are direct 3072,
125, and 103 bp repeats respectively (Cheung and Kieff
1982, Dambaugh and Kieff 1982); IR3 is a repeat array of
three nucleotide triplets, GGG, GCA and GGA (Heller et al.,
1982b); and U1, U2, U3, U4 and U5 are largely unique DNA
domains of 10, 3, 59, 40, and 30 kbp respectively. IR3 is
one part of the EBV genome which may have originated in
cell DNA since there are similar triplet nucleotide repeat
arrays in cell DNA (Heller et al., 1982a, 1985). Although
there is restriction endonuclease polymorphism among EBV
isolates, studies of EBV DNA from widely different
populations reveal little evidence for significant strain
differences (Heller et al., 1981; Raab-Traub et al., 1980;
Bornkamm et al., 1980). An exception is in the EBV U2
domain which varies widely among isolates (King et al.,
1982; Dambaugh et al., 1984). Even in this instance, the
variation does not correlate with the biologic or
geographic origin.

Following infection of B lymphocytes by EBV, in vitro
or in vivo, the virus persists in proliferating cells in a
latent state from which it can be reactivated. The
latently infected proliferating cells usually contain more
than one copy of the complete EBV genome (Nonoyama and
Pagano 1971). EBV DNA is heavily methylated in contrast to
virion DNA which is not methylated (Kintner and Sugden
1981). Most EBV DNA molecules in cells with multiple
copies are not linked to cell DNA (Nonoyama and Pagano
1972). These EBV DNA molecules are maintained as
covalently closed circular episomes (Lindahl et al., 1976).
The episomes are formed by covalent linkage between TR's at
each end of the DNA (Dambaugh et al., 1980). It is not
known if the process involves homologous recombination or
end ligation of TR's cleaved at a specific nucleotide
sequence. Defective EBV DNA molecules are also frequently
found in latently infected cell lines which have been
passaged extensively in vitro (Heller et al., 1981). These
defective molecules are characteristically deleted for
parts of the genome. It has not been established whether
the deleted molecules are maintained as episomes or are
integrated into cell DNA.

Analysis of cells for integrated EBV DNA is usually
complicated by the presence of episomes. Attempts to
demonstrate covalent linkage of viral and cell DNA by
isopycnic centrifugation and cell hybridization yielded
conflicting data. In situ chromosome cytological
hybridizations with cloned probes representative of most of
the EBV genome demonstrated that most or all of the genome
is linked to chromosomal sites; 4q25 on one homologue of
chromosome 5 of the IB4 cell line and 1P35 on one homologue
of chromosome 1 of the Namalwa cell line (Henderson et al.,
1983). Namalwa is a latently infected African Burkitt
tumor cell line which contains only one copy of EBV DNA.
The IB4 cell line, established by infection and growth
transformation of normal fetal lymphocytes with EBV,
contains several episomal copies of EBV DNA. The IR3
repeat which has homology to cell DNA did not mediate
integration in Namalwa or IB4 cells.

To investigate the organization of EBV DNA in Namalwa
and IB4 cells (Matsuo et al., 1984), labeled cloned EcoRi
fragments of EBV DNA representative of more than 90% of the
EBV genome were hybridized to Southern blots of EcoRI
fragments of Namalwa and IB4 DNA. The EBV EcoRI, A, B, C,

E, F, H, J and K fragment probes hybridize to only a single
EcoRI fragment in Namalwa and IB4 cells. EcoRI I also
hybridized to a single fragment in Namalwa DNA. The size
of the fragments identified with these probes is the size
expected for fragments of the standard EBV genome in the
case of Namalwa and for the deleted EBV genome in the case
of IB4. (IB4 is infected with the B95-8 isolate of EBV
which has a smaller EcoRI C fragment.) These data indicate
that these parts of the EBV genome in Namalwa and IB4 cells
are probably organized similarly to linear virion DNA; and,
that these fragments are unlikely to have recombined with
cell DNA. The EcoRI-D het fragment, which is formed by
covalent linkage of the terminal EcoRI fragments of
episomal EBV DNA, hybridizes to two Namalwa EcoRI fragments
and several IB4 EcoRI fragments. The Namalwa fragments are
18 and 5 kb, while the IB4 fragments range from 12-18 kb
and one fragment is 28 kb. Variation frequently occurs at
the ends of EBV DNA due to heterogeneity in the number of
copies of TR. Some of the IB4 EcoRI-D het fragments of 12-
18 kb could be fragments of episomal EBV DNA. However, TR
reiteration would be an unlikely explanation for the 28 kb
IB4 fragment since the TR unit is only 550 bp. These data
therefore suggest that the ends of EBV may have recombined
with cell DNAs.

Episome joined ends can be distinguished from free
ends or ends joined to cell DNA by examining the linkage of
U5 and U1. EcoRI, BamHI, BglII and PstI restriction
endonucleases do not cleave within TR and therefore
generate fragments from episomes in which part of U5 is
linked to TR and U1 (Figure 1). TR probe identified two
Namalwa fragments. One putative junction fragment (JU5)
hybridizes to U5 and not U1; while the second (JU1)
hybridizes to U1 and U5. The IB4 results are more
complicated. In this instance, all BglII, BamHI and PstI
fragments identified by TR also hybridized to U5 and to the
U1 probe, EcoRI I, indicating that TR had not recombined
with cell DNA. Further, although there is no BglII site in
EBV DNA between EcoRI I and EcoRI J, EcoRI I and J have
become unlinked in some IB4 EBV DNA molecules and not in
others. EcoRI I, TR and U5 hybridize to the same IB4 BglII
fragments. EcoRI J hybridizes to some of these fragments
as is expected for EBV episomes (designated Epi in Figure
1B); but not to others (designated JU5TRU1, for a putative
juncture fragment). Similar data were obtained with PstI
confirming a disassociation of EcoRI I and J. If the

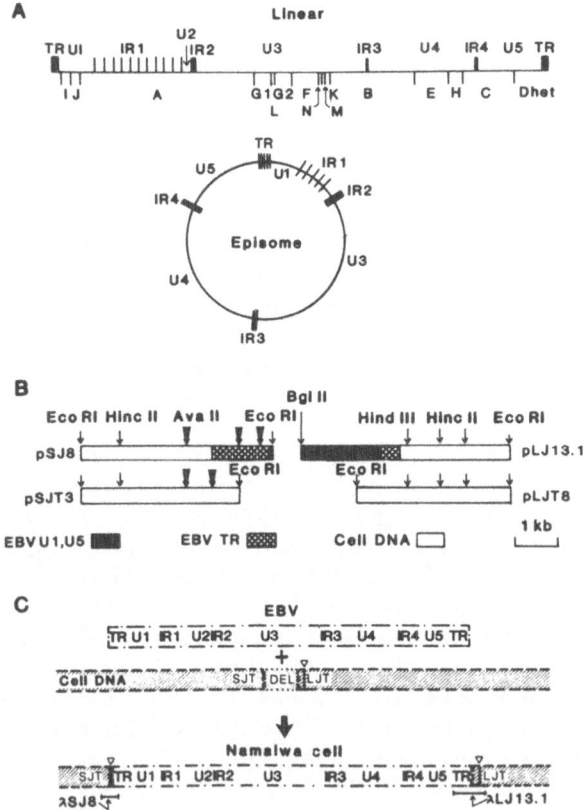

Fig. 1. (A) EBV DNA. TR, terminal repeat region; U, single—copy domains; IR, internal repeats; A to K, EcoRI fragments used as probes. (B) Recombinant clones. The EcoRI D probe includes the left and right terminal EcoRI fragments of linear virion DNA and hybridizes to two Namalwa EcoRI fragments designated LJ and SJ. The plasmid subclone pLJ 13.3 was derived from a lambda clone, LJ 13.1 of Namalwa LJ DNA. Similar designations are used for SJ and for clones of the homologous DNA's from the uninfected cells LJT and SJT. Ava II restriction endonuclease sites are indicated by thick arrows; all other restriction endonuclease sites are indicated by thin arrows. (C) The Namalwa—EBV integration. "Del" indicates cell DNA deleted at the site of EBV integration. The dark vertical lines and triangles indicate the 236 base pairs of cell DNA which becomes duplicated at an EBV cell DNA juncture site.

disassocation were due to recombination with cell DNA,
EcoRI I or J should hybridize to three fragments; a
standard EcoRI fragment from episomal EBV DNA molecules and
to the two parts of EcoRI I or J which had recombined with
cell DNA. EcoRI J hybridizes to only one fragment the size
of EcoRI J. EcoRI I hybridizes to at least two fragments
other than a standard EcoRI I fragment. A small fragment
is identified only with EcoRI I and could be a simple
junction with cell DNA. The other fragments are large
putative juncture fragments which hybridize to TR and U5 as
well as EcoRI I. Since there should be an EcoRI site
between TR and EcoRI, these data also indicate that there
has been a rearrangement of viral DNA within the large
putative juncture fragment.

Two independent lambda clones of the large putative
Namalwa junction fragment, four independent clones of the
small (designated LJ and SJ in Figure 1B) and at least
three independent clones of each of the homologous sites
from a library of human cell DNA clones (designated LJT and
SJT respectively in Figure 1B) were characterized so as to
determine the DNA sequences which had recombined and to
minimize possible cloning artifacts. Comparison of the
restriction endonuclease digests of the cloned Namalwa LJ
and SJ EcoRI fragments with digests of the cloned EcoRI
ends of EBV DNA indicates that the U5-TR portion of LJ DNA
and the U1-TR portion of SJ DNA are colinear with the
corresponding ends of linear EBV DNA. The rest of the LJ
or SJ DNA has no homology to EBV but is identical between
the two LJ clones or among the four SJ clones (Figure 2B).
This putative cell component of LJ hybridizes to a 3.5 kb
EcoRI fragment common among 4 of the 5 recombination target
site clones which were selected from a human cell DNA
library (Lawn et al., 1978) using the LJ probe. Restriction endonuclease analyses and blot hybridizations indicate
that the 3.5 kb cell target site EcoRI fragment, designated
LJT, is colinear with the cell DNA component of Namalwa LJ.
Similarly, restriction endonuclease and blot analyses
reveal that about 2.8 kb of the SJ target site EcoRI
fragment is colinear with SJ cell DNA.

The 3.5 kb LJT EcoRI fragment does not hybridize to
SJT DNA and the 3.7 kb SJT target site EcoRI fragment does
not hybridize to LJT DNA. Thus, the cell sequences with
which EBV has recombined are not immediately adjacent to
each other in uninfected cell DNA. In addition, LJT DNA

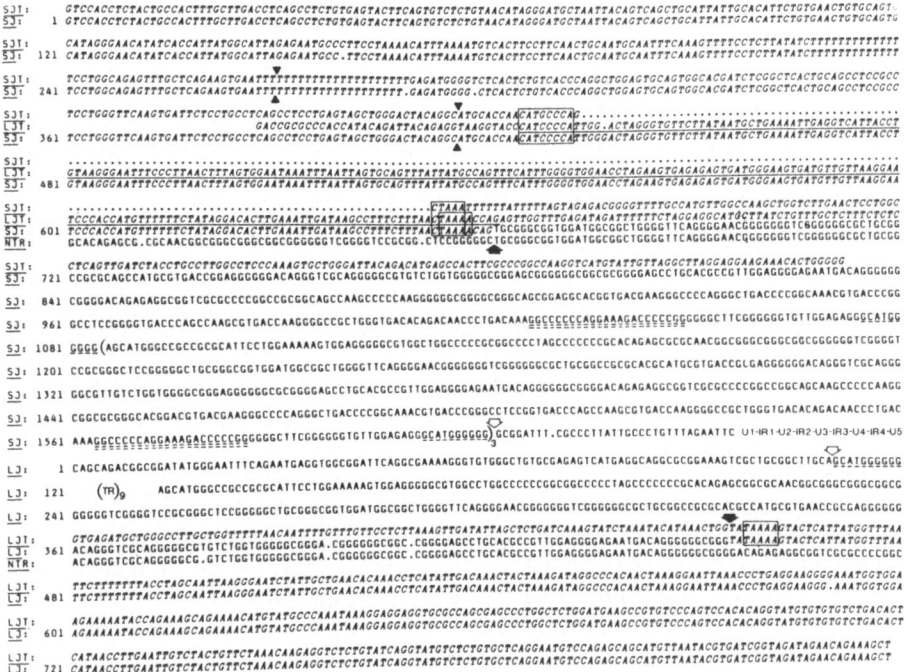

Fig. 2. Sequence of the Namalwa-EBV junction sites (SJ and LJ) and of the unrecombined sites for normal cells (SJT and LJT). Cell DNA is shown in italic letters. The sequence shown begins in SJ and SJT cell DNA, proceeds through EBV TR to an EcoRI site in EBV U1, begins again in EBV U5, proceeds through EBV TR, and ends at the HindIII cell DNA site in LJ and LJT. The dark triangles show the limits of an Alu homologous region in SJ and SJT. Continuous underlines indicate the 236-base-pair part of LJT, which is duplicated at the junction between EBV TR and SJ cell DNA. The dark arrows indicate junctions between EBV TR and cell DNA and open arrows indicate junctions between EBV TR and EBV U5 or U1. The single-dashed underlines indicate the 10-nucleotide direct repeat which brackets the EBV TR sequence and the double-dashed underlines indicate 22 nucleotides of TR which are homologous to a core sequence within the HSV "a" sequence. Dots indicate sites of variability among tandem copies of EBV TR in Namalwa. All sequences were analyzed for intra-and intersequence homology and compared to the Genbank databank.

does not hybridize to any of the surrounding EcoRI fragments in the lambda SJT clones, and SJT DNA does not hybridize to any fragments in the lambda LJT clones. The
average insert of human DNA in the phage library is 15-20
kb (Lawn et al., 1978). EBV integration in Namalwa has
therefore been accompanied by a deletion of more than 15 kb
of host DNA. Since there is no heterogeneity among the
Namalwa integration fragments in Southern analysis, the
deletion probably occurred during EBV insertion and was not
a subsequent event during continuous Namalwa cell culture.
Secondary events could account for some of the heterogeneity in putative TR-cell junction fragments in IB4.

The nucleotide sequence of the Namalwa EBV-cell DNA
junction sites (SJ and LJ) was compared to the prototype
EBV DNA sequence (Baer et al., 1984) and to the unrecombined sites (SJT and LJT) of normal cell DNA (Figure 2).
The nucleotide sequence of the cell DNA component of the
Namalwa small junction fragment was nearly identical to
that of the homologous site from cell DNA up to a point 236
nucleotides away from the juncture with TR. The sequence
in common between SJ and SJT includes a 146 bp homlogue of
the Alu family (bracketed by dark triangles in Figure 2).
The 236 bp SJ segment not common to SJT is also not homologous to EBV DNA nor to any DNA within the lambda SJT
clones. This 236 nucleotide sequence (indicated by under-
line in Figure 2) hybridizes to a LJT cell DNA sequence
which maps 1.2 kb away from the EBV-cell junction site
within an internal HincII fragment of the HindIII-EcoRI
cell DNA of LJ and LJT. At the point of recombination
between the 236 bp LJT sequence and SJT there are similar
nucleotide sequences, CAT(G or C)CCA. One nucleotide
further in SJT is a sequence CTAAA which is homologous to
the sequence CTAAAACAG at the end of the 236 bp DNA segment
where it joins EBV TR. Notably, "TAAA" is also in cell DNA
next to the site of joining with TR at the other end of EBV
DNA. The cell DNA for 50 nucleotides on both sides of the
integration site is unusually AT rich (75%). There is no
homology between the TR sequence and any of the LJT
sequence or SJT shown in Figure 2. The sequence of Namalwa
cell DNA joined to the Namalwa LJ TR is identical to the
same site in normal cell DNA (Figure 2). Since the
junctions of TR with cell DNA (indicated by black arrows in
Figure 2) are not at the same site in TR, it is likely that
part of TR was deleted during integration.

Relative to the start of TR at the U5-TR junction, the
first 11 nucleotides of TR are directly repeated at the TR-
U1 junction (junctions indicated by open arrows in Figure
2). The EBV TR sequence is similar in this regard to the
terminal repeat "a" sequence in herpes simplex (HSV) DNA
(Mocarski and Roizman, 1981, 1982a and b). Both sequences
are G-C rich, have oligo dG and oligo dC domains, and are
bracketed by short terminal direct repeats. The herpes
simplex terminal direct repeat, is the site of cleavage for
isomerization and packaging of HSV DNA (Mocarski and
Roizman, 1981, 1982a & b). Also, near the short EBV
terminal direct repeat is a sequence GGCCCCCAGGAAAGACCCCCGG
which has homology to a highly conserved domain of the
herpes simplex "a" sequence (Niza Frenkel, unpublished
observations).

These data and previously published in situ chromosome
hybridizations demonstrate that integration can be the sole
mechanism for intracellular virus persistence. Further,
these data suggest that EBV DNA integrated into Namalwa
cells from a linear format and that integration could be
site specific for sequence in TR. The EBV TR is similar to
the herpes simplex "a" sequence which is known to be specifically cleaved for isomerization and packaging. Further
evidence in support of site specificity within TR could
come from determination of the nucleotide sequences of
other integration sites and of the ends of virion EBV DNA.
Moreover, Namalwa EBV integration is by nonhomologous recombination as is characteristic of previously described
viral DNA integrations. A search of the TR and surrounding
Namalwa cell sequences revealed no significant dyad symme-
try as is noted in some DNAs involved in illegitimate recombination. There is also deletion and duplication of
cell DNA at the site of Namalwa EBV cell DNA recombination
as have been recognized near sites of integration of
smaller viral DNAs.

In IB4 cells, EBV DNA persists, both integrated on
chromosome 4Q25 and as episomes. The coexistence of episomes and integrated EBV DNA in IB4 cells is similar to
some cells infected with Papilloma viruses. Further, in
IB4, the EBV genome may have integrated from an episomal
format since unique DNAs from the left and right ends of
linear viral DNA are closely linked in integrated DNA as
they would be in episomes. These data also suggest that
EBV integration in IB4 cells is mediated by EcoRI I, not be

TR. However, the data are confounded by the rearrangement of EBV DNA EcoRI I and TR sequences at the site of integration.

Integration could be signficant to growth transforming and oncogenic properties of EBV if it has a cis effect on expression of adjacent cellular or viral genes as has been demonstrated with retroviruses. Chromosome 8 rearrangements altering expression of c myc are characteristic of malignant B cells. However, there is no apparent direct relationship between EBV integration on Namalwa chromosome 1 and the Namalwa 8:14 translocation. Further, although B lym is activated in Burkitt tumor cells and maps near the site of integration by chromosome cytological hybridizations (Diamond et al., 1983; Morton et al., 1984), we have demonstrated that B lym is not within 15 kb of the Namalwa integration site making it unlikely that integration has altered expression of this gene. Moreover, comparison of the Namalwa SJT and LJT sequences with cellular sequences in the Genbank databank as of March 1984 revealed no homology to known cell genes other than the Alu homology indicated in Figure 2. Thus the biologic significance of EBV integration will only be evident from further studies of EBV integration in human cells.

EXPRESSION OF LATENT EBV GENOMES

Replication of EBV is incompatible with persistent infection since EBV kills cells early in its replicative cycle. Lymphocytes whose growth in vitro has been enhanced by EBV infection usually show no evidence of virus replication. These cells contain a new intranuclear antigen, EBNA (Reedman and Klein 1973), and a new plasma membrane antigen, LYDMA (Svedmyr and Jondal 1975). EBNA is present in all Burkitt tumor lymphoblasts and in all NPC cells (Lindahl et al., 1974). Since detection of LYDMA is dependent on a functional test using cells growing in vitro, it is not known whether NPC cells express LYDMA. Three EBV genes are expressed in latently infected lymphocytes and probably also in NPC cells. These three genes are widely separate in the EBV genome and have been designated LT1, LT 2 and LT3 (Kieff et al., 1984, 1985). A fourth region encodes two small nonpolyadenylated polymerase 3 transcripts which are similar to adeno VA RNAs in structure and transcription (King et al., 1981; Rosa et

al., 1981) and function, in adenovirus replication (Bhat
and Thimmappaya 1983). These VA-like RNAs can replace adenovirus VA RNA. They could facilitate translation of EBV
RNAs in latently as well as productively infected cells.

The LT1 gene includes the EBV DNA IR1 and U2 domains
(Powell et al., 1979; King et al., 1980, 1981; van Santen
et al., 1981, 1983; Cheung and Kieff 1982; Dambaugh et al.,
1984; Hennessy and Kieff 1985). The principal RNA transcribed from this gene in latently infected cells includes
multiple copies of IR1 and terminates at a polyadenylation
site in U2. There is a promoter in each copy of IR1 which
could be the promoter for transcription of this RNA. The
cytoplasmic, putative messenger, RNA is a 3 kb spliced product of the primary transcript. There are approximately
three copies of this RNA in the cell cytoplasm. The RNA is
extensively spliced. Two contiguous and probably continuous RNA protected DNA segments have been defined in the 1.6
kb 3' to the polyadenylation site. This exon (assuming
continuity) coincides with a 1.5 kb open reading frame.
The open reading frame encodes most or all of a 88 kda
nuclear protein, EBNA2 (Dambaugh et al., 1984; Hennessy and
Kieff, 1983, 1985). Thus, differences among EBV isolates
in the length of a repeat region in the open reading frame
correlate with differences in size of a nuclear protein detected with EBV immune human sera; and, antiserum raised in
rabbits to a protein made in E. coli from the central part
of the open reading frame identify a 88 kda nuclear protein
in all latently infected cells. An "ATCATG" near the beginning of the open reading frame has the characteristics
of an initiation codon (Kozak 1984). Either of two potential splice acceptor sites 5' to the ATCATG but still within the open reading frame are likely to join the open reading frame to as yet unidentified 5' exons which could provide the amino terminus of the protein; or could be part of
a long 5' untranslated leader. The open reading frame 3'
to the putative initiation codon only encodes for a 53 kda
protein. However, the discrepancy between this and 88 kda
apparent size of EBNA2 is not unusual for a protein of such
high proline content as is encoded by the open reading frame.
The complete amino acid sequence of the open reading frame
of two EBV isolates B95-8 and AG876 beginning with the
initator methionine is shown in Figure 3. Interesting
features of the protein are (1) a polyproline domain, (2)
domains of positively charged AA alternating with noncharged
AA and (3) a negatively charged carboxy terminus.

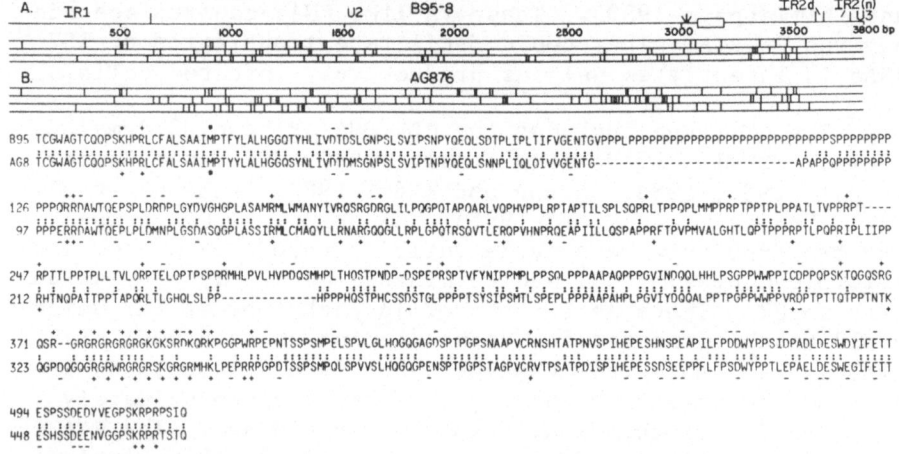

Fig. 3. Comparison of B95-8 and AG876 U2 reading frames and translation products. Panels A and B are schematic diagrams of the B95-8 and AG876 U2 regions indicating by vertical lines the positions of termination codons in the three possible reading frames for the 5' to 3' sense strand of DNA. The location of two consecutive polyadenylation signals is indicated by the vertical arrow. The open boxes show the position of an imperfect 105 bp repeat represented twice in AG876 U2 and once in B95-8 U2. IR2d represents the 38 bp partial copy of the IR2 sequence adjacent to U2. IR2(n) signifies a variable number of direct tandem copies of the IR2 seqence; in B95-8, n = 11; for AG876, n = 13. Panel C is the translation of the longest opening frame, the possible initiator methionine is indicated by an asterisk. Indicated underneath the B95-8 AA sequence is the AA predicted for the same open reading frame for AG876 EBV. Dashes indicate deleted amino acids after maximal alignment of the DNA and amino acid sequences. Coordinates indicated are for the B95-8 amino acid sequence.

The LT1 gene is important in initiation of growth transformation. First, P3HR-1 virus which is deleted for the U2 domain of LT1 is incapable of initiating growth transformation (Miller et al., 1974; Ragona et al., 1980). Recombinants between P3HR-1 and other EBV isolates which are capable of initiating growth transformation have all regained a U2 domain (Skare, Fresen and Strominger, personal communication). Second, transfections of lymphocytes with

U2 DNA transiently induces DNA synthesis (Volsky et al., 1984).

The LT2 gene includes the right end of U3, IR3 and the left end of U4 (Heller et al., 1982b; Summers et al., 1982; Hennessy et al., 1983; Hennessy and Kieff 1983). This gene encodes 3 copies of a 3.7 kb cytoplasmic RNA in each latently infected cell (van Santen et al., 1981; Heller et al., 1982b; Hennessy et al., 1983). The 3.7 kb RNA is spliced. A 2 kb exon is at its 3' end. The 5' exons have not been mapped. The 2 kb exon contains a single long open reading frame and a short untranslated 3' tail (Heller et al., 1982b; Baer et al., 1984). The open reading frame is translated into the EBNA 1 protein. Thus EBV immune human antisera which contain antibody against the EBNA protein react with part of the open reading frame expressed in E. coli (Hennessy and Kieff 1983); and, antibody raised in rabbits against the open reading frame expressed in bacteria react with the EBNA1 protein in latently infected cells (Hennessy and Kieff 1983). Since a DNA fragment which includes only 600 nucleotides 5' to the open reading frame when put into SV40 or polyoma expression vectors expresses a protein of the same size as EBNA1 (Summers et al., 1982; Fischer et al., 1984), it was likely that the open reading frame encodes the entire protein. This has been confirmed by fusing the HSV-1 alpha promoter directly to the open reading frame resulting in expression for EBNA1 protein of the same size as that found in latently infected cells (Hummel et al., unpublished observations).

The complete amino acid sequence of the EBNA1 protein is shown in Figure 4. The obvious features of this protein are domains of postive charged AA alternating with neutral AA, a glycine alanine copolymer domain encoded by the IR3 DNA repeat and a negatively charged carboxy terminus. These features are surprisingly similar to those of the EBNA2 protein.

The important biologic function of the EBNA1 gene is in maintaining EBV episomes (Yates et al., 1984; Yates et al., personal communication). A cis-acting repeat sequence has been defined in the U1 domain which is required for maintenance of episomes in latently infected cells (Yates et al., 1984). This putative origin of EBV episomal DNA replication preferentially survives in cells expressing EBNA1. Thus, EBNA1 may be similar to SV40 T antigen in its

```
                  10                                    20                                    30
MET SER ASP GLU GLY PRO GLY THR GLY PRO GLY ASN GLY LEU GLY GLU LYS GLY ASP THR SER GLY PRO GLU GLY SER GLY GLY SER GLY
                  40                                    50                                    60
PRO GLN ARG ARG GLY GLY ASP ASN HIS GLY ARG GLY ARG GLY ARG GLY ARG GLY ARG GLY GLY GLY ARG PRO GLY ALA PRO GLY GLY SER
                  70                                    80                                    90
GLY SER GLY PRO ARG HIS ARG ASP GLY VAL ARG ARG PRO GLN LYS ARG PRO SER CYS ILE GLY CYS LYS LYS THR HIS GLY GLY THR GLY
                  100                                   110                                   120
ALA GLY ALA GLY ALA GLY GLY ALA GLY ALA GLY GLY ALA GLY ALA GLY GLY ALA GLY ALA GLY GLY ALA GLY GLY ALA GLY GLY ALA GLY
                  130                                   140                                   150
ALA GLY GLY ALA GLY ALA GLY GLY GLY ALA GLY ALA GLY GLY GLY ALA GLY GLY ALA GLY GLY ALA GLY ALA GLY GLY GLY GLY ALA GLY ALA
                  160                                   170                                   180
GLY GLY GLY ALA GLY GLY ALA GLY ALA GLY GLY GLY ALA GLY GLY ALA GLY GLY ALA GLY ALA GLY GLY GLY ALA GLY ALA GLY GLY ALA GLY GLY
                  190                                   200                                   210
ALA GLY GLY ALA GLY ALA GLY GLY GLY ALA GLY GLY ALA GLY GLY ALA GLY ALA GLY GLY GLY ALA GLY GLY ALA GLY ALA GLY GLY ALA GLY ALA
                  220                                   230                                   240
GLY GLY ALA GLY ALA GLY GLY GLY ALA GLY GLY ALA GLY GLY ALA GLY GLY ALA GLY GLY ALA GLY GLY ALA GLY ALA GLY GLY ALA GLY ALA
                  250                                   260                                   270
GLY ALA GLY GLY ALA GLY ALA GLY GLY ALA GLY GLY ALA GLY GLY ALA GLY GLY ALA GLY GLY ALA GLY GLY ALA GLY GLY ALA GLY GLY ALA GLY
                  280                                   290                                   300
ALA GLY GLY GLY ALA GLY ALA GLY GLY ALA GLY ALA GLY GLY GLY ALA GLY GLY ALA GLY GLY ALA GLY GLY ALA GLY ALA GLY GLY ALA GLY ALA
                  310                                   320                                   330
GLY GLY ALA GLY ALA GLY GLY GLY ALA GLY ALA GLY ALA GLY GLY ALA GLY GLY ALA GLY ALA GLY GLY GLY ALA GLY ALA GLY GLY GLY GLY ARG GLY
                  340                                   350                                   360
ARG GLY GLY SER GLY GLY ARG GLY GLY ARG GLY GLY SER GLY GLY ARG GLY ARG GLY GLY GLY ARG GLY ARG ARG GLY ARG GLY ARG GLU ARG
                  370                                   380                                   390
ALA ARG GLY GLY SER ARG GLU ARG ALA ARG GLY ARG GLY ARG GLY ARG GLY GLU LYS ARG PRO ARG SER PRO SER SER GLN SER SER SER
                  400                                   410                                   420
SER GLY SER PRO PRO ARG ARG PRO PRO PRO GLY ARG ARG PRO PHE PHE HIS PRO VAL GLY GLU ALA ASP TYR PHE GLU TYR HIS GLN GLU
                  430                                   440                                   450
GLY GLY PRO ASP GLY GLU PRO ASP VAL PRO PRO GLY ALA ILE GLU GLN GLY PRO ALA ASP ASP PRO GLY GLU GLY PRO SER THR GLY PRO
                  460                                   470                                   480
ARG GLY GLN GLY ASP GLY GLY ARG ARG LYS LYS GLY GLY TRP PHE GLY LYS HIS ARG GLY GLN GLY GLY SER ASN PRO LYS PHE GLU ASN
                  490                                   500                                   510
ILE ALA GLU GLY LEU ARG ALA LEU LEU ALA ARG SER HIS VAL GLU ARG THR THR ASP GLU GLY THR TRP VAL ALA GLY VAL PHE VAL TYR
                  520                                   530                                   540
GLY GLY SER LYS THR SER LEU TYR ASN LEU ARG ARG GLY THR ALA LEU ALA ILE PRO GLN CYS ARG LEU THR PRO LEU SER ARG LEU PRO
                  550                                   560                                   570
PHE GLY MET ALA PRO GLY PRO GLY PRO GLN PRO GLY PRO LEU ARG GLU SER ILE VAL CYS TYR PHE MET VAL PHE LEU GLN THR HIS ILE
                  580                                   590                                   600
PHE ALA GLU VAL LEU LYS ASP ALA ILE LYS ASP LEU VAL MET THR LYS PRO ALA PRO THR CYS ASN ILE ARG VAL THR VAL CYS SER PHE
                  610                                   620                                   630
ASP ASP GLY VAL ASP LEU PRO PRO TRP PHE PRO PRO MET VAL GLU GLY ALA ALA ALA GLU GLY ASP ASP GLY ASP ASP GLY ASP GLY ASP
                  640
GLY ASP GLY ASP GLU GLY GLU GLU GLY GLN GLU
```

Fig. 4. Amino acid sequence of EBNAl projected from the nucleotide sequence and reading frame as described in text. Nucleotides encoding AA 1-383 are from Heller et al. 1982 and those encoding AA 384-642 are from Baer et al., 1984. The glycine alanine copolymer domain encoded by IR3, the basic glycine arginine domain and the acidic aspartic acid glycine domains are underlined.

specific recognition of an origin of DNA replication. Since EBNAl binds diffusely to chromatin, it may also have a direct effect on cellular origins of DNA synthesis.

The LT3 gene is the best characterized EBV gene expressed in latently infected cells since there are at least 60 copies of the messenger RNA in each cell (van Santen et al., 1981; 1983; Fennewald et al., 1984). The entire gene is within U5 near TR (Fennewald et al., 1984; Hennessy et al., 1984). The message is 2.9 kb and is spliced. Sixty nucleotides from the cap site of the RNA is an initiator ATG which begins a 1158 nucleotide open reading frame which spans three exons. The RNA has a 1700

235

nucleotide 3' tail after the end of the open reading frame
which is presumed to be untranslated in latent infection.
The important characteristics of the protein which is
encoded by the open reading frame (Figure 5) are six
markedly hydrophobic potentially trans membrane domains
linked by short peptides, predicted to be reverse turns; a
200 AA markedly acidic carboxy terminus; and, no identifiable translocation signal or N linked glycosylation sites.
The carboxy terminal 220 AA encoded by the open reading
frame were synthesized in bacteria, purified and used to
immunize rabbits (Hennessy et al., 1984). The immune rabbit
sera react with a 60 kda protein in the membrane of latently
infected cells. The protein is the same size as that
translated from the RNA in vitro. Immunofluorescent studies
with the rabbit antisera suggest that the carboxy terminus
of the protein is in the inner aspect of the plasma membrane
of latently infected cells. Thus, this protein could be
responsible for the generation of the LYDMA reactivity of
latently infected cells. A schematic diagram of the protein
and its proposed relationship to the phospholipid membrane
bilayer is shown in Figure 5. The 8 AA domains which are
predicted to be on the outer surface of the membrane could
be recognized by immune T cells. However, these domains are
not hydrophilic and may not extend sufficiently beyond the
membrane to be recognized by immune T cells. An alternative
possibility is that insertion of the EBV membrane protein
alters the conformation of an adjacent cell membrane
proteins creating the new LYDMA antigen.

The EBV LT3 membrane protein does not resemble these
membrane proteins of acute transforming retroviruses which
are partially homologous to growth factors or growth factor
receptors. Nor is it similar to the "ras" class of proteins. Rather, it resembles proteins with multiple membrane
spanning domains such as the rhodopsins, acetylcholine receptor, calcium pump protein and erythrocyte membrane band
3. The EBV protein might therefore have an ion transport
function similar to that of these other integral membrane
proteins. Growth transformed cells are known to have altered ion permability; and, calcium affects the proliferation of lymphocytes. The markedly acidic carboxy terminus
of the protein is also likely to have important biologic
properties by virtue of its binding to cellular cytoskeletal
or enzymatic components or by possessing ATPase or phosphoprotein kinase activity.

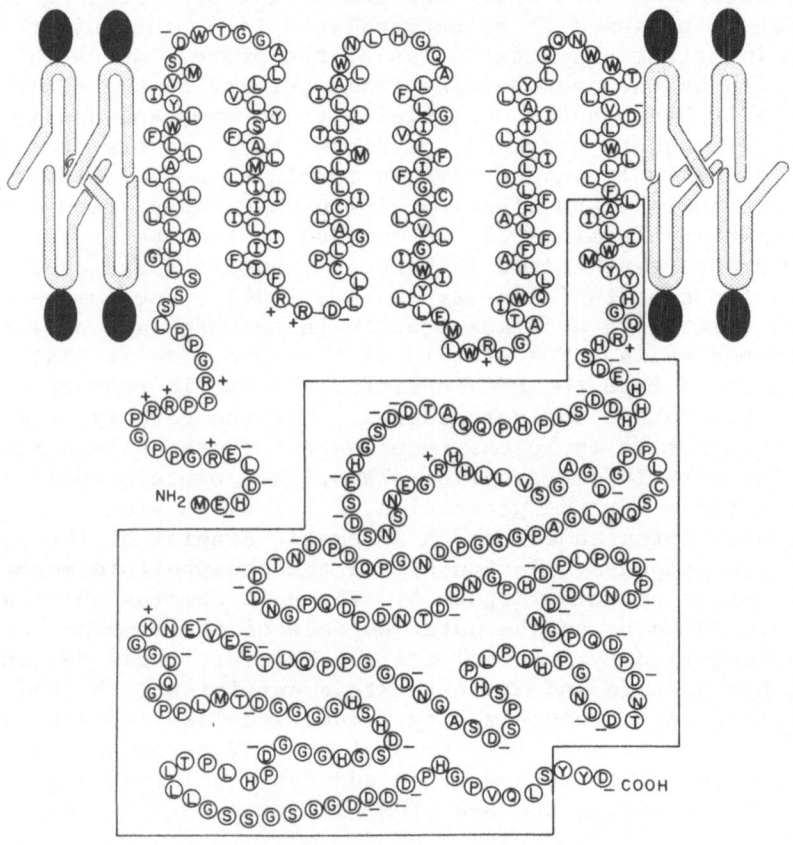

Fig. 5. A model of the EBV protein as it might be situated in the membrane. The portion of the U5 latent protein expressed in pKH548 in E. coli is boxed. Charged residues (+,-) and single letter amino acid designations are indicated.

CONCLUSION

EBV usually persists in cells as complete genome episomes, but may persist solely in transformed cells as a complete genome integrated into cell DNA. The finding of integrated EBV DNA raises the possibility that integration could have a cis effect on expression of cell or viral genes which are near the junctions between viral and cell DNA. Although initial data indicate that EBV does not in-

tegrate by homologous recombination or at a consistent
site, or near a known oncogene; this should be further investigated. While further investigation is not likely to
reveal a consistent association between integration and
transformation, it is likely to reveal effects which are
important in individual transforming events. It is highly
likely that integrative persistence is adequate for initiation of growth transformation. Three viral genes are
expressed in latently infected growth transformed cells
(Figure 6). These genes encode two intranuclear proteins
and a membrane protein. Mono specific antisera have been
developed and used to identify the proteins in latently
infected cells. Less is known about expression in tumor
tissue (Dambaugh et al., 1979; Raab-Traub et al., 1983),
especially NPC. The functions of these proteins vis a vis
growth transformation and latency are only partially understood. One function of EBNA1 is in trans on a putative EBV
ori sequence to permit episome replication. A function of
EBNA2 may be initial stimulation of cell DNA synthesis. In
specifying two nuclear and one membrane protein in transformed cells, EBV is similar to SV40, polyoma and adenoviruses. However, there is no homology between the EBNA1,
EBNA2 or LYDMA proteins and the proteins encoded by these
other viruses. It is likely therefore that identification
of the precise functions of the EBV proteins in lymphocyte
and epithelial cell transformation will lead to insight
into novel mechanisms of cell growth transformation.

The latent cell membrane protein is almost certainly
directly or indirectly responsible for immune cytolysis and
supression of growth of EBV infected lymphocytes by immune
T cells. Aspects of this recognition will require
considerable clarification. Nevertheless, identification
of the protein makes it possible to identify the LYDMA
epitope which should eventually make it feasible to
heighten immunity or to pre-immunize to prevent EBV
infection. This would be a quite novel approach at control
of herpes virus infection. Since LYDMA is usually
expressed on malignant B lymphocytes it may also be
possible to heighten tumor rejection. If the protein is
expressed on the surface of NPC cells, a similar approach
could be contemplated. Moreover, if the epitopes of the
EBV latent membrane protein protrude sufficiently from the
cytoplasmic membrane, they might react with specifically

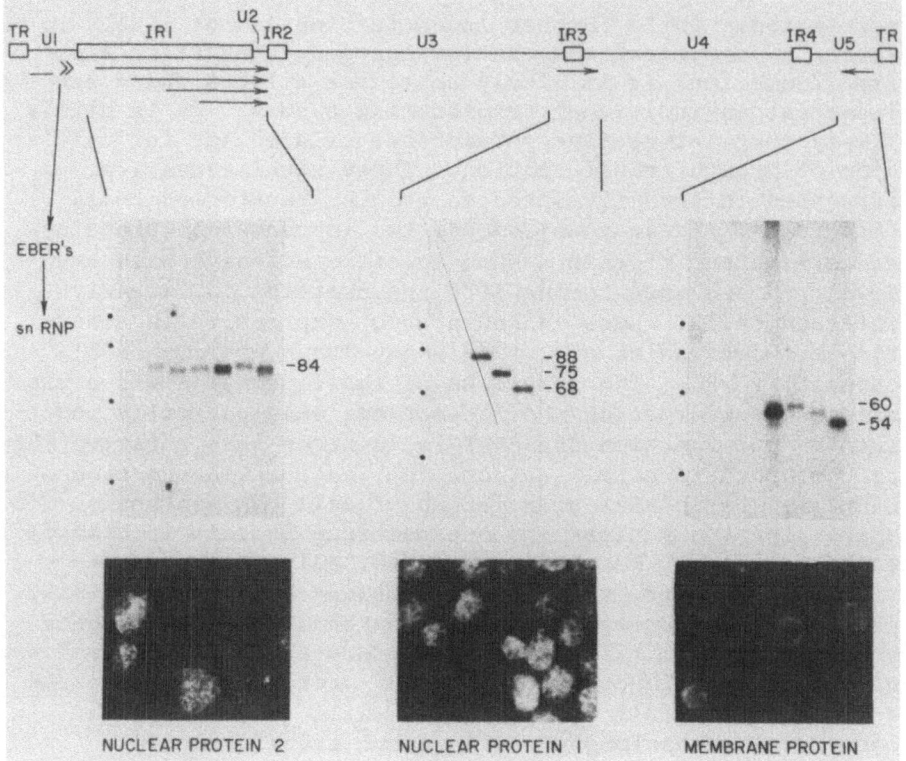

Figure 6. Summary of the major latent transcripts and their
protein products. Immediately below a schematic of the EBV
genome is the location of the transcripts characteristic of
EBV latent infection. Both polyadenylated and nonpolyadenylated transcripts are shown. The two nonpolyadenylated
small RNAs encoded by the U1 region are incorporated into
ribonuclear proteins (RNP). Representative Western blots of
proteins from different cell cell lines are shown stained
with antiserum specific to each particular gene product.
Size markers are indicated by dots beside each blot. Below
each blot a photograph of immunostained EBV-infected cells
indicates the products location within the cell.

directed antibody. If so, the tumor cell could be
specifically attacked by attaching toxins to the antibody.

ACKNOWLEDGEMENTS

This research was supported by Public Health Service
research grants CA 17281 and CA 19264 from the National
Institutes of Health and by American Cancer Society grant
ACS MV32J. K.H., S.F., M.H., L.P. and M.H. were supported
by National Research Service Awards GM 07183, AI 07182,
AI 07099, CA 09241 and AI 07182, respectively. T.M. was
supported by a Cancer Research Campaign International
Fellowship from the International Union Against Cancer.
E.K. is the recipient of a Faculty Research Award from the
American Cancer Society.

The contributions of colleagues and collaborators who
have contributed to these experiments or communicated
information and materials including G. Klein, W. Henle, G.
Pearson, H. Rabin, D. Thorley-Lawson, G. Miller, P.
Deininger, R. Baer, B. Barrell, B. Sugden, D. Volsky, P.
Levine, R. Pritchett, M. Dolyniuk, D. Hayward, D. Given, W.
King, A. Powell, V. van Santen and A. Cheung are greatly
appreciated.

REFERENCES

BAER, B., BANKIER, A., BIGGIN, M., DEININGER, P., FARREL,
P., GIBSON, T., HATFULL, G., HUDSON, G., SATCHWELL, S.,
SEGUIN, C., TUFFNELL, P., and BARRELL, B., DNA sequence and
expression of the B95-8 Epstein-Barr virus genome. Nature,
310, 207-211 (1984).

BHAT, R., and THIMMAPPAYA, B., Two small RNAs encoded by
Epstein-Barr virus can functionally substitute for the
virus-associated RNAs in the lytic growth of adenovirus 5.
Proc. Natl. Acad. Sci. U.S.A., 80, 4789-4793 (1983).

BORNKAMM, G.W., DELIUS, H., ZIMBER, U., HUDEWENTZ, J., and
EPSTEIN, M.A., Comparison of Epstein-Barr virus strains of
different origin by analysis of the viral DNAs. J. Virol.,
35, 603-618 (1980).

CHEUNG, A., and KIEFF, E., Epstein-Barr virus DNA XI: The nucleotide sequence of the large internal repeat in EBV DNA. J. Virol., 44, 286-294 (1982).

DAMBAUGH, T., NKRUMAH, F.K., BIGGAR, R.J., and KIEFF, E., Epstein-Barr virus RNA in Burkitt tumor tissue. Cell, 16, 313-322 (1979).

DAMBAUGH, T., BEISEL, C., HUMMEL, M., KING, W., FENNEWALD, S., CHEUNG, A., HELLER, M., RAAB-TRAUB, N., and KIEFF, E., Epstein-Barr virus DNA. VII. Molecular cloning and detailed mapping of EBV(B95-8) DNA. Proc. Natl. Acad. Sci. U.S.A., 77, 2999-3003 (1980).

DAMBAUGH, T., and KIEFF, E., Identification and nucleotide sequence of two similar tandem direct repeats in Epstein-Barr virus DNA. J. Virol., 44, 823-833 (1982).

DAMBAUGH, T., HENNESSY, K., CHAMNANKIT, L., and KIEFF, E., The U2 region of Epstein-Barr virus DNA may encode EBNA2. Proc. Natl. Acad. Sci. U.S.A., 81, 7207-7211 (1984).

DE THE, G., GESER, A., DAY, N., TUKEI, P., WILLIAMS, E., BERI, D., SMITH, P., DEAN, A., BORNKAMM, G., FEORINO, P., and HENLE, W., Epidemiological evidence for a causal relationship between Epstein-Barr virus and Burkitt's lymphoma: Results of the Ugandan prospective study. Nature, 274, 756-761 (1978).

DIAMOND, A., COOPER, G.M., RITZ, J., and LANE, M.-A., Identification and molecular cloning of the human Blym transforming gene activated in Burkitt's lymphomas. Nature, 305, 112-126 (1983).

EPSTEIN, M.A., and ACHONG, B.G., Introduction: discovery and general biology of the virus. In: M.A. Epstein and B.G. Achong (eds.), The Epstein-Barr Virus, pp. 1-22, New York, Springer-Verlag (1978).

FENNEWALD, S., VAN SANTEN, V., and KIEFF, E., Nucleotide sequence of an mRNA transcribed in latent growth-transforming virus infection indicates that it may encode a membrane protein. J. Virol., 51, 411-419 (1984).

FISCHER, D.K., ROBERT, M.F., SHEDD, D., SUMMERS, W.P., ROBINSON, J.E., WOLAK, J., STEFANO, J.E., and MILLER, G.,

Identification of Epstein-Barr nuclear antigen polypeptide in mouse and monkey cells after gene transfer with a cloned 2.9-kilobase-pair subfragment of the genome. Proc. Natl. Acad. Sci. U.S.A., 81, 43-47 (1984).

GIVEN, D., and KIEFF, E., DNA of Epstein-Barr virus. IV. Linkage map for restriction enzyme fragments of the B95-8 and W91 strains of EBV. J. Virol., 28, 524-542 (1978).

GIVEN, D., YEE, D., GRIEM, K., and KIEFF, E., DNA of Epstein-Barr virus. V. Direct repeats at the ends of EBV DNA. J. Virol., 30, 852-862 (1979).

HELLER, M., DAMBAUGH, T., and KIEFF, E., Epstein-Barr virus DNA. IX. Variation among viral DNAs. J. Virol., 38, 632-648 (1981).

HELLER, M., HENDERSON, A., AND KIEFF, E., Repeat array in Epstein-Barr virus DNA is related to cell DNA sequences interspersed on human chromosomes. Proc. Natl. Acad. Sci. U.S.A., 79, 5916-5920 (1982a).

HELLER, M., VAN SANTEN, V., and KIEFF, E., Simple repeat sequence in Epstein-Barr virus DNA is transcribed in latent and productive infections. J. Virol., 44, 311-320 (1982b).

HELLER, M., FLEMINGTON, E., Kieff, E. and DEININGER, P., Repeat arrays in cell DNA related to Epstein-Barr virus IR3 repeat. Molec. and Cell. Biol., in press (1985).

HENDERSON, A., RIPLEY, S., HELLER, M., and KIEFF, E. Chromosome site for Epstein-Barr virus DNA in a Burkitt tumor cell ine and in lymphocytes growth transformed in vitro. Proc. Natl. Acad. Sci. U.S.A., 80, 1987-1991 (1983).

HENLE, W., DIEHL, B., KOHN, G., ZUR HAUSEN, H., and HENLE, G., Herpes-type virus and chromosome marker in normal leukocytes after growth with irradiated Burkitt cells. Science, 157, 1065-1065 (1967).

HENLE, W., HENLE, G., HO, H.C., BURTIN, P., CACHIN, Y., CLIFFORD, P., DE SCHRYVER, A., DE THE, G., DIEHL, V., and KLEIN, G., Antibodies to EB virus in nasopharyngeal carcinoma, other head and neck neoplasms and control groups. J. Natl. Cancer Inst., 44:225-231 (1970).

242

HENNESSY, K., HELLER, M., VAN SANTEN, V., and KIEFF, E., A
simple repeat array in Epstein-Barr virus DNA encodes part
of EBNA. Science, 220, 1396-1398 (1983).

HENNESSY, K., and KIEFF, E., One of two Epstein-Barr virus
nuclear antigens contains a glycine-alanine copolymer
domain. Proc. Natl. Acad. Sci. U.S.A., 80, 5665-5669
(1983).

HENNESSY, K., FENNEWALD, S., HUMMEL, M., COLE, T. and
KIEFF, E., A membrane protein encoded by Epstein-Barr virus
in latent growth transforming infection. Proc. Natl. Acad.
Sci. U.S.A., 81, 7207-7211 (1984).

HENNESSY, K. and KIEFF, E., A second nuclear protein is
encoded by Epstein-Barr virus in latent infection.
Science, in press (1985).

HUANG, D., HO, J., and HENLE, G., Demonstration of
Epstein-Barr virus-associated nuclear antigen in
nasopharyngeal carcinoma cells from fresh biopsies. Int.
J. Cancer, 14, 580-588 (1974).

KIEFF, E., DAMBAUGH, T., HENNESSY, K., FENNEWALD, S.,
HELLER, M., MATSUO, T., and HUMMEL, M., Latent infection
and growth transformation by Epstein-Barr virus. In: G.
Giraldo and E. Beth (eds.), The Role of Viruses in Human
Cancer, Vol. 2, pp. 103-118, Elsevier Science Publishers
B.V. (1984).

KIEFF, E., HENNESSY, K., FENNEWALD, S., MATSUO, T.,
DAMBAUGH, T., HELLER, M., and HUMMEL, M., Biochemistry of
latent Epstein-Barr virus infection and associated cell
growth transformation. In: Proceedings of Burkitt's
Lymphoma: A Human Cancer Model, December 1983.
International Agency for Research on Cancer, in press
(1985).

KING, W., THOMAS-POWELL, A.L., RAAB-TRAUB, N., HAWKE, M.,
and KIEFF, E., Epstein-Barr virus RNA. V. Viral RNA in a
restringently infected, growth transformed cell. J.
Virol., 36, 506-518 (1980).

KING, W., VAN SANTEN, V., and KIEFF, E., Epstein-Barr virus
RNA. IV. Viral RNA in restringently and abortively
infected Raji cells. J. Virol., 38, 649-660 (1981).

KING, W., DAMBAUGH, T., HELLER, M., DOWLING, J., and KIEFF, E., Epstein-Barr virus DNA. XII. A variable region of the Epstein-Barr virus genome is included in the P3HR-1 deletion. J. Virol., 43, 979-986 (1982).

KINTNER, C., and SUGDEN, B., Conservation and progressive methylation of EBV DNA sequences in transformed cells. J. Virol. 38, 305-316 (1981).

KLEIN, G., GIOVANELLA, B.C., LINDAHL, T., FIALKOW, P.J., SINGH, S., and STEHLIN, J.S., Direct evidence for the presence of Epstein-Barr virus DNA and nuclear antigen in malignant epithelial cells from patients with poorly differentiated carcinoma of the nasopharynx. Proc. Natl. Acad. Sci. U.S.A., 71, 4737-4741 (1974).

KOZAK, M., Compilation and analysis of sequences upstream from the translational start site in eukaryotic mRNAs. Nucl. Acids Res., 12, 857-872 (1984).

LAWN, R., FRITSCH, E., PARKER, R., BLAKE, G., and MANIATIS, T., The isolation and characterization of linkeD delta and beta globin genes from a cloned library of human DNA. Cell, 15, 1157-1174 (1978).

LINDAHL., T., ADAMS, A., BJURSELL, G., BORNKAMM, G.W., KASCHA-DIERICH, G., and JEHN, U., Covalently closed circular duplex DNA of Epstein-Barr virus in a human lymphoid cell line. J. Mol. Biol., 102, 511-530 ((1976).

LINDAHL, T., KLEIN, G., REEDMAN, B.M., JOHANSSON, B., and SINGH, S., Relationship between Epstein-Barr virus DNA and the EBV-determined nuclear antigen (EBNA) in Burkitt's lymphoma biopsies and other lympho-proliferative malignancies, Int. J. Cancer, 13, 764-772 (1974).

MATSUO, T., HELLER, M., PETTI, L., O'SHIRO, E., and KIEFF, E., The entire Epstein-Barr virus genome integrates into human lymphocyte DNA. Science, 226, 1322-1325 (1984).

MILLER, G., ROBINSON, J., HESTON, L., and LIPMAN, M., Differences between laboratory strains of Epstein-Barr virus based on immortalization, abortive infection and interference, Proc. Natl. Acad. Sci. U.S.A., 71, 4006-4010 (1974).

MILLER, G., SHOPE, T., COOPE, D., WATERS, L., PAGANO, J., BORNKAMM, G., and HENLE, W., Lymphoma in cotton-top marmosets after inoculation with Epstein-Barr virus: Tumor incidence, histologic spectrum, antibody responses, demonstration of viral DNA, and characterization of viruses. J. Exp. Med., 145, 948-967 (1977).

MOCARSKI, E.S. and ROIZMAN, B., Site specific inversion sequence of herpes simplex virus genome: Domain and structural features. Proc. Natl. Acad. Sci., U.S.A., 78, 7041-7051 (1981).

MOCARSKI, E.S. and ROIZMAN, B., Herpesvirus-dependent amplification and inversion of cell-associated viral thymidine kinase gene flanked by viral a sequences and linked to an origin of viral DNA replication. Proc. Natl. Acad. Sci. U.S.A., 79, 5626-5630 (1982a).

MOCARSKI, E.S., and ROIZMAN, B., The structure and role of the herpes simplex virus DNA termini in inversion, circularization and generation of virion DNA. Cell, 31:89-97 (1982b).

MORTON, C.C., TAUB, R., DIAMOND, A., LANE, M.A., COOPER, G.M., and LEDER, P., Mapping of the human blym-1 transforming gene activated in Burkitt Lymphomas to chromosome 1. Science 223, 173-175 (1984).

NILSSON, K., High frequency establishment of human immunoglobulin producing lymphoblastoid lines from normal and malignant lymphoid tissue and peripheral blood. Int. J. Cancer, 8, 432-442 (1971).

NONOYAMA, M., HUANG, D.P., PAGANO, J.S., KLEIN, G., and Singh, S., DNA of Epstein-Barr virus detected in tissue of Burkitt's lymphoma and nasopharyngeal carcinoma, Proc. Natl. Acad. Sci. U.S.A., 70, 3265-3268 (1973).

NONOYAMA, M., and PAGANO, J.S., Detection of Epstein-Barr viral genome in non-productive cells. Nature (London) New Biol., 233, 103-106 (1971).

NONOYAMA, M., and PAGANO, J.S., Separation of Epstein-Barr virus DNA from large chromosomal DNA in non-virus producing cells. Nature (London) New Biol., 238, 169-171 (1972).

PEARSON, G., NEEL, H., WEILAND, L., MAHONEY, S.. TAYLOR, W., GOEPFERT, W., HUANG, A., LEVINE, P., LANIER, A., PILCH, B., and GOODMAN, M., Antibody dependent cellular cytotoxicity and disease course in North American patients with NPC; a prospective study. Int. J. Cancer, 33, 777-782 (1984).

POPE, J.H., HORNE, M.K., and SCOTT, W., Transformation of fetal human leucocytes in vitro by filtrates of a human leukemic cell line containing herpes-like virus. Int. J. Cancer, 3, 857-866 (1968).

POWELL, A., KING, W. and KIEFF, E., Epstein-Barr virus specific RNA. III. Mapping of the DNA encoding viral specific RNA in restringently infected cells. J. Virol., 29, 261-274 (1979).

PRASAD, V., Significance of metaplastic transformation in the pathogenesis of nasopharyngeal carcinoma. Clinical, histopathological, and ultrastructural studies. In: E. Grundmann, G.R.F. Krueger and D.V. Ablashi (eds.), Nasopharyngeal Carcinoma, Vol. 5, pp. 31-39, Gustav Fischer Verlag, (1981).

PRITCHETT, R.F., HAYWARD, S.D., and KIEFF, E., DNA of Epstein-Barr virus. I. Comparison of DNA of virus purified from HR-1 and B95-1 cells. J. Virol., 15, 556-569 (1975).

PRITCHETT, R., PEDERSEN, M., and KIEFF, E., Complexity of EBV homologous DNA in continuous lymphoblastoid cell lines, Virology 74, 227-231 (1976).

RAAB-TRAUB, N., DAMBAUGH, T., and KIEFF, E., DNA of Epstein-Barr virus. VIII. B95-8, the previous prototype, is an unusual deletion derivative. Cell, 22, 257-267 (1980).

RAAB-TRAUB, N., HOOD, R., YANG, C.-S., HENRY II, B., and PAGANO, J.S., Epstein-Barr virus transcription in nasopharyngeal carcinoma. J. Virol., 48, 580-590 (1983).

RAGONA, G., ERNBERG, I., and KLEIN, G., Induction and biological characterization of the Epstein-Barr virus

(EBV) carried by the Jijoye lymphoma line. Virology, 101, 553-557 (1980).

REEDMAN, B.M. and KLEIN, G., Cellular localization of an Epstein-Barr virus(EBV)-associated complement-fixing antigen in producer non-producer lymphoblastoid cell lines. Int. J. Cancer, 11, 499-520 (1973).

ROSA, M., GOTTLIED, E., LERNER, M., and STEITZ, J., Striking similarities are exhibited by two small Epstein-Barr virus-encoded ribonucleic acids and the adnovirus-associated ribonucleic acids VAI and VAII. Mol. Cell Biol., 1, 785-796 (1981).

SUMMERS, W., GROGAN, E., SHEED, D., ROBERT, M., LUI, C., and MILLER, G., Stable expression in mouse cells of nuclear neoantigen after transfer of a 3.4 megadalton cloned fragment of EBV DNA. Proc. Natl. Acad. Sci. U.S.A., 79, 5688-5692 (1982).

SVEDMYR, E., and JONDAL, M., Cytotoxic effector cells specific for B cell lines transformed by Epstein-Barr virus are present in patients with infectious mononucleosis. Proc. Natl. Acad. Sci. U.S.A., 72, 1622-1666 (1975).

VAN SANTEN, V., CHEUNG, A., and KIEFF, E., Epstein-Barr virus RNA. VII. Size and direction of transcription of virus-specific cytoplasmic RNA's in a transformed cell line. Proc. Natl. Acad. Sci. U.S.A., 78, 1930-1934 (1981).

VAN SANTEN, V., CHEUNG, A., HUMMEL, M., and KIEFF, E., RNA encoded by the IR1-U2 region of Epstein-Barr virus DNA in latently infected, growth transformed cells. J. Virol., 46, 424-433 (1983).

VOLSKY, D.J., GROSS, T., SINANGIL, F., KUSZYNSKI, C., BARTZATT, R., DAMBAUGH, T., and KIEFF, E., Expression of Epstein-Barr virus (EBV) DNA and cloned DNA fragmeents in human lymphocytes following Sendai virus envelope-mediated gene transfer. Proc. Natl. Acad. Sci. U.S.A., 81, 5926-5930 (1984).

WOLF, H., ZUR HAUSEN, H., KLEIN, G., BECKER, Y., HENLE, G., and HENLE, W., Attempts to detect virus-specific DNA sequences in human tumors. III. EBV DNA in nonlymphoid

nasopharyngeal carcinoma cells. Med. Microbiol. Immunol.,
161, 15-21 (1975)

YATES, J., WARREN, N., REISMAN, D., and SUGDEN, B., A
cis-acting element from the Epstein-Barr viral genome that
permits stable replication of recombinant plasmids in
latently infected cells. Proc. Natl, Acad. Sci. U.S.A.,
81, 3806-3810, (1984).

ZENG, Y., LIU, Y., LIU, C.H., CHEN, S., WEI, J., ZHU, J.,
and ZAI, H., Application of an immunoenzymatic and an
immunoautoradiographic method for mass survey of
nasopharyngeal carcinoma. Intervirology, 13, 162-168
(1980).

21

AN EPSTEIN-BARR VIRUS-DETERMINED NUCLEAR ANTIGEN
ENCODED BY A REGION WITHIN THE EcoRI A FRAGMENT OF
THE VIRAL GENOME

Lars Rymo, M.D.[1] and George Klein, M.D.[2]

[1]Department of Clinical Chemistry, Gothenburg
University, Sahlgren's Hospital, 413 45 Gothenburg,
and [2]Department of Tumour Biology, Karolinska
Institute, 104 01 Stockholm, Sweden

SUMMARY

Large Epstein-Barr virus (EBV) DNA restriction fragments
corresponding to regions transcribed in transformed, proliferating
cells were cloned into a cosmid derivative of the dominant-acting
selection vector pSV2-gpt. Recombinant vectors carrying the
EcoRI A fragment of EBV DNA were modified in the region
corresponding to the deletion of the virion DNA in the non-
transforming viral substrain P3HR-1, to create a series of recombinants lacking parts of this region. The recombinant vectors
were introduced into 3T3 mouse fibroblasts under selective conditions, and resistant clones shown to contain EBV DNA sequences
were analysed for the expression of EBV-related antigens detectable by direct, indirect, and anticomplement immunofluorescence techniques. Cells that contained the BamHI K fragment
expressed a nuclear antigen as expected. It is demonstrated here
that cells transfected with recombinant vectors containing the
major part of the EcoRI A fragment also express a nuclear
antigen detectable with certain anti-EBNA-positive human sera in
anticomplement immunofluorescence tests. This antigen is not
detected in cells transfected with EcoRI A derived vectors where
the BamHI H fragment has been deleted, nor in cells transformed
with vectors carrying the BamHI H fragment alone. Direct and
indirect immunofluorescence did not reveal the presence of
antigens associated with productive infection in any of the EBV
DNA transfected fibroblast clones.

In this study we address the question whether EBV genome regions transcribed in transformed, non-virusproducing cells, other than the BamHI K fragment region, are also involved in the induction of EBV-associated nuclear antigens. EBV DNA fragments representing these regions were introduced into 3T3 mouse fibroblast using transducing selectable vectors. Cells transformed stably with EBV DNA were selected and characterised with regard to the expression of EBV-related antigens as detected by immuno-fluorescence techniques. We show here, that cells transfected with recombinant vectors containing the major part of the EcoRI A fragment of EBV DNA express a nuclear antigen that reacts with certain anti-EBNA-positive human sera in ACIF tests. This antigen is not present in cells transfected with EcoRI A carrying vectors where the BamHI H fragment has been deleted.

MATERIAL AND METHODS

Cell Culture, DNA Transfection and Plasmids

NIH 3T3 mouse fibroblasts were obtained from Rudolf Jaenisch, Hamburg, West Germany. The cells were maintained in Dulbecco-modified Eagle medium (DMEM) supplemented with 10% fetal calf serum, penicillin and streptomycin.

Cells from subconfluent monolayers were transfected in suspension with the appropriate recombinant vector DNA (10^6 cells/10 µg DNA/ml) using the calcium phosphate-DNA precipitation technique of Graham and van der Eb (1973) as modified by Shen et al. (1982). After 3 days in DMEM containing 10% fetal calf serum, the cells were plated at a density of 2×10^5 cells per 9 cm Petri dish, and the medium was replaced by DMEM containing 10% fetal calf serum, xanthine (250µg/ml), mycophenolic acid (8 µg/ml), hypoxanthine (15 µg/ml), aminopterin (2 µg/ml), and thymidine (10 µg/ml). The medium containing these supplements was changed the next day and thereafter every 3-4 days. Mycophenolic acid resistant colonies were isolated with cloning cylinders after 14-21 days. The cloned cells were maintained in Iscove's modification of Dulbecco's medium (GIBCO) with the same supplements.

Recombinant plasmids containing the EcoRI or BamHI cleavage fragments of B95-8 EBV DNA were described previously (Arrand et al., 1981). The plasmids pSV2-gpt and pSV2-gpt/BglIIdel were obtained from Paul Berg, Stanford University, CA. pSV2-gpt/BglIIdel is a variant of the pSV2-gpt plasmid where the first 121 nucleotides of the gpt segment including the BglII cleavage site have been deleted (Mulligan and Berg, 1981). A cosmid derivative of pSV2-gpt, designated pSV2-gpt·cos2, was constructed as described (Rymo and Klein, submitted for publication).

Demonstration of EBV-associated Antigens

Early antigen (EA), viral capsid antigen (VCA), and Epstein-Barr virus-determined nuclear antigen (EBNA) were demonstrated on fixed cell smears as described earlier (Klein and Dombos, 1973; Reedman and Klein, 1973).

RESULTS

Construction of Transfection Vectors Containing EBV DNA Sequences

A library of cloned restriction enzyme fragments of EBV DNA covering the whole genome was established earlier (Arrand et al., 1981). Plasmids containing the EcoRI B and BamHI H, K and M fragments of B95-8 EBV DNA and the EcoRI D_{end} fragment of circular Raji EBV DNA, respectively, were digested with the appropriate restriction endonucleases, and the excised fragments were purified on agarose gels and recloned in the dominant-acting selection vectors pSV2-gpt or pSV2-gpt·cos2 by standard techniques. The EcoRI fragments B and D$_{end}$ containing vectors were designated pEB-gpt and pED$_{end}$-gpt(Raji) and the BamHI H, K and M containing vectors pBH-gpt, pBK-gpt and pBM-gpt, respectively.

We wanted to construct a series of deletions in the EcoRI A fragment involving sequences corresponding to the region which is deleted in virion DNA from the non-transforming P3HR-1 strain. Thus a plasmid carrying the EcoRI A fragment was partially digested with BamHI under conditions which resulted in cleavage of one BamHI site per molecule on the average. The plasmid DNA was then digested to completion with EcoRI and the resulting BamHI—EcoRI fragments were directionally cloned in the vector pSV2-gpt·cos2 between the EcoRI and the BamHI sites. Recombinant clones that hybridised to the appropriate BamHI fragments of EBV DNA and to SV40 DNA were characterised further by small-scale isolation of plasmid DNA and restriction enzyme analysis. Five clones designated pE△A1-gpt, pE△A2-gpt, pE△A3-gpt, pE△A4-gpt, and pE△A5-gpt were selected on the basis of their EBV DNA BamHI fragment composition and used for transfection experiments. The pE△A1-gpt clone contains the left part (on the conventional map) of the EcoRI A fragment up to and including the BamHI F fragment. The pE△A2-gpt clone has lost the BamHI F fragment, the pE△A3-gpt clone the BamHI F and H fragments, and the pE△A4-gpt clone the BamHI F, H and Y fragments as compared to the pE△A1 clone. The pE△A5-gpt clone consists of the right end of the EcoRI A fragment and lacks the BamHI C part.

Tranformation of NIH 3T3 Mouse Fibroblasts with EBV DNA-containing Selection Vectors

The recombinant vectors listed in Table I were introduced into 3T3 mouse fibroblasts by a modification (Shen et al., 1982) of the calcium phosphate technique of Graham and van der Eb (1973). Transformants were selected for growth in mycophenolic acid-containing medium also supplemented with xanthine, hypoxanthine, aminopterin, and thymidine. Clones were isolated with cloning cylinders. Small-scale preparations of cellular DNA from a large number of clones were analysed with dot hybridisation (Thomas, 1980) for the presence of EBV DNA sequences. Five representative EBV DNA-containing clones for each recombinant selection vector were chosen for analysis of the expression of EBV-related cellular antigens, except for pE△A3-gpt and pE△A5-gpt transfected cells where only two and three positive clones had been obtained, respectively.

Table I summarises the results. Cells transfected with the BamHI K fragment expressed a nuclear antigen as expected (Summers et al., 1982). We have also found that cells stably transformed with the recombinant vectors pE△A1-gpt, pE△A2-gpt, and pE△A5-gpt express a nuclear antigen detectable by certain anti-EBNA-positive human antisera in ACIF tests. The nuclear staining pattern of the transformed cells was diffuse and finely granular, similar to the EBNA pattern of conventionally stained EBV-transformed lymphoblastoid cell lines and mouse fibroblasts stably transfected with the BamHI K fragment. Between 10 and 70% of the cells expressed the nuclear antigen in the different EcoRI A transfected clones. The intensity of the nuclear fluorescence varied considerably between different cells within a certain clone.

The nuclear antigen could not be detected when an antiserum against a chemically synthesised 14 residue copolymer of glycine and alanine, the structure of which was deduced from the internal repeat sequence of the BamHI K fragment (Dillner et al., 1984), was used in the ACIF test. This antiserum identifies the BamHI K encoded nuclear antigen and gave a positive staining reaction with our BamHI K-transfected clones.

Cells transfected with recombinant vectors where EBV DNA sequences corresponding to the BamHI H fragment (pE△A3-gpt) or the BamHI H and Y fragments (pE△A4-gpt) had been deleted, did not contain antigen in the nucleus, nor did cells transfected with vectors carrying the BamHI H fragment alone. Cells stably transfected with the EcoRI D$_{end}$ fragment, which includes the third region of the EBV genome transcribed in transformed, non-virusproducing cells, did not contain nuclear antigens detectable with the ACIF test.

Table I. EBV-associated antigens in cells tranformed with recombinant selection vectors containing EBV DNA fragments

Recombinant vector	Antigen containing clones[a] number of positives/number tested		
	EA	VCA	EBNA
pE△A1-gpt (CWYHF)[b]	0/5	0/5	4/5
pE△A2-gpt (CWYH)[b]	0/5	0/5	5/5
pE△A3-gpt (CWY)[b]	0/2	0/2	0/2
pE△A4-gpt (CW)[b]	0/5	0/5	0/5
pE△A5-gpt (WYHFQU)[b]	0/3	0/3	1/3
pBH-gpt	NT	NT	0/5
pEB-gpt	0/5	0/5	0/5
pBK-gpt	0/5	0/5	4/5
pBM-gpt	NT	NT	0/5
pED$_{end}$-gpt (Raji)	0/5	0/5	0/5

EA   = early antigen
VCA = viral capsid antigen
EBNA = EBV-determined nuclear antigen
NT   = not tested

[a]At least two EBNA antibody positive and two EBV-antibody negative sera were tested against each clone. Positive reactions were only seen with anti-EBNA antibody positive, but not with EBV-antibody negative sera. Each clone was tested between two and five times in independent repeat tests.

[b]Letters within parenthesis denote the BamHI fragment sequences of EBV DNA present in the recombinant. The number of BamHI W repeats in the different vectors has not been determined.

Direct immunofluorescence with an anti-EA anti-VCA antibody positive FITC conjugated human IgG and indirect immunofluorescence with an anti-VCA+EA- human serum did not reveal the presence of antigens associated with productive infection in any of the EBV DNA-transfected fibroblast clones.

To demonstrate the presence of EBV DNA sequences in cells transfected with EcoRI A-carrying recombinant vectors that expressed the nuclear antigen, high molecular weight cellular DNA was prepared and cleaved with restriction endonucleases EcoRI and BamHI. The resulting DNA fragments were separated by electrophoresis in agarose gels, transferred to nitrocellulose and analysed by hybridisation. The results show that fragments corresponding to the BamHI W, Y and H fragments of B95-8 EBV DNA were present in the cells.

DISCUSSION

The major new finding of our study concerns the EBV-specific nuclear fluorescence detected by the ACIF reaction in cells which have received the major part of the EcoRI A fragment of EBV DNA. The nuclear fluorescence was not induced when the BamHI H fragment was deleted from the transducing vector. The BamHI C, F, Q, and U fragment regions of the EcoRI A fragment were not neccessary for the expression of the nuclear antigen and the BamHI W and Y region did not seem to be sufficient. Furthermore, the BamHI H fragment could not induce the nuclear antigen by itself. This implies that the coding sequence for the antigen is within the BamHI W, Y, and H fragment region.

The results are in line with the fact that a major transcript is generated from this area of the genome in EBV-transformed cells. A model for the generation of a cytoplasmic polyadenylated 3 kb mRNA has been provided by van Santen et al. (1983) from RNA hybridisation data. The primary transcript is synthesised in a left to right direction on a standard physical map. It is spliced from a large primary transcript and consists of small segments from the BamHI W repeat and larger, possibly continuous, exons encoded by the BamHI Y and H fragments, plus a polyadenylate tail.

The transcription model is supported by DNA sequence data identifying strong promotor sequences (CCAAT and TATAAA) and potential splice sites in BamHI W, long open reading frames in BamHI W, Y, and H, and an AATAAA polyadenylation signal in BamHI H (Cheung and Kieff, 1982; Jones and Griffin, 1983; Jones et al., 1984).

It is clear from the present study that the nuclear antigen induced by the EcoRI A fragment in mouse cells is different from the BamHI K encoded EBNA, since the induced antigen does not react with an antiserum raised against a chemically synthesised glycine-alanine peptide that has been shown to identify EBNA1 (Dillner et al., 1984). The EcoRI A induced antigen might be related to the other previously described EBNA subtype, EBNA2 (Strnad et al., 1982; Hennessy and Kieff, 1983), or it might represent still another EBNA species. However, the absence of a detectable 81-kdalton antigen in P3HR-1 cells (Strnad et al., 1981) is indicative of a relationship between the EcoRI A induced antigen and EBNA2. Furthermore, it has recently been demonstrated by immunoblotting techniques that monkey COS cells transfected with pEΔA2-gpt DNA express an EBNA polypeptide of a similar size as the EBNA2 polypeptide in Raji cells (Ricksten and Rymo, in preparation).

ACKNOWLEDGEMENT

We are grateful to Paul Berg for providing the pSV2-gpt vector, to Barbro Ehlin-Henriksson, Jane Löfvenmark, Ingeli Paul and Ann Christine Synnerholm for excellent technical assistance, and to Eli Lilly Sweden AB for providing mycophenolic acid. This work was supported by the Swedish Medical Research Council, project no 05667, and the PHS grant no 5 R01 CA 28380-03, awarded by the National Cancer Institute, DHHS.

REFERENCES

ARRAND,J.R., RYMO,L., WALSH,J.E., BJÖRK,E., LINDAHL,T., and GRIFFIN,B., Molecular cloning of the complete Epstein-Barr virus genome as a set of overlapping restriction endonuclease fragments, Nucleic Acids Res., 9, 2999-3014 (1981).

CHEUNG,A., and KIEFF,E., Long internal direct repeat in Epstein-Barr virus DNA. J. Virol., 44, 286-294 (1982).

DILLNER,J., STERNÅS,L., KALLIN,B., ALEXANDER,H., EHLIN-HENRIKSSON,B., JÖRNVALL,H., KLEIN,G., and LERNER,R., Antibodies against a synthetic peptide identify the Epstein-Barr virus-determined nuclear antigen. Proc. Natl. Acad. Sci., USA, 81, 4652-4656 (1984).

GRAHAM,F.L., and VAN DER EB,A.J., A new technique for the assay of infectivity of human adenovirus 5 DNA. Virology, 52, 456-467 1973).

HENNESSY, K., HELLER, M., VAN SANTEN, V., and KIEFF, E., Simple repeat array in Epstein-Barr virus DNA encodes part of the Epstein-Barr nuclear antigen. Science, 220, 1396-1398 (1983).

JONES,M.D., and GRIFFIN,B.E., Clustered repeat sequences in the genome of Epstein-Barr virus. Nucleic Acids Res., 11, 3919-3937 (1983).

JONES,M.D., FOSTER,L., SHEEDY,T., and GRIFFIN,B., The EB virus genome in Daudi Burkitt's lymphoma cells has a deletion similar to that observed in a non-transforming strain (P3HR-1) of the virus. EMBO J., 3, 813-821 (1984).

KLEIN,G., and DOMBOS,L., Relationship between the sensitivity of EBV-carrying lymphoblastoid lines to superinfection and the inducibility of the resident viral genome. Int. J. Cancer, 11, 327-337 (1973).

MULLIGAN,R.C., and BERG,P., Factors governing the expression of a bacterial gene in mammalian cells. Mol. Cell. Biol., 1, 449-459 (1981).

REEDMAN,B.M., and KLEIN,G., Cellular localization of an Epstein-Barr virus-associated complement-fixing antigen in producer and non-producer lymphoblastoid cell lines. Int. J. Cancer, 11, 599-620 (1973).

SHEN,Y.-M., HIRSCHHORN,R.R., MERCER,W.E., SURMACZ,E., TSUTSUI,Y., SOPRANO,K., and BASERGA,R., Gene transfer: DNA microinjection compared with DNA transfection with a very high efficiency. Mol. Cell. Biol., 2, 1145-1154 (1973).

STRNAD,B.C., SCHUSTER,T.C., HOPKINS III,R.F., NEUBAUER, R.H., and RABIN,H., Identification of an Epstein-Barr virus nuclear antigen by fluoroimmunoelectrophoresis and radioimmuno-electrophoresis. J. Virol., 38, 996-1004 (1981).

SUMMERS,W., GROGAN,E., SHEDD,D., ROBERT,M., LUI,C., and MILLER,G., Stable expression in mouse cells of nuclear neoantigen after transfer of a 3.4-megadalton cloned fragment of Epstein-Barr virus DNA. Proc. Natl. Acad. Sci. USA, 79, 5688-5692 (1982).

THOMAS,P., Hybridization of denatured RNA and small DNA fragments transferred to nitrocellulose. Proc. Natl. Acad. Sci USA, 77, 5201-5206 (1980).

VAN SANTEN,V., CHEUNG,A., and KIEFF,E., Epstein-Barr virus RNA VII. Size and direction of transcription of virus specified cytoplasmic RNA in a cell line transformed by EBV. Proc. Natl. Acad. Sci. USA, 78, 1930-1934 (1981).

VAN SANTEN,V., CHEUNG,A., HUMMEL,M., and KIEFF,E., RNA encoded by the IR1-U2 region of Epstein-Barr virus DNA latently infected, growth-transformed cells. J. Virol., 46, 424-433 (1983).

22

AN EBV RNA WITH A REPETITIVE SPLICED STRUCTURE

BODESCOT, M., CHAMBRAUD, B. and PERRICAUDET, M.

Institut de Recherches Scientifiques sur le

Cancer, C.N.R.S., 94800 VILLEJUIF, FRANCE

SUMMARY

We are studying the Epstein-Barr virus genes expressed in the Burkitt's lymphoma latently infected Raji cells. We describe here a cDNA representing a spliced RNA transcribed rightward from the IR1-U2 region. The cDNA contains several repeats of two exons, 66 and 132 bp, which are transcribed from the IR1 repeats, and four exons transcribed from U2. The longest open reading frame of the cDNA presumably corresponds to the carboxy-terminal 261 amino acids of a polypeptide containing several repeats of a 66 amino acid sequence. Since part of this coding region is deleted in the P3HR-1 non-immortalizing virus, this polypeptide might be involved in the process of growth-transformation of B-lymphocytes.

INTRODUCTION

The Epstein-Barr virus (EBV) genome is nearly 170×10^3 bp. Several clusters of repeated sequences, designated TR, IR1, IR2, IR3 and IR4 (Given et al., 1979 ; Kintner and Sugden, 1979 ; Cheung and Kieff, 1982 ; Dambaugh and Kieff, 1982 ; Heller et al., 1982a, 1982b ; Jones and Griffin, 1983) divide the genome into the five U1, U2, U3, U4 and U5 regions (Figure 1A). Studies about the P3HR-1 non-immortalizing virus suggest that the IR1-U2 region of the viral genome is implicated in immortalization of B-lymphocytes (Bornkamm et al., 1982 ; Hayward et al., 1982 ; King et al., 1982 ; Rabson et al., 1982 ; Jeang and Hayward, 1983 ; Stoerker and Glaser, 1983 ; Stoerker et

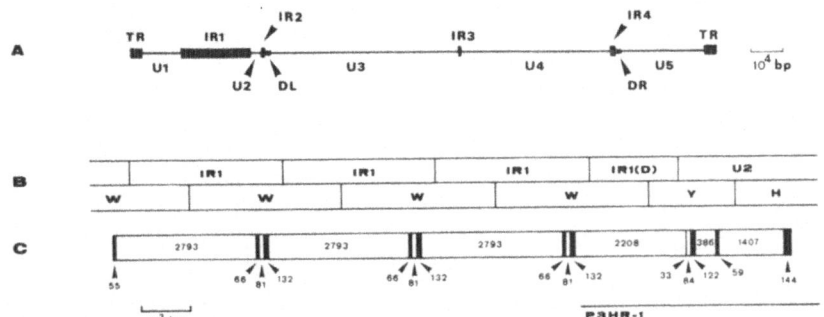

Figure 1 : (A) Structure of the EBV genome according to
Kieff et al. (1983). (B) An enlargement of the IR1-U2
region. A non-integral number of the IR1 repeat separates
U1 and U2. IR1(D) is the incomplete copy. The BamHI-W, Y,
and H fragments are indicated. (C) Structure of the exons
and introns which are described by the cDNA. The sizes are
indicated in bp. A part of the deleted region of the non-
immortalizing P3HR-1 virus is shown. (Reprinted with
permission).

al., 1983). This region is transcribed into polyadenylated
RNAs in latently infected cells (Rymo, 1979 ; Thomas-Powell
et al., 1979 ; King et al., 1980, 1981 ; Van Santen et al.,
1981, 1983 ; Arrand and Rymo, 1982 ; Heller et al., 1982b ;
Weigel and Miller, 1983). We are using cDNA cloning and
sequencing to elucidate the structural organization of the
RNAs and proteins which are expressed from this region in
the Burkitt's lymphoma latently infected Raji cells.

MATERIAL AND METHODS
Cell culture
Raji cells were grown in RPMI 1640 medium
supplemented with 10 % fetal calf serum (Gibco
laboratories).
Preparation of cytoplasmic RNA
Cytoplasmic RNA was extracted according to
Brawerman et al. (1972). Cells were lysed in 10 mM Tris-HCl
(pH 7.8), 150 mM NaCl, 5 mM $MgCl_2$ containing 0.6 %
Nonidet-P40. The cytoplasmic fraction was phenol extracted
in the presence of 0.5 % SDS and 1 mM EDTA, and RNA was
ethanol precipitated.

cDNA cloning
        cDNA was cloned according to Rougeon and Mach
(1976), and to Perricaudet et al. (1979). First strand cDNA
was synthesized with avian myeloblastosis virus reverse
transcriptase from cytoplasmic RNA using oligo(dT) as a
primer. After alkaline hydrolysis of the RNA, self priming
of the cDNA enabled synthesis of the opposite strand. The
hairpin loop was digested with the S1 nuclease. The
double-stranded cDNA was inserted into the PstI restriction
site of the pBR322 plasmid after dG/dC tailing with
terminal transferase. E.coli strain HB101 was transformed
according to Mandel and Higa (1970).
In situ hybridization
        The BamHI-W, Y, and H fragments of the B95-8
viral genome were prepared from the plasmids described by
Dambaugh et al. (1980), and nick-translated according to
Rigby et al. (1977). In situ hybridization was performed
according to Cami and Kourilsky (1978).
DNA sequencing
        The cDNA was sequenced according to Sanger et al.
(1977 and 1980). The mp8 and mp9 derivatives of the M13
phage (Messing and Viera, 1982) were used as vectors.

                        RESULTS
        cDNA copies of Raji cytoplasmic RNAs were
inserted into the PstI restriction site of the pBR322
plasmid. About 30,000 colonies were screened by in situ
hybridization. One clone hybridized to the three BamHI-W,
Y, and H fragments of the B95-8 viral genome (Figure 1B).
The PstI insert is nearly 1 kb. The locations of three SmaI
and three HinfI restriction sites suggested the presence of
a repeated sequence (Figure 2A). The sequencing strategy
used for the PstI insert is shown in Figure 2B, and the
sequence is given in Figure 3. Comparison with the
nucleotide sequence of the BamHI-W (Cheung and Kieff,
1982 ; Jones and Griffin, 1983), BamHI-Y (Baer et al.,
1984), and BamHI-H (Baer et al., 1984 ; Jones et al., 1984)
fragments showed that the PstI insert describes a spliced
RNA transcribed rightward from the IR1-U2 region (Figures
1C and 2C). Two exons, 66 and 132 nucleotides, start at
nucleotides 1340 and 1487 of the BamHI-W fragment. These
exons constitute a 198 nucleotide unit which is tandemly
repeated three times. These repeats start at nucleotides
56, 254, and 452 of the cDNA. The 5' end of the cDNA is
made up of an incomplete copy of the unit. Three exons, 33,
122, and 59 nucleotides, are transcribed from the BamHI-Y

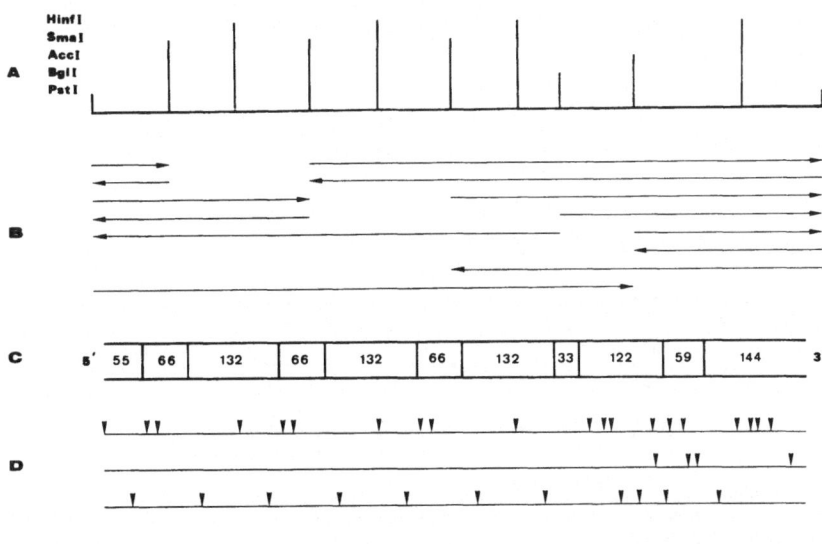

Figure 2 : (A) Restriction map of the PstI insert. (B) Sequencing strategy. The arrows indicate the restriction fragments which where cloned into the M13 mp8 and mp9 derivatives, and the direction of synthesis from primer. (C) The different exons described by the cDNA. The sizes are indicated in bp. (D) Map of the stop codons in the three reading frames.

fragment. These exons start at nucleotides 755, 872, and 1380 of the BamHI-Y fragment, and at nucleotides 650, 683, and 805 of the cDNA. One exon, at least 144 nucleotides, is transcribed from the BamHI-H fragment. This exon starts at nucleotide 1005 of the BamHI-H fragment, and at nucleotide 864 of the cDNA. The sizes of the introns, from the 5' to the 3' end, are 2793, 81, 2793, 81, 2793, 81, 2208, 84, 386, and 1407 nucleotides relative to the B95-8 viral genome (Figure 1C). The junctions between exons and introns follow the GT-AG rule (Breathnach and Chambon, 1981 ; Mount, 1982). The longest open reading frame extends from the 5' end to nucleotide 784 (Figure 2D and 3), and presumably corresponds to the carboxy-terminal 261 amino acids of a polypeptide. This would be rich in proline, arginine, and glycine, and would contain several repeats of a 66 amino acid sequence.

        Northern blot experiments were performed with Raji cytoplasmic polyadenylated RNA. A single-stranded

```
            10        20        30        40        50    ▼  60        70
5'(G)₁₆ TAGAAGGGTCCTCGTCCAGCAAGAAGAGGAGGTGGTAAGCGGTTCACCTTCAGGGCCTAGGGGAGACCGAAGT
        R  R  V  L  V  Q  Q  E  E  E  V  V  S  G  S  P  S  G  P  R  G  D  R  S

        80        90       100       110       120 ▼     130       140       150
GAAGGCCCTGGACCAACCCGGCCCGGCCCCCCGGTATCGGGCCAGAGGGTCCCCTCGGACAGCTCCTAAGAAGGCACCGG
E  G  P  G  P  T  R  P  G  P  P  G  I  G  P  E  G  P  L  G  Q  L  L  R  R  H  R

       160       170       180       190       200       210       220       230
TCGCCCAGTCCTACCAGAGGGGGCCAAGAACCCAGACGAGTCCGTAGAAGGGTCCTCGTCCAGCAAGAAGAGGAGGTGGTA
S  P  S  P  T  R  G  G  Q  E  P  R  R  V  R  R  R  V  L  V  Q  Q  E  E  E  V  V

       240       250 ▼    260       270       280       290       300       310
AGCGGTTCACCTTCAGGGCCTAGGGGAGACCGAAGTGAAGGCCCTGGACCAACCCGGCCCGGGCCCCCCGGTATCGGGCCA
S  G  S  P  S  G  P  R  G  D  R  S  E  G  P  G  P  T  R  P  G  P  P  G  I  G  P

      ▼      330       340       350       360       370       380       390
GAGGGTCCCCTCGGACAGCTCCTAAGAAGGCACCGGTCGCCCAGTCCTACCAGAGGGGGCCAAGAACCCAGACGAGTCCGT
E  G  P  L  G  Q  L  L  R  R  H  R  S  P  S  P  T  R  G  G  Q  E  P  R  R  V  R

      400       410       420       430       440       450 ▼     460       470
AGAAGGGTCCTCGTCCAGCAAGAAGAGGAGGTGGTAAGCGGTTCACCTTCAGGGCCTAGGGGAGACCGAAGTGAAGGCCCT
R  R  V  L  V  Q  Q  E  E  E  V  V  S  G  S  P  S  G  P  R  G  D  R  S  E  G  P

      490       500       510        ▼      530       540       550
GGACCAACCCGGCCCGGGCCCCCCGGTATCGGGCCAGAGGGTCCCCTCGGACAGCTCCTAAGAAGGCACCGGTCGCCCAGT
G  P  T  R  P  G  P  P  G  I  G  P  E  G  P  L  G  Q  L  L  R  R  H  R  S  P  S

      570       580       590       600       610       620       630       640
CCTACCAGAGGGGGCCAAGAACCCAGACGAGTCCGTAGAAGGGTCCTCGTCCAGCAAGAAGAGGAGGTGGTAAGCGGTTCA
P  T  R  G  G  Q  E  P  R  R  V  R  R  R  V  L  V  Q  Q  E  E  E  V  V  S  G  S

         ▼     660       670       680 ▼    690       700       710       720
CCTTCAGGGCCACTACGGCCACGTCCCCGGCCTCCAGCTCGGTCTCTTAGAGAGTGGCTGCTACGCATTAGAGACCACTTT
P  S  G  P  L  R  P  R  P  R  P  P  A  R  S  L  R  E  W  L  L  R  I  R  D  H  F

      730       740       750       760       770       780       790       800
GAGCCACCCACAGTAACCACCCAGCGCCAATCTGTCTACATAGAAGAAGAAGAGGATGAAGACTAAGTCACAGGCTTAGCC
E  P  P  T  V  T  T  Q  R  Q  S  V  Y  I  E  E  E  E  D  E  D

    ▼   810       820       830       840       850       860    ▼ 870       880
AGTAACCCAGCACTGGCGTGTGACGTGGTGTAAAGTTTTGCCTGAACCTGTGGTTGGGCAGACCCCGCAGACTTAGACGAA

      890       900       910       920       930       940       950       960
AGTTGGGATTACATTTTTGAGACAACAGAATCTCCTAGCTCAGATGAAGATTATGTGGAGGGACCCAGTAAAAGACCTCGC

      970       980       990      1000
CCCTCCATCCAGTAAAAACCCTTGCCCTCTCCAGCAAACAATG (C)₁₇ 3'
```

Figure 3 : Nucleotide sequence of the PstI insert and amino acid sequence corresponding to the longest open reading frame. The junctions between exons and introns are indicated by arrows.

probe was synthesized from an M13 recombinant containing the largest SmaI-PstI fragment of the cDNA. This probe contained the entire cDNA sequence and was complementary to the RNA corresponding to the cDNA. Four RNA species were detected. The approximate sizes was 1.5, 3.0, 3.2 and 4.2 kb. The 3.0 and 3.2 kb species appear to be the most abundant.

DISCUSSION

A cDNA library has been constructed from the cytoplasmic RNAs of the Burkitt's lymphoma latently infected Raji cells. We report here the characterization of a cDNA which describes a spliced RNA transcribed rightward from the IR1-U2 region of the viral genome. The cDNA contains several tandem repeats of two exons, 66 and 132 bp, transcribed from IR1, and four exons transcribed from U2.

As no ATG is found in the longest open reading frame, nor a poly(A) sequence at the 3' end, the cDNA is an incomplete copy of a larger RNA. An AATAAA sequence is located on the B95-8 viral genome 8 bp downstream from the 3' end of the cDNA (Baer et al., 1984 ; Jones et al., 1984). Thus, at the 3' end, the RNA should contain some 30-40 additional nucleotides and a poly(A) tail (Proudfoot and Brownlee, 1976 ; Le Moullec et al., 1983). The location of the 5' end of the RNA might be inferred from the Northern blot analysis. In Raji cells, the viral genomes contain eight copies of the IR1 repeat (Polack et al., 1984). If the eight copies are transcribed and spliced as described here, a piece of RNA of 2.1 kb would be generated. Thus if we assume that the cDNA corresponds to one of the three largest RNAs, the promoter and additional exons should be located within U1. Alternatively, the 5' end might be located within the IR1 region. Each IR1 repeat of the B95-8 viral genome contains a sequence which promotes transcription in a Hela cell extract (Van Santen et al., 1983). In this hypothesis, translation could be initiated only if the 5' end of the precursor were processed in a different way than described here.

A part of this RNA is transcribed from a region which is deleted in the non-immortalizing P3HR-1 virus (Figure 1C). Marker rescue experiments have shown that this region is implicated in the immortalization process (Stoerker and Glaser, 1983 ; Stoerker et al., 1983). Thus the polypeptide described here might play a role in immortalization.

ACKNOWLEDGEMENTS

We are indebted to Dr. B. Barrell for providing us with the nucleotide sequence of the B95-8 BamHI-Y and H fragments before publication. We thank Dr. E. Kieff for the gift of the plasmids containing the B95-8 BamHI-W, Y, and H fragments. This investigation was supported by grants from the "Fondation pour la Recherche Médicale Française" and from the "Centre National de la Recherche Scientifique".

REFERENCES

ARRAND, J., and RYMO, L., Characterization of the major Epstein-Barr virus-specific RNA in Burkitt lymphoma-derived cells. J. Virol., 41, 376-389 (1982).

BAER, R., BANKIER, A., BIGGIN, M., DEININGER, P., FARRELL, P., GIBSON, T., HATFULL, G., HUDSON, G., SATCHWELL, S., SEGUIN, C., TUFFNELL, P., and BARRELL, B., DNA sequence and expression of the B95-8 Epstein-Barr virus genome. Nature, 310, 207-211 (1984).

BORNKAMM, G., HUDEWENTZ, J., FREESE, U., and ZIMBER, U., Deletion of the nontransforming Epstein-Barr virus strain P3HR-1 causes fusion of the large internal repeat to the DS_L region. J. Virol., 43, 952-968 (1982).

BRAWERMAN, G., MENDECKI, J., and LEE, S., A procedure for the isolation of mammalian messenger ribonucleic acid. Biochemistry, 11, 637-641 (1972).

BREATHNACH, R., and CHAMBON, P., Organization and expression of eucaryotic split genes coding for proteins. Ann. Rev. Biochem., 50, 349-383 (1981).

CAMI, B., and KOURILSKY, P., Screening of cloned recombinant DNA in bacteria by in situ colony hybridization. Nucleic Acids Res., 5, 2381-2390 (1978).

CHEUNG, A., and KIEFF, E., Long internal direct repeat in Epstein-Barr virus DNA. J. Virol., 44, 286-294 (1982).

DAMBAUGH, T., BEISEL, C., HUMMEL, M., KING, W., FENNEWALD, S., CHEUNG, A., HELLER, M., RAAB-TRAUB, N., and KIEFF, E., Epstein-Barr virus DNA VII : molecular cloning and detailed mapping. Proc. Natl. Acad. Sci. USA, 77, 2999-3003 (1980).

DAMBAUGH, T., and KIEFF, E., Identification and nucleotide sequences of two similar tandem direct repeats in Epstein-Barr virus DNA. J. Virol., 44, 823-833 (1982).

GIVEN, D., YEE, D., GRIEM, K., and KIEFF, E., DNA of Epstein-Barr virus V : direct repeats of the ends of Epstein-Barr virus DNA. J. Virol., 30, 852-862 (1979).

263

HAYWARD, S., LAZAROWITZ, S., and HAYWARD, G., Organization of the Epstein-Barr virus DNA molecule II : fine mapping of the boundaries of the internal repeat cluster of B95-8 and identification of additional small tandem repeats adjacent to the HR-1 deletion. J. Virol., 43, 201-212 (1982).

HELLER, M., HENDERSON, A., and KIEFF, E., Repeat array in Epstein-Barr virus DNA is related to cell DNA sequences interspersed on human chromosomes. Proc. Acad. Sci. USA, 79, 5916-5920 (1982 a).

HELLER, M., VAN SANTEN, V., and KIEFF, E., Simple repeat sequence in Epstein-Barr virus DNA is transcribed in latent and productive infections. J. Virol., 44, 311-320 (1982 b).

JEANG, K., and HAYWARD, S., Organization of the Epstein-Barr virus DNA molecule III : location of the P3HR-1 deletion junction and characterization of the Not I repeat units that form part of the template for an abundant 12-O-tetradecanoylphorbol-13-acetate-induced mRNA transcript. J. Virol., 48, 135-148 (1983).

JONES, M., FOSTER, L., SHEEDY, T., and GRIFFIN, B., The EB virus genome in Daudi Burkitt's lymphoma cells has a deletion similar to that observed in a non-transforming strain (P3HR-1) of the virus. The EMBO J., 3, 813-821 (1984).

JONES, M., and GRIFFIN, B., Clustered repeat sequences in the genome of Epstein-Barr virus. Nucleic Acids Res., 11, 3919-3936 (1983).

KIEFF, E., DAMBAUGH, T., HUMMEL, M., and HELLER, M., Epstein-Barr virus transformation and replication. In : Klein G. (ed.), Advances in viral oncology, pp. 133-182, Raven Press, New York (1983).

KING, W., DAMBAUGH, T., HELLER, M., DOWLING, J., and KIEFF, E., Epstein-Barr virus DNA XII : a variable region of the Epstein-Barr virus genome is included in the P3HR-1 deletion. J. Virol., 43, 979-986 (1982).

KING, W., THOMAS-POWELL, A., RAAB-TRAUB, N., HAWKE, M., and KIEFF, E., Epstein-Barr virus RNA V : viral RNA in a restringently infected growth-transformed cell line. J. Virol., 36, 506-518 (1980).

KING, W., VAN SANTEN, V., and KIEFF, E., Epstein-Barr virus RNA VI : viral RNA in restringently and abortively infected Raji cells. J. Virol., 38, 649-660 (1981).

KINTNER, C., and SUGDEN, B., The structure of the termini of the DNA of Epstein-Barr virus. Cell, 17, 661-671 (1979).

LE MOULLEC, J., AKUSJARVI, G., STALHANDSKE, P.,PETTERSSON, U., CHAMBRAUD, B., GILARDI, P., NASRI, M., and PERRICAUDET, M., Polyadenylic acid addition sites in the adenovirus type 2 major late transcription unit. J. Virol., 48, 127-134 (1983).

MANDEL, M., and HIGA, A., Calcium-dependent bacteriophage DNA infection. J. Mol. Biol., 53, 159-162 (1970).

MESSING, J., and VIEIRA, J., A new pair of M13 vectors for selecting either DNA strand of double digest restriction fragments. Gene, 19, 269-276 (1982).

MOUNT, S., A catalogue of splice junction sequences. Nucleic Acids Res., 10, 459-472 (1982).

PERRICAUDET, M., AKUSJARVI, G., VIRTANEN, A., and PETTERSSON, U., Structure of two spliced mRNAs from the transforming region of human subgroup C adenoviruses. Nature, 281, 694-696 (1979).

POLACK, A., DELIUS, H., ZIMBER, U., and BORNKAMM, G., Two deletions in the Epstein-Barr virus genome of the Burkitt lymphoma nonproducer line Raji. Virology, 133, 146-157 (1984).

PROUDFOOT, N., and BROWNLEE, G., 3' non-coding region sequences in eucaryotic messenger RNA. Nature, 263, 211-214 (1976).

RABSON, M., GRADOVILLE, L., HESTON, L., and MILLER, G., Non-immortalizing P3J-HR-1 Epstein-Barr virus : a deletion mutant of its transforming parent, Jijoye. J. Virol., 44, 834-844 (1982).

RIGBY, P., DIECKMANN, M., RHODES, C., and BERG, P., Labelling deoxyribonucleic acid to high specific activity in vitro by nick-translation with DNA polymerase I. J. Mol. Biol., 113, 237-251 (1977).

ROUGEON, F., and MACH, B., Stepwise bio-synthesis in vitro of globin genes from globin mRNA by DNA polymerase of avian myeloblastosis virus. Proc. Natl. Acad. Sci. USA, 73, 3418-3422 (1976).

RYMO, L., Identification of transcribed regions of Epstein-Barr virus DNA in Burkitt lymphoma-derived cells. J. Virol., 32, 8-18 (1979).

SANGER, F., COULSON, A., BARRELL, B., SMITH, A., and ROE, B., Cloning in single-stranded bacteriophage as an aid to rapid DNA sequencing. J. Mol. Biol., 143, 161-178 (1980).

SANGER, F., NICKLEN, S., and COULSON, A., DNA sequencing with chain-terminating inhibitors. Proc. Natl. Acad. Sci. USA, 74, 5463-5467 (1977).

STOERKER, J., and GLASER, R., Rescue of transforming Epstein-Barr virus (EBV) from EBV genome-positive epithelial hybrid cells transfected with subgenomic fragments of EBV DNA. Proc. Natl. Acad. Sci. USA, 80, 1726-1729 (1983).

STOERKER, J., HOLLIDAY, J., and GLASER, R., Identification of a region of the Epstein-Barr virus (B95-8) genome required for transformation. Virology, 129, 199-206 (1983).

THOMAS-POWELL, A., KING, W., and KIEFF, E., Epstein-Barr virus-specific RNA III : mapping of DNA encoding viral RNA in restringent infection. J. Virol., 29, 261-274 (1979).

VAN SANTEN, V., CHEUNG, A., HUMMEL, M., and KIEFF, E., RNA encoded by the 1R1-U2 region of Epstein-Barr virus DNA in latently infected growth-transformed cells. J. Virol., 46, 424-433 (1983).

VAN SANTEN, V., CHEUNG, A., and KIEFF, E., Epstein-Barr
virus RNA VII : size and direction of transcription of
virus-specified cytoplasmic RNAs in a transformed cell
line. Proc. Natl. Acad. Sci. USA, <u>78</u>, 1930-1934 (1981).

WEIGEL, R., and MILLER, G., Major EB virus-specific
cytoplasmic transcripts in a cellular clone of the HR-1
Burkitt lymphoma line during latency and after induction
of viral replicative cycle by phorbol esters. Virology,
<u>125</u>, 287-298 (1983).

23

CHARACTERIZATION OF AN EBV-ASSOCIATED PROTEIN KINASE

Elizabeth Fowler

Dept. of Microbiology and Immunology
University of South Alabama
Mobile, Alabama 36688

SUMMARY

Biologically active Epstein-Barr virions have been isolated in a two-step procedure using affinity chromatography on ricin-agglutinin Sepharose followed by sedimentation in non-ionic density gradients of Nycodenz. A protein kinase activity co-purifies with the virions through both steps. Nucleocapsids prepared by extraction of purified virions with 2% NP-40, 0.05% Na deoxycholate, 1M urea retain more than 85% of the activity present in virions indicating that the kinase is an intrinsic component of the nucleocapsid. SDS gel electrophoresis of autophosphorylated virions and nucleocapsids reveals that the same proteins are phosphorylated in both preparations. The primary phosphorylated proteins have molecular weights of 100,000, 76,000, 52,000 amd 46,000 daltons. Analysis of the phosphoamino acids shows a high level of phosphoserine with a trace of phosphothreonine; no tyrosine phosphorylation is detectable. These data indicate that Epstein-Barr virus, like other herpes viruses, has an endogeneous protein kinase activity contained in its nucleocapsid.

INTRODUCTION

Several herpes viruses have been reported to possess protein kinase activity (Lemaster and Roizman, 1980; Randall et al., 1972; Rubenstein et al., 1972; Mar et al., 1981) which is contained in the nucleocapsid and phosphorylates virion structural proteins. We have recently developed a two step purification for Epstein-Barr virus (EBV) which yields intact virions that are biologically active in four different assays including transformation of B lymphocytes (Fowler et al., 1984). The extent of contamination of the purified virions with cellular proteins is less than 3%. In order to determine whether EBV is similar to other herpes viruses in having an endogenous kinase activity, we have measured kinase activity during the purification of virions and nucleocapsids. We find that protein kinase activity cosediments with EB virions and nucleocapsids in Nycodenz gradients. This activity phosphorylates the same proteins in both nucleocapsids and virions. Serine is the principal phosphorylated amino acid.

MATERIALS AND METHODS

EBV was obtained from cell-free supernatants of cultures of the EBV-producing B95-8 cell line (Miller et al., 1972) by centrifugation at 17,000 x g for 90 min. The pellet from 3 l of culture fluid was suspended in 8 ml 0.15M NaCl, 3.4mM KCl, 9.4mM NaHPO$_4$, 1.7mM KH$_2$PO$_4$, 1mM CaCl$_2$, 0.5mM MgCl$_2$, 100 µg/ml bacitracin, pH 7.4. This preparation is referred to as crude virus. Preparations were routinely labeled by culturing cells in the presence of 2 µC/ml ^{3}H-thymidine. CsCl density gradient centrifugation showed that 95% of the ^{3}H-DNA in the virus suspension had the expected viral density. Virions were purified by affinity chromatography on ricin-agglutinin Sepharose and non-ionic density gradient centrifugation using a 15 - 35% gradient of Nycodenz (Accurate Chemical Company) as described previously (Fowler et al., 1984). Nucleocapsids were prepared by treatment of purified virions with 2% NP-40, 0.5% Na deoxycholate, 1M urea for 1h at 4°C followed by pelleting through 30% glycerol, 0.05M Tris HCl, pH 8.0 or banding in a 30 - 50% Nycodenz gradient.

The number of EBV genomes in each preparation was determined by dot hybridization analysis using a ^{32}P-labeled BamH1 W fragment (Dambaugh et al., 1980). The intensity of each spot was measured by densitometry and compared to a standard curve constructed using known amounts of unlabeled BamH1 W fragment. EBV was assayed for its capacity to bind to Raji cells using the indirect immunofluorescent assay described by Simmons et al. (1983).

Protein kinase activity was assayed using a standard reaction mixture containing 0.05M Tris HCl, pH 8.0, 0.05M MgCl$_2$, 1mM dithiothreitol, 1% NP-40, [γ-^{32}P]ATP (10 uC/reaction, specific activity adjusted to 1 C/mole). Reactions were incubated at 37° C for 1h and terminated by the addition of 1/2 volume 0.2M EDTA, 1mM Na vanadate, 50mM NaF, 2 mg/ml lysozyme. Quadruplicate aliquots were precipitated with 10% TCA saturated with Na pyrophosphate. Precipitates were collected on 0.8 μ nitrocellulose filters and measured by liquid scintillation counting using Betafluor (National Diagnostics, Somerville, NJ).

Viral proteins were analyzed by SDS gel electrophoresis on 12.5% polyacrylamide gels (Laemmli, 1970). Gels were stained with silver (Wray et al., 1981), dried and autoradiographed using Kodak X-Omat AR film. The molecular weights of the viral bands were estimated from a standard curve constructed with proteins of known molecular weights.

For phosphoamino acid analysis virions were phosphorylated under standard conditions. The TCA precipitable material was subjected to partial acid hydrolysis in 6N HCl for 2h at 110°C. The hydrolysates were applied to thin layer cellulose plates along with unlabeled standards of phosphoserine, phosphothreonine and phosphotyrosine and subjected to two dimensional electrophoresis at pH 1.9 and pH 3.5 (Avruch et al., 1982). The plates were stained with 1% ninhydrin in acetone to reveal the standards and subjected to autoradiography using Kodak X-Omat AR film.

RESULTS AND DISCUSSION

EBV was subjected to affinity chromatography on ricin-agglutinin Sepharose followed by sedimentation in a 15 - 35% gradient of Nycodenz. Under these conditions a protein kinase activity co-purified with the EB virions. As shown in Figure 1, a peak of kinase activity co-sedimented in Nycodenz with the virions detected by dot hybridization and by the immunofluorescent binding assay. Table 1 shows that 70% of the kinase activity in the starting viral suspension was retained in the Nycodenz purified material. Much of the activity that was removed during purification was apparently non-viral since SDS gel electrophoresis of the phosphorylated proteins demonstrated significant phosphorylation of non-viral material in the crude preparations (data not shown). The kinase activity was precipitated by a monoclonal antibody to membrane antigen. The activity required magnesium but was not enhanced by cAMP. Furthermore, the activity in purified virions phosphorylated virion proteins and was not significantly enhanced by addition of exogenous substrates. From these data we conclude that EB virions contain a cyclic AMP independent protein kinase activity which primarily phosphorylates endogenous virion proteins.

Table 1

Protein Kinase Activity of EBV Preparations

	^{32}P incorporation[a] EBV genome	% Activity
1. Crude virus	159	100
2. Ricin eluate	118	74
3. Nycodenz purified virions	112	70
4. Nucleocapsids	98	62

[a] Standard reactions were performed as described in methods. The amount of ^{32}P incorporated was corrected for the number of EBV genomes determined by dot hybridization.

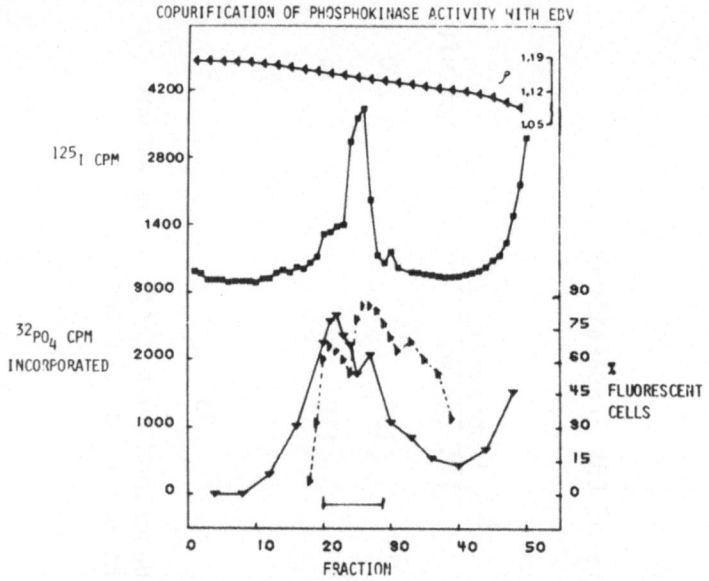

DOT HYBRIDIZATION ACROSS NYCODENZ GRADIENT

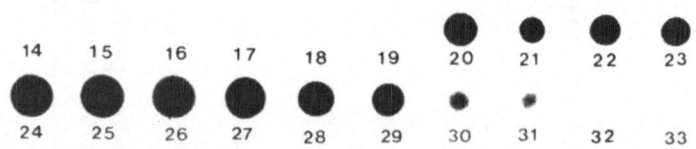

Figure 1. Copurification of protein kinase activity with EBV. ^{125}I-EBV (0.5 ml) was layered on a 15-35% Nycodenz gradient and centrifuged for 2h at 102,000 x g in a Ty65 rotor. Fractions (0.2 ml) were collected and the TCA precipitable counts (—■—) were measured in a gamma spectrometer. Aliquots were taken for dot hybridization and assayed for the ability to bind to Raji cells (—▶—). Protein kinase activity (—▼—) was measured on aliquots of a parallel gradient of unlabeled EBV. Densities were calculated from the refractive indices (—◀—). Lower panel: Dot hybridization of 10 μl aliquots of each fraction to a ^{32}P BamH1 W probe. The bound radioactivity was detected by autoradiography. All fractions from the gradient were tested; only fractions 20-31 showed detectable hybridization.

The kinase activity was greatly enhanced in the presence of non-ionic detergents consistent with the hypothesis that the activity resides in the nucleocapsid. To test this hypothesis, nucleocapsids were prepared from purified virions by extraction with 2% NP-40, 0.05% Na deoxycholate, 1M urea. The extracted material banded in Nycodenz at a density of 1.24 g/cc. Electron microscopy of the material recovered from the gradient revealed typical nucleocapsid morphology with little adherent tegument material. The purified nucleocapsids retained 88% of the kinase activity present in the Nycodenz purified virions (Table 1). This result indicates that the EBV-associated kinase activity, like the kinase activity in herpes simplex, is tightly associated with the nucleocapsid and is an integral viral protein rather than an adsorbed cellular contaminant.

The polypeptides which were phosphorylated by the EBV associated kinase were examined by SDS gel electrophoresis followed by autoradiography. Four phosphorylated polypeptides were consistently observed in both the purified virions and the nucleocapsids (Figure 2). The 46,000, 52,000 and 76,000 dalton polypeptides are prominent components of ^{125}I-labeled virions and correspond to nucleocapsid proteins identified by others (Dolyniuk et al., 1976) in ^{35}S-methionine labeled preparations. The 100,000 dalton polypeptide was undetectable in the ^{125}I-labeled preparations although a protein of this size has been visualized in ^{35}S-labeled virions, but not nucleocapsids. The phosphorylated 100,000 dalton polypeptide was present at equal intensity in the virion and nucleocapsid preparations. This phosphoprotein is unlikely to be a contaminating envelope protein since the density of the nucleocapsids and their appearance in the electron microscope indicated complete removal of the envelope. Furthermore analysis of the polypeptides by SDS gel electrophoresis on 7.5% gels showed that membrane antigen was undetectable in the nucleocapsid preparation supporting the contention that the envelope had been removed. If the phosphorylated 100,000 dalton polypeptide present in the nucleocapsid preparation were a residual envelope protein, it should be present with significantly less intensity compared to the level observed in virions. Since this was not the case, we conclude that the 100,000 dalton peptide is a minor component of nucleocapsids which was not detectable in ^{35}S- or ^{125}I-labeled preparations.

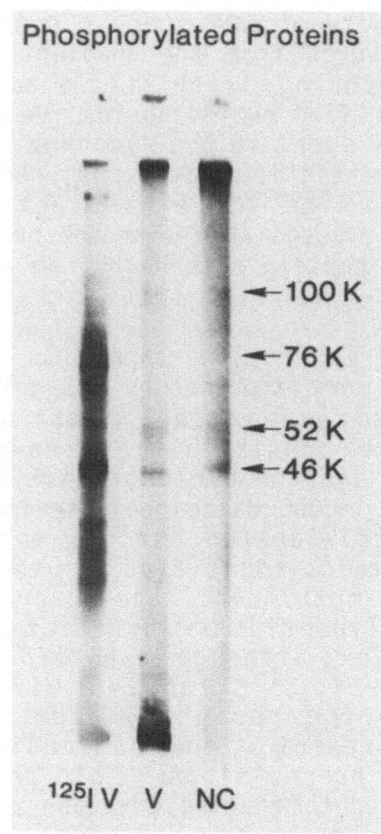

Figure 2. SDS gel electrophoretic analysis of proteins phosphorylated in virions and nucleocapsids. Virions and nucleocapsids were isolated from Nycodenz gradients and phosphorylated in the standard reaction. A second sample of purified virions was labeled with ^{125}I. Proteins were separated by SDS-PAGE on 12.5% gels using the Laemmli buffer system and visualized by autoradiography of the dried gel. The molecular weights were calculated using a standard curve constructed with proteins of known molecular weight run on the same gel. Left: ^{125}I-virions; center: phosphorylated virions; right: phosphorylated nucleocapsids.

In order to determine which amino acids were phosphorylated, virions were incubated in the standard reaction and the TCA precipitable material was subjected to partial acid hydrolysis. The phosphoamino acids were then separated by two dimensional electrophoresis. As shown in Figure 3, serine is the predominant phosphorylated species. A trace of phosphothreonine is visible but there is no evidence for tyrosine phosphorylation.

The experiments described above demonstrate that EBV has a intrinsic protein kinase activity. This activity phosphorylates the same four polypeptides in nucleocapsids and enveloped virions; this phosphorylation primarily involves serine residues. The EBV kinase activity is similar to the activity found in herpes simplex in that it is intimately associated with the nucleocapsids, it primarily phosphorylates endogeneous proteins and it is independent of cyclic AMP but dependent on magnesium.

ACKNOWLEDGMENTS

We greatly appreciate the many contributions of Scott Hall to this work and thank Dr. Sharish Shenolikar for helpful discussions. This work was supported by Public Health Service grant CA36776 awarded by the National Cancer Institute.

Phospho Amino Acid Analysis

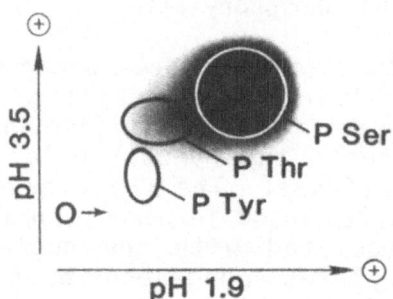

Figure 3. Phosphoamino acid analysis. TCA precipitable
material in phosphorylated EB virions was hydrolyzed for 2h
at 110°C. The hydrolysate was applied to a cellose plate
for two dimensional electrophoresis at pH 1.9 and pH 3.5.
Labeled phosphoamino acids were detected by
autoradiography. Unlabeled standards were run on the same
plate and visualized by ninhydrin staining. The positions
of the standards are denoted by the circles superimposed on
the autoradiograph.

REFERENCES

Avruch, J., Nemenoff, R.A., Blackshear, P.J., Pierce, M.W.,
and Osathanondh, R., Insulin-stimulated tyrosine
phosphorylation of the insulin receptor in detergent
extracts of human placental membranes. J. Biol.
Chem. **257**, 15162-15166 (1982).
Dambaugh,T., Beisel, C., Hummel, M., King,W., Fennewald,
S., Cheung, A., Heller, M., Raab-Traub, N. and Kieff, E.,
Epstein-Barr virus DNA. VII. Molecular cloning and
detailed mapping of EBV (B95-8) DNA, Proc. Natl. Acad.
Sci. USA **77**, 2999-3003 (1980).

277

Dolyniuk, M., Wolff, E., and Kieff, E., Proteins of Epstein- Barr virus. II. Electrophoretic analysis of the polypeptides of the nucleocapsid and the glucosamine- and polysaccharide-containing components of enveloped virus, J. Virol. **18**, 289-297 (1976).
Fowler, E., Raab-Traub, N., and Hester, S., Purification of biologically active Epstein-Barr virus by affinity chromatography and non-ionic density gradient centrifugation, J. Virol. Methods (1984) in press.
Laemmli, U., Cleavage of structural proteins during the assembly of the head of bacteriophage T_4, Nature **227**, 680-685 (1970).
Lemaster, S., and Roizman, B., Herpes simplex virus phosphoproteins. II. Characterization of the virion protein kinase and of the polypeptides phosphorylated in the virion, J. Virol., **35**, 798-811 (1980).
Mar, E.-C., Patel, P.C., and Huang, E.-S., Human cytomegalovirus-associated DNA polymerase and protein kinase activities. J. Gen. Virol. **57**, 149-156 (1981).
Miller, G., Shope, T., Lisco, H., Still, D., and Lipman, M., Epstein-Barr virus: Transformation, cytopathic changes and viral antigens in squirrel monkey and marmoset leukocytes, Proc. Natl. Acad. Sci. USA **69**,383-387 (1972).
Randall, C.C, Rodgers, H.W., Downer, D.N., and Gentry, G.A., Protein kinase activity in equine herpes virus, J. Virol. 9, 216-222 (1972).
Rubenstein, A.S., Gravell, M., and Darlington, R., Protein kinase in enveloped herpes simplex virions, Virology **50**, 287-290 (1972).
Simmons, J.G., Hutt-Fletcher, L.M., Fowler, E., and Feighny, R.J., Studies of the Epstein-Barr virus receptor found on Raji cells. I. Extraction of receptor and preparation of anti-receptor antibody, J. Immunol. **130**, 1303-1308 (1983).
Wray, W., Boulikas, T., Wray, V.P. and Hancock, R. Silverstaining of proteins in polyacrylamide gels. Anal. Biochem. 118, 197-203 (1981).

24

CHARACTERIZATION OF THE GENES WITHIN THE BamH1 M FRAGMENT

OF EPSTEIN-BARR VIRUS DNA THAT MAY DETERMINE THE FATE OF

VIRAL INFECTION

Jeff Sample[1], Akiko Tanaka[2], Gerald Lancz[1], and Meihan Nonoyama[2]

[1]Department of Medical Microbiology and Immunology, The University of South Florida College of Medicine, 12901 North 30th Street, Tampa, FL 33612

[2]Showa University Research Institute for Biomedicine in Florida, 10900 Roosevelt Boulevard North, St. Petersburg, FL 33702

SUMMARY

The mRNAs transcribed from the Epstein-Barr virus (EBV) BamHI M restriction fragment were analyzed by nuclease S1 and exonuclease VII mapping. This type of structural characterization led to the determination of which portion of these genes is present in the defective genome of the P3HR-1 strain of EBV. Three unspliced mRNAs of the immediate-early or early class were identified and found to have common 3' termini. Mapping the 5' termini of these three mRNAs indicated that a 4.0-kilobase transcript is initiated within the adjacent BamHI a fragment and is terminated within the BamHI M fragment and a 2.65 and a 1.9-kilobase transcript each are encoded entirely within the BamHI M fragment. Less than 500 bases of DNA encoding the 3' portion of these mRNAs were found to be present in the defective P3HR-1 virus genome. Thus, production of these mRNAs must occur by transcription of DNA in the standard genome during the disruption of latency caused by superinfection with the P3HR-1 strain of EBV.

INTRODUCTION

When lymphoblastoid cells that are latently infected
with Epstein-Barr virus (EBV) are superinfected with the
P3HR-1 strain of EBV, activation of the EBV replication
cycle results (Miller et al., 1974; Menezes et al., 1975;
Yajima and Nonoyama, 1976; Seigneurin et al., 1977). This
requires superinfection with a population of P3HR-1 EBV
that contains standard and defective genomes (Rabson et al.,
1983; Miller et al., 1984). P3HR-1 EBV that contains a
standard genome is incapable of disrupting latency. The
mechanism(s) by which virus-containing defective DNA
activates the productive cycle of virus replication is
unknown. It seems likely that at least one function of the
co-infecting defective virus is to mediate expression of
genes that are important for initiating the replicative
cycle but that are normally repressed.

Following superinfection, the first detectable mRNAs
synthesized in association with the viral replicative cycle
are transcribed from the BamHI M restriction fragment of
the EBV genome (Sample et al., 1984). This suggests that
genes within this region of the genome are involved in the
initiation of virus replication. DNA from the BamHI M
restriction fragment is present in the defective genomes
of P3HR-1 EBV (Cho et al., 1984a, Miller et al., 1984). It
is possible that the presence of these genes in a transcriptionally active form may contribute to the ability of
P3HR-1 virus with defective genomes to activate EBV
replication in latently infected cells. This was tested
by characterizing the genes within the BamHI M fragment by
nuclease mapping of the three mRNAs transcribed in this
region. Analysis of cloned fragments of DNA from the
defective genome containing sequences from the BamHI M
fragment indicated that less than 500 bases of the 3'
portion of these genes are present in the defective genome.
Thus, if expression of these potentially key replicative
genes is mediated by the defective virus, this probably
occurs by trans-activation of genes in the standard genome.

RESULTS

Transcription of the BamHI M Region

Figure 1 illustrates the different species of mRNAs

transcribed from the BamHI M region of the EBV genome in
various productively infected cells. In Raji cells
superinfected with P3HR-1 EBV, major polyadenylated
transcripts of 4.0, 2.65 and 1.9-kilobases (kb) were
detected (lane 1). A 2.0-kb transcript, which was
detected in 12-0-tetradecanoylphorbol-13-acetate (TPA) -
induced B95-8 cells (lane 3), may also have been present in
superinfected cells, but was indistinguishable on Northern
blots due to the intensity of hybridization to transcripts
in the 1.9 to 2.0-kb size range. Inhibition of protein
synthesis in superinfected cells by treatment with
cycloheximide from the time of superinfection to cell
harvest (15 hr postinfection), did not prevent transcription
of the BamHI M DNA (lane 2). The level of the 4.0-kb
transcript did appear to be lower in cycloheximide-treated
cells. The concentration of cycloheximide employed (5µg/ml)
was sufficient to inhibit expression of EBV early antigen
following superinfection. These data suggest that the
BamHI M mRNAs are of the immediate early or early class. A
5.5-kb transcript was also detected in TPA-induced P3HR-1
cells (lane 4) and to a lesser extent in superinfected
Raji cells (lane 1). This mRNA may have been transcribed
from defective DNA.

Nuclease Mapping of the BamHI M mRNAs

     To characterize the structure of the BamHI M mRNAs
and map the genes within the BamHI M fragment, nuclease S1
and exonuclease VII mapping was performed. ^{32}P-end-labeled
subfragments of the BamHI M fragment were hybridized to
cytoplasmic RNA from TPA-induced B95-8 cells, the DNA-RNA
hybrids were then digested with nuclease S1 or exonuclease
VII and the products of nuclease digestion were subjected
to electrophoresis in alkaline agarose gels as described
(Berk and Sharp, 1978). The results of these experiments
are shown in Figure 2 and are summarized as follows. The
4.0, 2.65 and 1.9-kb transcripts are transcribed in a
rightward direction and are unspliced. The 5' portion of
the gene encoding the 4.0-kb transcript lies within the
BamHI a restriction fragment, whereas the 2.65 and 1.9-kb
transcripts are encoded entirely within the BamHI M fragment.
All three mRNAs share a common 3' transcription termination
site that is approximately 100 bases to the right of the

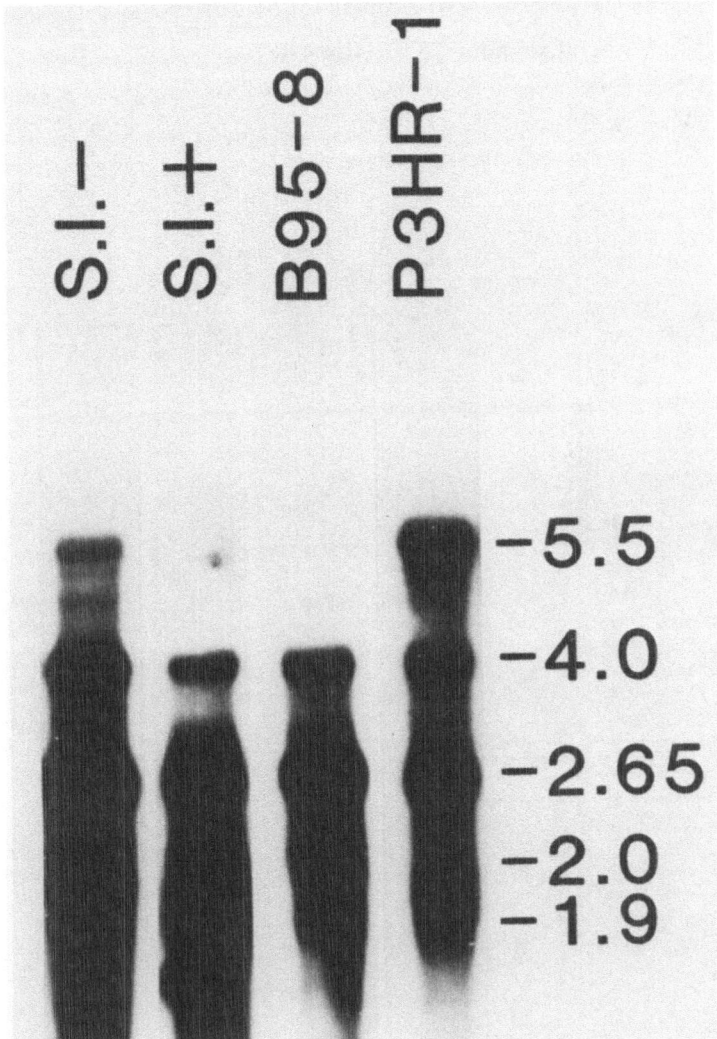

Figure 1 - Hybridization of ^{32}P-labeled BamHI M DNA to nitrocellulose filters containing polyadenylated RNA from Raji cells superinfected (S.I.) with P3HR-1 EBV in the absence (-) and presence (+) of cycloheximide; TPA-induced B95-8 cells; or TPA-induced P3HR-1 cells. Sizes are indicated in kilobases (kb).

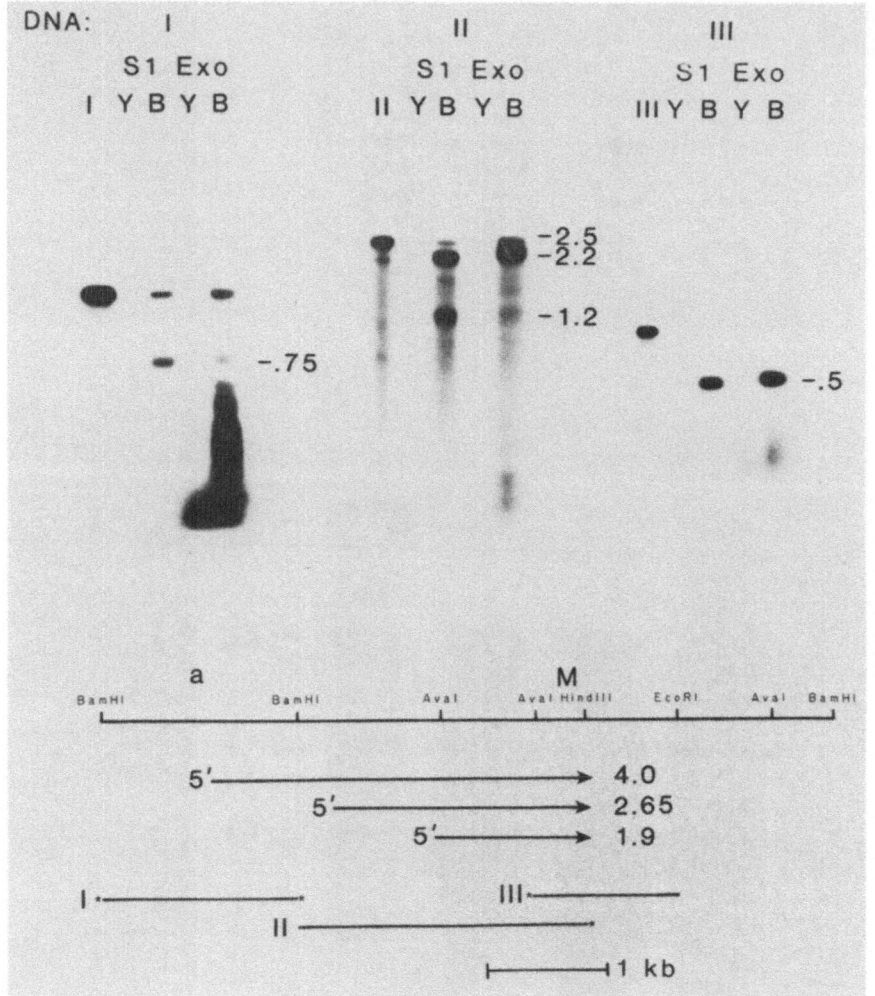

Figure 2 - Nuclease SI and Exonuclease VII mapping of the BamHI M mRNAs. Lanes containing the full-length DNA fragment (I,II,III), DNA that was hybridized to yeast RNA (Y) or B95-8 cell cytoplasmic RNA (B) are indicated. All numbers indicate size in kb, including 3'-poly(A) tails of the mRNAs (horizontal arrows). Asterisks indicate the positions of 5' (Iand II) or 3' (III) ^{32}P.

HindIII restriction site within the BamHI M fragment.
Although the 2.0-kb transcript has not been completely
characterized, preliminary data suggest that this mRNA is
encoded by the right half of the BamHI M fragment and is
transcribed in a leftward direction. It is not known
whether this mRNA is spliced or not.

Structure of the Defective Viral DNA Containing DNA from
the BamHI M Fragment

Analysis of P3HR-1 viral DNA, containing both defective
and nondefective molecules, by Southern blot hybridization
indicated that restriction fragments of the defective genome
contain sequences from the BamHI M fragment and the adjacent
BamHI S fragment (Figure 3). The structure of the defective
DNA that hybridized to DNA from the BamHI M fragment was
characterized to determine whether the BamHI M genes were
intact in the defective genome. Two cloned EcoRI fragments
of 21 and 18-kilobase pairs (kbp) with homology to BamHI M
DNA were characterized by Southern blot hybridization. The
21-kbp EcoRI fragment was found to contain less than 500
bases of the 3' coding region for the 4.0, 2.65 and 1.9-kb
mRNAs. This was determined by hybridization of the ^{32}P-
labeled plasmid DNA (pJS184) containing the 21-kbp fragment
to blots of restricted plasmid DNA containing the standard
BamHI M fragment (Figure 4A). The left-most portion of
the BamHI M fragment which hybridized to ^{32}P-pJS184 was a
400-bp HindIII-AvaI subfragment (see Figure 2 for a
restriction map of BamHI M). The 21-kbp EcoRI fragment
also contained DNA from the BamHI B', W'I ' and A fragments.
This was determined by hybridization of the ^{32}P-labeled
21-kbp EcoRI fragment to dot blots of the cloned BamHI
fragments of B95-8 EBV DNA (data not shown) and by
hybridization to Southern blots of BamHI- and EcoRI-
restricted P3HR-1 viral DNA (Figure 4B). The linkage
map of the entire 21-kbp EcoRI fragment and the terminal
portion of the left adjacent 18-kbp EcoRI fragment, which
contains the right-most EcoRI-BamHI subfragment of BamHI M
and most of BamHI S, is illustrated in Figure 4D. Note
also that the BamHI M-S DNA in the defective genome is in
the opposite orientation relative to the standard genome.
These data are in agreement with those of Cho et al.(1984a)
who recently reported the complete structure of the
defective genome of P3HR-1 EBV.

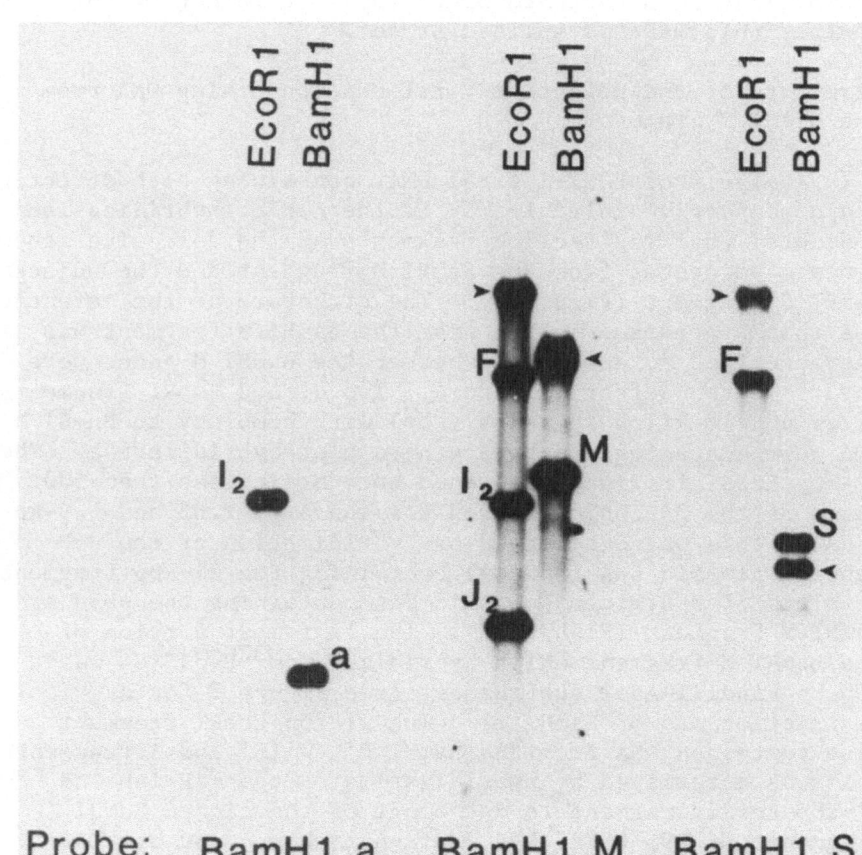

Figure 3 – Southern blots of EcoRI–or BamHI–restricted P3HR-1 EBV DNA hybridized to ^{32}P–labeled BamHI a, M or S DNA fragments. The standard restriction fragments are labeled whereas fragments from defective genomes are indicated by an arrowhead.

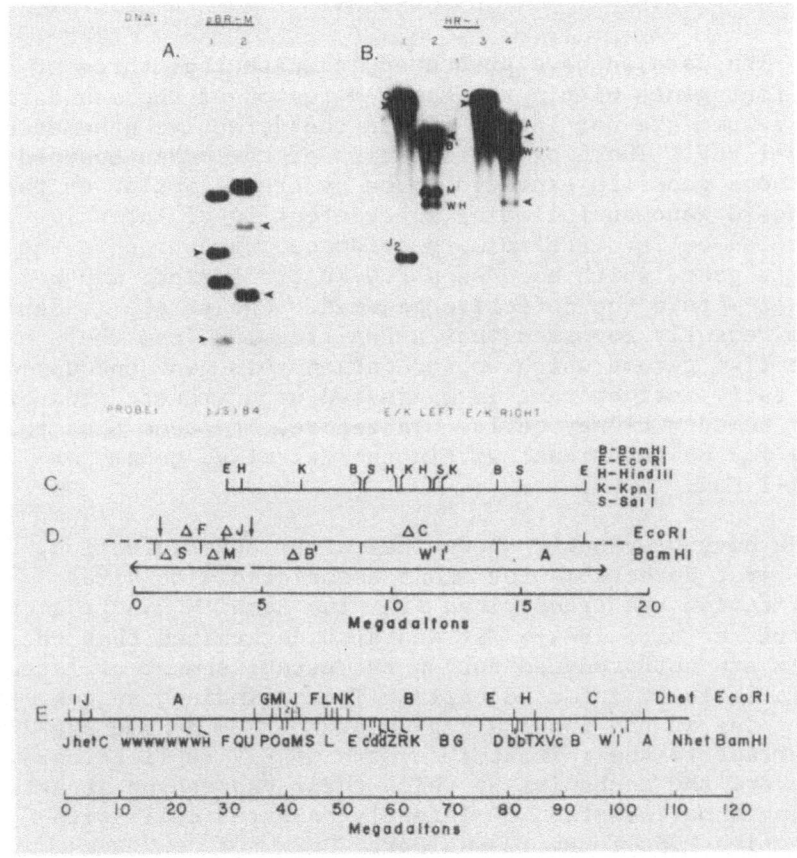

Figure 4 - Characterization of the defective P3HR-1
EBV DNA containing BamHI M sequences.   (A) Southern blots
of BamHI, AvaI and HindIII (lane 1) or BamHI, PstI and
HindIII (lane 2) restricted pBR-M (containing BamHI M);
arrowheads indicate viral DNA bands. (B) Southern blots of
BamHI (lanes 2 and 4) or EcoRI (lanes 1 and 3) restricted
P3HR-1 viral DNA; letters indicate standard or homologous
(WH) fragments, arrowheads DNA from defective molecules.
(C) Restriction map of the 21-kbp EcoRI fragment.   (D)
Structure of the defective DNA.   (E)   Standard P3HR-1 EBV
genome. See text for further explanation.

DISCUSSION

The data we have presented indicate that three of the four genes within the BamHI M region of the standard EBV genome are not intact within the defective genomes of P3HR-1 EBV. Therefore, production of the mRNAs encoded by these genes is expected to be by transcription of the standard genomes following superinfection of latently infected cells. Preliminary evidence suggests that the fourth gene, which encodes a 2.0-kb transcript, may be intact within the defective genomes. Cho et al. (1984b) have recently reported that a DNA fragment from the defective genome which would contain this gene encodes an EBV early antigen that is expressed upon transfection of baby hamster kidney cells. Therefore, it seems that this gene may be functional within the defective genome of P3HR-1 EBV.

We have previously shown that after superinfection, the first detectable EBV mRNAs associated with virus replication are transcribed from the BamHI M DNA fragment (Sample et al., 1984). It was also determined that these mRNAs are not produced during the establishment of latency or in latently infected cells. These findings suggest that the genes in the BamHI M region of the EBV genome are important to the initiatory events of EBV replication. However, the mechanism by which their expression is activated during superinfection of latently infected cells with defective P3HR-1 EBV is unclear.

Although the gene encoding the 2.0-kb mRNA may be intact and in a transcriptionally active form within the defective genome, it seems unlikely that its gene product activates transcription of the other BamHI M genes since they all appear to be of the same temporal class. Alternatively, activation of these genes may be mediated by an intermediate function of the defective virus. Whatever the mechanism, understanding how defective P3HR-1 virions disrupt latency will provide valuable insight into the regulatory events responsible for the establishment and maintenance of EBV latent infections.

ACKNOWLEDGMENTS

This work was supported by grant R01 CA 31950 from the National Institutes of Health and a grant from Sigma Xi, the Scientific Research Society.

287

REFERENCES

Berk, A.J., and Sharp, P.A., Spliced early mRNAs of simian
virus 40. Proc. Natl. Acad. Sci. USA, 75, 1274-1278
(1978).

Cho, M.-S., Bornkamm, G.W., and zur Hausen, H., Structure
of defective DNA molecules in Epstein-Barr virus
preparations from P3HR-1 cells. J. Virol., 51, 199-207
(1984a).

Cho, M.-S, Gissmann, L., and Hayward, S.D., Epstein-Barr
virus (P3HR-1) defective DNA codes for components of
both the early antigen and viral capsid antigen complexes.
Virology, 137, 9-19 (1984b).

Menezes, J., Liebold, W., and Klein, G., Biological
differences between Epstein-Barr virus (EBV) strains
with regard to lymphocyte transforming ability, superinfection and antigen induction. Exp. Cell Res., 92,
478-484 (1975).

Miller, G., Robinson, J., Heston, L., and Lipman, M.,
Differences between laboratory strains of Epstein-Barr
virus based on immortalization, abortive infection, and
interference. Proc. Natl. Acad. Sci. USA 71, 4006-4010
(1974).

Miller, G., Rabson, M., and Heston, L., Epstein-Barr virus
with heterogeneous DNA disrupts latency. J. Virol., 50,
174-182 (1984).

Rabson, M., Heston, L., and Miller, G., Identification of
a rare Epstein-Barr virus variant that enhances early
antigen expression in Raji cells. Proc. Natl. Acad. Sci.
USA, 80, 2762-2766 (1983).

Sample, J., Tanaka, A., Lancz, G., and Nonoyama, M.,
Identification of Epstein-Barr virus genes expressed
during the early phase of virus replication and during
lymphocyte immortalization. Virology, 139, 1-10 (1984).

Seigneurin, J.-M., Viullaume, M., Lenoir G., and de-The',
G., Replication of Epstein-Barr virus: ultrastructural
and immunofluorescent studies of P3HR-1-superinfected
Raji cells. J. Virol., 24, 836-845 (1977).

Yajima, Y., and Nonoyama, M., Mechanisms of infection with
   Epstein-Barr virus I. Viral DNA replication and
   formation of noninfectious virus particles in superinfected Raji cells. J. Virol., 19, 187-194 (1976).

25

EFFECTS OF TUNICAMYCIN ON BINDING OF EPSTEIN-BARR VIRUS

N. Balachandran and L.M. Hutt-Fletcher

Dept. of Comp & Exp. Pathology, University of Florida

Gainesville, Florida 32610

SUMMARY

The effects of tunicamycin were examined on the
production of lytic and transforming strains of Epstein-
Barr virus (EBV). The drug altered expression of EBV
membrane antigens on the surface of producer cells and
markedly reduced the yield of virus. The small amount of
virus that was released was still able to bind to receptor
positive cells, but its ability to induce immunoglobulin
synthesis in normal B cells was compromised.

INTRODUCTION

The membrane of EBV contains four major polypeptides
of approximate molecular weights 350/300K, 220/200K, 140K
and 85K (Thorley-Lawson et.al., 1982). All, except the
140K molecule, are glycoproteins containing relatively
large amounts of carbohydrate. The 350/300K and 220/200K
glycoproteins, (gp350/300 and gp220/200) share sequences
and are reported to contain both 'N'- and 'O'-linked
sugars; the 85K glycoprotein (gp85) contains 'N'-linked
sugars (Strnad et.al., 1983). Antibodies to gp350/300 and
gp220/200 can inhibit virus binding and antibodies to gp85
can also neutralize virus. We therefore decided to use
inhibitors of glycosylation to examine the possible role
of the carbohydrate portions of these molecules in the
selective adsorption and penetration of EBV into B cells.
We report here on the effects of tunicamycin, an inhibitor
of 'N'-linked glycosylation (Klenk and Schwarz, 1982).

MATERIALS AND METHODS

The superinducible P3HR1(cl.13) line and the MCUV5 marmoset line (both a gift of Dr. George Miller, Yale University) were grown in RPMI with 10% fetal calf serum. Before each experiment, cells were washed, adjusted to 2×10^6/ml, preincubated for 3h with different concentrations of tunicamycin (1,2,5µg) and induced with phorbol ester (TPA 30ng/ml). Untreated cells with and without TPA were included as controls. To label DNA, ^{3}H-thymidine (1µCi/ml) was added simultaneously with TPA. Three days post-induction, virus was harvested from clarified spent culture medium by centrifugation at 20,000 g (ECV). Intracellular virus for binding assays was harvested by lysing cell pellets with 2 cycles of freeze/thawing and clarification of the lysates at 400 g (ICV). Such preparations were also used to determine the amount of acid insoluble ^{3}H-thymidine and EBV DNA in extranuclear cell fractions.

Surface antigen expression on P3HR1(cl.13) cells and EBV binding to EBV receptor positive (EBVR$^+$) and receptor negative (EBVR$^-$) cells were measured by indirect immunofluorescence (Simmons et.al., 1983) on cells fixed with 0.1% paraformaldehyde. Fluorescence intensity and distribution were assessed visually or by analysis of 30,000 cells in a FACS-II apparatus. ^{3}H-virus was allowed to bind to fixed cells for 60 min at 37°C and assayed by comparing acid precipitable counts bound with the total precipitable counts added.

EBV DNA in extracellular and intracellular virus was measured by dot hybridization (Kafotis et.al. 1979) with a-^{32}P-labeled Bam Hl W fragment cloned in pBR322 (Dambaugh et.al., 1980). Total cellular EBV DNA was measured by cell blot hybridization with the same probe (Brandsma and Miller, 1980). Relative amounts of DNA were determined by integrating scans of autoradiographs of hybridizations and expressing values as a percentage of those obtained for samples treated with TPA alone.

Induction of immunoglobulin synthesis in fresh peripheral B cells was measured as previously described (Hutt-Fletcher et.al., 1983).

RESULTS

P3HR1(cl.13) cells were induced with TPA in the presence or absence of tunicamycin. At two days post induction the expression of EBV membrane antigens on the cell surface was determined by immunofluorescence. Both the number of cells expressing antigen and the intensity of fluorescence were reduced by tunicamycin although the viability of the cells was not significantly altered (Table I).

Table I. EBV antigen expression in membranes of TPA-induced P3HR1 cells grown in tunicamycin (Tu) and tested at 2 days p.i. by immunofluorescence.

Treatment	% Cell viability	Membrane fluorescence with	
		Human anti-EBV	Monoclonal anti-gp350/220
No drug	77	++++[1]/35[2]	++++/40
2 μg Tu	74	++/15	++/10
5 μg Tu	71	++/9	+/9

[1] Intensity of fluorescence
[2] Percent fluorescent cells

If cells were induced in the presence of ^{3}H-thymidine and virus was harvested from spent culture medium after three days, tunicamycin markedly reduced the acid-precipitable counts associated with the ECV (Table II). In contrast, it slightly increased the acid-precipitable counts associated with the extranuclear fraction of cell lysates.

The apparent reduction in extracellular virus was corroborated by measurement of EBV DNA in similar extra-cellular and intracellular preparations. The relative amount of DNA in extracellular virus fell after treatment with tunicamycin (Fig. 1); the relative amount of extra-nuclear, intracellular EBV DNA increased at concentrations

Table II. Effect of tunicamycin (Tu) on yield of acid
insoluble ^{3}H in (ECV) and (ICV) from P3HR1(cl.13) cells
and binding of counts to EBVR$^+$ cells

Cells grown in:	Relative yield of cpm		% of cpm bound to EBVR$^+$ cells[1]	
	ECV	ICV	ECV	ICV
No drug	100	100	22	4
1 µg Tu	10	129	19	1
2 µg Tu	8	144	16	2
5 µg Tu	5	148	18	1

[1] No counts bound to receptor negative cells

of up to 5µg tunicamycin, at which point, although there
was still more DNA inside than outside the cell, the
amount of intracellular DNA fell. Analysis of intracellular EBV DNA by cell blot hybridization (Fig. 1; right
panel) indicated that this reduction reflected a drop in
total intracellular EBV DNA and not merely in that which
was extranuclear and possibly encapsidated. Similar
results were obtained using virus from MCUV5 cells (not
shown).

Although the total yield of virus from cells was
reduced by tunicamycin, the virus that was released was
still able to bind to cells. The percentage of acid
precipitable counts in pellets of spent culture medium
which bound to EBVR$^+$ and not EBVR$^-$ cells remained constant
(Table II).

Comparable results were also obtained by measuring
virus binding to EBVR$^+$ cells by indirect immunofluorescence using human anti-EBV antibody. Fluorescence
intensity was quantitated in a FACS II apparatus.
Although the amount of bindable virus harvested from
tunicamycin treated cells was less than that from
untreated cells either with or without induction with TPA,
measurable amounts of virus were repeatably found (Fig.2).

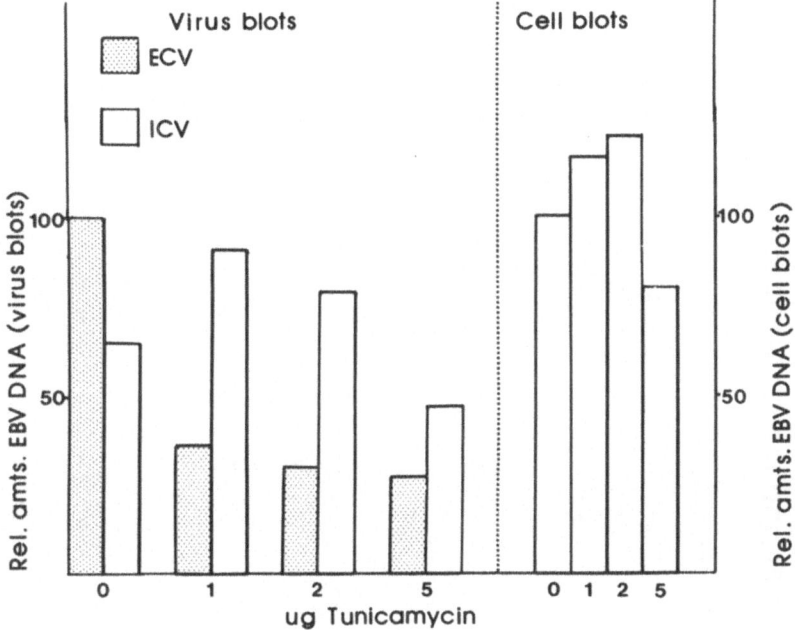

Fig. 1. Relative amounts of EBV DNA in ECV and ICV (left panel) and in whole cells (right panel) treated with tunicamycin (no drug = 100).

In the absence of tunicamycin, more bindable virus was present outside than inside the cell (Fig. 3; panel A); in its presence, more was found inside than out (Fig.3; panel B). However, although the amount of bindable extracellular virus was markedly less in the presence of tunicamycin than in its absence (Fig. 3; panel C) the total amount of bindable virus inside the cell was not increased in the presence of drug (Fig. 3; panel D). In all these experiments, the fluorescence intensity profiles of virus from cells treated with 2 or 5μg of tunicamycin were superimposable.

Although there was no discernable difference in the yield of bindable virus from MCUV5 cells treated with 2 or 5μg of tunicamycin, the virus obtained from MCUV5 cells treated with 5μg was not able to induce immunglobulin synthesis in freshly isolated B cells (Table III).

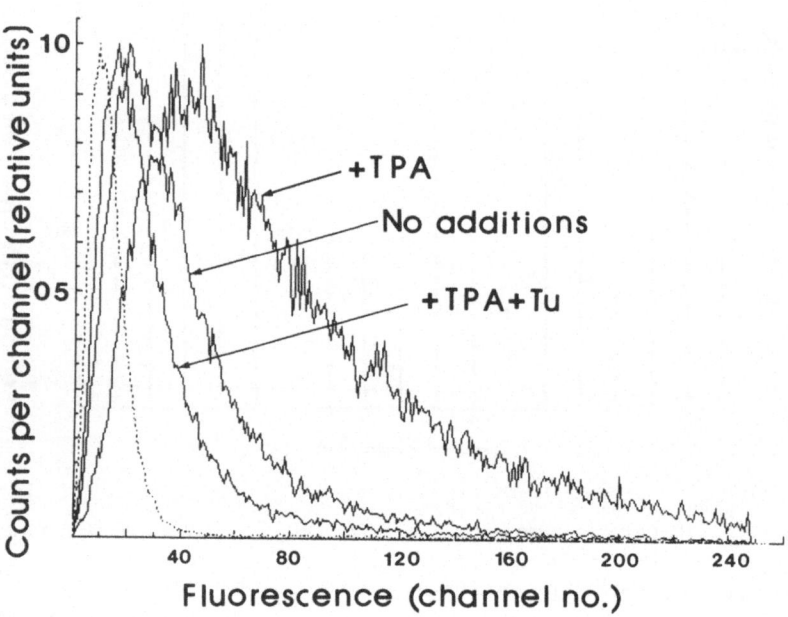

Fig.2.   FACS-II fluorescence intensity profiles of binding of virus from TPA and tunicamycin treated cells (┄┄┄┄no virus control)

DISCUSSION

Inhibition of co-translational addition of 'N'-linked sugars by tunicamycin has been shown to affect the intracellular transport, assembly, antigenicity and biological functions of several viruses (Klenk and Schwarz, 1982). Small amounts of enveloped herpes simplex virus lacking carbohydrate are produced in the presence of tunicamycin. However, although such virus has been shown to bind to

Table III. Immunoglobulin production by peripheral
blood lymphocytes infected with EBV (MCUV5) grown in
the presence of tunicamycin (Tu).

Virus grown in presence of:	Immunoglobulin conc (ng/ml)	
	ECV	ICV
No drug	7990	6312
1 μg Tu	6226	4874
2 μg Tu	5779	2214
5 μg Tu	564	516
No virus	526	

cells, its infectivity is reduced and a failure to penetrate cells has been suggested.

Our results to date with EBV suggest that this virus
is similarly affected. First, we can conclude that
'N'-linked glycosylation probably plays an important role
in the egress of mature EBV from B cells. Not only is
there a reduction in the yield from tunicamycin-treated
cells of properly enveloped virus capable of binding
specifically to B cells, there is also a loss of encapsidated EBV DNA. In addition there is an apparent impairment
of the insertion of membrane antigens into the infected
cell membrane. The reduction in the number of surface
antigen positive cells, as detected by a polyvalent high
titer human antibody as well as by a monoclonal anti-
gp350/220, suggests that this represents a real failure in
transport of protein and not merely loss of antigenic
sites by changes in sugar residues or in conformation.
The possibility that tunicamycin interferes with the
induction of the productive cycle of EBV replication has
also been considered. However, preliminary experiments
indicate that, although the drug does have some effect on
virus induction, the reduction in the number of cells
expressing early antigen is only about 20% (data not
shown) and is not sufficient to account for our results.

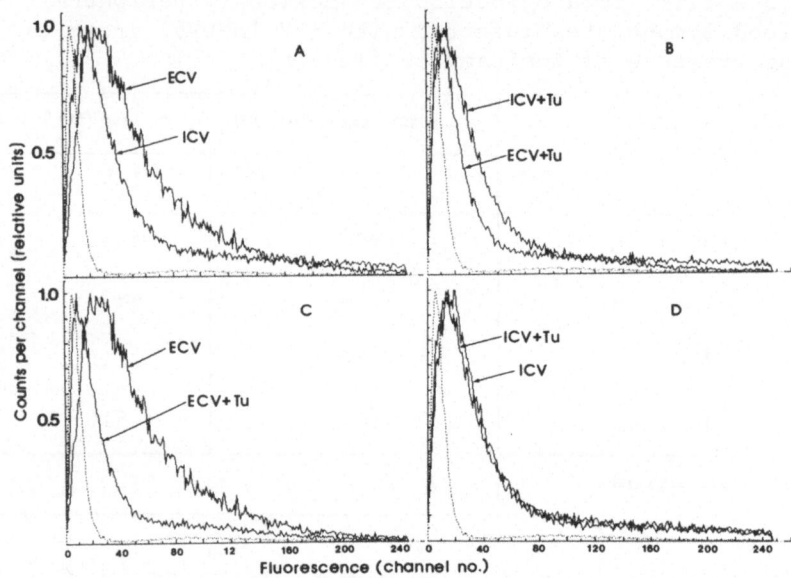

Fig.3. FACS-II fluorescence intensity profiles of binding of ECV and ICV from TPA and tunicamycin treated cells (····· no virus control).

Secondly, we have some indication that tunicamycin may interfere with complete virus assembly and envelopment. The loss in extracellular virus that binds to receptor positive cells is not accompanied by an increase in similar virus inside the cell even at concentrations that are accompanied by an increase in the total amount of intracellular EBV DNA.

Thirdly, as with herpes simplex virus, a small amount of EBV is released in the presence of tunicamycin with apparently unimpaired ability to bind to receptor positive cells. This indicates either that some small amounts of normally glycosylated virus are still being made or that

'N'-linked sugars play no role in binding EBV to the B
cell receptor. We are as yet unable to distinguish
conclusively between these two possibilities. We have
been unable to immunoprecipitate fully glycosylated
species from extracellular virus made in the presence of
tunicamycin, but we have not been able to identify
glycoprotein precursors either. The failure of virus made
in the presence of tunicamycin to induce immunoglobulin
synthesis in normal B lymphocytes, argues against the
normalcy of its glycoproteins and for its inability to
penetrate cells to which it binds. However concentrations
of 5µg/ml of tunicamycin appear to have an inhibitory
effect on viral DNA synthesis and it is conceivable that
at this concentration a lower proportion of extracellular
enveloped virus contains DNA.

A dissociation of the functions of binding and
penetration does however have interesting implications and
is supported by our findings (unpublished) that monoclonal
antibodies that neutralize virus do not necessarily
inhibit virus binding. The possibility that 'N'-linked
sugars are relevant to successful virus penetration will
be further explored.

ACKNOWLEDGEMENTS

This work was supported by grant number AI20662 from
the National Institutes of Health. We thank Susan Turk
and Ann McNicol for technical help and Dr. P.A. LeBlanc
for help with the FACS-II analysis.

REFERENCES

BRANDSMA, J., and MILLER, G., Nucleic acid spot
hybridization: rapid quantitative screening of lymphoid
cell lines for Epstein-Barr virus DNA. Proc. Natl. Acad.
Sci. USA. 77, 6851-6855 (1980).

DAMBAUGH, T., BEISEL, C., HUMMEL, M., KING, W., FENNEWALD,
S., CHEUNG, A., HELLER, M., RAAB-TRAUB, N., and KIEFF, E.,
Epstein -Barr virus (B95-8) DNA VII. Molecular cloning and
detailed mapping of EBV (B95-8) DNA. Proc. Natl. Acad.
Sci., USA. 77, 2999-3003 (1980).

HUTT-FLETCHER, L.M., BALACHANDRAN, N., and ELKINS, M.H., B cell activation by cytomegalovirus. J. Exp. Med. 158, 2171-2176 (1983).

KLENK, H.D., and SCHWARZ, R.T., Viral glycoprotein metabolism as a target for anti-viral substances. Antiviral Res. 2, 177-190 (1982).

KAFOTIS, F.C., JONES, C.W., and EFSTRATIADIS, A., Determination of nucleic acid sequence homologies and relative concentrations by dot blot hybridizatin procedure. Nucl. Acids Res. 7,1541-1552 (1979).

SIMMONS, J.G., HUTT-FLETCHER, L.M., FOWLER, E., and FEIGHNY, R.J., Studies of the Epstein-barr virus receptooe found on Raji cells. I. Extraction of receptor and preparation of anti-receptor antibody. J. Immunol. 130. 1301-1308 (1983).

STRNAD, B.C., ADAMS, M.R., and RABIN, H., Glycosylation pathways of two major Epstein-Barr virus membrane antigens. Virology. 127, 168-176 (1983).

THORLEY-LAWSON, D.A., EDSON, C.M., and GEILINGER, K., Epstein-Barr virus antigens - a challenge to modern biochemistry. Adv. Cancer. Res. 36, 295-348 (1982).

26

STRUCTURE AND EXPRESSION OF THE EPSTEIN-BARR VIRUS GENOME

P.J. Farrell, J. Dyson, P. Tuffnell, M. Biggin, T. Gibson', A. Bankier', G. Hudson', G. Hatfull', S. Satchwell' and B. Barrell'.
Ludwig Institute for Cancer Research and 'Laboratory of Molecular Biology, MRC Centre, Hills Road, Cambridge CB2 2QH, England

SUMMARY

Protein sequence comparison has revealed an analogue of the HSV glycoprotein B in the EBV genome. Another region of the EBV genome has apparently arisen by tandem gene duplication. Weak transcription enhancers have been detected in EBV and more mRNAs have been mapped. Some of the latent cycle mRNAs have exons derived from both the short and long unique regions.

RESULTS AND DISCUSSION

Because of the difficulty of analysing the EBV genome by classical genetic methods we have taken a structural approach to understanding the organisation and control of the viral gene expression. The complete DNA sequence of the B95-8 strain of EBV has been established using the random M13 cloning/dideoxynucleotide sequencing method (Baer et al, 1984). Computerised analysis was used to identify the major open translational reading frames that might be encoded by the DNA sequence. The principal reading frames were chosen on grounds of length, position relative to likely transcriptional signals and analysis of codon usage. The biased G+C composition of the EBV genome (59.94%) made it possible to use a simple form of codon usage analysis in many regions of the genome. By this type of procedure we identified at least 80 likely protein coding regions within the viral DNA and we expect the total number of genes in EBV to lie in the range of 80-100. We name reading frames according to the BamHI restriction fragment in which they begin and their direction on the genome. So BKRF1 is a Frame starting in the Bam K fragment going Rightward on the genome.

Although it was obvious that the pattern of open reading frames must reflect the pattern of genes in the virus we neither knew the identities of most of these genes nor the way in which they might be expressed. Earlier experiments had located most or all of the protein coding part of the EBNA-1 protein to the BKRF1 reading frame (Fischer et al, 1984; Hennessy and Kieff, 1983). Work by several labs (Hummel et al 1984, Biggin et al, 1984) has located the genes for the gp350 and gp220 components of membrane antigen (MA) to the BL-LF1 reading frame.

In order to identify the functions of more of the EBV genes we compared the protein sequences of polypeptides encoded by all the major reading frames in EBV with computer libraries of sequences of known proteins. From this analysis we found that several polypeptides encoded by the EBV genome are highly homologous to Herpes Simplex Virus genes of known function. The reading frames BORF2 and BaRF1 are homologous to the ribonucleotide reductase genes of HSV1, HSV2 and also the large subunit of the E. Coli enzyme. Similarly the reading frame BALF5 is homologous to the HSV1 DNA polymerase and the frame BALF4 is homologous to the glycoprotein B of HSV1. In the glycoprotein B homologues important structural elements of the sequence are conserved even though the degree of identity of sequence is quite modest in some regions. Thus 10 out of 11 cysteine residues in the two proteins are in similar positions. Also a hydrophobic region in the HSV enzyme which is thought to anchor the protein in the cell membrane is matched by a similar sequence in the EBV protein sequence. No further protein sequence homologies have been detected between EBV and HSV but as more HSV genes are sequenced, further homologies may emerge. The occurrence of such homologous regions and the similarity of the overall virus organisation and morphology supports the idea of a common evolutionary origin for EBV and HSV. Some regions of the EBV genome seem to have arisen by tandem duplication followed by subsequent mutational drift. The region between 92 - 102kb on the genome map may have been generated by a triplication of a progenitor sequence because the protein sequence of parts of the BERF1, BERF2b and BERF4 reading frames are all homologous to each other.

We have determined the structures of genes in EBV by physically mapping viral mRNAs on to the viral genome. This has been done using a combination of Northern blotting, S1 mapping and primer extension experiments on RNA from B95-8 cells. These experiments have been made much easier by using the very large collection of well characterised, small M13 clones of the EBV genome generated during the sequencing program. By these methods 35 mRNAs have been accurately mapped at the DNA sequence level and 27 transcription promoters have been identified so far.

These RNAs have been categorised into three regulatory groups, though in the future this will probably prove to be an oversimplification. These classes are called latent cycle, early productive cycle and late productive cycle. RNAs of the different classes are distinguished by comparing their levels in untreated B95-8 cells, cells treated for 3 days with 30-50 mg/ml of 12-o-tetradecanoyl -phorbol-13-acetate (TPA) and cells treated thus with TPA in the presence of 125 µg/ml phosphonoacetic acid (PAA). TPA induces cells into the productive cycle and PAA blocks viral DNA synthesis, thus distinguishing the early and late productive cycle mRNAs. Control B95-8 cells are mostly in the latent viral cycle with about 0.2 - 0.5% spontaneously in productive infection. After TPA treatment up to 30% of the cells switch to the productive cycle so the levels of productive cycle RNAs are strongly increased (20-50 fold) by TPA treatment. In contrast the proportion of latent cycle cells is hardly affected and the level of latent cycle RNAs is only modestly changed by TPA treatment. PAA prevents the TPA induction of late RNAs but not the early ones. This classification is generally simple and reliable but we have observed a few RNAs which are only partially affected by PAA treatment of the cells and this may indicate a greater complexity of the control of expression.

All the transcription promoters we have mapped appear to be of the RNA polymerase II type. This was demonstrated for the first 12 promoters we mapped using an RNA polymerase II in vitro transcription system to show that transcription from the promoters was inhibited by low levels of α-amanitin, specific for RNA polymerase II. More recently we have not directly tested every promoter by in vitro transcription but note that sequences upstream of the mapped transcription starts always contain a sequence homologous to the TATA box of polymerase II promoters.

Although we have so far probably only mapped about 1/3 of the EBV mRNAs, already some general features of EBV transcription are apparent. There are many examples of overlapping RNAs. A very frequent arrangement is to have 3' coterminal sets of RNAs starting at different promoters but sharing a common 3' end and poly A addition site. A similar arrangement has been seen in HSV. Usually in such RNAs the likely protein coding sequences do not overlap, the longer RNAs from such families presumably having very long 3' untranslated regions. The different RNAs in such overlapping sets are not necessarily in the same regulatory class. In principle some such RNAs could act as polycistronic mRNAs but the fact that RNAs of different regulatory classes overlap seems to argue against this. There are examples of truly overlapping genes where the protein coding sequence of one RNA overlaps another either in different translational phases (e.g. BLLF1 and BLLF3) or in the same translational phase (see below).

Splicing of EBV RNAs is not uncommon and an interesting dichotomy in splicing is emerging. All of the latent cycle RNAs identified so far seem to be spliced but only one out of 33 productive cycle RNAs is spliced. This is the RNA for the gp220 MA protein and even in this case the corresponding unspliced RNA is also translated. The reason for this difference is unclear. Intervening sequences might permit the virus RNAs to appear more similar to cellular RNAs during latency. Perhaps in the in vivo productive cycle, production of viral RNAs is too rapid to be hindered by the cellular splicing machinery or occurs in a cell where the cellular splicing mechanism is damaged.

It is not yet known whether the regulation of expression of EBV genes is at the transcriptional level or in the processing and stability of viral RNAs. There is a precedent for transcriptional control in Herpes simplex virus. Because transcriptional enhancers can have a profound effect on the level of expression from a promoter, we have tried to map enhancers in the B95-8 EBV genome. These experiments were in collaboration with W. Schaffner and F. Weber and used a defective SV40 "enhancer trap" system which they developed (Weber et al, 1984). SV40 lacking the 72bp repeat enhancers will not grow in CV1 cells. By cotransfecting fragments of the EBV genome with this defective SV40, recombinant SV40 genomes which grow and which have acquired EBV DNAs can be recovered. Two such recombinants were obtained and analysed. They contained short regions of EBV DNA around 110,800 and 133,950 on the B95-8 map within the SV40 sequence.

Although these recombinant sequences behave as true enhancers since they will increase transcription when placed in either orientation downstream of a test gene, their activity is extremely weak. In the standard test of enhancing a β-globin gene transfected into HeLa cells they only caused a 3-fold increase in transcription of the globin gene. Sometimes enhancer activity is cell type specific and we have not yet screened for enhancers specific for lymphoid cells; unfortunately no rapid enhancer trap system has yet been developed for lymphoid cells. It may be interesting that the enhancer at 133,950 lies within the region which (Griffin and Karren, 1984) apparently immortalises monkey and marmoset epithelial cells.

The expression of the latent cycle RNAs has only partially been unravelled. The EcoR1-Dhet region encodes the most abundant latent cycle EBV transcript in B95-8 cells. We have determined the structure of this RNA and shown it to consist of two short 5' exons joined to a longer 3' exon. The exons have been accurately mapped and shown to contain a reading frame for a 45k membrane protein. We and others have previously considered the possibility that this protein might be LYDMA, the antigen recognised by cytotoxic T cells that specifically kill EBV infected cells. It may be possible to test this hypothesis by attempting to convert LYDMA negative cells to a LYDMA positive phenotype by introduction of the gene for this protein. Even if the protein encoded by this gene is LYDMA it presumably also has another function more directly helpful to the virus. Since the gene is apparently always expressed in EBV - transformed B-lymphocytes and is a membrane protein, it may be involved in the action of the B cell growth factor secreted by such cells and required for their growth (Gordon et al, 1984). Interestingly, a second RNA is expressed, during the late productive cycle, which overlaps part of the gene for the latent protein. It would apparently encode a protein identical to the C terminal part of the latent cycle protein. This would retain two of the six membrane-spanning sequences and all the large C terminal globular domain present in the latent cycle protein. It is intriguing to know what the function of this protein might be.

The region of the viral genome around the major internal repeats is also expressed during the latent virus cycle. Because the transformation-defective strains P3HR1 and Daudi have deletions here, within the Bam HI W,Y,H region, it seems likely that genes localised here may be directly involved in the immortalisation process in B lymphocytes. Using a combination of Northern blotting and S1 mapping we have found at least two rightward latent cycle RNAs here of sizes 3.7 kb and 2.4 kb. Some of this mapping was done by M. Bodescot and M. Perricaudet who also constructed a partial cDNA clone of one of the equivalents of these RNAs in Raji cells (Bodescot et al, 1984). This RNA has a very highly spliced structure. Both the 3.7 kb and the 2.4 kb RNA appear to start at a promoter in the short unique region just to the left of the major internal repeats. We have mapped the boundaries of this 5' exon on to the DNA sequence and also mapped the common 3' end of the two RNAs in the Bam HI H fragment. The RNAs are very highly spliced and their genes cover almost 40 kb of the EBV genome. The RNAs differ in their structure near their 3' ends, the 3.7 kb having most of the reading frame BYRF1 within it, the 2.4 kb RNA lacking most of this. The 3.7 kb RNA may encode the nuclear antigen EBNA-2.

The RNA for the EBNA-1 protein, which is about 4 kb in length presents an interesting puzzle. It is clear from transfection experiments that most and probably all of the coding sequence of the EBNA-1 protein lies in the reading frame BKRF1 (Fischer et al, 1984). S1 mapping indicates that the EBNA-1 RNA stops at a poly A addition signal just downstream of this frame and that the 2.0 kb exon containing BKRF1 is spliced to the rest of the mRNA. This seems to suggest that the RNA has a very long 1.8 kb 5' untranslated region, a very unusual situation in eukaryotes. The natural promoter for the EBNA-1 gene has not yet been located.

Two other latent cycle RNAs (about 2.0 kb and 1.8 kb in length) are derived from the short unique region. These are transcribed rightward on the genome, terminating at a poly A addition signal at 5.85 kb on the map and have a spliced structure. Their translation products are unknown.

It should be possible to construct a detailed map of the regulated expression of EBV genes in the relatively near future. This offers many possibilities for future research. Understanding the structure of genes active in the latent virus cycle will contribute to solving the mechanism by which the virus immortalises cells. It is important to remember that we have only studied the gene expression in lymphoid cells and the gene expression may differ in some respects in epithelial cells. An analysis of the mechanism by which the controlled expression of different sets of viral genes is achieved should allow us to understand the mechanism of latency and entry into the lytic virus cycle.

REFERENCES

Baer R., Bankier A.T., Biggin M.D., Deininger P.L., Farrell P.J., Gibson T.J., Hatfull G., Hudson G.S., Satchwell S.C., Seguin C., Tuffnell P.S. and Barrell B.G. DNA sequence and expression of the B95-8 Epstein-Barr Virus genome. Nature 310 207-211 (1984).

Biggin M.D., Farrell P.J. and Barrell B.G. Transcription and DNA sequence of the Bam HI L region of B95-8 Epstein-Barr Virus. EMBO J. 3 1083-1090 (1984).

Bodescot M., Chambraud B., Farrell P. and Perricaudet M. Spliced RNA from the IR1-U2 region of Epstein-Barr virus: presence of an open reading frame for a repetitive polypeptide. EMBO J. 3 1913-1917 (1984).

Fischer D.K., Robert M.F., Shedd D., Summers W.P., Robinson J.E., Wolak J., Stefano J.E. and Miller G. Identification of Epstein-Barr nuclear antigen polypeptide in mouse and monkey cells after gene transfer with a cloned 2.9kb subfragment of the genome. Proc. Natl. Acad. Sci. USA 81 43-47 (1984).

Gordon J., Ley S.C., Melamed M.D., English L.S. and Hughes-Jones N.C. Immortalised B lymphocytes produce their own B cell growth factor. Nature 310 145-147 (1984).

Griffin B.E. and Karren L. Immortalisation of monkey epithelial cells by specific fragments of Epstein-Barr virus DNA. Nature 309 78-82 (1984).

Hennessy K. and Kieff E. One of two Epstein-Barr virus nuclear antigens contains a glycine-alanine copolymer domain.Proc. Natl. Acad. Sci. USA 80 5665-5669 (1983).

Hummel M., Thorley-Lawson D.A. and Kieff E. An Epstein-Barr virus DNA fragment encodes messages for the two major envelope glycoproteins (gp350/300 and gp250/200). J. Virol. 49 413-417 (1984).

Weber F., de Villiers J. and Schaffner W. An SV40 enhancer
   trap incorporates exogenous enhancers or generates
   enhancers from its own sequences.
   Cell 36 983-992 (1984).

27

BRIEF COMMUNICATION

ANTIBODY RESPONSE TO EPSTEIN-BARR VIRUS-SPECIFIC DNase IN

THIRTEEN PATIENTS WITH NASOPHARYNGEAL CARCINOMA

Jen-Yang Chen[1], R Palmer Beasley[1,2], Chia-Siang
Chien[3], and Czau-Siung Yang[1]
[1]Department of Bacteriology, College of Medicine
National Taiwan University; [2]University of Washington Medical Research Unit, Taipei; and
[3]Government Employees' Clinical Center, Taipei,
Taiwan, Republic of China

Nasopharyngeal carcinoma (NPC) is one of the major neoplastic malignancies of Southern Chinese males. In Taiwan,
it has been reported as the most common cancer of males (Yeh,
1966). A recent long-term study established that the survival
rate over a 10-year period could be as high as 77% if the
treatments were initiated at stage I of the disease (Hsu et
al, 1981) and therefore an early diagnosis method is required.

Recently, an Epstein-Barr virus (EBV)-specific DNase has
been described (Clough, 1979; Cheng et al, 1980a). It has also
been reported that the majority of sera from Chinese NPC
patients neutralized EBV DNase activity (Cheng et al, 1980b).
In another study of antibody to EBV DNase activity in sera
obtained from NPC patients in Taiwan, it was found that
most of the sera contained high levels of antibody as early
as the stage I of the disease (Chen et al, 1982).

During a prospective study for hepatoma, 13 individuals
were found to have a previous history of, or to be suffer-
ing from, NPC. Serum samples obtained from this group were
examined in this study for anti-EBV DNase activity.

Materials and Methods

Preparation of cellular extract. The preparation of

extracts from 5-iodo-2'-deoxyuridine (IUDR)-treated P3HR1 cells has been described previously (Chen et al, 1982).

Patient sera. Sera obtained from 13 patients with NPC, found in a prospective study of hepatoma in 22,707 Chinese males in Taiwan (Beasley et al, 1981) were examined for the presence of anti-EBV DNase activity.

Enzyme assay and detection of antibody to EBV DNase activity. The assay procedure for EBV DNase activity was the same as that described for the exonuclease (Cheng et al,1980a). One unit of DNase activity is defined as the amount of the enzyme that converts 1 μg of double-stranded DNA to acid-soluble material in 10 minutes at 37°C. The procedure for the detection of antibody was as described previously (Cheng et al, 1980b) except that 0.05 units instead of 0.1 units of enzyme in the cellular extract was added to the assay tubes.

Results

Table 1 shows the results of EBV DNase neutralization tests in sera from 13 NPC patients. The sera obtained from patients 1, 2 and 3 before diagnosis of the disease contained

TABLE 1-ANTI-EBV DNase ANTIBODY IN THIRTEEN NPC PATIENTS

Patient	Date of blood collection	Months before/after diagnosis	EBV DNase units neutralized	Death (months after blood collection)
1	6/76	-12	3.6	NPC(24)
2	3/78	- 9	5.8	NPC(19)
3	2/78	-28	8.4	NPC(31)
4	6/77	-29	1.4	NPC(50)
5	4/77	+76	8.0	NPC(27)
6	10/76	+36	8.1	NPC(17)
7	4/78	+ 5	7.5	NPC(10)
8	2/78	+74	8.7	NPC(18)
9	3/77	+ 9	3.2	NPC(33)
10	1/77	+24	2.0	NPC(36)
11	11/76	+15	5.7	NPC(62)
12	8/77	+15	0.5	NPC(42)
13	3/77	+53	0	Pneumonia (71)

significantly elevated levels of antibody. A single serum
sample was also obtained from a fourth patient 29 months
prior to diagnosis of NPC, but did not exhibit specific neutralization by our criteria. Sera obtained after diagnosis
of NPC from 9 other patients were also examined for the antibody. However, the levels of antibody in sera obtained at
the remissison stage of the disease, ie, patients 9 to 13
with the possible exception of one patient (patient 11) were
either very low or undetectable.

Discussion

Antibody to EBV-specific DNase has been demonstrated in
most of the sera obtained from Chinese NPC patients (Cheng
et al, 1980b). This antibody has also been found to be elevated as early as stage I of the disease (Chen et al, 1982).
Thus, it is necessary to determine the presence of this antibody in serum before the diagnosis of the disease to evaluate
the hypothesis that this antibody could be a useful marker
for the early diagnosis of NPC. In this study, sera from 13
NPC patients were examined for the antibody. Most of the
sera taken from the four individuals before diagnosis have
significant to high levels of the antibody.

Recently, serial serum samples from 17 juvenile NPC
patients were examined for their capacity to neutralize the
activity of EBV-specific DNase (Tan et al, 1982). The
results suggested that anti-EBV DNase might be a useful
marker for the prognosis of the disease. In this study, in
the cases of the sera taken from patients 9, 10, 11, 12 and
13, most of whom had survived for more than 30 months after
blood collection, contained low or insignificant levels of
the antibody, whilst most sera taken from patients 5, 6, 7
and 8 contained high levels of the antibody and the patients
survived less than 30 months. These results are compatible
with the hypothesis that patients with low levels of antibody in their sera may have a better prognosis.

In conclusion, the results suggest that antibody to
EBV DNase activity is a useful marker for both the early
diagnosis and prognosis of NPC.

Acknowledgments

The authors thank Dr T J Harrison and Dr K N Tsiquaye of Department of Medical Microbiology, London School of Hygiene and Tropical Medicine, Keppel Street, London WC1E, UK for their kindly review of the manuscript. The help of Professor S M Tu, Dr L Y Hwang, Mr C J Lai, Ms S F Hsu, Ms S M Tsoi, Ms M Y Liu and Ms T Gidman is very much appreciated. This study was supported partly by a grant NSC 72-0412-B-002-33 from National Science Council, Taipei, Taiwan, R.O.C.

References

BEASLEY,R.P., HWANG,L.Y., LIN,C.C. and CHIEN,C.S. Hepato-cellular carcinoma and hepatitis B virus: A prospective study of 22,707 men in Taiwan. Lancet 2,1129-1133 (1981).
CHEN,J.Y., LIU,M.Y., LYNN,T.C. and YANG,C.S. Antibody to Epstein-Barr virus-specific DNase in patients with nasopharyngeal carcinoma. Chinese J. Microbiol. Immunol., 15, 255-261 (1982).
CHENG,Y.C., CHEN,J.Y., HOFFMAN,P.J. and GLASER, R. Studies on the activity of DNase associated with the replication of Epstein-Barr virus. Virology, 100,334-338 (1980a).
CHENG,Y.C., CHEN,J.Y., GLASER,R. and HENLE,W. Frequency and levels of antibodies to Epstein-Barr virus-specific DNase are elevated in patients with nasopharyngeal carcinoma. Proc. Natl. Acad. Sci. USA, 77,6162-6165 (1980b).
CLOUGH,W. Deoxyribonuclease activity found in Epstein-Barr virus producing lymphoblastoid cells. Biochemistry, 18, 4517-4521 (1979).
HSU,M.M., LYNN,T.C., HSIEH,T., HUANG,S.C. and TU,S.M. Factors affecting the survival of patients with nasopharyngeal carcinoma. J. Formosan Med. Assoc., 80,1296-1306 (1981).
TAN,R.S., CHENG,Y.C., NAEGELE,R.F., HENLE,W., GLASER,R. and CHAMPION,J. Antibody responses to Epstein-Barr virus-specific DNase in relation to the prognosis of juvenile patients with nasopharyngeal carcinoma. Int. J. Cancer, 30,561-565 (1982).
YEH,S. Some geographic aspects of most common diseases in Taiwan. Part two, Int. Pathol., 7,24-28 (1966).

28

Recent Developments in Nucleic Acid Hybridization

Wolf, H., Gu, S., Haus, M. and Leser, U.

Max-von-Pettenkofer Institute, Munich, FRG.

SUMMARY

Nucleic acid hybridization is widely used for scientific applications but essentially restricted to specialized laboratories. The use of recombinant m13 phages as hybridization probes offers a considerable advantage over the commonly used recombinant plasmids as the preparation of the probe DNA is very simple. This suggested probe-DNA can be labelled efficiently even with longer half-life isotopes like ^{125}I and used in the hybridization reaction. In a new sandwich technique, the first step is a specific hybridization with an unlabelled recombinant m13 DNA carrying an insert of the desired specificity. In a second step a universally usable labelled probe directed against the m13 part of the recombinant phage DNA is applied. In combination with simple devices for the collection of clinical specimens (Richter et al,1983) and protocols for rapid sample preparation, nucleic acid hybridization should become acceptable for routine diagnostic laboratories.

INTRODUCTION

The most important requirement for nucleic acid hybridization concerns the preparation of hybridization probes. In vivo labeling of nucleic acids by metabolic pathways has been used to obtain probes. The main difficulties are the usually limiting low-specific activity (Frenkel et al 1976), the difficulty of getting enough labeled material, and the problem of contaminating or cross-reacting sequences. These sequences may appear due to incomplete purification of the desired gene, for example from cell lysates, or can be due to sequences within the genome used as probe. DNA of the various herpes group viruses are a typical example as several areas of the viral genomes cross-hybridize with DNA from uninfected human cells (Peden et al,1982). Recombinant DNA techniques have been most valuable to overcome the problems of limited availability of probes and allow the use of only selected fractions of the viral genome without cross-reacting areas.

METHODS

Iodination of nucleic acids

To 10µg of ml3 lyophilized ssDNA were added 1µl of water, 2µl of .25 M Na-acetate buffer pH 4.65, 125^I (Amersham ImS30 100Ci/ml), and 2µl of Thallium III chloride)100 mM in buffer). The tightly closed tubes were incubated for 30 min at 60°C, chilled on ice and the reaction stopped by the addition of 150µl stop mix (100 mM Tris pH8, 10 mM EDTA, 10mM Na_2So_3 and 100µg/ml Poly A). The probe was purified on a 5 ml Sephadex G50 column with 10 mM Tris pH8, 1mM EDTA. Fractions of approx. 300µl were collected and the first peak of radioactivity was pooled, heated 10 min at 60°C and stored at -20°C in the dark in a lead container.

Sandwich hybridization

a. Prehybridization
The nitrocellulose filter (10 x 10 cm) containing the DNA fragments is prehybridized for 3 h at 68°C in 25 ml prehybridization mix: 6 x SSC, 0.5% SDS, 5 x Denhard solution (1x: 0.02% Ficoll, 0.02% Polyvinylpyrrolidone, 0.02% BSA), 100 µg/ml calf thymus DNA (preheated at 100°C for 10 min. and chilled to 4°C).

b. Hybridization (sequence specific step)
17.5ml prehybridization mix are removed from the bag and discarded. 185µl 0.4 M EDTA and 1µg single-stranded m13 DNA carrying an insert of the desired gene are added to the remaining 7.5ml. The resealed bag is incubated for 16 h at 68°C. The filter is washed in 2 x SSC, 0.5% SDS for 5

min. at room temperature and then in 0.1 x SSC, 0.5% SDS 2 x 20 min. at 68°C.

c. Posthybridization (common step for different sequences)
Posthybridization was carried out with denatured ^{32}P nick-translated DNA probe prepared from m13 mp8 double-stranded replicative form of bacteriophages without an insert. The specific activity of the nick-translated probe was 10^8 cpm/µg. The filter was incubated with 7.5ml $_{32}$hybridization solution as in step a. 10^7 cpm of the ^{32}P nick-translated probe (preheated at 100°C for 5 min., cooled to 4°C) are added and the bag is incubated for 16 h at 63°C.
The washing procedure is similar as above, but the temperature is 63°C. After washing the paper is dried and exposed below -60°C with Kodak x-omat S film and lightning plus amplifying screen.

RESULTS

Preparation of labeled probes

Various procedures have been used to introduce labels into nucleic acids. DNA can be transcribed in vitro in the presence of radioactive ribonucleoside triphosphates into cRNA using RNA polymerases. In analogy, RNA can be reverse-transcribed into cDNA. The original template can be destroyed with appropriate nucleases. Although of great importance in earlier studies, these techniques have been largely replaced by other methods.
The most frequently used protocol is based on the use of E. coli DNA-polymerase I. This enzyme acts on DNase introduced nicks in double-stranded DNA as a 5'-3' exonuclease and a 5'-3' synthetase which allows the efficient and random introduction of label. With ^{32}P labeled nucleotides specific activities between 10^8 and 10^9 cpm/µg DNA can be obtained.
Wherever the plasmid part introduces a danger of unwanted signals, it can be replaced by another vector or removed with restriction enzyme digestion and electrophoretic separation. The desired sequence can be eluted from the gel with special procedures (Langridge et al 1980; Vogelstein and Gillespie 1979), which abolish otherwise observed inhibitory effects of remaining contaminants from the gel matrix. Recently the single-stranded DNA phage m13

has been used as a cloning vehicle for probes (Messing 1983). This approach is very helpful as m13 sequences rarely occur in natural specimens and do not have to be removed and the recombinant phage DNA can be purified in large amounts without the use of ultracentrifuge or other specific equipment. Single-stranded DNA can be labeled by primer-directed synthesis of a second strand which spans all or part of the m13 sequences or by chemical reactions (Hu and Messing 1982; Gu et al 1983).

Single-stranded nucleic acids and, with reduced efficiency, also double-stranded nucleic acids can be labelled by chemical introduction of ^{125}I into the cytosines of DNA or RNA (Commerford 1971; Gu et al 1983); Han and Harding 1983; Prensky 1976). The specific advantage of chemical modification is easy scaling up for the mass production of labeled probes.

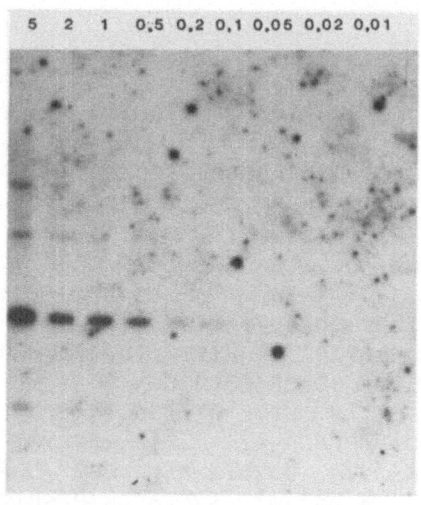

Fig. 1: Variable amounts of Raji cell (50 EBV genomes per cell) DNA given in the top line of the figure in µg was digested with Bam H1 restriction enzyme, separated on an agarose gel, blotted to a nitrocellulose membrane and used for sandwich hybridization. For the first hybridization step, 0.13 µg/ml (total 1µg/blot) of a m13 phage containing a fragment of the Bam W piece of EBV was hybridized, for the second step 10^7 cpm of a ^{32}P labelled probe prepared by nick translation of replicative form of wild type m13 was used.

Indirect "sandwich" hybridization

It was only recently (Wolf et al 1984) that a new approach was developed which overcomes the need to introduce label in each specific hybridization probe. The technique takes advantage of the recombinant DNA technique and links the specific sequence 1 to another sequence 2. Sequence 2 can conveniently be m13 DNA. These probes are used in a first hybridization step unlabeled and in high concentrations which favors fast and complete hybridization. After removal of excess probe, a second probe is added which is homologous to sequence 2 and ideally contains both orientations of the DNA strands. This second probe can be universally applied and under specific conditions, forms a network on top of sequence 2. An example is given in Fig. 1 where with appropriate exposure time as little as 200 fg of the specific sequences represented in the probe were found in the test DNA. We have used this technique to detect a fragment of EBV DNA in Southern blots; with ^{32}P labeled double-stranded replicative form of m13 for second step hybridization we could detect 0.5 pg of the fragment of EBV (Fig. 1). The sandwich hybridization yielded an amplification of the hybridization signal of up to 100-fold when nonradioactive "second probes" were utilized.

DISCUSSION

Simplified hybridization procedures are necessary if the potential of this technique should be exploited on a wider scale for clinical problems. Viruses which cannot be cultivated in vitro e.g. Rota and Hepatitis B, or where cultivation takes too long e.g. Cytomegalovirus, Dengue, Hepatitis A and particularly where viruses linked to proliferative diseases of man, like EBV, Papillomaviruses or HTLV require reliable and fast tests for the presence of the respective genomic materials. The growing variety of probes raises prohibitory stock-holding problems for routine labs if short-lived isotope labels are used. Chemical iodination of recombinant m13 DNA is a very fast and simple way to prepare hybridization probes. Sandwich hybridization especially if non-radioactive second probes are used is only slightly more complex but reduces logistic problems considerably while allowing excellent sensitivity.

REFERENCES

Commerford, S. Iodination of nucleic acids in vitro.
Biochem. 10:1993-1999 (1971).

Frenkel, N., Locker, H., Cox, B., Roizman, B., and Rapp,
F. Herpes simplex virus DNA in transformed cells:
Sequence complexity in five hamster cell lines and one
derived hamster tumor. J. Virol. 18:885-893 (1976).

Gu, S., Wolf, H., Zeng, Y. Cloning fragments of EBV-DNA
in single-stranded phage ml3 mp8. I. Preparation and
identification of cloned DNA. Cancer (China), 129-277
(1983).

Han, J., and Harding, J. Using iodinated single-
stranded ml3 probes to facilitate rapid DNA sequence
analysis-nucleotide sequence of a mouse-lysine tRNA
gene. Nucleic Acid Res. 11:7 (1983).

Hu, N., and Messing, J. The making of strand-specific
ml3 probes. Gene 17:271-277 (1980).

Langridge, J., Langridge, P., and Bergquist, P.
Extraction of nucleic acids from agarose gels.
Analytical Biochem. 103:264-271 (1980).

Messing, J. New ml3 vectors for cloning. Methods in
Enzymology 101:20 (1983).

Peden, K., Mounts, P., and Hayward, G., Homology
between mammalian cell DNA sequences and human herpesvirus genomes detected by a hybridization procedure
with high-complexity probes. Cell 31:71-80 (1982).

Prensky, W. The radioiodination of RNA and DNA to high
specific activities. Methods in Cell Biology 13:121-
152 (1976).

Richter, W., Gu, S., Seibl, R., and Wolf, H. A new
method for examination of carcinomas of the nasopharynx. Nasopharyngeal carcinoma: Current concepts
(1983).

Vogelstein, B., and Gillespie, D. Preparative and
analytical purification of DNA from agarose. Proc.
Natl. Acad. Sci., 76:615-619 (1979).

MECHANISMS OF EBV TRANSFORMATION

29

THE HOST-CELL RANGE OF THE EPSTEIN-BARR VIRUS

Ronald Glaser, Ph.D.

The Ohio State University

College of Medicine and Comprehensive

Cancer Center

Columbus, Ohio   43210

SUMMARY

The Epstein-Barr virus (EBV) has been classified as a
B-lymphocyte tropic virus, primarily based on the host-
range of EBV under laboratory conditions. Since it is now
generally accepted that the EBV is associated with the
epithelial cells of NPC, an important issue in the pathogenesis of EBV is the host range of cells that can be
infected. Data have accumulated from several laboratories
showing that EBV can infect certain epithelial cells directly, including epithelial explant cell cultures prepared
from NPC biopsy specimens, epithelial tumor cell lines,
normal epithelial nasopharyngeal cells from squirrel monkeys, and, more recently, primary human cervical epithelial
cells. In addition, important information has come from
studies in which laboratory manipulations have been used
to insert EBV or EBV DNA into a variety of cell types.
These procedures include transfection, microinjection, and
receptor implantation experiments, as well as the preparation of epithelial hybrid cells. The ability to demonstrate routine direct infection of epithelial cells, and
more importantly, to demonstrate that they can be transformed by EBV, will help clarify the association of EBV
with NPC.

INTRODUCTION

The Epstein-Barr virus (EBV) is a human oncogenic herpesvirus that is the etiologic agent for infectious mononucleosis (IM) (Henle et al., 1968). It has been very closely associated with several human malignant diseases, including African Burkitt's lymphoma (BL) (Epstein et al., 1964), nasopharyngeal carcinoma (NPC) (zur Hausen et al., 1970), and B-cell lymphoma, particularly in immune-suppressed patients (Saemundsen et al., 1981). Historically, the "laboratory host-range" for EBV has been B-lymphocytes of human origin, as well as B-lymphocytes from certain non-human primates (Miller et al., 1971; Miller and Lipman, 1973). However, when it was recognized that the epithelial cells of NPC tumors were EBV genome positive, rather than infiltrating lymphocytes (Wolf et al., 1973), the search for an explanation of a mechanism(s) through which EBV becomes associated with these cells intensified. This overview will address some of the studies which clarify the association of EBV with epithelial cells, and which ultimately led to proving that EBV can infect certain epithelial cells. In addition, the expression and replication of EBV in cells other than B-lymphocytes or lymphoblasts "infected" with EBV using a variety of laboratory manipulations will also be explored.

HISTORICAL REVIEW

Latently Infected Epithelial Cells In Vivo

The first report to link EBV with epithelial cells of NPC in tumor tissue was reported by Wolf et al. (1973), using an in situ hybridization procedure. This was later confirmed by Klein et al. (1974a) in which EBNA was shown to be associated with the epithelial cells of NPC as well. Several years later, Lemon et al. (1977) found evidence for EBV replication in normal nasopharyngeal epithelial cells exfoliating from the oropharynx of patients with IM. A more recent study has confirmed this original observation and hypothesized that the cells within which EBV is productively replicating in the oropharynx of IM patients may include oropharyngeal epithelial cells (Sixbey et al., 1984). This is consistent with the previously reported hypothesis suggesting that similar cells might be the site of infection in the nasopharynx of squirrel monkeys (Glaser et al., 1978).

In an attempt to search for the presence of EBV DNA or EBNA in "normal" nasopharyngeal epithelial cells in patients who were at risk to NPC in China (as determined by the presence of VCA IgA antibody levels) tissues from persons were examined for the presence of latent EBV (Desgranges et al., 1982). Of 14 normal individuals tested for EBV DNA using biopsies obtained from the nasopharynx, 10 had EBV DNA internal repeats and 11 specimens had EBNA positive cells. The same individuals in both groups also had significant antibody titers to EA IgG. No other gross changes were noted in these normal individuals which suggested that they had NPC. It is possible that EBV was latently associated with the cells in the normal mucosa. It is of interest that several VCA IgA positive individuals were negative for EBV DNA and EBNA in biopsies obtained from their nasopharynx. The significance of this observation is not clear.

There is also evidence that EBV may be associated with other regions of the nasopharynx, including the area designated as Waldeyer's Ring. It has been suggested that EBV might be involved in the etiology of carcinomas of this region. The EBV antibody spectrum in patients with poorly differentiated carcinomas of this anatomical region of the throat was similar to that of those reported of NPC patients (Wolf et al., 1983). In a preliminary report using in situ hybridization, Wolf and co-workers have found that EBV DNA was limited to cells in poorly differentiated carcinomas that appeared to be epithelial cells. It is not known whether epithelial cells obtained from this anatomical region of the nasopharynx, or from carcinomas arising from Waldeyer's Ring, have receptors for EBV and are infectable with the virus.

EBV and Epithelial Cells; In Vitro Studies

From 1964, when EBV was first isolated, until 1971, there were no published data showing that EBV could infect or replicate in any cell other than cells of B-cell lineage, though a variety of cells had been tested to study the host-range of EBV. The first study examining the replication of EBV in a cell other than a B-lymphocyte or a B-lymphoblastoid cell line was performed by preparing epithelial/somatic cell hybrids of BL cell lines (Glaser and O'Neill, 1972). Epithelial/BL hybrid cells, such as D98/

HR-1, were used as a model to study the expression and regulation of the EBV genome in cells with an epithelial morphology (Glaser and Rapp, 1972; Glaser et al., 1973a, 1973b,
1976a; Glaser and Nonoyama, 1974). These cells, grown as a
monolayer, were morphologically compatible with epithelial
cells, but were limited by the fact that they were still
hybrid cells. Indeed, one study showed that at least some
markers of the phenotype of the BL parent were expressed
on the membrane of the epithelial hybrid cells (Glaser et
al., 1977). Nevertheless, these early studies clearly
showed that EBV could remain latent in epithelial-like
cells for many years, and that the cells expressed the
Epstein-Barr virus associated nuclear antigen (EBNA) similar to EBV positive lymphoid cell lines. The epithelial/BL
hybrid cells were also used in early studies that demonstrated that after treatment with certain drugs such as
iododeoxyuridine (IUdR) the virus genome could be induced
to replicate, either in part or in total, depending upon
the conditions (Glaser and O' Neill, 1972; Glaser and Rapp,
1972). The results of these studies suggested that if one
could bypass what was presumably a lack of an appropriate
receptor for EBV on an epithelial-like cell and get the
virus genome into such a cell, virus latency could be
established. In one of these papers (Glaser and Rapp,
1972), an hypothesis was made, based on those data: "An
important conclusion can be drawn from the data presented.
Namely, the EBV genome can persist and can be activated
to synthesize virus in a cell type (epithelial) other than
a lymphoblastoid cell. The possibility exists that the
lymphoblastoid cells, in which the EBV is found in vivo,
may not be the sole site of infectious virus replication,
an observation already made in the case of the Marek's
disease herpes virus (MDHV)." In 1973, after this paper
was published, it was shown that the EBV genome was in the
epithelial cells of NPC tumors. These observations, however, did not resolve the question as to how the EBV became associated with epithelial cells under natural conditions.

Several years after the initial studies with epithe-
lial/BL hybrid cells, already described, an experiment was
performed which demonstrated that the same cells (D98/HR-1)
could be directly superinfected with virus derived from the
HR-1 cell line (lytic virus) (Glaser et al., 1976b). These
data suggested that the EBV receptor was being expressed
on the surface of the D98/HR-1 cells, and that the genetic

information for that receptor came from the HR-1 BL parent cell line. Moreover, it demonstrated that if one found an epithelial cell with the appropriate receptor for EBV, the virus could attach, penetrate, and replicate in a lytic manner to produce at least early antigen (EA). Following this observation, superinfection studies were performed using explanted epithelial tumor tissue obtained from NPC biopsy specimens (Glaser et al., 1976b). Epithelial explant cell cultures were prepared from NPC biopsies and infected with concentrated HR-1 EBV. The cells were found to be positive for EBV antigens as determined by immunofluorescence (IF), confirming the presence of EBV receptors on the cells, and demonstrating susceptibility to EBV. It is possible that the receptor for EBV was expressed in the epithelial tumor cells after the normal epithelial progenitor cells were in some way latently infected with EBV. It is also possible that the normal progenitor nasopharyngeal epithelial cells possessed receptors for EBV and were infected in the nasopharynx at some point earlier in life. The data from these experiments did not rule out either conclusion. Nevertheless, it was the first observation which showed that epithelial cells of any kind, other than epithelial/hybrid cells, had receptors for EBV and could be directly infected.

In an unconfirmed study (Huang et al., 1977), data were obtained suggesting that normal human nasopharyngeal epithelial explant cultures could be stimulated to grow after exposure to transforming virus obtained from the B95-8 cell line. In that study, explant epithelial cell cultures prepared from fresh biopsy specimens from non-neoplastic nasopharyngeal epithelial mucosa, NPC tumors, other tumors of the head and neck, as well as freshly removed tonsils, were compared to noninfected control explant cultures 14 days after infection. A significant enhancement of growth was found in normal nasopharyngeal epithelial cells exposed to B95-8 virus when compared to the other cell cultures, or mock-infected cultures. In addition, several of the nasopharyngeal epithelial explant cell cultures showed increased growth characteristics, as well as cellular morphological changes. The implication was that the normal nasopharyngeal epithelial cells had been infected with EBV; however, no evidence, e.g., EBV antigens or the presence of virus DNA, was reported.

Additional data to support the hypothesis that normal nasopharyngeal epithelial cells can be infected with EBV came from a study in which nasopharyngeal epithelial cell explant cultures prepared from squirrel monkey tissue were used. It had been previously shown that squirrel monkeys exposed to B95-8 EBV by application of virus concentrates to the nasopharynx by pipetting virus into the nose and throat, seroconverted (EA, VCA and EBNA) for EBV in the absence of disease (Glaser et al., 1978). This demonstrated that an immune response could be induced with EBV administered by the nasopharyngeal route without the necessity of directly inoculating the virus by injecting tissue with a syringe. These results strongly suggested that there were cells in the nasopharynx of squirrel monkeys (presumably on the surface) which had receptors for EBV and in which EBV could lytically replicate. The infected cells were probably B-lymphocytes, but might also have included epithelial cells.

Attempts were then made to determine if such tissue did contain epithelial cells with receptors for EBV. It was subsequently shown that normal nasopharyngeal explant cell cultures from squirrel monkeys had EBV receptors and could be infected, showing the ability of the virus to bind to the surface of receptor-positive cells (demonstrated by electron microscopy) and to express EBV EA after infection, as determined by IF (Glaser et al., 1980).

In a subsequent study, it was demonstrated that a human epithelial tumor cell line, designated U, could be infected with the non-lytic transforming B95-8 strain of EBV (Ben-Bassat et al., 1982). Cells infected with B95-8 EBV were EBNA positive and showed an increase in cellular DNA synthesis. Mock-infected cells were negative, as were cells infected with HR-1 EBV.

Thus, a series of studies, beginning with the use of epithelial/hybrid cells and ultimately including a variety of different kinds of epithelial cells, have now conclusively demonstrated that EBV can infect and replicate in cells of epithelial origin and that the host range for EBV is not limited to B-lymphocytes. These studies, as well as speculations regarding the reasons for the problems found in attempts to infect epithelial cells with EBV have been explored in previous reviews (Simons and Shanmugaratnam [eds.], 1982; 1982; Glaser et al., 1984 [in press]).

RECENT STUDIES ON INFECTION OF EPITHELIAL CELLS
WITH EBV

As already discussed, it is clear that EBV receptors
are expressed on the surface of certain epithelial tumor
cells, as well as normal nasopharyngeal epithelial cells
from the nasopharynx of squirrel monkeys. It has been
very difficult to demonstrate a similar phenomenon using
normal human nasopharyngeal tissue for reasons which are
not clear (Glaser, 1984, in press). In a recent study, it
was possible to show that primary human epithelial cells
prepared from cervical tissue have receptors for EBV and
can be infected with EBV derived from throat washings from
acute IM patients (Sixbey et al., 1983). Infected cells
showed EA and VCA associated with viral replication, and
EBV DNA could be detected in the cells. In addition, EBNA
was also expressed in the EBV infected cells. Supernatants
prepared from EBV infected cell cultures were used in an
attempt to transform cord blood lymphocytes; however, no
evidence for the production of transforming EBV under
these conditions was found.

Another recent report (preliminary note) has suggested
that normal human nasopharyngeal epithelial cells might be
infectable with EBV (Thompson et al., 1983). Cell cultures
derived from several normal tissue specimens were exposed
to EBV by adding the virus to medium for several weeks, or
by co-cultivation with lethally X-irradiated EBV producing
lymphoblastoid cells. Cells sloughing from cell sheets
which grew out from the tissue showed VCA staining by IF
in the cultures that had been co-cultivated with virus producing cells. Virus particles were detected by electron
microscopy in cells showing bundles of keratin, supporting
the association of EBV with the epithelial cells. The
cells being shed from the monolayers were described as undergoing differentiation, leaving the basal layer of undifferentiated cells behind; cells in the basal layer were EBV-
VCA negative. It is of interest that virus replication
observed in these two studies seemed to be limited to ter-
minally differentiated epithelial cells. Both of these
studies are consistent with the data showing replication
of the MDHV in differentiated epithelial cells in feather
follicles of infected chickens as previously discussed
(Calnek et al., 1970).

It is interesting to speculate that if EBV becomes
associated with undifferentiated epithelial cells in the
nasopharynx, the virus genome is repressed and a latent
infection results. Virus could remain in these cells for
years without malignant transformation, or, in certain
individuals might induce NPC. Perhaps it is at this level
that cofactors such as chemical carcinogens could interact
with latently infected epithelial cells. The significance
of EBV replicating lytically in differentiating epithelial
cells needs further clarification and should be an area
for further studies.

EXPRESSION OF THE EBV GENOME AFTER TRANSFECTION

In order to study the replication and expression of
the EBV genome in cells lacking a receptor for EBV, research
was initiated using the calcium precipitation technique in
which intact EBV DNA or cloned fragments were transfected
into epithelial cells. Several studies have shown that
EBV DNA obtained from several different virus isolates can
be successfully transfected into different cell types,
including epithelial cells. In one study, EBV DNA was
obtained from nontransforming HR-1 EBV and transfected
into two different epithelial tumor cell lines. The EBV
DNA was also obtained from a transforming strain of EBV,
AG-876, and transfected into the same cells (Stoerker et
al., 1981). The results showed that successful transfection was accomplished by the expression of EBV specific
EA/VCA in a small percentage of cells, as well as the
presence of EBV DNA. Two interesting observations were
made from those experiments. First, DNA obtained from
both transforming and nontransforming EBV produced lytic
replication in both epithelial cell types under the same
conditions, and second, that the expression of HR-1 DNA
was restricted to EA only in one of the epithelial cell
lines and not the other. This latter observation is consistent with several published studies showing the importance of the host cell on the expression and replication
of EBV, described earlier.

It was also demonstrated that human placenta cells
(fibroblasts) could be transfected with EBV DNA, and these
data suggested that the virus could replicate in such
cells to produce infectious transforming virus (Miller et
al., 1981). Subsequent experiments showed that EBV DNA

could be used to successfully transfect Vero, BSC-1 and
owl monkey kidney cells, and that EA could be expressed in
these cells (Grogan et al., 1981).

In a recent study, transfection techniques were used
to transform epithelial cells (Griffin and Karron, 1984).
African green monkey kidney (AGMK) cells were transfected
with different regions of the EBV genome. Using DNA from
cosmid clones, designated P13/P33 and P-31, AGMK cells
were obtained which had extended growth characteristics.
The transformed cells grew in medium with low serum concentrations, as well as in soft agar. The cells, however,
did not induce tumors in nude mice, and therefore did not
appear to be oncogenically transformed. The cells were
negative by IF for all EBV antigens, including EBNA. Whether transformation of the AGMK cells was due to the genetic information associated with the transfected regions
of the EBV genome or to an endogenous retrovirus oncogene
enhanced by an interaction with EBV DNA is not known. It
will be of interest to see if this observation can be confirmed and extended so that the ability to study transformation of epithelial cells by EBV can finally be explored
and delineated.

MICROINJECTION

Another method used successfully to study the expression and replication of EBV in non-lymphoblastoid cells
has been microinjection with EBV DNA. The first report of
successful microinjection of EBV DNA into cells came from
Graessman et al. (1980) who successfully transfected human
diploid fibroblast cells, rat fibroblasts, and AGMK cells.
The only EBV marker expressed in these cells was EA. The
apparent absence of VCA expression may have been associated
with a restriction of the EBV genome by the microinjected
cells; similar results were obtained in a study after
transfection (Stoerker et al., 1981).

In a second and more recent study it was shown that
two different epithelial tumor cell lines could be microinjected with intact EBV DNA as well as cloned fragments.
The procedure is presently being used for functional mapping studies of the EBV genome (Glaser et al., 1983; Boyd
et al., 1985, this volume).

STUDIES OF EBV REPLICATION IN CELLS WHICH
HAVE TRANSPLANTED EBV RECEPTORS

Infection of epithelial cells which have had EBV receptors implanted on the cell membrane have resulted in the expression of EA, VCA and EBNA. It was also possible to show virus replication in several other cell types, including T-lymphocytes derived from human and mouse origin, and mouse fibroblasts (Volsky et al., 1980; Shapiro and Volsky, 1982). Similar studies from another laboratory have also shown the usefulness of this procedure in studying the replication of EBV in receptor-negative cells (Khelifa and Menezes, 1983a). Their data suggest that the binding of EBV to cell membranes could be mediated by the Sendai virus hemagglutinin-neuraminidase molecule. The authors proposed that one possible mechanism that might allow or enhance penetration of EBV into normal nasopharyngeal epithelial cells might be the presence of viruses which produce hemagglutinin-neuraminidase, which could become associated with the surface of epithelial cells and enhance infection.

CONCLUSIONS

This overview has addressed the specific issue of infectability of epithelial cells by EBV, and the importance of these data for both latency and the induction of NPC. This issue has been controversial and it is this author's opinion that the number of studies from different laboratories using different cells and different EBV isolates should settle the issue; the host range of EBV clearly includes certain epithelial cells, as well as B-lymphocytes. The overview also included data from studies in which cells have been manipulated in various ways to enhance the infectability of EBV or to bypass EBV receptor-negative cell membranes. While many of these studies are artificial, and may have little to do with processes under natural conditions, they do yield information which may ultimately be important in understanding the replication of the EBV genome in nonlymphoblastoid cell types, as well as the expression of the virus genome vis-a-vis the cell genome under a variety of conditions. As these data are put into a better perspective, some of the studies may yield important information in learning about EBV and NPC.

The approaches outlined in this overview are also important for studying EBV in general, particularly in the
areas of cellular and molecular biology, and for that reason are worth pursuing. However, the ultimate experiment,
in this author's opinion, that will clarify the essential
role of EBV in NPC from a virological standpoint, is one
that will show both that normal human nasopharyngeal epithelial cells can be infected with EBV and transformed
after infection with properties consistent with oncogenically transformed tumor cells.

REFERENCES

Ben-Bassat, H., Mitrani-Rosenbaum, S., and Goldblum, N.,
Induction of Epstein-Barr virus nuclear antigen and DNA
synthesis in a human epithelial cell line after Epstein-
Barr virus infection. J. Virol., 41:703-708 (1982).

Boyd, V.A., Stoerker, J., Holliday, J.E., and Glaser, R.,
Mapping of Epstein-Barr virus antigens by microinjection
of human cells. This volume (in press).

Calnek, B.W., Adlinger, H.K., and Kahn, D.E., Feather-
follicle epithelium: A source of envelope and infectious
cell-free Herpesvirus of Marek's disease. Avian Dis., 14:
219-233 (1970).

Desgranges, C., Bornkamm, W., Zeng, Y., Wang, P.C., Zhu,
J.S., Shang, M., and de-The, G. Detection of Epstein-Barr
viral DNA internal repeats in the nasopharyngeal mucosa of
Chinese with IgA-EBV specific antibodies. Int. J. Cancer
29:87-95 (1982).

Epstein, M.A., Achong, B.G., and Barr, Y.M., Virus particles in cultured lymphoblasts from Burkitt's lymphoma.
Lancet, 1:702-703 (1964).

Glaser, R., and O'Neill, F.J., Hybridization of Burkitt
lymphoblastoid cells. Science, 176:1245-1247 (1972).

Glaser, R., and Rapp, F., Rescue of Epstein-Barr virus from
somatic cell hybrids of Burkitt lymphoblastoid cells.
J. Virol., 10:288-296 (1972).

330

Glaser, R., Decker, B., Farrugia, R., Shows, T., and Rapp, F., Growth characteristics of Burkitt somatic cell hybrids in vitro. Cancer Res., 33:2026-2029 (1973a).

Glaser, R., Nonoyama, M., Decker, B., and Rapp, F., Synthesis of Epstein-Barr virus antigens and DNA in activated Burkitt somatic cell hybrids. Virology, 55:62-69 (1973b).

Glaser, R., and Nonoyama, M., Host cell regulation of induction of Epstein-Barr virus. J. Virol., 14:174-176, (1974).

Glaser, R., Farrugia, R., and Brown, N., Effect of the host cells on the maintenance and replication of Epstein-Barr virus. Virology, 69:132-142 (1976a).

Glaser, R., de-The, G., Lenoir, G., and Ho. J.H.C., Superinfection of epithelial nasopharyngeal carcinoma cells with Epstein-Barr virus. Proc. Nat. Acad. Sci., 73:960-963, (1976b).

Glaser, R., Lenoir, G., Ferrone, S., Pellegrino, M., and de-The, G., Cell surface markers on epithelial/Burkitt hybrid cells superinfected with Epstein-Barr virus. Cancer Res. 37:2291-2296 (1977).

Glaser, R., Lee, K.J., Lang, C.M., and Levy, B., Seroconversion against Epstein-Barr virus in two non-human primate species infected by the oropharyngeal route. J. Infec. Dis. 138:695-698 (1978).

Glaser, R., Lang, C.M., Lee, K.J., Schuller, D.E., Jacobs, D., and McQuattie, C., Attempt to infect non-malignant nasopharyngeal epithelial cells from humans and squirrel monkeys with Epstein-Barr virus. J. Nat. Cancer Inst., 64: 1085-1090, (1980).

Glaser, R., Boyd, A., Stoerker, J., and Holliday, J., Functional mapping of the Epstein-Barr virus genome: Identification of sites coding for the restricted early antigen, the diffuse early antigen and the nuclear antigen. Virology, 129:188-198 (1983).

Glaser, R., Noyes, I., and Milo, G., Pathogenesis of Epstein-Barr virus infection: The Host-Range of EBV now includes epithelial cells. J. Cell. Biochem. (in press).

Graessmann, A., Wolf, H., and Bornkamm, G.W., Expression of Epstein-Barr virus genes in different cell types after microinjection of viral DNA. Proc. Nat. Acad. Sci. USA, 77:433-436 (1980).

Griffin, B.E., and Karran, L., Immortalization of monkey epithelial cells by specific fragments of Epstein-Barr virus DNA. Nature, 309:78-82 (1984).

Grogan, E., Miller, G., Henle, W., Rabson, M., Shedd, D., and Niederman, J.C., Expression of Epstein-Barr viral early antigen in monolayer tissue cultures after transfection with viral DNA and DNA fragments. J. Virol., 40:861-869 (1981).

Henle, G., Henle, W., and Diehl, V., Relation of Burkitt tumor associated herpes-type virus to infectious mononucleosis. Proc. Nat. Acad. Sci. USA, 59:94-101 (1968).

Huang, D.P., Ho, H.C., Ng, M.H., and Lui, M., Possible transformation of nasopharyngeal epithelial cells in culture with Epstein-Barr virus from B95-8 cells. Brit. J. Cancer, 35:630-634 (1977).

Khelifa, R., and Menezes, J., Sendai virus envelopes can mediate Epstein-Barr virus binding to and penetration into Epstein-Barr virus receptor-negative cells. J. Virol. 46:325-322 (1982).

Klein, G., Giovanella, B.C., Lindahl, T., Fialkov, P.J., Singh, S., and Stehlen, J., Direct evidence of the presence of EBV in cells from patients with anaplastic carcinoma of the nasopharynx. Proc. Nat. Acad. Sci. 71:4737-4740 (1974a).

Lemon, S.M., Hutt, L.M., Shaw, J.E., Li, J-L.H., and Pagano, J.S., Replication of EBV in epithelial cells during infectious mononucleosis. Nature, 268:268-270 (1977).

Miller, G., Lisco, H., Kohn, H.I., and Stitt, D., Establishment of cell lines from normal adult human blood leukocytes by exposure to Epstein-Barr virus and neutralization by human sera with Epstein-Barr virus antibody. Proc. Soc. Exp. Biol. Med., 137:1459-1465 (1971).

Miller, G., and Lipman, M., Release of infectious Epstein-Barr virus by transformed marmoset leukocytes. Proc. Nat. Acad. Sci. USA. 70:190-194, (1973).

Miller, G., Grogan, E., Heston, L., Robinson, J., and Smith, D., Epstein-Barr viral DNA: Infectivity for human placental cells. Science, 212:452-455 (1981).

Saemundsen, A.K., Purtillo, D.T., Sakamoto, K., Sullivan, J.L., Synnerholm, A.C., Hanto, D., Simmons, R., Anvret, M., Collins, R., and Klein, G., Documentation of Epstein-Barr virus infection in immunodeficient patients with life-threatening lymphoproliferative diseases by Epstein-Barr virus complementary RNA-DNA and virus DNA-DNA hybridization. Cancer Res., 41:4237-4242 (1981).

Shapiro, I.M., and Volsky, D.J., Infection of normal epithelial cells by Epstein-Barr virus. Science, 219:1225-1228 (1982).

Simons, M.J., and Shanmugaratnam, K. (eds.), The Biology of Nasopharyngeal Carcinoma, Report No. 16, U.I.C.C. Technical Report Series, Vol. 71, U.I.C.C., Geneva (1982).

Sixbey, J.W., Vesterinen, E.H., Nedrud, J.G., Raab-Traub, N., Walton, L.A., and Pagano, J.S., Replication of Epstein-Barr virus in human epithelial cells infected in vitro. Nature, 306:480-483 (1983).

Sixbey, J.W., Nedrud, J.T., Raab-Traub, N., Hanes, R.A., and Pagano, J.S., Epstein-Barr virus replication in oropharyngeal epithelial cells. N. Eng. J. Med., 310:1225-1230 (1984).

Stoerker, J., Parris, D., Yajima, Y., and Glaser, R., Pleiotropic expression of Epstein-Barr virus DNA in human epithelial cells. Proc. Nat. Acad. Sci. USA, 78:5852-5855 (1981).

Thompson, J.L., Epstein, M.A., Achong, B.G., Chen, J.J., Production of EB virus by normal human nasopharyngeal epithelial cells exposed to the virus in vitro. Ann. Virol. (Inst. Pasteur), 134E:573-579 (1983).

Volsky, D.J., Shapiro, I.M., and Klein, G., Transfer of
Epstein-Barr virus receptors to receptor-negative cells
permits virus penetration and antigen expression. Proc.
Nat. Acad. Sci. 77:5453-5457 (1980).

Wolf, H., zur Hausen, H., and Becker, V., EBV virus genomes
in epithelial nasopharyngeal carcinoma cells. Nature New
Biol., 244:245-247, (1973).

Wolf, H., Hause, M., and Wilms, E., On the viral etiology
of carcinomas of Waldeyer's Ring. In Nasopharyngeal Carcinoma: Current Concepts, Prasad, U., Ablashi, D.V.,
Levine, P.H., and Pearson, G.R. (eds.), University of
Malaysia, Kuala Lumpur, p. 277-279 (1983).

zur Hausen, H., Schulte-Holthausen, H., Klein, G., Henle,
W., Henle, G., Clifford, P.,and Santesson, L., EB virus DNA
in biopsies of Burkitt tumors and anaplastic carcinomas of
the nasopharynx. Nature, 228:1056, (1970).

30

EPSTEIN-BARR VIRUS (EBV) GROWTH TRANSFORMATION IS ASSOCIATED
WITH AN ALTERATION IN c-myc CHROMATIN STRUCTURE

William H. Schubach[1], Bart H. Steiner[1], and
Mark Birkenbach[2]

Section of Medical Oncology, Department of
Medicine[1], Department of Laboratory Medicine
and Pathology[2], University of Minnesota,
Minneapolis, MN 55455, USA

SUMMARY

Because the c-myc gene is involved in translocations in
Burkitt's lymphoma, and because of the association of this
neoplasm with EBV infection, we have studied the chromatin
structure of the c-myc gene in EBV growth-transformed cells
and in an isogenic population of cells enriched for B-cells.
No correlation between the location of DNaseI-hypersensitive
sites and translocation breakpoints in c-myc was found. A
novel DNaseI-hypersensitive site appears within the second
c-myc intron of growth-transformed cells that is absent in
nontransformed cells. In contrast, the pattern of nuclease
sensitivity around the pro-α-2-collagen gene is identical in
both cells. Thus some gene product encoded or induced in
EBV transformed cells may serve to alter chromatin structure
in c-myc. This alteration might play a role in controlling
c-myc expression in growth-transformed cells.

INTRODUCTION

The c-myc gene is the cellular progenitor of the
transforming gene of the avian myelocytomatosis virus
(Sheiness et. al.,1980). Altered expression of c-myc has
been found in a variety of human and animal neoplasms (Eva
et. al.,1982). In murine plasmacytomas and human Burkitt's
lymphoma a class of rearrangements brings the c-myc gene in
proximity with an immunoglobulin gene (reviewed in Perry,

1983 and Robertson, 1983). Although this translocation is correlated with malignant transformation, the role of c-myc in lymphomagenesis is undefined. In Burkitt's lymphoma the translocation may result in enhanced levels of c-myc expression (Westin et. al,1982), sometimes employing an immunoglobulin enhancer (Hayday et. al. 1984). It may result in synthesis of an altered c-myc mRNA (Saito et. al., 1983), or it may result in alteration of the cycle-specific regulation of c-myc expression that characterizes nontransformed cells (Kelly et. al.,1983).

Alteration of chromatin structure, as determined by the appearance of DNaseI hypersensitive sites, accompanies transcriptional activation of specific genes. DNaseI-hypersensitive sites are relatively local, double-stranded cuts that occur often, but not always, near the 5' end of actively transcribed genes (reviewed in Elgin, 1981). DNaseI-hypersensitive sites also correlate with chromatin structures that mediate tissue-specific transcriptional enhancement (Shermoen and Beckendorf, 1982; Parslow and Granner, 1983). In addition, DNaseI-hypersensitive sites may coincide with sites of region-specific recombination as has been observed in the yeast mating type locus (Nasmyth, 1982) and at a site of switch recombination in immunoglobulin heavy chain genes (Mills et. al., 1983). In avian bursal lymphomas, sites of avian leukosis virus provirus intergration are clustered in regions that correspond to DNaseI-hypersensitive sites that are present in the unrearranged allele (Schubach and Groudine, 1984).

Because of the association of antecedent EBV infection with endemic Burkitt's lymphoma (De The et. al., 1978) and because of the above-noted translocations involving c-myc, we wanted to determine whether EBV growth transformation correlated with alterations of the chromatin structure in c-myc that might in turn be correlated with translocation breakpoints. In addition, because alteration of chromatin structure in the region of a gene can be mediated by trans-acting factors (Emerson and Felsenfeld, 1984; Green et. al., 1983) we wished to determined whether EBV transformation was associated with alterations in c-myc chromatin structure. Such an alteration in chromatin structure could mediate changes in the expression of a cellular gene such as c-myc, whose function may play a role in growth transformation.

MATERIALS AND METHODS

Isolation of Cells and Nuclei: 10^9 peripheral blood mononuclear cells were harvested from a normal, EBNA-negative individual (1944) in a Fenwall CS3000 leukopheresis apparatus. This preparation was enriched for B-cells by separation on Ficoll-hypaque, adsorption to plastic to remove macrophages, and rosetting T-cells with AET-treated sheep red blood cells (Torok-Storb et. al., 1983). Peripheral blood mononuclear cells from the same individual were transformed with the B95-8 strain of EBV and grown in RPMI supplemented with 10% fetal calf serum. The cells used were grown in mass culture and contained approximately 50 EBV genomes per cell. Nuclei were isolated by lysis of cells in NP40 as previously described (Schubach and Groudine, 1984).

DNase I Digestion and Blot Hybridization: Nuclei were suspended on ice a at nucleic acid concentration of 0.6 mg/ml in RSB (10 mM NaCl, 3mM $MgCl_2$,10mM TRIS pH 8.3), made 10^{-4} M $CaCl_2$, and digested with various amounts of bovine pancreatic DNaseI (Sigma) for 10 minutes at 37^0.
DNA was prepared from DNaseI-treated nuclei as previously described (Schubach and Groudine, 1984). For blot hybridization, 20 µg of DNA were digested with 30 units of the indicated restriction enzyme(s) overnight. DNA was electrophoresed in neutral agrose gels, blotted onto nitrocellulose (Southern, 1975), and probed with the indicated ^{32}P-labelled probes.

Preparation of Probes: Segments of the human c-myc region were subcloned into M13mp18. The identity of the fragment was verified by dideoxy sequence determination (Sanger et. al., 1977). The insert was isolated on a neutral acrylamide gel, eluted, and nick translated. Hybridization and wash conditions were as previously described (Schubach and Groudine, 1984).

RESULTS

We have compared the pattern of DNaseI-hypersensitive sites in the region of the c-myc gene that is found in EBV transformed B-cells with that found in a population of nontransformed cells. The non-transformed cells in this study were isolated from a healthy, EBNA-negative person by leukopheresis. Mononuclear cells were partially enriched for a population of B-cells and null cells by ficoll-hypaque separation, adsorption to plastic to remove macrophages, and rosetting T-cells by adsorption to sheep RBCs. This enrichment scheme resulted in a population of cells that was 25-35% polyvalent immunoglobulin positive and 20-30% sRBC-rosette positive. Less than 5% of the cells were macrophages as judged by morphologic staining. Thus, between 70% and 80% of the cells were a mixture of B and null cells. These cells (1944NL) were used as the source of nuclei.

Peripheral blood mononuclear cells from the same individual were growth transformed by the B95-8 strain of EBV (1944 EBV) and used to prepare nuclei. The transformed cells contained approximately 50 EBV genomes per cell. Because minor DNA sequence polymorphisms can alter the appearance of DNaseI-hypersensitive sites (McGinnis et. al, 1983) we chose to use cells from the same individual for this comparative analysis.

We identified the location of DNaseI-hypersensitive sites by digesting nuclei from the normal (1944NL) and EBV-transformed cells (1944EBV) with increasing concentrations of DNaseI. DNA was isolated from these nuclei, cleaved with restriction enzymes, blotted onto nitrocellulose, and hybridized to cloned, ^{32}P-labelled DNA fragments of c-myc. Figure 1A shows the results of one such experiment. DNA was cleaved with ClaI, separated on a 0.8% agarose gel, blotted, and probed with the 0.4 kb PstI fragment designated in Figure 1B. The probe detects a "parent" fragment of 6.5 kb. In addition the blot demonstrates subbands, one end of which is generated by the restriction enzyme and the other end of which is generated by cleavage by DNaseI. The subbands correspond to hypersensitive sites (I and II) which lie at the sites designated on the restriction map in Figure 1B. Within the level of detection of this study, no differences in the patterns of DNaseI-hypersensitive sites were found that distinguish 1944NL from 1944EBV cells in the region extending 3kb upstream from the first (non-coding) c-myc exon to the beginning of the second exon.

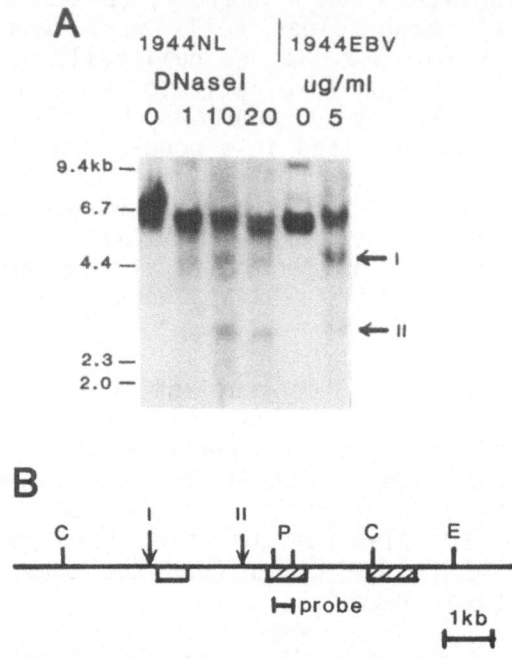

Figure 1: Identification of DNaseI-hypersensitive sites upstream from the second c-myc exon. (A) Nuclei were digested with DNaseI at the concentrations indicated, DNA isolated, cleaved with ClaI, electrophoresed in a 0.8% agarose gel, blotted onto nitrocellulose, and probed with the 0.4kb PstI fragment denoted in (B). Sub-bands (I and II) are designated by arrows. (B) Location of DNaseI hypersensitive sites in the region. Identical subbands are found in 1944NL and 1944EBV cells. Restriction enzymes are abbreviated as follows: C=ClaI; S=SmaI; P=PstI; E=EcoRI. Molecular weight marker locations are at the left.

The hypersensitive sites downstream from the second exon were studied using the indirect end-labelling technique (Wu, 1980). In the example shown in Figure 2, DNA was isolated from DNaseI-treated nuclei, digested with SmaI and ClaI, electrophoeresed on a 1.1% agrose gel, and blotted onto nitrocellulose using as a probe the 0.39kb SmaI-PstI fragment depicted in Figure 2B. Using this probe, which hybridizes to the end of the restriction fragment, all the DNaseI hypersensitive sites downstream from the SmaI site can be directly located. Several hypersensitive sites (III, IV, V, VI and VII) are located in this region in both 1944NL

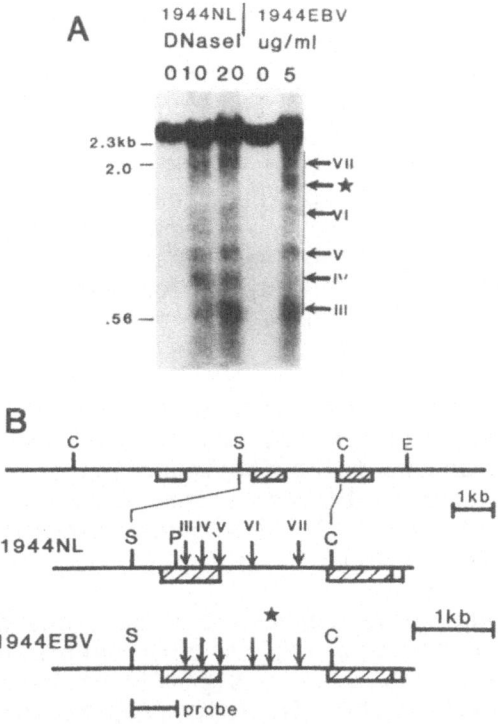

Figure 2: Identification of DNaseI-hypersensitive sites in the second c-myc exon and intron. (A) The protocol was the same as in Figure 1, except that the DNA was cleaved with SmaI and ClaI, electrophoresed in a 1.1% gel, and the blot probed with the .39kb PstI-SmaI fragment designated in B. Subbands (III, IV, V, VI, and VII) are indicated by arrows. A subband unique to 1944EBV cells is denoted by a star. (B) Location of DNaseI-hypersensitive sites in this region in 1944NL and 1944EBV cells.

and 1944EBV cells. A new hypersensitive site in the middle of the second intron is present in 1944EBV cells that is absent in 1944NL cells. This difference persists in 1944EBV cells treated with TPA to induce virus production (Lin et. al., 1979). No other distinguishing hypersensitive sites were found in the region studied, which encompasses 8kb downstream from the first ClaI site in Figure 1. By contrast, the pattern of DNaseI-hypersensitive sites around the human pro-α-2-collagen gene were identical in 1944NL and 1944EBV cells (data not shown). The novel hypersensitive site seen in 1944EBV cells was not found in fetal liver cells (data not shown).

DISCUSSION

We have studied the distribution of DNaseI-hypersensitive sites in the c-myc gene in a population of EBV growth-transformed cells and in an isogenic population of cells partially enriched for normal B-cells. The original impetus for this inquiry was to determine whether there was a correlation between the location of hypersensitive sites and the translocation breakpoints in the c-myc region found in various Burkitt's lymphoma cell lines. No distinguishing hypersensitive sites were found in the first intron or in the region 2kb upstream from the first exon, a region where the majority of translocation breakpoints have been mapped.

A novel DNaseI-hypersensitive site was found in the second c-myc intron in EBV transformed cells during both latent infection and following induction of virus production by TPA. By contrast, no differences in the chromatin structure were found in the pro-α-2-collagen region of 1944NL and 1944EBV cells.

The 1944NL cells represent a heterogeneous population of mononuclear cells which is only partially enriched for normal B-cells. There is a significant percentage of both T-cells and null cells in this preparation. Thus it is possible that the distinguishing hypersensitive site in 1944EBV cells is a feature of the B-cell lineage, and not a direct or indirect result of EBV transformation.

Another possible interpretation of this finding is that an EBV-encoded protein or a cellular protein induced in EBV transformed cells binds to the c-myc region creating the observed hypersensitive site in the second intron. Such a factor (or factors) might be involved in a subtle alteration of c-myc expression in EBV transformed cells. Both mitogen stimulation and EBV transformation enable B-cells to traverse critical phases of the cell cycle and acquire transformation "competence" (DeFranco et. al., 1982). It has been shown that mitogen stimulation causes a 20-fold increase in c-myc transcription independent of new protein synthesis (Kelly et. al.,1983) It is tempting to speculate that EBV and mitogen stimulation alter expression of specific genes that in turn alter the growth properties of the cell. It should be noted, however, that a region which binds a nuclear factor in B-cells and has been proposed as a regulatory region for c-myc expression (Siebenlist et. al., 1984), was not included in the region encompassed by this study.

The possible role of EBV encoded proteins in altering chromatin structure is testable by construction of selectable vectors containing EBV fragments, such as that encoding EBNA (Fischer et. al.,1984), transfection into appropriate cells, and analysis of the resultant patterns of nuclease sensitivity.

ACKNOWLEDGEMENTS

We wish to thank K. Gajl-Peczalska for surface marker studies and Iris Cochran for typing the manuscript. This work was supported by NIH grant CA36977 from the NCI, by ACS grant IN-13-W-15, and by grants from the Minnesota Medical Foundation and the Graduate School of the University of Minnesota.

REFERENCES

DEFRANCO, A.L., KUNG, J.T., and PAUL, W.E., Regulation of growth and proliferation in B cell populations. Immunol. Rev. 64, 161-182 (1982).

DE THE, G. GESER, A., DAY, N.W., TUKEI, P.M., WILLIAMS, E.H., BERI, D.P., SMITH, P.G., DEAN, A.G., BORNKAMM, G.W., FEORINO, P., and HENCE, W., Epidemiological evidence for causal relationship bewteen Epstein-Barr virus and Burkitt's lymphoma from Ugandan Prospective Study. Nature 274, 756-761 (1978).

ELGIN, S.C.R., DNAase I-hypersensitive sites of chromatin. Cell 27, 413-415 (1981).

EMERSON, B.M., and FELSENFELD, G., Specific factor conferring nuclease hypersensitivity at the 5' and of the chicken adult β-globin gene. Proc. Nat. Acad. Sci. USA 81, 95-99 (1984).

EVA, A., ROBBINS. K.C., ANDERSEN, P.R., SRINIVASAN, A., TRONICK, S.R., REDDY, E.P., NELSON, W.E., GALEN, A.T.,

LAUTENBERGER, J.A., PAPAS, T.S., WESTIN, E.H., WONG-STAAL, F., GALLO, R.C., and AARONSON, S.A., Cellular genes analogous to retroviral oncogenes are transcribed in human tumour cells. Nature 295, 116-119 (1982).

FISCHER, D.,., ROBERT, M.F., SHEDD, D., SUMMERS, W.P., ROBINSON, J.E., WOLAK, J., STEFANO, J.E., and MILLER, G., Identification of Epstein-Barr nuclear antigen polypeptide in mouse and monkey cells after gene transfer with a cloned 2.9-kilobase-pair subfragment of the genome. Proc. Nat. Acad. Sci. USA., 81, 43-47 (1984).

GREEN, M.R., TREISMAN, R., and MANIATIS, T., Transcriptional activation of cloned human β-globin genes by viral immediate-early gene products. Cell 35, 137-148 (1983).

HAYDAY, A.C., GILLIES, S.D., SAITO, H., WOOD, C., WIMAN, K., HAYWARD, W.S., and TONEGANA, S., Activation of a translocated human c-myc gene by an enhancer in the immunoglobulin heavy-chain locus. Nature 307, 334-340 (1984).

KELLY, K., COCHRAN, B.H., STILES, C.D., and LEDER, P., Cell-specific regulation of the c-myc gene by lymphocyte mitogens and platelet-derived growth factor. Cell 35, 603-610 (1983).

LIN, J-C., SHAW, J.E., SMITH, M.C., and PAGANO, J.S., Effect of 12-O-tetradelanoyl-phorbol-13-acetate on the replication of Epstein-Barr virus I. Characterization or viral DNA. Virology 99, 183-187 (1979).

MCGINNIS, W., SHERMOEN, A.H., HEEMSKERK, J., and BECKENDORF, S.K., DNA sequence changes in an upstream DNaseI-hypersensitive region are correlated with reduced gene expression. Proc. Nat. Acad. Sci. USA 80, 1063-1067 (1983).

MILLS, F., FISHER, M., KURODA, R., FORD, A., and GOULD, H., DNAaseI hypersensitive sites in the chromatin of human mu immunoglobulin heavy-chain genes. Nature 306, 809-812 (1983).

NASMYTH, K.A., Regulation of yeast mating-type chromatin structure by SIR: An action at a distance affecting both transcription and transposition. Cell 30, 567-578 (1982).

PARSLOW, T., and GRANNER, D. Structure of a nuclease-sensitive region inside the immunoglobulin kappa gene: evidence for a role in gene regulation. Nucl. Acids. Res. 11, 4775-4792 (1983).

PERRY, R.P., Consequences of myc invasion of immunoglobulin loci: facts and speculations. Cell 33, 647-479 (1983).

ROBERTSON, M., Paradox and paradigm: the message and meaning of myc. Nature 306, 733-736 (1983).

SAITO, H., HAYDAY, A.C., WIMAN, D., HAYWARD, W.S., and TONEGANA, S., Activation of the c-myc gene by translocation: A model for translational control. Proc. Nat. Acad. Sci. USA, 80, 7476-7480 (1983).

SANGER, F., NICKLEN, S., and COULSON, A.R., DNA sequencing with chain terminating inhibitors. Proc. Nat. Acad. Sci. USA, 74, 5463-5467 (1977).

SCHUBACH, W., and GROUDINE, M., Alteration of c-myc chromatin structure by avian leukosis virus integration. Nature 307, 702-708 (1984).

SHEINESS, D.K., HUGHES, S.H., VARMUS, H.E., STUBBLEFIELD, E., and BISHOP, J.M., The vertebrate homologue of the putative transforming gene of avian myelocytomatosis virus: characteristics of the DNA locus and its RNA transcript. Virology 105, 415-424 (1980).

SHERMOEN, A.W., and BECKENDORF, S.K., A complex of interacting DNAaseI-hypersensitive sites near the Drosophila glue protein gene, Sgs4. Cell 29, 601-607 (1982).

SIEBENLIST, U., HENNIGHAUSEN, L., BATTEY, J., and LEDER, P., Chromatin structure and protein binding in the putative regulatory region of the c-myc gene in Burkitt's lymphoma. Cell 37, 381-391 (1984).

SOUTHERN, E.M., Detection of specific sequences among DNA fragments separated by agarose gel electrophoresis. J. Mol. Biol. 98, 503-515 (1975).

TOROK-STORB, B., NEPOM, G.T., NEPOM, B.S., and HANSEN, J.A., HLA-DR antigens on lymphoid cells differ from those on myeloid cells. Nature 305, 541 (1983).

WESTIN. E.H., WONG-STAAL, F., GELMANN, E.P., DALLA-FAVERA, R., PAPAS, T., LAUTENBERGER, J.A., EVA, A., REDDY, E.P., TRONICK, S.R., AARONSON, S.A., and GALLO, R.C., Expression of cellular homologs of retroviral oncogenes in human hematopoietic cells. Proc. Nat. Acad. Sci. USA 79, 2490-2492 (1982).

WU, C., The 5' ends of Drosophila heat shock genes in chromatin are sensitive to DNAaseI. Nature 286, 854-860 (1980).

31

STUDY OF NUCLEOSOMAL ORGANIZATION OF CHROMATIN IN EBV

PRODUCER AND NON-PRODUCER CELLS

V.Zongza* and S.D.Kottaridis**

*Lecturer, Biochemistry and Mol. Biology Depart-
ment, School of Sciences, University of Athens,
Athens 157 01, Greece.
**Head of the Virology Department, Papanikolaou
Research Center, Hellenic Anticancer Institute,
171 Alexandras Ave., Athens 11522, Greece.

INTRODUCTION

The chromatin of higher and lower eukaryotes consists
of nucleoprotein units, the nucleosomes (Felsenfeld, 1978).
Investigations of the chromatin structure with nucleases
have revealed that the DNA length per nucleosome varies
between different organisms, different tissues, cells of the
same tissue, and is correlated with the transcriptional
activity of the cells and their state of differentiation
(Noll, 1976; Lohr et al., 1977; Ord and Stocken, 1979).

The experiments of Mintz and Ilmensee (1975) and of
Fahmy and Fahmy (1980) have refocused attention on the possibility that the mechanisms of neoplastic transformation
may involve aberrant differentiation. The heritable alterations in gene-expression (Stein et al., 1978) and the anomalous gene-expression (Weinhouse, 1972), which often accompa-
ny neoplasia suggest that altered gene regulation plays an
important role in many, if not all, examples of neoplastic
transformation. Thus, the study of chromatin structural
organization as related to its function in neoplastic cells,
seems to be very important.

In the present study we examined the accessibility to
micrococcal nuclease of the nuclei derived from two lymphoma cell lines, P_3HR-1 and Raji, which have the Epstein-
Barr Virus (EBV) integrated (zur Hausen and Schulte-Holthausen, 1970; Nonoyama and Pagano, 1971), and from tonsil

lymphocytes, and we estimated the DNA length per nucleosome in these cells.

The kinetics of chromatin digestion by micrococcal nuclease showed that Raji nuclei were more sensitive than P$_3$HR-1 nuclei to the nuclease. It is also noteworthy that the DNA-repeat length of Raji nuclei (in which the EBV follows a lysigenic cycle), is shorter than that of P$_3$HR-1 nuclei (which have a lytic cycle of EBV) (Nonoyama and Pagano, 1971; zur Hausen and Schulte-Holthausen, 1970).

MATERIALS AND METHODS

Micrococcal nuclease and restriction fragments of bacteriophage ΦX174 RF DNA with Hae III restriction endonuclease, were purchased from Bethesda Research Laboratories, Inc. Proteinase K was purchased from Sigma. The rest of the chemicals were either from Sigma or from Serva.

Lymphocytes and Preparation of Nuclei

Raji and P$_3$HR-1 cells were grown in RPMI medium with 15% fetal calf serum.

Cells were obtained from the culture medium by centrifugation at 500 g for 10 min. They were dispersed in 10 volumes of reticulocyte standard buffer (RSB, 0.1 M NaCl, 1.5 mM MgCl$_2$, 0.01 M Tris-HCl, pH 7.0), (Pederson and Pavis, 1980) and homogenized with a hand-driven Teflon pestle (clearance 0.25 mm) with 10 strokes. The homogenate was centrifuged at 1000 g for 10 min, the supernatant was discarded and the nuclear pellet was washed several times in 10 vol. of RSB until the supernatant appeared clear.

Nuclei from freshly removed tonsils were prepared as following: Minced tissue was homogenized in 7 vol. of 0.34 M sucrose, 3 mM MgCl$_2$, 0.01 M Tris-HCl (pH 7.4), in a Potter-Elvehjem homogenizer with a motor-driven Teflon pestle (clearance 0.25 mm) with 15 strokes at 1700 rev./min. The homogenate was filtered through nylon bolting cloth and centrifuged at 1000 g for 10 min. The nuclear pellet was resuspended in 2.1 M sucrose, 1 mM MgCl$_2$, 0.01 M Tris-HCl (pH 7.4), layered over an equal volume of the same buffer and centrifuged at 70.000 g for 1 h. The nuclear pellets were finally suspended in Digestion Buffer (10 mM NaCl, 3 mM MgCl$_2$, 10 mM Tris-HCl, ph 7.4) in a concentration of 1 mg of nuclear DNA/ml. The analysis of DNA content of samples was performed according to Burton (1956).

Nuclease Digestion and DNA Electrophoresis
In nuclear samples 10 mM CaCl$_2$ was added to a final
concentration of 1 mM. Nuclei were digested with micrococcal nuclease (10 unit/100 µg of DNA) at 37°C for various
times. The reaction was terminated by the addition of 100 mM.
EDTA to a final concentration of 5 mM and by chilling on
ice.The determination of the amount of DNA that was rendered acid soluble, was made as described by Zongza and
Mathias (1979). DNA was purified (Zongza and Mathias,1979)
and purified DNA was dissolved in Tris-acetate buffer containing 15% glycerol.

The electrophoresis of DNA was carried out in vertical
2.5% acrylamide (acrylamide/bisacrylamide; 19:1), 0.5%
agarose slab gels (12 cm long) with Tris-acetate buffer system (0.4 M Tris-HCl (pH 7.8), 0.2 M sodium acetate, 20 mM
EDTA), (Loening, 1967; Peacock and Dingman, 1968). 5 µg of
DNA was applied to each slot and the electrophoresis was
carried out for about 4 h at 20 mA constant current. Gels
were stained in a solution of 3 µg of Ethidium bromide/ml
for 30 min. DNA was visualized in the gels under U.V. light
and photographed.

RESULTS

a) Kinetics of Digestion by Micrococcal Nuclease
The kinetics of digestion of chromatin with micrococcal
nuclease, based on the determination of the amount of DNA
that is rendered acid soluble (Fig. 1), has shown that Raji
nuclei, compared with P$_3$HR-1 nuclei are more easily attacked
by micrococcal nuclease. Nuclei derived from tonsils seem
to be the less-accessible to the nuclease. These results
suggest that the chromatin in Raji cells is in a more open
and accessible to nuclease form than the chromatin in the
other two cell types. The experiments have been repeated
four times and show good reproducibility.

b) Estimation of the DNA Repeat Length
For the measurement of the DNA-repeat length, nuclei
were digested with micrococcal nuclease, as described in the
Experimental section, to the same extent, usually to the
point at which 4% of the original DNA had become acid soluble. To ensure strict comparability it is essential to examine on gels samples digested to the same degree because the
amount of material in a band may affect the migration of the

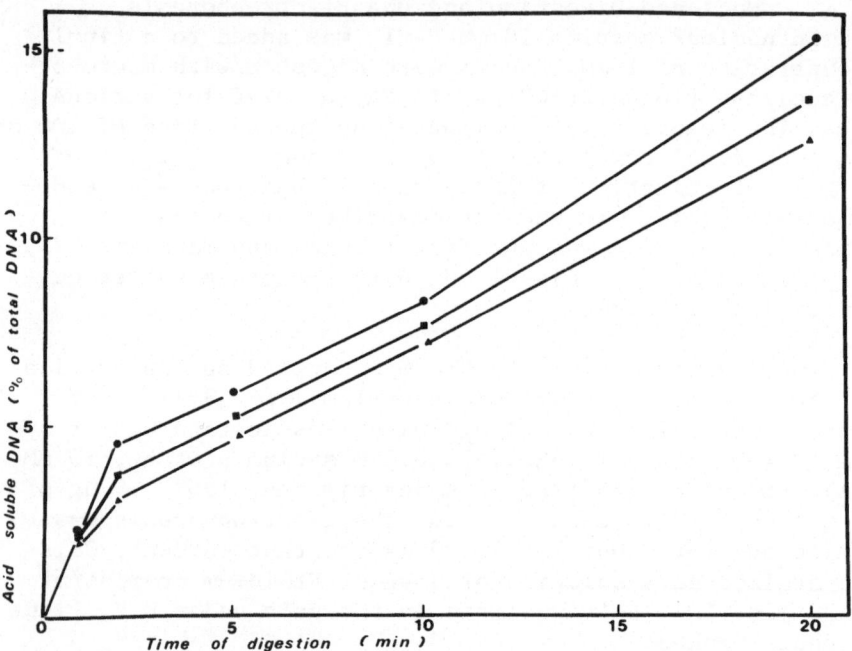

Fig. 1. <u>Time course of digestion of nuclei by micrococcal nuclease.</u>
Nuclei were digested with micrococcal nuclease for various times, and the amount of DNA that was rendered acid-soluble was determined as described in the experimental section. The values shown are the means for duplicates with the control values subtracted. Symbols: ●, Raji nuclei; ■, P$_3$HR-1 nuclei; ▲ lymphocytes nuclei.

neighbouring bands and lengthening the time of digestion may shorten the DNA fragment.

The band sizes were obtained from a calibration graph constructed with the use of restriction fragments of bacteriophage ΦX174 RF DNA run in the same slab gel (Fig. 2). The values for the DNA repeat lengths in the various nuclei were measured by taking the slope of a graph of band size (in base pairs) against band number (Fig. 3) as in method of Noll and Kornberg (1977). In this way the effect of over-digestion is eliminated. Table 1 gives the band sizes of multiples up to a unit size of 4 for the samples shown in Fig. 2 and also the nucleosomal DNA repeat lengths for the same nuclear samples. As Table 1 shows the nucleosomal DNA

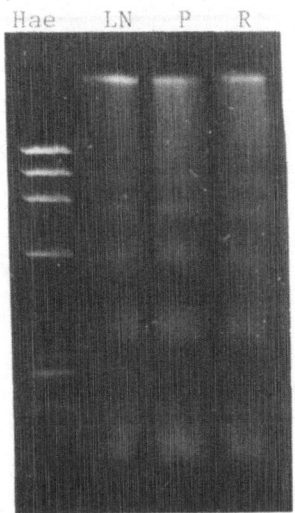

Fig. 2. <u>A photograph of a typical gel for the determination</u>
<u>of the size of DNA fragments obtained from digestion by mi-</u>
<u>crococcal nuclease of Raji nuclei, P_3HR-1 nuclei and lympho-</u>
<u>cyte nuclei.</u>
Nuclei were digested until 4% of the original DNA was rendered acid soluble. DNA was purified as described in the Experimental section and electrophoresed in 2.5% acrylamide/0.5%
agarose slab gels (12 cm long) with the Tris-acetate buffer
system.Electrophoresis was carried out at room temperature
at 20 mA for 4 h. Abbreviations: LN, lymphocyte nuclei;Hae,
Hae III restriction endonuclease digest of bacteriophage
ΦX174 RF DNA; R, Raji nuclei; P, P_3HR_1 nuclei.

length varies in the three types of nuclei. Raji nuclei
appear to have shorter DNA-repeat length than P_3HR-1 nuclei
and tonsil nuclei. It has been suggested that the short DNA
repeat length could be a consequence of the packaging of
chromatin in "active nuclei" (Morris, 1976; Thomas and
Thompson, 1977). The estimated DNA repeat lengths for the
three nuclear types examined are in agreement with the accessibility of the same nuclei to micrococcal nuclease (Fig.1).
So the more accessible to nuclease Raji nuclei have also
the shorter repeat length, meaning that these cells have
a more relaxed form of chromatin than the other two cell
types.

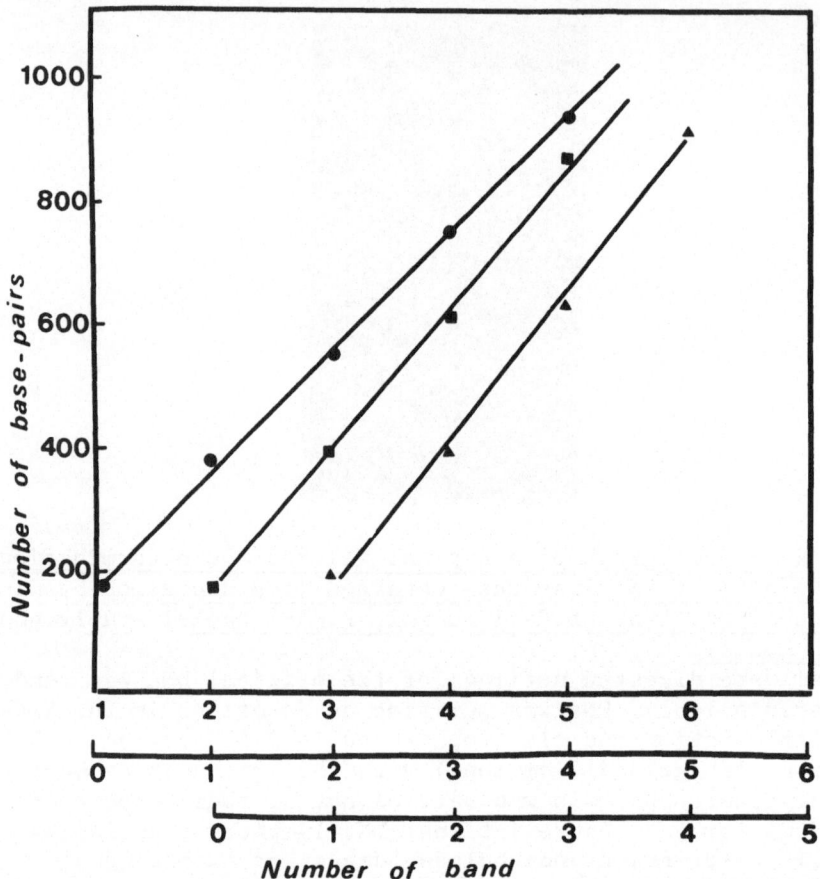

Fig. 3. <u>Three typical graphs for the determination of DNA-</u>
<u>repeat lengths in the Raji, P₃HR-1 and lymphocyte nuclei.</u>
The sizes of DNA bands were estimated from the calibration
curve constructed with the use of restriction fragments of
bacteriophage ΦX174 RF DNA run in the same slab gel (Fig.2).
The distances of migration were measured to the midpoints
of the bands on the gels. The DNA-repeat length is determined from the slope of the graph. Raji nuclei (●, upper ab-
scissa scale). P₃HR-1 nuclei (■, middle-abscissa scale);
lymphocyte nuclei (▲, lower-abscissa scale).

Table 1. Sizes (in base pairs) of DNA fragments produced by micrococcal nuclease digestion of nuclei studied, and DNA-repeat lengths of the same nuclei.

The sizes of the DNA fragments in the polyacrylamide gels, one of which is shown in Fig. 2, were estimated from a calibration curve as described in the text. Mobilities were measured from the origin to the midpoints of the bands on gel photographs. Repeat lengths were determined by linear regression as described in the text and Fig. 3.

	Size of DNA fragment (number of base pairs)		
oligomer size	Raji	P_3H-R-1	lymphocytes
1	165	174	176
2	365	389	400
3	546	575	656
4	735	851	871

Average repeat lengths		
Raji	P_3H-R1	lymphocytes
185±1	214±3	233±3

DISCUSSION

The present results show that nuclei from two lymphoma cell lines have a different packaging of chromatin than normal lymphocytes. The two lymphoma cell lines, Raji and P_3HR-1 are also different in the same respect. Thus, Raji nuclei have a shorter DNA-repeat length than nuclei derived from the other two cell types examined. The DNA-repeat length is measured in the overall products of the digestion and consequently the DNA-repeat length obtained is a mean value for the repeat lengths in the genome.

The explanation of our results should be seeked in the known differences between Raji and P_3H_3-1 cells. Raji cells have the EBV integrated in the genome and do not produce EBV particles (zur Hausen and Schulte-Holthausen, 1970; Novoyama and Pagano, 1971). In the same cells it has been found a translocation involving the c-myc oncogene, which is translocated to the chromosome carrying the immunoglobulin genes (Klein, 1983) and it has been suggested that the above translocation is responsible for the neoplastic phenotype of these cells. In P_3HR-1 cells the EBV is also integrated, but a certain region of the EBV genome is missing (Klein, 1984). Possibly the absence of this region is responsible for the lytic viral cycle occuring in these cells.

However, in order to understand better the relation
of chromatin structure and gene regulation, we must examine
the chromatin structure of regions flanking the integrated
E.B. viral genome by DNase I, an enzyme, which specifically
recognizes and digests regions of chromatin preceding
active in trascription genes (Elgin, 1981; Schubach and
Groudine, 1984).

The estimation of repeat lengths in newly transformed
lymphocytes would also enable us to understand the alteration in the packaging of chromatin, that is observed between
normal lymphocytes and cells from lymphoma cell lines. Work
concerning the formentioned two points is under progress.

The change of DNA length per nucleosome observed could
also be attributed to histone modifications, such as acety-
lation and phosphorylation (Wallace et al., 1977; Letnasky,
1978), since histones with lower positive charge could be
the cause of a shorter DNA repeat length. Alternatively,
the role of non-histone proteins should be also considered,
and specially the role of HMG proteins that have been correlated with transcriptional activity and chromatin structure
(Goodwin et al., 1978; Weisbrod et al., 1980). We believe
that the study of non-histone proteins in the cells we have
examined would be very useful in an understanding of the
organization of chromatin in these cells.

REFERENCES

Burton, K. A study of the conditions and mechanisms of the
    diphenylamine reaction for the calorimetric estimation
    of deoxyribonucleic acid. Biochem. J. 62, 315-323 (1956).
Elgin, S.C.R. DNase. I- Hypersensitive sites of chromatin.
    Cell 27, 413-415 (1981).
Fahmy, M.J. and Fahmy, O.G. Intervening DNA insertions and
    the alteration of gene expression by carcinogens. Cancer
    Res. 40, 3327-3382 (1980).
Felsenfeld, G. Chromatin. Nature (London) 271, 115-122
    (1978).
Goodwin, G.H., Walker, J.M. and Johns, E.W. The High Mobility
    Group (HMG) nonhistone Chromosomal Proteins in the "Cell
    Nucleus" (Busch H. ed.) vol. 6, p.p. 181-219, Academic
    Press N.Y. (1978).
zur Hausen, H. and Schulte-Holthausen, H. Presence of EB
    virus nucleic acid homology in a "virus-free" line of
    burkitt tumour cells. Nature (London) 227, 245-248 (1970).

Klein, G.,Specific chromosomal translocations and the genesis of B-cell-derived tumors in mice and men. Cell 32, 311-315 (1983).

Klein, G., Burkitt lymphoma and nasopharyngeal carcinoma: Transformation related aspects. Abstract book "1st International Symposium on EBV and Associated Malignant Diseases", 1984.

Letnasky, K., Nuclear proteins in genetically active and inactive parts of chromatin. FEBS Lett. 89, 93-97 (1978).

Loening, U.E., The fractionation of high-molecular-weight ribonucleic acid by polyacrylamide-gel electrophoresis. Biochem. J. 102, 251 (1967)

Lohr, D., Kovacic, R.T. and Van Holde, K.E., Quantitative analysis of the digestion of yeast chromatin by Staphylococcal nuclease. Biochemistry 16, 463-471 (1977).

Mintz, B. and Illmensee, K., Normal genetically mosaic mice produced from malignant teratocarcinoma cells. Proc. Nat. Acad. Sci. (USA) 72, 3585-3589 (1975).

Morris, N.R., A comparison of the structure of chicken erythrocyte and chicken liver chromatin. Cell 9, 627-632 (1976).

Noll, M., Differences and similarities in chromatin structure of Neurospora crassa and higher eucaryotes. Cell 8, 349-355 (1976).

Noll, M. and Kornberg, R.D., Action of micrococcal nuclease on chromatin and the location of histone H_1. J. Mol. Biol. 109, 393-404 (1977).

Nonoyama, M. and Pagano, J., Detection of Epstein-Barr viral genome in nonproductive cells. Nature (London) New Biol. 233, 103-106 (1971).

Ord, M.G. and Stocken, L.A., Nucleosomes from normal and regenerating rat liver. Biochem. J. 178, 173-185 (1979).

Peacock, A.C. and Dingman, C.W., Molecular weight estimation and separation of ribonucleic acid by electrophoresis in agarose- acrylamide composite gels. Biochemistry 7, 668-674 (1968).

Pederson, T. and Pavis, N.G., Messanger RNA processing and nuclear structure: isolation of nuclear ribonucleoprotein particles containing b-globin messenger RNA precursors. J. Cell Biol. 87, 47-54 (1980).

Schubach, W. and Groudine, M., Alteration of C-myc chromatin structure by avian leukosis virus integration. Nature (London) 307, 702-708 (1984).

Stein, G.S., Stein, J.L. and Thompson, J.A., Chromosomal proteins in transformed and neoplastic cells: A review. Cancer Res. 38, 1181-1201 (1978).

Thomas, J.O. and Thompson, R.J., Variation in chromatin structure in two cell types from the same tissue: a short DNA repeat length in cerebral cortex neurons. Cell 10, 633–640 (1977).

Wallace, B.R., Sargent, T.D., Murphy, R.F. and Bonner, J., Physical properties of chemically acetylated rat liver chromatin. Proc. Natl. Acad. Sci. (USA) 74, 3244–3243 (1977).

Weinhouse, S., Glycolysis, Respiration and anomalous gene expression in experimental hepoctomas. Cancer Res. 32, 2007–2016 (1972).

Weisbrod, S., Groudine, M. and Weintraub, H., Interaction of HMG 14 and 17 with actively transcribed genes. Cell 19, 289–301 (1980).

Zongza, V. and Mathias, A.P., The variation with age of the structure of chromatin in three cell types from rat liver. Biochem. J. 179, 291–298 (1979).

NOVEL BIOLOGICAL FUNCTIONS ASSOCIATED WITH EPSTEIN-BARR

VIRUS DNA

Beverly E. Griffin[1], David King[2] and Loraine Karran[2]

Royal Postgraduate Medical School, Hammersmith Hospital,
London W12 and[2] Imperial Cancer Research Fund, Lincoln's
Inn Fields, London WC2, England.

SUMMARY

Transfection of primary kidney cells of non-human primates with specific EBV DNA fragments from a recombinant
DNA cosmid library, has allowed two functions associated
with the virus to be mapped on to the viral genome. One,
designated the "p31 function" is associated with cellular
immortalisation, and the other, designated the "p5 function"
appears to play a role in cellular differentiation. Transfection experiments with fragment p31 show that in vitro,
epithelial cells of both AGMK, marmoset and human origin are
selected for growth proliferation in preference to fibroblasts. Poorly differentiated epithelial cell lines have
been established. In the presence of EBV DNA fragments
p31 + p5, "dome" producing cells from primary marmoset kidneys have been obtained. A working hypothesis regarding
EBV-induced transformation of epithelial cells is put forward.

INTRODUCTION

In spite of considerable evidence that supports the notion
of a virally-encoded transforming gene within EBV, the localisation of this gene and its identity have not been unambig-
uously resolved. Our attempts to provide a solution are
straight-forward and involve transfection of fragments of
viral DNA into primary cells, then monitoring phenotypic
alterations. Initial trials were carried out using a cosmid
'library' composed of partially digested Bam HI fragments of

EBV and a mixture of cell types (both fibroblast and epithelial cells) derived from the kidneys of an African green monkey (AGMK cells). These experiments, which have been published (Griffin and Karran. 1984), demonstrated that immortalisation of epithelial cells could be induced in response to two specific overlapping regions of the viral genome carried within the cosmid p13 (p33) and p31 (see Figure 1) It is worth noting that in the presence of the EBV DNA, epithelial cells were stimulated to outgrow fibroblasts under ordinary tissue culture conditions (E4 media + 5% foetal calf serum), strongly suggestive of intervention in normal cell growth by some viral function.

The established AGMK cell lines obtained in this fashion have the following properties:

a. Morphologically, they resemble the flat, cuboidal epithelium.

b. Immunofluorescence, using a monoclonal antibody specific for epithelial cells (LE 61; Lane, 1982), reveals tonofilaments characteristic of this class of cell.

c. 'Footprints' of EBV DNA in the cellular chromosome alter their sequence arrangements during the continuous passaging of cells in culture, ultimately reaching a stable pattern which persists after more than two years in culture.

d. Cells grow to relatively high density in vitro.
None of the cell lines. however, displayed fully-transformed phenotypes. That is, their growth in either low serum (1%) or semi--solid media (soft agar) was limited, and they did not produce tumours in athymic (nu/nu) mice. The in vitro experiments are thus in accord with the hypothesis that EB virus provides a function that stimulates cells to proliferate continuously, but tumour formation depends on 'cofactors' or 'genetic accidents', such as those that induce or succeed chromosomal translocations (Klein, 1983).

IMMORTALISATION OF MARMOSET (AND HUMAN) CELLS

The study of chromosomal alterations as possibly necessary events accompanying the expression of a fully--transformed phenotype would be difficult with AGMK cells since they have 60 chromosomes (diploid number), and many of those are large and of similar size (Hsu and Benirschke, 1967). For this reason, as well as to determine whether the capacity to immortalise epithelial cells is a general property of the EBV DNA fragments, experiments similar to those described above were carried out with primary kidney cells derived from

common marmosets (Callithrix jacchus, chromosome diploid no.
46; Hsu and Benirschke, ibid.).   Again, in these marmoset
studies, epithelial cells were found to outgrow fibroblasts
in response to one of the two (p31) 'immortalising fragments'
used in the AGMK experiments (Griffin et al., in press).
The properties of the marmoset cells. which have now been in
culture for more than a year, are in most respects similar
to those already described above for AGMK cells.   At passage
20 (nearly eight months in culture), the p31-transfected
marmoset cells were still diploid.   Morphologically, they
grew in a more or less ordered manner, but reached higher
densities (about twice) than normal cells by close-packing
on a dish; they did not overgrow to form foci.   The cells
appeared to reach a quiescent state and could remain so for
a long time in culture.   They had gross features reminiscent
of cells derived from undifferentiated renal carcinomas
(Bennington and Beckwith, 1979).

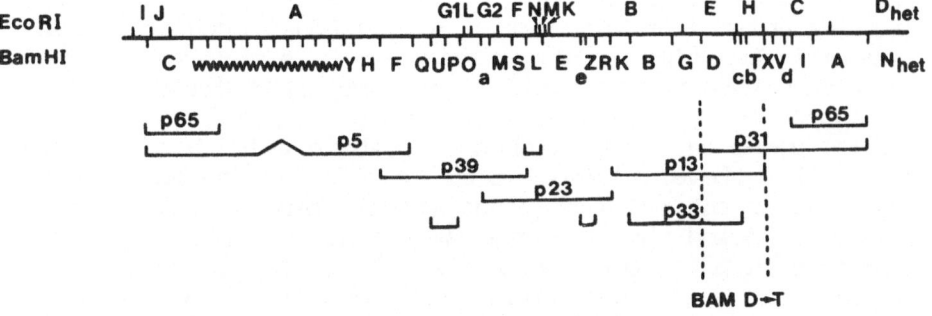

Figure 1
   Schematic diagram showing the location of the EBV DNA cosmid library in relation to the EcoRI and BamHI physical maps
of the viral genome, strain B95-8. (Adapted from Griffin
and Karran, 1984).

Several further studies were carried out on the primary marmoset kidney cells: (a). One involved transfections using p31 DNA that had been cleaved with a variety of restriction endonucleases. This experiment had the two-fold purpose of providing circumstantial evidence in support of a virally-encoded 'immortalising function' and further localisation of it on to the viral genome. Preliminary results (D.K., unpublished) suggest that a number of enzymes, including BamHI, EcoRI (see Figure 1 as well as BglII and HindIII, destroyed the function. These data support the notion of an EBV encoded immortalising gene since the functional behaviour of transfecting DNA could be altered by a variety of enzymes. (b). The second type of experiment sought to ask whether another gene, complementary for transformation might also exist within EBV. Thus, fragment p31 was used together with other individual EBV DNA clones in transfection experiments with marmoset cells, and morphological changes associated with fully-transformed cells (dense focus formation etc.) were monitored. In the instance where p31 and p5 were used in concert, an unusual and unexpected alteration was observed (Griffin et al., in press). That is, after several months in culture, cells that were allowed to grow to confluence and rest in culture were found to be highly polarised and to produce "domes" or "hemi-cysts"characteristic of actively transporting epithelium (Handler et al., 1980). Such cells can generally be stimulated to form domes by a variety of agents that also show potent activity in inducing a differentiated phenotype in other systems, such as Friend erythroleukemia cells; domes can be abolished by ouabain which binds to transport sites (Lever, 1980). Similar effects were observed in the case of the p31 + p5 transfected marmoset cells. It should be noted that no domes were observed in cells transfected with p31 alone, or in control cells. Thus, it would appear that cell polarisation and concomitant dome formation were related to the expression of some function associated with information contained in the cosmid, p5. Morphologically, p31/p5 transfected cells had gap junctions, etc., indicative of highly differentiated cells.

Although it may be premature to ascribe any specific significance to the latter experiments, the facts that the immortalised cells are undoubtedly sensitive to agents that induce differentiation and that domes have been postulated to arise in vitro by a process similar to cell differentiation (Lever, 1981, and references therein), have led us to evoke a working hypothesis regarding EBV (Griffin, in press). This assumes that EBV encodes (within p31) a function associated

with immortalisation, and(within p5)a function associated
with differentiation.   Further it presupposes that subpopulations of Eb viruses exist which, lacking the 'p5 function', can immortalise a cell via the 'p31 function' and make
it thus susceptible to subsequent events that generate the
fully-transformed phenotype.

Analogous experiments have also been carried out on human
cells.   It has been observed that transfection with p31 results in growth alteration of mammary cells (Griffin et al.
in press), although no cell line has yet been established.
On the other hand, primary foetal human kidney cells transfected with p31 have been in culture for nearly seven months
and epithelial cells in the population are still growing
well under normal tissue culture conditions.   Thus it would
appear that a function capable of immortalising a sub-set of
epithelial cells from a variety of sources may exist within
the EB viral genome.

CONCLUSIONS

For practical reasons, all our transfection studies to date
have been carried out on epithelial cells; it is by no means
clear,however, that our data could be extrapolated to allow
conclusions to be drawn about immortalisation of B lymphocytes.   Indeed, the deletion localised on to the BamHI W, Y,
and H fragments (see Figure) in the non-transforming P3HR-1
viral strain have focussed on this region (which lies within
p5) as a transformation-associated area of the EBV genome
with regard to B-lymphocytes (Rabson et al., 1982; Jeang and
Hayward, 1983; and Stoerker and Glaser, 1983).   Further, since
experimentation on other 'transforming' viruses have pointed
to a variety of mechanisms whereby viruses induce continuous
proliferation of cells (including viral gene expression, in-
sertion-promotion, mutation, etc.) it will not be surprising
if a virus as large as EBV can interact with epithelial cells
and B-lymphocytes in different manners.

REFERENCES

BENNINGTON, J.L. and BECKWITH, J.B. In: H.I. Firminger (ed),
Tumors of the kidney, renal pelvis and ureter, Atlas of tumor
pathology vol. 12, pp 289, 302, Am. Cancer Society and Armed
Force Inst. Path. (1979).

GRIFFIN, B.E. and KARRAN, L. Immortalisation of monkey epithelial cells by specific fragments of Epstein-Barr virus DNA, Nature 309, 78-82 (1984).

GRIFFIN, B.E. Transformation of cells by Epstein-Barr virus: An hypothesis. In: N.A. Mitchison and M. Feldmann (eds.) 16th Int. Leucocyte Culture Conf., Humana Press Inc., in press.

GRIFFIN, B.E., KARRAN, L., KING, D., AND CHANG, S.E. Immortalising gene(s) encoded by Epstein-Barr virus. In: P.W. Rigby and N.A. Wilkie (eds), Viruses and Cancer. Soc. Gen. Microbiol., vol. 37, Cambridge Univ. Press in press.

HANDLER, J.S., PERKINS, F.M. and JOHNSON, J.P. Studies of renal cell function using cell culture techniques. Am. J. Physiol. 238, F1-F9 (1980).

HSU, T.C. and BENIRSCHKE, K. In: An Atlas of Mammalian Chromosomes, vol. 1, folio 1, pp.46,48. Springer Verlag, Berlin (1967).

JEANG, K.T. and HAYWARD, S.D. Organization of the Epstein-Barr virus DNA molecule. III. Location of the P3HR-1 deletion junction and characterization of the NotI repeat units that form part of the template for an abundant 12 O-tetradecanoyl phorbol-13-acetate-induced mRNA transcript. J.Virol. 48, 135-148 (1983).

KLEIN, G. Specific chromosomal translocations and the genesis of B cell-derived tumors in mice and man. Cell 32, 311-315 (1983).

LANE, E.B. Monoclonal antibodies provide specific intramolecular markers for the study of epithelial tonofilament organisation. J. Cell Biol. 92, 665-676 (1982).

LEVER, J.E. Regulation of dome formation in kidney epithelial cell cultures. Ann. N.Y. Acad. Sci. 372, 371-383 (1981).

RABSON, M., GRADOVILLE, L., HESTON, L. and MILLER, G. Non-immortalizing P3J-HR-1 Epstein-Barr virus: A deletion mutant of its transforming parent, Jijoye. J. Virol. 44, 834-844 (1982).

STOERKER, J. and GLASER, R. Rescue of transforming Epstein-Barr virus (EBV) from EBV-genome positive epithelial hybrid cells transfected with subgenomic fragments of EBV DNA. Proc. Natl. Acad. Sci., U.S.A. <u>80</u>, 1726-1729 (1983).

33

EBV DNA CONTENT AND EXPRESSION IN NASOPHARYNGEAL CARCINOMA

N. Raab-Traub[1], D. Huang[2], C.S. Yang[3], and G. Pearson[4]

[1]Dept. of Microbiology and Immunology, Univ. of North Carolina, Chapel Hill, NC, U.S.A.; [2]Medical and Health Dept., Queen Elizabeth Hospital, Kowloon, Hong Kong; [3]National Taiwan University, Taipei, Taiwan; [4]Dept. of Microbiology, Georgetown Univ. School of Medicine, Washington, D.C., U.S.A.

SUMMARY

We have analyzed EBV DNA content in nasopharyngeal carcinoma tissue (NPC) samples from endemic and nonendemic regions. The samples were histopathologically classified as 1) squamous cell carcinomas, 2) nonkeratinizing carcinomas, or 3) undifferentiated carcinomas. Southern blots, prepared from DNA purified from NPC tissues, were hybridized to ^{32}P-labeled cloned restriction enzyme fragments of EBV. Twenty-six samples of all three histopathologic classifications regions were positive for EBV DNA. We have also attempted to identify the state of viral expression within the tumor at either the level of transcription or by identification of viral polypeptides.

The EBV sequences which encode mRNA in latently infected lymphocytes are transcribed in most tumor specimens. Some NPC tumor specimens and tumors grown in nude mice contain RNA encoded by additional sequences. We have identified by Northern blot analyses some of the EBV mRNAs encoded by these sequences. Many are similar in size to EBV mRNAs which are believed to encode early functions.

To investigate the possibility of viral activation, immunoblots of tumor tissue lysates were prepared and reacted with high-titer EBV-specific antisera and a monoclonal antibody to the diffuse component of the early antigen (EAd). Proteins which may be viral specific were detected in some specimens; however EAd could not be detected in twenty-two tumor samples including those which had activated transcription.

INTRODUCTION

Nasopharyngeal carcinomas are classified into three histopathologic subtypes by the World Health Organization: 1) squamous cell carcinomas, 2) nonkeratinizing carcinomas and 3) undifferentiated carcinomas. Types 2 and 3 are associated with the Epstein-Barr virus in that EBV DNA is detected in the tumor tissue and patients have elevated titers to the diffuse component of the EBV early antigen and to the viral capsid antigen (VCA) (Andersson-Anvret et al., 1977). These antibody titers can be correlated to tumor mass and disease progression (Henle et al., 1973); therefore it is of interest to determine if the antibody response reflects the state of viral expression within the tumor tissue.

Our studies have concentrated on detecting the Epstein Barr viral genome in NPC tissue and determining the state of viral expression either by analyzing the viral sequences which are transcribed or by identifying viral polypeptides within the tumor tissue. The viral sequences which are transcribed in latently infected lymphocytes which only express EBNA are also transcribed in several WHO 3 carcinomas from patients with elevated titers to EA and VCA. In contrast, a WHO 1 carcinoma contained EBV DNA and had activated transcription similar to abortively infected Raji cells (Raab-Traub et al., 1983). Similar transcription was detected in NPC tissue grown in nude mice.

MATERIALS AND METHODS

Analysis of Nucleic Acid in Tumor Tissue

The tissue specimens are homogenized in 4 M guanidine thiocyanate, made into a dilute CsCl solution (0.5 gm/ml), layered over a 5.7 M CsCl cushion, and centrifuged at 80,000 X g for 24 hours. After centrifugation, the DNA layer is pulled from the cushion interface, dialyzed, digested with proteinase K, and extracted with phenol:chloroform twice. The DNA is usually digested with BamHI, transferred to nitrocellulose after electrophoresis, and hybridized to ^{32}P labelled BamHI V fragment of EBV, representing the large internal repeat sequence, IR1 (Dambaugh et al., 1980). This procedure unequivocally identified a 2 X 10^6 dalton fragment.

The RNA fraction forms a clear translucent pellet under the cushion and is separated into polyadenylated RNA by oligo dT chromatography. The polyadenylated RNA is copied into a ^{32}P labelled cDNA with avian myeloblastosis virus reverse transcriptase. This labelled cDNA is then hybridized to Southern blots of recombinant EBV fragments spanning the genome and to blots of the oncogenes. In some cases, RNA is subjected to electrophoresis through a denaturing formaldehyde agarose gel, transferred to nitrocellulose, and hybridized to the labelled EBV fragments which were identified as transcribed in the cDNA analysis.

Preparation of Immunoblots

Tumor tissue is homogenized in 0.15 M NaCl, 0.57 Triton X-100, 0.5% Na-deoxycholate, 1mM PMSF, 50 mM Tris pH 7.3 and the protein content is determined by a Folin–Lowry reaction. Protein lysates are subjected to electrophoresis through a 7.5% polyacrylamide gel and transferred to nitrocellulose by diffusion. The sheets are treated with high titer human antisera or a monoclonal antibody to EA(d) (Pearson et al., 1983). Bound antibody is detected using peroxidase – tagged goat anti-human or anti-mouse IgG.

RESULTS

Detection of EBV DNA

The hybridization to Southern blots of multiple NPC DNAs is shown in Figure 1. In most cases hybridization with BamHI V is readily detectable particularly in the NPC WHO Type 3 undifferentiated carcinomas and the NPC WHO 2 non-keratinizing carcinomas. The hybridization to the WHO-1 differentiated tumors is usually much weaker or borderline detectable. For example in NPC 62, (Figure 1) hybridization with BamHI V weakly identified a fragment at 2×10^6 d. This was confirmed by hybridization with ^{32}P-labelled EcoRI B which identified the BamHI fragments B,E,G,K, and R, which are contained within EcoRI B. Almost all of the NPC samples, Table 1, which have been analyzed by this method have been positive for EBV DNA. These NPC samples are from American, Alaskan, African, and Taiwanese sources. All three of the histologic subtypes have been positive for EBV DNA although the comparatively weak hybridization in the NPC Type 1 may reflect low copy number or mixed histology.

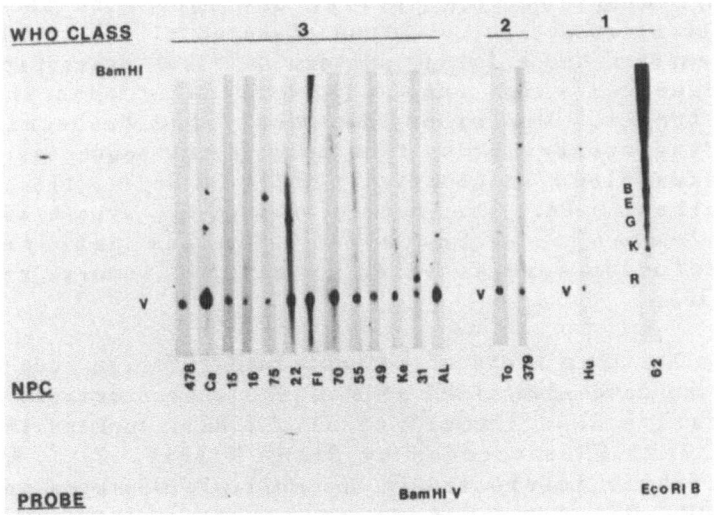

Fig 1: Hybridization of EBV DNA Fragments to BamHI digested DNA purified from NPC biopsies.

TABLE 1

Detection of EBV DNA in Nasopharyngeal Tissues

HISTOLOGY	# SPECIMENS	# POSITIVE
WHO3—undifferentiated	20	19
WHO2—nonkeratinizing	3	3
WHO1—differentiated	3	3

Analysis of EBV Transcription

Our previous studies have identified the EBV sequences which encode RNA in NPC human biopsy material as well as NPC grown in nude mice (Raab-Traub et al., 1983). By comparison with data obtained from lymphoblastoid cell lines we can ascertain the state of viral infection and begin to identify functions which may be particularly important in NPC.

NPC biopsies from several patients with WHO 3 or undifferentiated carcinomas and elevated titers to the EBV early antigen had a latent pattern of viral transcription. The sequences which encode EA were not transcribed in these tumors. One biopsy had activated transcription including transcription from most of the sequences which are transcribed in abortively infected Raji cells, which do synthesize EA. This patient was an NPC type 1 and had no detectable titer to EA. Similar patterns of transcription were detected in all NPC tumors grown in nude mice.

On Northern blots of RNA obtained from the nude mouse tumors we have identified some of the characteristic early replicative RNAs (Hummel et al., 1982), such as the 2.7 and 1.9 kb RNAs encoded by BamHI H (Fig. 2). We are particularly interested in the multiple messages encoded by EcoRI DIJ het because it is the most abundantly transcribed fragment in both the biopsies and the tumors grown in nude mice.

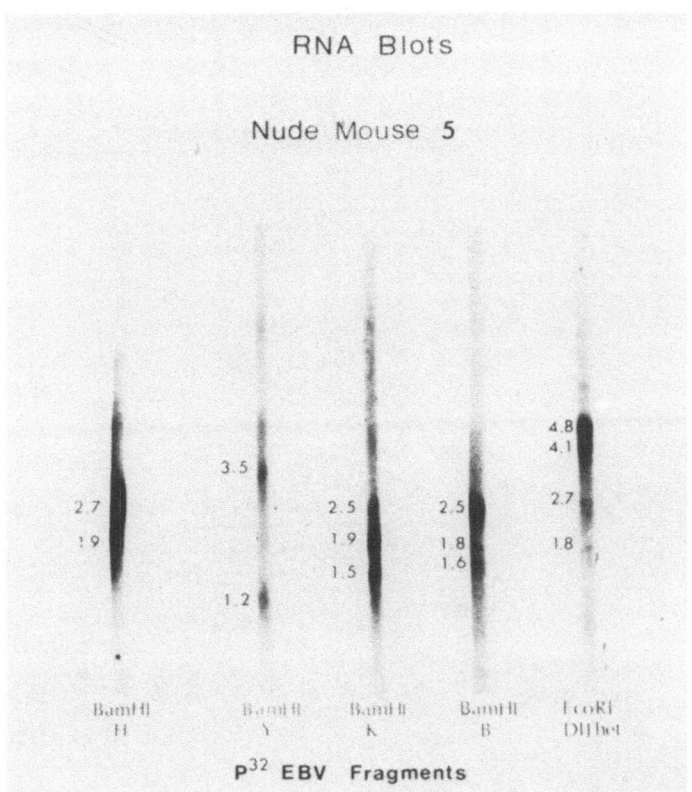

RNA Blots

Nude Mouse 5

P^{32} EBV Fragments

Fig 2: Identification of EBV mRNAs encoded by BamHI, H,Y,K,B or EcoRI DIJ het in NPC grown in nude mice.

Screening for Oncogene Expression

We have begun to establish a library of recombinant DNA clones of identified oncogenes and have utilized this material to screen NPC tissue for elevated transcription of a particular oncogene. Recombinant DNAs which include the following oncogenes have been used in this screening: v-myb, v-myc, Harvey v-ras, Kirsten v-ras, Harvey c-ras, v-abl, v-mos, mouse-mos, human-mos, v-erb, v-fos, v-ros, v-sis, and v-sarc. The labelled cDNAs to NPC RNA have been hybridized to Southern blots of the oncogene library (Fig. 3). Although we have not yet detected any

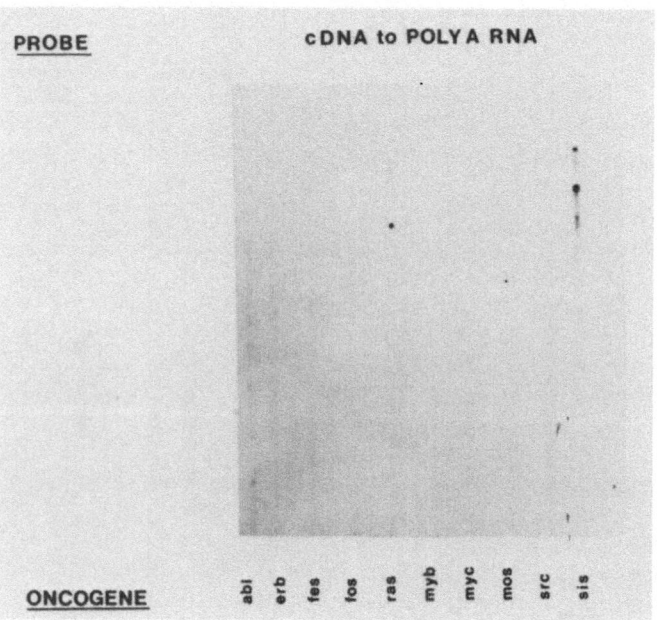

PROBE · cDNA to POLY A RNA

ONCOGENE · abl erb fes fos ras myb myc mos src sis

Fig 3: Hybridization of ^{32}P-labelled cDNA of polyadenylated RNA from NPC tissue grown in nude mice to a library of recombinant DNAs containing various oncogenes.

hybridization with material prepared from biopsies, with cDNA from 4 NPCs grown in nude mice there is apparently abundant transcription to simian sarcoma virus, v-sis. In addition a control carcinoma of the ethmoid grown in nude mice also strongly hybridized to sis.

Screening for Viral Proteins in NPC

Many biopsy samples are too small to obtain nucleic acid; therefore we have prepared immunoblots of protein lysates from these samples to identify viral proteins and to determine if the tumors are latently or abortively infected. On duplicate filters the protein lysates of 5 NPCs were compared with proteins from Raji cells superinfected with HR1 (Fig. 4). For most screenings we have used one extremely high titer serum EA 1:160,000, VCA 1:640,000, the gift of Dr. G. Lenoir. The protein lysate from NPC 18 had several proteins which may be virus-specific. Other high titer EA sera, usually from patients with NPC, were also used for screening. Using this type of analysis we have compared the reactivities of a few NPC sera on SIRC proteins. For example the high-titer sera reacts strongly with a 110K protein whereas the NPC sera does not but reacts strongly with the 49K component of EA(d). Of the few sera we have compared there does not appear to be consistent reactivity with particular viral proteins in superinfected Raji cells.

Very few proteins reacted in the NPC lysates and it is impossible to identify positively any of these as viral proteins using human sera with multiple reactivities. Therefore as monospecific and monoclonal antibodies become available to individual viral proteins we wish to screen duplicate immunoblots of these lysates. Because of the characteristic high antibody titers to EA(d) in patients with NPC, it was of particular interest to screen tumors for the presence of EA(d) using a monoclonal antibody. This antibody reacts extremely well on immunoblots and identifies a heterogeneous family of proteins in SIRC ranging from 49–55 K. It can be seen in Figure 5 that the reactivity with the monoclonal serum in contrast to the high titer sera is much greater. We have screened 22 NPCs by this method and have not detected a trace of this EA(d). Interestingly EA(d) was not detected in NPC 18 which had a low level of transcription from the fragment which encodes this component of EA. This suggests that either the protein is present at such a low level to be undetectable, that another unrelated protein is encoded by these sequences, or that the sequences are transcribed but not translated.

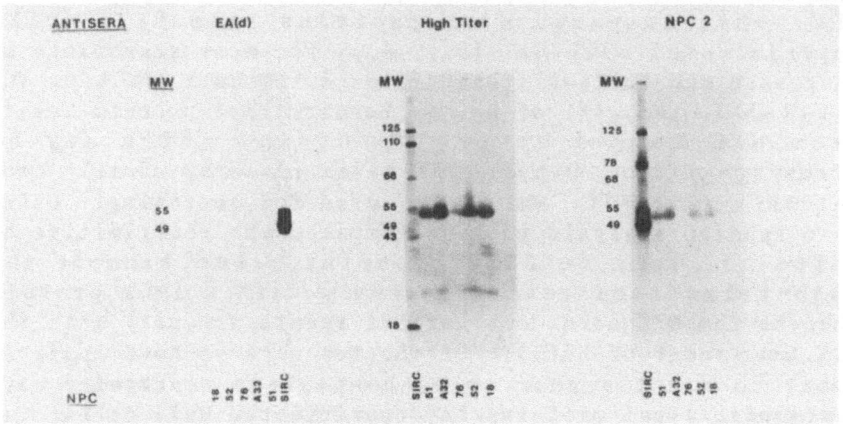

Fig 4: Immunoblots of polypeptides extracted from NPC tissues or superinfected Raji cells reacted with high titer human sera or a monoclonal to EA(d).

DISCUSSION

It is apparent that all NPC from endemic and nonendemic regions of all histologic subtypes contain EBV DNA. The hybridization to differentiated tumors is weak but unequivocal in that multiple EBV fragments can be identified on Southern blots. In addition, one differentiated NPC, NPC 18, contained detectable EBV DNA and extensive transcription of viral RNA. This patient had a low titer to VCA and no detectable titer to EA. Many of the sequences which were transcribed in this specimen and in all of the tumors grown in nude mice are believed to encode components of EA. However, analysis of the protein content of NPC 18 and the tumors grown in nude mice on immunoblots revealed that these tumors did not contain the component of EA(d) recognized by the monoclonal antibody. It is possible that the sequences are transcribed but not translated or that they encode a different class of viral functions.

All tumors grown in nude mice contain relatively abundant RNA homologous to the simian sis oncogene. It is possible that infiltrating mouse stroma may contain active endogenous retroviruses and that the cDNA may be hybridizing to the viral non-oncogene sequences. However, several of the oncogene clones including the Harvey and Kirsten v-ras clones also contain viral sequences which are more likely to be homologous to the endogenous mouse retroviruses than those in the simian v-sis clone.

It is difficult to assess the relation of the v-sis transcription to the malignant state of the cells. All of the positive tumors have been carcinomas which are not believed to express the receptor for platelet-derived growth factor (PDGF). One of the proposed mechanisms of action of v-sis requires binding of excreted v-sis or PDGF to the receptor on the malignant cell; therefore these carcinoma cells presumably would not be susceptible to v-sis. Nevertheless, it is tempting to speculate that growth in nude mice either activates transcription of sis or PDGF or provides a selection for tumors which are expressing sis.

REFERENCES

Andersson-Anvret, M., Forsby, N., Klein, G., and Henle, W., Relationship between Epstein-Barr virus and undifferentiated nasopharyngeal carcinoma: Correlated nucleic acid hybridization and histopathologic examination. Int. J. Cancer, 20, 486-494 (1977).

Dambaugh, T., Beisel, C., Hummel, M., King, W., Fennewald, S., Cheung, A., Heller, M., Raab-Traub, N., and Kieff, E., Epstein-Barr virus (B95-8) DNA. VII. Molecular cloning and detailed mapping of EBV (B95-8) DNA. Proc. Nat. Acad. Sci. (Wash), 77, 2999-3003 (1980).

Henle, W., Ho, H.C., and Kwan, H.C., Antibodies to Epstein-Barr virus related antigens in nasopharyngeal carcinoma. Comparison of active cases with long term survivors. J. Nat. Cancer Inst., 51, 361-369 (1973).

Hummel, M., and Kieff, E., Epstein-Barr virus RNA. VIII. Viral RNA in permissively infected B95-8 cells. J. Virol., 43, 262-272 (1982).

Pearson, G.R., Vroman, B., Chase, B., Sculley, T., Hummel, M., and Kieff, E., Identification of polypeptide components of the Epstein-Barr virus early antigen complex using monoclonal antibodies. J. Virol., 47, 193-201 (1983).

Raab-Traub, N., Hood, R., Yang, C.S., Henry, B., and Pagano, J., Epstein-Barr virus transcription in nasopharyngeal carcinoma. J. Virol., 48, 580-590 (1983).

34

TRANSFORMATION OF HUMAN LYMPHOCYTES BY COINFECTION

WITH EBV DNA AND TRANSFORMATION-DEFECTIVE VIRIONS

David J. Volsky, Barbara Volsky, Mona Hedeskog,
Faruk Sinangil and Thomas G. Gross
Department of Pathology and Eppley Institute for
Research in Cancer
University of Nebraska Medical Center
Omaha, NE 68105

SUMMARY

EBV DNA from B95-8 strain was introduced into human cord blood lymphocytes (CBL) using reconstituted Sendai virus envelopes (RSVE) as gene transfer vehicles. The DNA-treated CBL expressed EB virus nuclear antigen (EBNA) and synthesized cellular DNA at an increased rate but were not immortalized. However, when EBV DNA transfer was followed by exposure to UV-inactivated virions, full transformation was achieved. The resulting lymphoblastoid cell lines, termed NEB, were of B-cell origin, contained 25-50 EBV genome equivalents/cell, but were not capable of expressing EA or VCA, or releasing infectious virus. Permanent cell lines were also established when CBL were coinfected with purified EBV DNA and EBV of the HR-1 strain. These cells, termed HBD, were of T-cell origin, did not express EBNA, but contained EBV genomes as determined by nucleic acid hybridization. The HBD cells also expressed the early and virus capsid antigens of EBV as determined by immunofluorescence and radioimmunoprecipitation. The NEB and HBD cell lines will be useful for analyzing the mechanism of cell transformation by EBV.

INTRODUCTION

Studies on the cell transforming function of EBV have been hampered by the lack of viral mutants, host range restriction to B lymphocytes and restricted

Transformation of human lymphocytes by EBV DNA

permissivity of B cells for virus replication (reviewed
by Miller, 1980; Kieff et al., 1983). Attempts to
extend the host range of the virus by DNA transfection,
microinjection or virus receptor transplantation resulted
in virus replication rather than transformation (Graessman
et al., 1980; Volsky et al., 1980; Miller et al., 1981).
An alternative approach to identify and study the transforming genes of EBV is by following the function of
segments of EBV genome in normal human B lymphocytes.
In order to achieve efficient DNA transfer into lymphocytes, we have recently applied reconstituted Sendai
virus envelopes (RSVE) as gene transfer vehicles. In
principle, EBV DNA is first trapped within RSVE during
envelope reconstitution. It is then introduced with
high efficiency into recipient cells by vesicle-cell
fusion (Shapiro et al., 1981, Volsky et al., 1983; Volsky
et al., 1984a,b). RSVE-transferred cloned BamHI K fragment
of EBV DNA induced EBNA in 4% of human B cells but it
did not have any effect on lymphocyte proliferation
(Volsky et al., 1983, 1984a). Other segments of the
EBV genome, such as the cloned BamHI Dl fragment, induced
transient stimulation of lymphocyte DNA synthesis but
no EBNA (Volsky et al., 1984a). However, no lymphocyte-
transforming EBV DNA fragment has been as yet identified.
Curiously, even purified EBV DNA did not promote cell
transformation, although it induced EBNA in 1% of the
RSVE-transferred B cells and stimulated cellular DNA
synthesis (Volsky et al., 1983, 1984a). The purpose
of these experiments was to determine conditions permitting
transformation of human lymphocytes by purified EBV
DNA. We have found that UV-inactivated EBV facilitates
transformation of B lymphocytes by RSVE-transferred
DNA. Coinfection of cord blood lymphocytes with EBV
DNA and EBV of the HR-1 strain also resulted in the
establishment of lymphoblastoid cell lines, but of the
T cell origin.

MATERIALS AND METHODS

Sendai virus (SV) was propagated in 10-day old fertilized
eggs, purified and checked for fusogenic activity as
previously described (Volsky et al., 1980). For reconstitution and DNA entrappment, isolated SV envelopes (1 mg
of viral protein) in Triton X-100 were mixed with EBV
DNA (20-30 μg) and reconstituted by dialysis. The reconstituted, DNA-loaded vesicles (RSVE/DNA) were treated

Transformation of human lymphocytes by EBV DNA

with DNAse (1 mg/ml)to remove untrapped DNA. RSVE-mediated
DNA transfer (Volsky et al., 1983, 1984a) was performed
by fusing RSVE/DNA vesicles with lymphocytes at 37°C
in a buffer containing 160 mM NaCl/10 mM Tris-HCl, pH
7.5. The cells were then washed in RPMI 1640 medium
(+15% FBS), exposed to UV-inactivated virions or HR-1
EBV as described, and cultured under standard conditions.
HR-1 virus of the HH 514-15 strain EBV (Heston et al.,
1982) was obtained from starving HH 514-16 cells grown
in the presence of TPA (20 ng/ml) at 33°C. The supernatants were concentrated 500 fold. B95-8 virus DNA was
obtained from nuclear extracts of B95-8 cells and purified
by repetitive CsCl equilibrium centrifugation. The
procedure for isolation, propagation and characterization
of cloned EBV DNA fragments was as described (Dambaugh,
et al., 1980).

RESULTS

I. Establishment of the NEB and HBD Cell Lines

Two strategies were used to achieve transformation
of human cord blood lymphocytes with purified EBV DNA.
In the first approach, the cells were coinfected with
purified EBV DNA and UV-inactivated virions. In confirmation to our previous results (Volsky, et al., 1984a),
purified EBV DNA induced transient stimulation of CBL
DNA synthesis and EBNA induction, but not permanent
cell immortalization. When the RSVE-mediated EBV DNA
transfer was followed by an application of UV-inactivated
B95-8 or W91 EBV virions, cellular DNA synthesis increased
exponentially, and the cells became immortalized (Table
1). The UV-treated virions themself did not induce
any EBNA or stimulation of cellular DNA synthesis (Table
1). Six different cell lines, designated NEB, were
established by coinfection with different preparations
of EBV DNA and UV-inactivated virions. The cells have
been in continuous culture for more than a year.

The second experimental approach was to use purified
EBV DNA for the functional complementation of the non-
transforming P3HR-1 strain of EBV. One hour after the
RSVE-mediated EBV DNA transfer, lymphocytes were exposed
to P3HR-1 virus of the HH 514-16 strain (HH-EBV) (Heston
et al., 1982). We have chosen this viral isolate because
it was shown to induce little or no EA upon superinfection

Transformation of human lymphocytes by EBV DNA

of Raji cells (Heston et al., 1982), and thus it could
be expected to be less cytolytic. The DNA-transferred,
HH-EBV exposed cells were cultured on irradiated feeder
layer of mouse 3T3 fibroblasts and tested weekly for
EBNA induction and morphologic transformation. The
coinfected cells rapidly proliferated during the first
two weeks, then declined to few living cells about four
weeks after infection, and finally grew out into permanently transformed cell lines about two months from
the beginning of the experiment (Table 1). The cells
exposed to HH-EBV alone died within two weeks after
viral infection (Table 1). Three cell lines, termed
HBD, have been established by this approach. The cells
have been in continuous culture for over a year.

II. Characterization of the NEB and HBD cell lines

The principal features of NEB-11 and HBD-1 cell lines
are summarized in Table 2. NEB-11 cells contained B-1[+],

Table 1. Establishment of lymphoblastoid cell lines by coinfection
of human cord blood lymphocytes with purified EBV DNA and UV-inactivated
virions or EBV of HR-1 strain.

System	EBNA induction[1]	Stimulation of celluler DNA synthesis[2]	Immortalization into LCL	LCL code
B-EBV	+	+	+	LCL
UV-B-EBV	–	–	–	
RSVE/DNA	+	+	–	
RSVE/DNA+UV-B-EBV	+	+	+	NEB
HH-EBV	–	–	–	
RSVE/DNA+HH-EBV	n.t.	n.t.	+	HBD

[1] Tested by anticomplement immunofluorescence 3 days after infection/
coinfection; [2] Tested by [^{3}H]-thymidine uptake into total cellular
DNA 7 days after infection/coinfection; B-EBV: EBV of B95-8 strain;
UV-B-EBV: UV-inactivated B-EBV (22 min under 15 W GE germicidal lamp;
10 erg/mm^2/sec); HH-EBV: HR-1 EBV of the HH-514-16 strain (Heston et
al., 1982); RSVE/DNA: reconstituted Sendai virus envelopes containing
purified EBV DNA from B95-8 cells.

Transformation of human lymphocytes by EBV DNA

Ig$^+$, EBV-R$^+$, OKT-11$^-$, OKT-3$^-$ lymphocytes, indicating
that the cells belong to a B cell lineage (Table 2). All
of the NEB cell lines established in our laboratory
expressed surface IgM (Volsky, et al., 1984b), confirming
the known tropism of EBV to IgM-expressing lymphocytes.
All NEB-11 cells expressed EBNA; however, the cells
were completely negative for EA and VCA. TPA and N-butyrate treatment had no effect on the antigen induction
in these cells, neither could infectious virus be collected
from cell supernatants after the treatment (Table 2).
Nucleic acid hybridization with EBV DNA fragments as
probes confirmed that NEB-11 cells contained multiple
copies of the viral genome (Fig. 3A).

In contrast to NEB-11 cell line, the cell surface
characterization of HBD-1 cells showed that they contained
predominantly OKT-11$^+$, OKT-3$^+$, Leu-1$^+$, B-1$^-$, Ig$^-$, EBV-R$^-$
lymphocytes (Table 2), indicating that the cells belong
to a T cell lineage. Examination of Wright's-stained
HBD cell smears revealed mature and immature T cell
morphology (not shown). Cytogenetic analysis revealed
that HBD lines have normal karyotypes as expected in
cells obtained from healthy donors. In spite of their
T cell origin, HBD cells contain and actively express
EBV. This has been demonstrated in several ways: 1)
up to 30% of cells expressed early antigen (EA) and
virus-capsid antigen (VCA), as detected by indirect
immunofluorescence on methanol-fixed slides using monoclonal anti-EA and anti-VCA antibodies. About 40% of
cells bound monoclonal anti-MA antibodies as measured
on living cells by flow cytometry. In spite of this
active expression of several EBV-specific antigens,
the HBD cells were found to be negative for the only
antigen normally expressed in EBV-transformed human
B lymphocytes, EBNA (Table 2); 2) mature and immature
herpesvirus particles have been detected by electron
microscopy; 3) multiple viral genomes have been detected
by DNA-DNA hybridization with a specific viral probe
(Fig. 3A); 4) numerous EBV-specific polypeptides have
been identified in extracts of (^{35}S)-labelled HBD cells
by immunoprecipitation with a high titer EA$^+$/VCA$^+$ human
serum (Fig. 3B). The HBD cells were negative for the
only other human virus capable for transforming T lymphocytes, HTLV, as the cells did not express ATLV-determined
antigens (ATLA), did not contain any C-type retroviral
particles when examined by EM, and were negative for

Transformation of human lymphocytes by EBV DNA

Table 2. Properties of the NEB and HBD cell lines.

Marker/feature	NEB-11	HBD-1	Method of Determination
Leu-1	N.T.	+	Flow cytometry
OKT-11	-	+	Flow cytometry
OKT-3	-	+	Flow cytometry
M-1	-	-	Flow cytometry
B-1	+	-	Flow cytometry
Ig	+	-	Flow cytometry
EBV-R	+	-	Flow cytometry
EBV-MA	-	+	Flow cytometry
EBNA	+	-	ACIF
EA	-	+	IF
VCA	-	+	IF
ATLA	-	-	IF
RT	-	-	RT assay
Herpes virus particles	+	+	EM
EBV genome	+	+	Hybridization
Inducibility with TPA or n-butyrate	-	-	IF
Retrovirus particles	-	-	EM

One year-old (continuous cultures) NEB and HBD cell lines were analyzed
for cell surface phenotype by an Ortho system 50H cytofluorograph (flow
cytometry) with a Lexel argon ion laser tuned to 488 nm wavelength at
0.5 W. The monoclonal markers were purchased from Ortho Diagnostic
Systems, Inc. (OKT-3, OKT-11), Coulter Electronics, Inc. (B-1), Becton-
Dickinson (Leu-1), Tago, Inc. (Ig). EBV receptors (EBV-R) were assayed
using fluorescein-isothiocyanate labelled-virions and flow cytometry
as previously described. EBV-membrane antigen (EBV-MA) was detected
flow cytometrically using mouse monoclonal anti-MA antibodies. EBNA
was assayed by an anticomplement immunofluorescence (ACIF) test; EA,
VCA and ATLA were determined by indirect _in situ_ immunofluorescence
(IF) using smears of methanol fixed cells and mouse monoclonal anti-EA
and anti-VCA antibodies (Biotech Res. Laboratories, Maryland), or serum
from an ATL patient, respectively. Poly(A)-dependent DNA polymerase
activity (reverse transcriptase assay, RT) was measured in 50 μl of
50 mM Tris HCl, pH 7.5/5 mM dithiothreitol/100 mM KCL/10 mM MgCl$_2$/ 10 μ
M (^{3}H)dTTP/0.1% Triton X-100 containing 2 μg of poly (A), 0.4 μg of
(dT)$_{12-18}$ and 20 μl of cell extracts. For electron microscope examinations (EM), cells were fixed in 1% glutaraldehyde, processed according
to standard methods and observed under Phillips EM300. DNA-DNA hybridization was performed as described in legend to Fig. 1.

Transformation of human lymphocytes by EBV DNA

Fig. 1. Detection of EBV genome and EBV-specific polypeptides in the
T-lymphoblastoid cell line HBD-1.

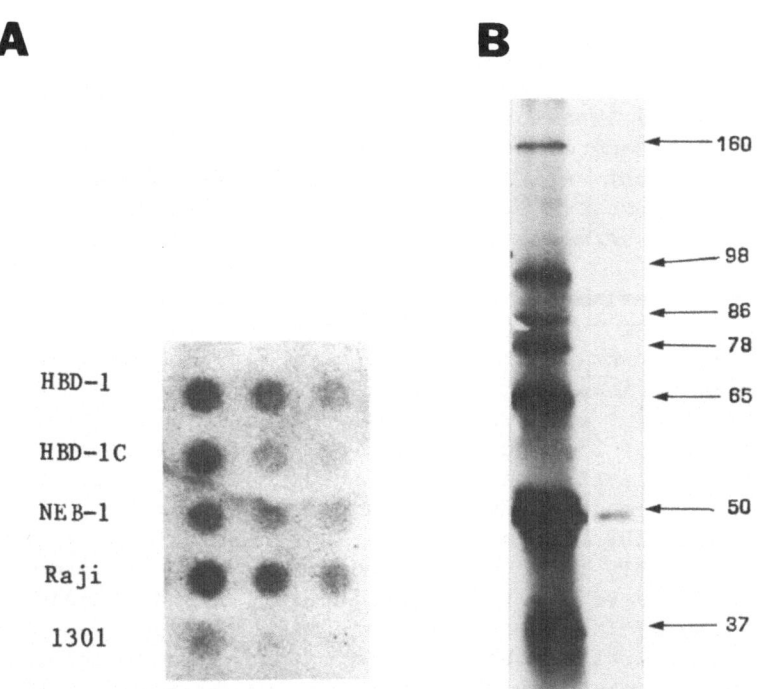

A: Detection of EBV genome in HBD cells by nucleic acid hybridization.
The cells (in 2-fold dilutions starting from 0.5 x 10^6) were washed
and blotted on nitrocellulose filters in a manifold apparatus using
the Quick-Blot procedure of Bresser et al (1983). The hybridization
cocktail contained 6 x SSC/50% formamide/0.05 M Na$_3$PO$_4$ buffer, pH 7.0/1%
SDS, and denatured EBV DNA fragments EcoRI J, EcoRI G$_2$, and EcoRI C+H
labelled with [^{32}P] by nick translation (2 x 10^7 cpm/μg DNA). The
hybridization was for 48 h at 42°C, followed by posthybridization washes
and autoradiography. B: Polyacrylamide gel electropherogram of EBV-
specific polypeptides immunoprecipitated from [^{35}S]-labelled HBD-1 cells.
The cells were pulsed with [^{35}S]methionine (New England Nuclear, 1166.5
Ci/mmol; 50 μCi/ml) in methionine-free medium for 6 h, then washed,
extracted, and immunoprecipitated with an EA$^+$/VCA$^+$ serum (titer: EA
1:5120, VCA 1:20480; EBNA 1:40; lane 1) or EBV negative (C.K.) serum
(lane 2). The immunoprecipitates were resolved on 10% SDS-polyacrylamide
gels, followed by autoradiography using Kodak XAR-2 paper and intensifying
screens. The numbers referring to EBV-determined proteins are their
mol. wt. x 10^{-3}.

Transformation of human lymphocytes by EBV DNA

the reverse transcriptase activity (Table 2).

DISCUSSION

In this work we report on the establishment of human lymphoblastoid cell lines by coinfection with purified EBV DNA and transformation-defective virions. Conditions have been determined which permit efficient introduction and permanent expression of purified EBV DNA in normal human lymphocytes, paving the way for the identification of EBV DNA segments involved in B cell transformation.

Two experimental strategies have been successful in achieving cell transformation after RSVE-mediated transfer of EBV DNA: a subsequent application of UV-inactivated virions from the transforming strains of EBV, or subsequent exposure to the nontransforming HH-514-16 substrain of EBV (Table 1). The possible contribution of UV-inactivated virions to the process of cell transformation by isolated EBV DNA is presently unclear. The amplified EBV DNA in NEB cell lines seems to originate predominantly from transfected DNA rather than from UV-inactivated virus (Volsky, et al., 1984b), indicating that the UV-inactivated virions have a helper function in the process of cell transformation. The following mechanisms might be considered: a) stimulation of the cellular DNA repair process by UV-treated EBV DNA, which could promote integration of the transfected DNA; b) the UV-inactivated and RSVE transferred DNA could undergo recombination in which the tranforming but not lytic genes of the virus would be rescued; c) the RSVE-transferred EBV DNA could be supplemented by a function which has been destroyed/ removed by DNA purification, but which was present in the UV-inactivated virions.

The establishment of EBV genome-positive T cell lines (HBD) was unexpected. This is the first time that EBV genome is found associated with permanently growing hematopoietic cells of non-B lineage origin. Preliminary studies show that the HBD cells contain both the HR-1 and B95-8 virus DNA, indicating that the two genomes complement each other in sustaining the transformed state. That the HBD cells are permissive to EBV replication, as reflected by the high expression of EA, VCA and MA antigens (Table 1), indicates that T lymphocytes are less successful than B cells in controlling EBV

Transformation of human lymphocytes by EBV DNA

replication. This could also explain the elimination
of B lymphocytes from the EBV DNA/HH-virus coinfected
cultures.

One of the most intriguing features of the HBD cell
lines is the complete lack of expression of EBNA. Not
only is EBNA not detected by immunofluorescence (Table
1), but there seems to be a block of the transcription of
the EBV DNA region that encodes for the antigen (Sinangil
and Volsky, unpublished results). The lack of EBNA
expression in HBD cells suggests that this antigen might
not be an essential indicator of EBV infection. EBV
replication in vivo might actually occur in EBNA-negative
cells, which were never expected to harbor the virus.

ACKNOWLEDGEMENTS

We are grateful to Dr. E. Kieff for supplying the
recombinant EBV DNA plasmids, to Dr. W. Henle for supplying
the high titer EA^+/VCA^+ human sera and to Dr. G. Miller
for sending the HH-514-16 cell line. We are also indebted
to L. Pertile and C. Kuszynski for excellent technical
assistance, and to S. Blum for typing the manuscript.
This work was supported by PHS NIH grants CA 33386 and
CA 37465, and by the Nebraska Department of Health (LB506).

REFERENCES

Bresser, J., Doering, J., and Gillespie, D., Selective
mRNA or DNA immobilization from whole cells. **DNA 2**,
243-254 (1983).

Dambaugh, T., Beisel, C., Hummel, M., King, W., Fennewald,
D., Cheung, A., Heller, M., Raab-Traub, N., and Keiff,
E., Epstein-Barr virus (B95-8) DNA. VII: Molecular Cloning
and Detailed Mapping. **Proc. Natl. Acad. Sci. USA 77**,
2999-3003 (1980).

Graessman, A., Wolf, H., and Bornkamm, G.W., Expression
of Epstein-Barr virus genes in different cell types
after microinjection of viral DNA. **Proc. Natl. Acad. Sci.,
USA 77**, 433-436 (1980).

Heston, L., Rabson, M., Brown, N., and Miller, G., New
Epstein-Barr virus variants from cellular subclones
of P3J-HR-1 Burkitt lymphoma. **Nature 295**, 160-163 (1982).

Transformation of human lymphocytes by EBV DNA

Kieff, E., Dambaugh, T., Hummel, M., and Heller, M.,
Epstein Barr virus transformation and replication. **In:**
G. Klein (ed.), **Advances in Viral Oncology, Vol. 3,**
p. 133-182, Raven Press, New York (1983).

Miller, G., Biology of Epstein-Barr Virus. **In:** G. Klein
(ed.),**Viral Oncology,** p. 713-738, Raven Press, New York
(1980).

Miller, G., Grogan, E., Heston, L., Robinson, J. and
Smith, D., Epstein-Barr viral DNA: Infectivity for human
placental cells. **Science 212,** 452-455 (1981).

Shapiro, I.M., Klein, G., and Volsky, D.J., Epstein-Barr
virus coreconstituted with Sendai virus envelopes infects
Epstein-Barr virus receptor-negative cells. **Biochem
Biophys Acta 676,** 19-24 (1981).

Volsky, D.J., Shapiro, I.M., and Klein, G., Transfer
of Epstein-Barr virus receptors to receptor-negative
cells permits virus penetration and antigen expression. -
Proc. Natl. Acad. Sci. USA 77, 5453-5457 (1980).

Volsky, D.J., Sinangil, F., Gross, T., Shapiro, I.,
Dambaugh, T., King, W., and Kieff, E., Functional mapping
of the Epstein-Barr virus (EBV) genome using Sendai
virus envelope mediated gene transfer. **In:** D.W. Golde
and P.A. Marks (eds.), **Normal and Neoplastic Hematopoiesis,
UCLA Symposia on Molecular and Cellular Biology, New
Series, Vol. 9,** p. 425-434, Alan R. Liss, Inc., New
York (1983).

Volsky, D.J., Gross, T., Sinangil, F., Kuszynski, C.,
Bartzatt, R., Dambaugh, T. and Kieff, E., Expression
of Epstein-Barr virus (EBV) DNA and cloned DNA fragments
in human lymphocytes following Sendai virus envelope-mediated
gene transfer. **Proc. Natl. Acad. Sci. USA 81,** 5926-5930
(1984a).

Volsky, D.J., Bartzatt, R., Kuszynski, C., Sinangil,
F., and Gross, T., Transformation of human cord blood
lymphocytes by coinfection with UV-inactivated Epstein-Barr
virus (EBV) and purified EBV DNA. **In:** Fred Rapp (ed.),
**Herpesvirus, UCLA Symposia on Molecular and Cellular
Biology, New Series, Vol. 21,** in press, Alan R. Liss,
Inc., New York (1984b).

35

EPSTEIN-BARR VIRUS-ACTIVATING SUBSTANCE(S) FROM SOIL

Yohei Ito[1], Harukuni Tokuda[1], Hajime Ohigashi[2],
Koichi Koshimizu[2] and Yi Zeng[3]

[1]Department of Microbiology, Faculty of Medicine
and [2]Department of Food Sciences and Technology,
Faculty of Agriculture, Kyoto University, Kyoto
606, Japan; and [3]Institute of Virology, National
Centre for Preventive Medicine, China

SUMMARY

The soil samples collected from under the plants Euphori-
biaceae and Thymeleaeceae, which are known to contain
Epstein-Barr virus (EBV)-activating and tumor-promoting
diterpene ester compounds, yielded ether extracts which
induced EBV early antigen (EA). Such findings were extended
to field studies in southern areas of China where one of
the EBV-associated diseases, nasopharyngeal carcinoma (NPC),
is endemic. The soil samples collected from such areas,
particularly those from under the Chinese tung oil trees
(Aleurites fordii), which contain active diterpene ester
HHPA (12-0-hexadecanoyl-16hydroxyphorbol-13-acetate), also
exerted positive reactivity. When plants which do not contain diterpene esters were transplanted into a petri dish
with such active substances or were grown in soil in which
they had accumulated, even those portions of the plants
which were relatively free of such substances yielded
extracts which activated EBV early antigen (EA) in non-
producer human lymphoblastoid cells (Raji).

INTRODUCTION

Many, if not all, species of plants belonging to the families of Euphorbiaceae and Thymeleaeceae contain irritating compounds. Such irritants are known to be diterpene esters based on the skeleton structures of tigliane, daphnane and ingenane (Evans and Schmidt, 1980). The chemical extracts from these plants have long been recognized to possess tumor-promoting capacity (Diamond et al., 1980). The linkage between such tumor-promoting substances and EBV-EA induction was shown in recent studies (zur Hausen et al., 1979). It was also revealed that many species of such plants are currently used as folk remedies in areas where two EBV-associated diseases, Burkitt's lymphoma (BL) and NPC, are endemic (Hirayama and Ito, 1981; Ito et al., 1981a). The Chinese tung oil tree, a member of Euphorbiaceae, is popular all over the southern provinces of of China, where it is cultivated chiefly for industrial purposes. The plant itself is also used as a source of herbal drugs. The essential diterpene ester of this plant with EBV-activating and tumor-promoting capacity is HHPA (Ito et al., 1983).

To determine how the EBV-activating substances gain access to the human system, we decided to test the hypothesis that such substances were in the soil in which the plants were grown. This phenomenon of higher plants affecting each other through the release of chemicals through their plant bodies, known to botanists and agriculturists, is termed allelopathy (Muller and Chou, 1972). After assaying the soil extracts, it became evident that such was the case (Ito et al., 1983). Extracts of soil samples obtained from under the plants with EBV-activating diterpenes exerted similar activities and their potency was comparable to those of the plant extracts. The uptake of EBV-activating substances by plants which primarily lack such substances was also observed. The implications of these findings to the possible etiology of BL and NPC is discussed.

MATERIALS AND METHODS

Cells and cell culture

The Raji cell line, an EBV-nonproducer cell containing multiple copies of EBV genomes, was cultivated in RPMI 1640

medium containing 10% fetal calf serum, 100 units of penicillin and 250 µg/ml of
streptomycin. Under these conditions, Raji cells showed a
spontaneous rate of induction of EBV EA of less than 0.01%.

Assay method for EA induction

The synergistic assay for EA induction in Raji cells was
employed (Ito et al., 1981a). The cells were adjusted to a
density of 1 x 10^6 cells/ml and were incubated with 4 mM
n-butyrate and various test extracts at varying concentrations. 12-0-tetradecanoylphorbol-13-acetate (TPA), at a
concentration of 10-20 ng/ml, served as a positive positive
control, and cultures treated only with n-butyrate as a
negative control. The results were read after 48 hrs incubation of the cells at 37^oC, using the indirect immunofluorescence (IF) technique (Henle and Henle, 1966). Cell
smears were prepared on glass slides, air-dried, and fixed
with acetone at room temperature for 10 min. Activated
Raji cells expressing EBV EA were stained with EA+-virus
capsid antigen (VCA+) high titer serum from an NPC patient,
kindly provided by Prof. H. Hattori, Kobe University
School of Medicine. The untreated cultures served as controls. In each assay, at least 500 cells were counted
randomly and the EA+ cells were determined. The number of
viable cells in the culture was determined by the methylene-
blue exclusion test (Ito et al., 1981b).

Soil extracts

Soil samples (20 g each) were collected from under plants
of Euphorbiaceae and Thymeleaeceae, and also from plants of
other species (controls). The collections were made at
about 0.5 cm from the base of the plant stems at a soil
depth of 0.2 cm. The samples were extracted with an equal
volume of ether for 20 min at room temperature (20^oC).
After evaporating the solvent, the crude extracts were
weighed and redissolved in dimethylsulfoxide (DMSO) as a
stock solution of 10 mg/ml. The extracts were prepared in
final concentractions of 100, 20 and 4 ng/ml and tested for
induction of EA in Raji cells.

RESULTS AND DISCUSSION

Detection of EBV-activating potency in extracts from soil
samples under plants containing active diterpene esters

The phenomenon of allelopathy, described above, is accomplished by chemicals being released from plants by rain-wash,
root excretion and decay; by pollens, decomposition of fallen
leaves, flowers and fruit, bark, etc. In the case of the
Euphorbiaceae and Thymeleaeceae plants, it seemed probable
that the EBV-activating substances might also be found in
the soil surrounding those plants. This was found to be
the case. Table 1 shows EBV-EA induction (5.9% - 23.8%) in
Raji cells by ether extracts of soil under plants containing
diterpene esters, suggesting excretion of EBV-activating
compounds; while those from under plants free of such substances, selected randomly from areas of our university
campus, did not induce EA.

TABLE 1
EBV-ACTIVATING PRINCIPLES IN SOIL EXTRACTS

Samples taken from soil under plants		Soil Extracts (μg/ml) #	EBV EA-positive cells (%)
Species	Family		
Sapium sebiferum	Euphorbiaceae	20	23.8
Codiaeum variegatum	Euphorbiaceae	20	12.0
		4	9.8
Euphorbia lathyris	Euphorbiaceae	20	3.6
		4	7.5
Daphne odora	Thymeleaceae	20	15.6
		4	10.5
Edgeworthia papyrifera	Thymeleaceae	20	5.9
Vinca rosea	Apocynaceae	20	0.1*
		4	0.1
Castanea crenata	Fagaceae	20	0.1
Control Ground		20	0.1
Commercial soil		20	0.1

#Dissolved in DMSO and used with n-butyrate (4 mM).
*Represents figure less than or equal to 0.1%.

387

Comparison of EBV-activating potency of soil extracts with those of "parental" plants

It is of interest to determine whether the active substances detectable in the soil around the affected plants are related to the active chemical(s) in the plant bodies per se. We are currently in the process of isolating and purifying the compounds from the soil. The chromatographic data, although still too preliminary to be definitive, indicate that the compounds are at least of the diterpene ester-type. In Table 2, data of the EBV EA-inducing capacity of extracts from various portions of the plant bodies, as compared with those of the soil extracts beneath them, are shown. It may be noteworthy that the EBV-induction activity of extracts derived from the soil is comparable to, if not exceeding, that of the plant-derived extracts, suggesting that such compounds are actively released in large quantities in the soil.

TABLE 2

COMPARISON OF EBV-ACTIVATING POTENCY OF EXTRACTS FROM PLANTS
AND FROM SOIL UNDER THE PLANTS

Plant species	Concentration	EBV EA-positive cells (%)
Codiaeum variegatum		
Flowers	10	11.4
	2	7.5
Stems	10	13.4
	2	10.2
Soil	20	12.0
	4	9.8
Daphne odora		
Leaves	10	31.0
	2	29.7
Soil	20	15.6
	4	10.5
Sapium sebiferum		
Leaves	2	24.5
	0.4	28.7
Soil	20	19.3
	10	10.3
Control		
TPA*		31.1
n-butyrate		0.1

*used with n-butyrate (4 mM).

EBV-activating factors extracted from plants growing in media or soil containing diterpene esters and possibly related active substances.

The next question posed was whether the active substances accumulated in the soil could be absorbed by plants of other species which normally do not possess such compounds. A model experimental system was designed by. culturing germinating plants in medium containing EBV-activating compounds. Extracts were prepared from the upper portion of the plant were assayed for EBV-EA induction after three days.

Bean sprouts and other plants, which are all common vegetables sold at the market, were the plants chosen to be grown in medium containing EBV-EA-inducing compounds. When 50 µg of TPA and a 5% acetone solution were added to the medium (cotton bed on which the germinating plants were transplanted and kept for 3 days), the EBV-EA induction by the plant extract was 7.3% and 10.2%, respectively; whereas the positive control, with 20 ng/ml of TPA in the synergistic assay system, showed 22.6% positive cells.

The next set of experiments were carried out on a species of vegetable, Zingider mioga, a popular Japanese delicacy. We selected a location in a private garden (Dr. Y's), where the vegetable was cultivated under a Sapium sebferum tree, a parental plant producing active diterpene ester HHPA. The location where the vegetable was grown is illustrated in Fig. 1, and the results of the assay of the extracts for EBV-activation are shown in Table 3. The portion of Z. mioga eaten grows underground (root). The results show that considerable EA-inducing activity (1% - 24%) was found in the vegetable, normally free of such diterpene esters.

FIGURE 1

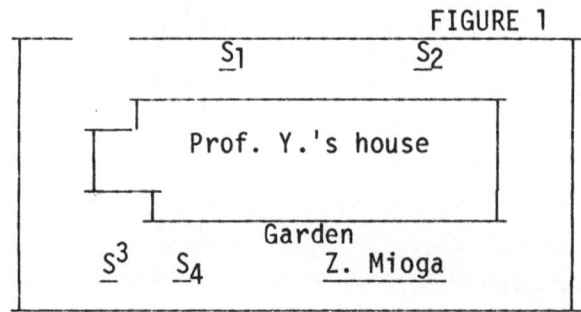

S: Sapium sebiferum (4 trees)

Z. mioga: where vegetables are cultivated (2.0 m from S_4.

TABLE 3
EBV-ACTIVATING POTENCY OF EXTRACTS OF SOILS AND VEGETABLES
(Zingider mioga) FROM PROF. Y.'S GARDEN

Samples	Concentration (μg/ml)	EBV EA+ Cells (%)
Soil		
between S_1 & S_2	50	1.0
	10	21.0
	2	2.0
between S_3 & S_4	50	24.3
	10	11.0
	2	7.2
Control soil	50	0.1
	10	0.1
	2	0.1
Plant (Z. mioga)		
Leaves	50	0.1
	10	0.1
	2	0.1
Roots #1[a]	50	0.1
	10	2.3
	2	1.6
#2	50	1.1
	10	5.6
	2	6.1
#3 (pickled)	50	2.1
	10	4.3
	2	2.1
Control		
TPA		27.8
n-butyrate		0.1

[a]Roots are the underground portion of the vegetable eaten as delicacy in Japanese cooking.

The implications of the uptake of active substances by non-active plants may have profound significance. However, the evidence is not sufficient at the present time to suggest any link to human neoplasia. Further careful studies must be carried out to draw any conclusions.

Field study for EBV-activating substances from soil under
trees of Aleurites fordii (Chinese tung oil tree) and other
plants in NPC endemic areas of southern China

A survey of soil samples collected from under Chinese tung
oil trees and other Euphorbiaceaes was carried out (Z.Y.)
in the southern provinces of China to confirm our findings.
The rate of EBV EA induction by such samples was as high as
59.5%. Although the data are still qualitative rather than
quantitative, it appears EBV-activating substances do exist
in the soils of endemic NPC areas (Zeng et al., 1984) and
may be a contributing factor in the complex etiology of
NPC.

REFERENCES

Diamond, L., O'Brien, T.G., and Barid, W.M. Tumor promoters
and mechanism of tumor promotion. Adv. Cancer Res., 32,
1-74 (1980).

Evans, F.J., and Schmidt, R.J. Plants and plant products
that induce contact dermatitis. J. Med. Plant Res., 38,
291-316 (1980).

Henle, G. and Henle, W. Immunofluorescence in cells
derived from Burkitt's lymphoma. J. Bacteriol., 91,
1248-1256 (1966).

Hirayama, T. and Ito, Y. A new view of the etiology of
nasopharyngeal carcinoma. Rev. Med., 10, 614-622 (1981).

Ito, Y., Kishishita, M., Morigaki, T., Yanase, S., and
Hirayama, T. Induction and intervention of Epstein-Barr
virus expression in human lymphoblastoid cells: A simulation model for study of cause and prevention of nasopharyngeal carcinoma and Burkitt's lymphoma. In: E. Grundmann,
G.R.F. Krueger and D.V. Ablashi (eds.), Nasopharyngeal
Carcinoma, Cancer Campaign, 5 (1981a), pp. 255-255-262.

Ito, Y., Yanase, S., Fujita, J., Hirayama, T., Takashima,
M., and Imanaka, H. A short-term in vitro assay for promoter
substances using human lymphoblastoid cells latently infected
with Epstein-Barr virus. Cancer Letters, 13, 29-37 (1981b).

Ito, Y., Yanase, S., Tokuda, H., Kishishita, M., Ohigashi, H., Hirota, M., and Koshimizu, K. Epstein-Barr virus activation by tung oil, extracts of Aleurites fordii and its diterpene ester HHPA. Cancer Letters, 18, 87-95 (1983).

Muller, C.H. and Chou, C.H. Phytotoxins: an ecological phase of phytochemistry. In: J.B. Harborne (ed.). Phytochemical Ecology, Academic Press, London (1972), pp. 201-206.

Zeng, Y., Miao, X.C., Jaio, B., Li, H.Y., Ni, H.Y., and Ito, Y. Epstein-Barr virus activation in Raji cells with ether extracts of soil from different areas in China. Cancer Letters, 23, 53-59 (1984).

zur Hausen, H., Bornkamm, G.W., Schmidt, R., and Hecker, H. Tumor initiators and promoters in induction of Epstein-Barr virus. Proc. Natl. Acad. Sci. USA, 76, 782-785 (1979).

36

HYDROCORTISONE ENHANCEMENT OF BOTH EBV REPLICATION AND

TRANSFORMATION OF HUMAN CORD LYMPHOCYTES

D. V. Ablashi[1], J. Whitman[2], J. Dahlberg[1],
G. Armstrong[3], and J. Rhim[1]

[1]National Cancer Institute, Bethesda, MD
20205, [2]Advanced Biotechnologies, Inc., Silver
Spring, MD 20904, [3]Division of Virology,
Bureau of Biologics, Bethesda, MD 20205

SUMMARY

We investigated the interaction of hydrocortisone (HC)
and EBV. The treatment of P3HR-1 cells (propagated at 34°C
and 37°C) with various concentrations of HC for 7 and 21
days resulted in enhanced levels of antigen positive cells
with a maximum increase at 21 days. Virus harvested from
HC-treated P3HR-1 cells grown at 34°C had a 1-2 log higher
titer in Raji cells when compared to control virus.

Treatment of AG876 EB virus producer cells grown at
34°C with 5 and 10 μg/ml HC for 7 and 21 days resulted in
up to a 3-fold higher level of IF membrane positive cells.
Cells treated for 21 days with 10 μg/ml of HC exhibited
a 3-5-fold increase in VCA positive cells. When human cord
blood mononuclear cells were infected with AG876 EBV and
maintained in HC, earlier transformation was observed.

These data suggest that hydrocortisone is able to
enhance the expression of the EBV genomes present in human
cells and leads to increased levels of antigen expression
and virus production. The mechanism by which this glucocorticoid hormone modulates EBV expression remains to be
determined.

INTRODUCTION

Ringold et al. (1984) have recently summarized current research on glucocorticoid-inducible genes involving regulatory sequences of the MMTV LTR. Rhim (1983) has shown that glucocorticoids enhanced the transformation of mammalian cells caused by the Kirsten strain of murine sarcoma virus. Other studies showed that steroid hormones led to enhanced yields of polyoma virus (Morhenn et al., 1973) and further increased the production of endogenous type C virus induced by 5-iodo2'-deoxyuridine from mouse fibroblast cells (Paran et al., 1973). Long-term suspension growth of normal immature myeloid cells from human peripheral blood was first accomplished in medium containing hydrocortisone (Saladhuddin et al., 1982). Thompson et al. (1983) were able to grow normal human nasopharyngeal epithelial cells up to 146 days when the culture medium included hydrocortisone. These cells also could be subcultured up to 50 passages, thus providing a method for culturing nasopharyngeal epithelial cells in quantities suitable for extensive experimental work with Epstein-Barr virus. Armelin et al. (1983), after treatment of a rat glioma cell line with hydrocortisone, demonstrated that the morphological alterations caused by this hormone were accompanied by the induction of an endogenous type C RNA tumor virus. Our own studies have shown that higher yields of Herpesvirus saimiri (HVS) can be obtained when owl monkey kidney cells were treated with hydrocortisone immediately after virus infection (unpublished observation).

Based on the these observations, we were interested in investigating the effect of hydrocortisone on EBV. Although Magrath et al. (1979) reported on the effect of low temperature and corticosteriods on EBV producer cells, they did not study enhancement of virus transformation, EA induction or specific yields of EBV.

MATERIALS AND METHODS

Chemicals

Hydrocortisone was purchased from Sigma Chemical Co., St. Louis, Missouri, USA, and a stock solution was prepared in ethanol as described by Rhim (1983). This stock solution was diluted into growth medium (RPMI-1640 supplemented with 10% fetal calf serum and 50 µg/ml gentamycin; Advanced Biotechnologies Inc., Silver Spring, MD, USA) just prior to use.

Cells

EB virus producer (P3HR-1, AG876) and nonproducer (Raji) cell lines were used for measuring EBV antigens (EA, VCA, MA) by immunofluorescence by previously described assays. The P3HR-1 and AG876 cells were propagated in RPMI 1640 medium containing 10% fetal calf serum (FCS) and cell cultures were carried at 35°C or 37°C. Human cord blood mononuclear cells, used for transformation with the AG876 strain of EBV, were grown in the same medium.

Virus

Purified EBV from P3HR-1 and AG876 EBV cells was prepared by Advanced Biotechnology, Inc. Briefly, lots of virus were clarified by centrifugation at 10,000 xg for 10 min., concentrated by continuous flow zonal centrifugation in a sucrose gradient, pelleted, and resuspended in 0.2% of the original volume in complete growth medium. Concentrated P3HR-1 strain of EBV contained 10^5 EA units/ml in Raji cells and the concentrated strain of AG876 virus had 10^3 transforming units/ml in human cord blood lymphocytes.

To test the effect of hydrocortisone (HC) on EBV replication, cultures were treated for 7-21 days with 1-50 µg/ml of HC. These cultures were then tested for altered levels of EA and VCA production by immunofluorescence, and culture fluids were tested for their level of infectious virus.

Virus titration

Infectivity of P_3HR-1 virus was carried out by a standard procedure involving superinfection of Raji cells by serial dilutions of virus and measuring the resultant induction of EA by immunofluorescence 48 hours later. Titers of the transforming strain AG876 were determined by infecting human cord blood lymphocytes with serial 10-fold dilutions of virus and determining the limiting dilution capable of causing colonies of transformed cells by 8 weeks (Faggioni et al., 1983).

Detection of EBV antigens

EBV antigen detection was carried by IF using previously described assays. To detect MA, two monoclonal antibodies kindly supplied by Dr. Pearson, Dept. of Microbiology, Univ. of Georgetown, Washington, D.C. were employed (Qualtiere et al., 1982) using AG876 cell cultures. Human sera were used to detect EA and VCA antigens.

EA was induced either by treatment of cells with TPA and N-butyric acid (Faggioni et al., 1983) or by superinfection of Raji cells with P3HR-1 EBV for 48 hours.

RESULTS AND DISCUSSION

Hydrocortisone-mediated enhancement of EBV antigens in P3HR-1 and AG876 cells

Hydrocortisone levels ranging from 1-50 µg/ml were maintained for 7 days in P3HR-1 cultures prior to testing these cells for the level of EBV VCA/EA antigens.

As can be seen in Table 1, 5 µg/ml of hydrocortisone doubled the number of antigen producer cells, while 1 µg/ml was somewhat less effective, and higher levels exhibited toxicity. This enhancement may be useful in obtaining higher yields of antigen expressing cells for routine EBV serology and suggested that higher yields of virus might be obtainable. Table 2 shows that when P3HR-1 and AG876 cells were treated continuously with 5 or 10 µg/ml of hydrocortisone over a period of 21 rather than 7 days at 35°C, up to 35-40% of the P3HR-1 cells and 25-30% of the AG876 cells were VCA/EA positive. In contrast to the P3HR-1 cells, 10 µg/ml of hydrocortisone was not toxic to the AG876 cells and this level induced a four-fold-higher level of VCA/EA positive cells than was present in the control culture.

Table 1

Enhancement of EBV-VCA/EA by hydrocortisone in P3HR-1 cells

Dose of hydro-cortisone	% cells positive for EBV-VCA/EA	Average increase over control
1 µg/ml	< 20-22	9%
5 µg/ml	> 25-27	12%
10 µg/ml	< 20 (some toxicity)	8%
50 µg/ml	toxic (>85% dead cells by 24 hours)	-
No HC	< 12	-

The cells were seeded at 1×10^6/ml at 35°C in the presence or absence of HC for 7 days and were examined for VCA.

Yields of EBV from hydrocortisone treated cells

Both P3HR-1 and AG876 cells were grown in the presence of 5 or 10 µg per ml of hydrocortisone at 35°C. After 7 days, the cultures were fed by adding fresh medium containing 5 µg/ml of HC and the cells were allowed to age for another 7 days prior to titering the culture fluids for virus. When P3HR-1 was titered by its ability to induce EA in Raji cells, one or more logs of additional infectious virus was routinely obtained from hydrocortisone treated P3HR-1 cells while approximately two additional logs of transforming AG876 virus, as determined by its ability to transform human cord blood lymphocyes, was regularly observed (data not shown).

Enhancement of EBV-EA with hydrocortisone

Raji cells were used to compare the ability of HC and TPA to affect EA induction either alone or following super-infection with P3HR-1 virus. Table 3 shows that Raji cells were slightly stimulated (up to 2% EA positive cells) by hydrocortisone alone. TPA alone, at 20 ng/ml, induced 10-12% of the cells to produce EA, and HC did not enhance the TPA mediated EA induction. Of greater interest was the observation that 5 µg/ml HC led to an approximate doubling

Table 2

Effect of HC on EBV-VCA/EA in P3HR-1 or AG876 cells continuously grown in 5 and 10 μg/ml HC for 21 days at 35°C

Hydrocortisone dose	% EBV-VCA/EA(+) cells[*]	Average increase of VCA/EA
A. P3HR-1 cells		
1 μg/ml	>25 but less than 30	>10%
5 μg/ml	>35 but less than 40	>20%
No HC	⩾15	-
B. AG876 cells		
1 μg/ml	>12	6%
5 μg/ml	⩾20	>15%
10 μg/ml	>25 but less than 30	⩽20%
No HC	5-7	-

*Average of three experiments. The cells were changed with
 medium containing HC twice during duration of experiment.

of the EA level obtained by superinfection of Raji cells
with P3HR-1 EBV (Table 3). This effect was highly dependent
on the concentration of HC used, with both 1 μg/ml and
10 μg/ml producing only modest augmentation. The reason
for this is not known. The mechanisms by which HC and TPA
induced EA appear to be independent. Our data indicate
that HC can significantly increase the ability of infectious virus to induce EA antigens.

Enhancement of EBV-membrane antigens in the presence of
hydrocortisone

The data in Fig. 1 show that both monoclonal antibodies to EBV-MA detected a higher percentage of antigen
positive cells in the presence of 10 μg/ml of hydrocortisone. The optimal level of cells expressing MA occurred
after 21 days of continuous HC treatment. Since these

Table 3

Effect of HC on EBV-EA expression in Raji cells

Treatment*	μg/ml HC	EBV-EA	Average Enhancement
A. Raji	0	-	-
	1	-	0
	5	-	1%
	10	-	$\leq$2%
B. Raji + P3HR-1/EBV (48 hrs)	0	>45	-
	1	>55%	10%
	5	<80%	35%
	10	<50%	5%
C. Raji + 20 ng/ml TPA	0	10-12%	-
	5	10-12%	0

*Raji cells were grown at 37°C. 1×10^6 cells were treated for 7 days. Group C cultures were treated with TPA for 2 hrs and then medium containing 5 μg/ml HC was added to half the cultures, and incubated for 7 days.

monoclonals recognize different glycoproteins (Qualtiere et al., 1982) it can be concluded that HC enhances the yield of both glycoproteins.

Enhancement of transformation by EB virus with hydrocortisone treatment

Human cord blood mononuclear cells at 5×10^5 cells/ml were PHA stimulated with 5 μg/ml 24 hours prior to EBV infection. Immediately after EBV infection, medium containing 5 or 10 μg/ml of hydrocortisone was added to the cells. Treated and untreated cultures were incubated at 37°C and were periodically tested for the presence of EBNA positive cells and colonies of transformed cells. Treatment of these cultures with 10 μg/ml HC led to a greater than 2 log enhancement in the efficiency of transformation by EB virus, as defined by the reciprocal of the limiting of virus leading

**ENHANCEMENT OF EBV-MEMBRANE ANTIGEN IN AG-876 STRAIN
OF EBV IN THE PRESENCE OF HYDROCORTISONE*†**

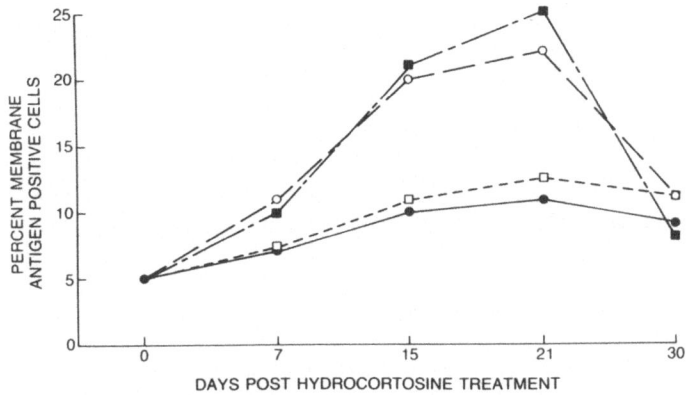

DAYS POST HYDROCORTOSINE TREATMENT

* 1 Two Monoclonal Antibodies (2F5.6 and 2L10) to EBV
Membrane were used in Detection of Membrane
Antigen Positive Cells (Supplied by Dr. Gary Pearson)

† 2 The AG876 Cells were Split at a Ratio of 1 x 10⁶/ml.
The Medium was Changed once a Week and Cells
were incubated at 34/35°C

○- -○ 10 ug/ml Hydrocortisone. Tested the Cells with 2F 5.6 Antibody
■- -■ 10 ug/ml Hydrocortisone. Tested the Cells with 2L10 Antibody
●—● No Hydrocortisone was used. Tested the Cells with 2F5.6
□- -□ No Hydrocortisone. Tested the Cells with 2L10 Antibody

to transformation of mononuclear cells. In addition, transformed colonies appeared earlier when HC was present (10-25
days compared to 14-45 for control cultures). These results
are similar to those obtained by Faggioni et al. (1983) using
MNNG and TPA, although there is no data to support the possibility that the enhanced rate of transformation by EBV
mediated by HC, carcinogens (Henderson and Fronko, 1984)
or by tumor promotors (Mizuno et al., 1983) is mediated by
a common mechanism.

The data presented here clearly suggest that the interaction of hydrocortisone with human cells producing EBV not
only results in a higher level of antigen positive cells
but also enhances the production of infectious virus. In
addition, HC leads to enhanced transformation of cells by
a transforming strain of EBV.

The development of neoplastic disease has classically
been considered a multistep process involving a series of
progressive changes rather than a single-step conversion
of a normal cell to a highly malignant neoplastic cell.
The role of activated cellular proto-oncogenes in the pathogenesis of many neoplams is well documented (Cooper, 1984).

In the case of Burkitt's lymphoma, in which infection of
B cells by EBV is a crucial early step, activation of c-myc
by nondisjucntion is also a critical event. It is well
known that glucogenic steroids can affect a wide spectrum
of cellular functions, some of which could be capable of
altering or enhancing EBV mediated replication or transformation. Thus, development of BL and NPC may not only
involve infection of appropriate target cells by EBV in
an appropriate immunological milieu, but also be greatly
influenced by other enviromental factors such as chemical
carcinogens, tumor promotors and hormonal imbalance.

REFERENCES

ARMELIN, M.C., GARRIDO, J., and ARMELIN, H.A., RNA tumor
virus production accompanies the transformed phenotype
change by hydrocortisone hormone in rat glioma cells.
Cell Biol. Int. Rep. 7, 689-696 (1983).

COOPER, G.M., Activation of cellular transforming genes in
neoplasms. In: F. Bresciani, R.J.B. King, M.E. Lipman,
M. Namer and J.O. Raymand (eds.), Progress In Cancer Research and Therapy, vol. 31, pp. 1-6 (1984).

FAGGIONI, A., ABLASHI, D.V., ARMSTRONG, G., DAHLBERG, J.,
SUNDAR, S.K., RICE, J.M., and DONOVAN, P.J., Enhancing
effect of N-methyl-N-nitrosoquanidine (MNNG) on Epstein-Barr
virus (EBV) replication and comparison of continuous and
discontinuous TPA treatment of EBV nonproducer and producer
cells for antigen induction and/or stimulation. In: U.
Prasad, D.V. Ablashi, P.H. Levine and G.R. Pearson (eds),
Nasopharyngeal Carcinoma: Current Concepts, pp. 333-345,
Univ. of Malaya Kuala Lumpur (1983).

HENDERSON, E.E. and FRONKO, G., Studies on chemical carcinogen enhancement of Epstein-Barr virus induced transformation of human neonatal and adult peripheral blood
lymphocytes. Int. J. Cancer 33, 331-338 (1984).

MAGRATH, I.T., PIZZO, P.A., NOVIKOVS, L., AND LEVINE, A.S.,
Enhancement of Epstein-Barr virus in producer cells by
combination of low temperature and corticosteroids.
Virology 97, 477-481 (1979).

MIZUNO, F., KOIZUMI, OSATO, T., KORWARO, J.O., and ITO, Y., The enhancing effect of extracts from Chinese and African Euphorbiaceae plants on transformation by Epstein-Barr virus. In: U. Prasda, D.V. Ablashi, P.H. Levine and G.R. Pearson (eds.), Nasopharyngeal Carcinoma: Current Concepts, p. 329-332, Univ. of Malaya Kuala Lumpur, Malaysia (1983).

MORHENN, V., RABINOWITZ, A., and TOMKINS, G.M., Effect of adrenal glucocorticoids on polyma virus replication. Proc. Natl. Acad. Sci. USA 70, 1088-1089 (1973).

PARAN, M., GALLO, R.C., RICHARSON, L.S., and WU, A.M., Adrenal corticosteroids enhance production of type C virus induced by 5-iodo-2′-deoxyuridine from cultured mouse fibroblasts. Proc. Natl. Acad. Sci. USA 70, 2391-2395 (1973).

QUALTIERE, L.F., CHASE, R., VROMAN, B., and PEARSON, G.R., Identification of Epstein-Barr virus strain differences with monoclonal antibody to a membrane glycoprotein. Proc. Natl. Acad. Sci. USA 79, 616-620 (1982).

RHIM, J.S., Glucocorticoids enhance viral transformation of mammalian cells. Proc. Soc. Exptl. Bio. Med. 174, 217-223 (1983).

RINGOLD, G., COSTELLO, M., DOBSON, D., FRANKEL, F., HALL, C., and LEE, F., Glucocorticoid regulation of gene expression. In: F. Bresciani, R.J.B. King, M.E. Lipman, M. Namer and J.P. Raymand (eds.), Progress In Cancer Research and Therapy, vol. 31, pp. 7-18, Raven Press, New York (1984).

SALAHUDDIN, S.Z., MARKHAM, P.D., and GALLO, R.C., Establishment of long-term monocyte suspension cultures from normal human peripheral blood. J. Exptl. Med. 155, 1842-1857 (1982).

THOMPSON, J.L., EPSTEIN, M.A., ACHONG, B.G., and CHEN, J.J., A culture method giving substantial yields of normal nasopharyngeal epithelial cells for work with Epstein-Barr virus. J. Virol. Methods 6, 319-328 (1983).

37

BRIEF COMMUNICATION

DETECTION OF EBNA AND RESCUE OF TRANSFORMING EBV IN

MEGAKARYOCYTE CELLS ESTABLISHED IN CULTURE

D. Morgan[1] and D. V. Ablashi[2]

[1]Department of Haematology and Oncology,
Hahnemann University, Philadelphia, PA, USA;
[2]National Cancer Institute, Bethesda,
Maryland 20205, USA.

INTRODUCTION

For the first time, cell lines of human megakaryocytes
have been obtained from the circulating blood of normal
donors and patients with various blood disorders (Morgan and
Brodsky, 1984). The cultures consist predominantly of small
"lymphoid" cells accompanied by giant multinucleated cells
which spontaneously accumulate (Fig. 1, A and B). These
giant cells have morphological properties common to well-
defined megakaryocytes (Morgan and Brodsky, 1984). These
"lymphoid" cells have no surface markers specific for lymphocytes, monocytes or granulocytes. Most significantly,
greater than 80% of the cells do express antigens specific
for and/or associated with megakaryocytes and platelets
(Table 1).

Besides the B-lymphoblastoid cells, only the epithelial
cells from nasopharyngeal carcinoma (NPC) have been found
to contain EBV, suggesting a very narrow cell specificity
(Glaser, 1983). We were interested in investigating whether
megakaryocytes, the blood platelet precursors (Kapff and
Jandle, 1981), contain EBV, particularly those derived from
EBV-antibody-positive normal individuals, patients with
immunologic abnormalities and patients with certain cancers.

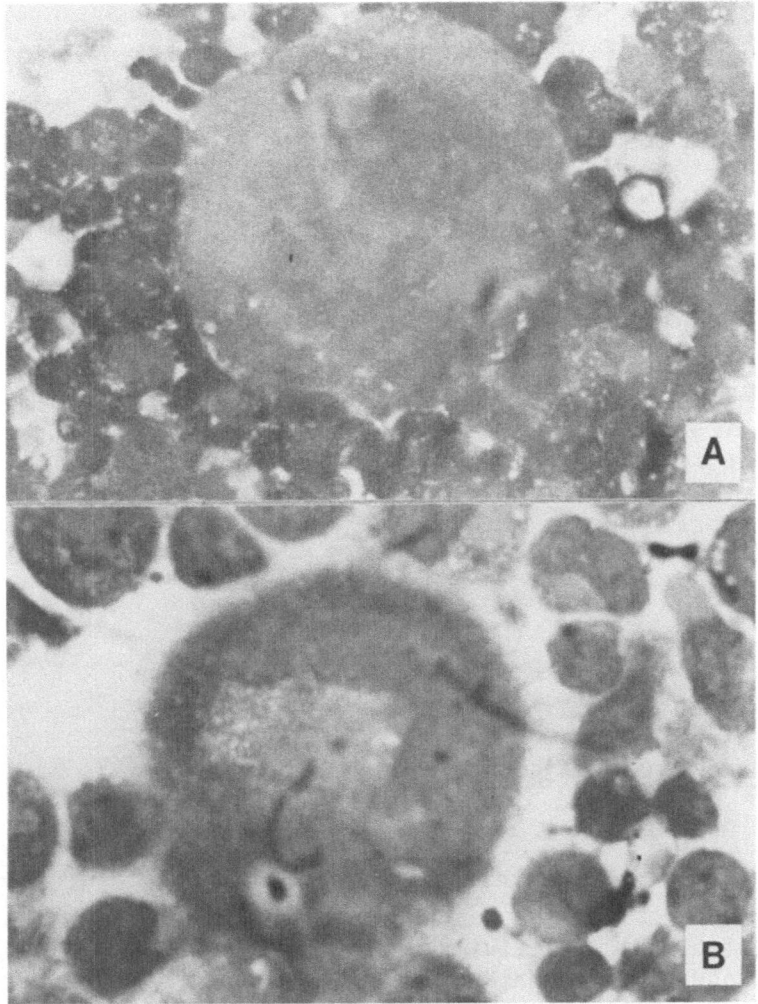

Fig. 1: Megakaryocyte Cell Cultures after Cytocentri-
         fugation and Morphological Staining.

(A) Note the predominant diploid cells surrounding
cell with at least 16 nuclei.

(B) Amultinucleated cell showing cytoplasmic
spicules.

The "lymphoid" cells are diploid megakaryocytes
as defined by platelet-specific reagents (oil
immersion).

Table 1

Indirect IF of Cell Lines Using Antibodies Against
Platelet-Related Proteins

Platelet antigens detected by:	Membrane (% cells)	Cytoplasm (% cells)
Monospecific antisera		
Fibrinogen	<10+	Negative
Fibronectin	Negative	<20+
B-thromboglobulin	Negative	<10 weakly+
Factor VIII-Ag	<10+	<10 weakly+
PF4*	>90+	10-80+
PDGF*	>90+	10-20+
Gp*IIb	>80+	ND
GPIIIa	>80+	ND
Monoclonal antibody		
GpIIb-IIIa	>80+	ND

Membrane labeling was performed on cell suspensions and
cytoplasmic reactivity was detected on fixed cells by
standard IF techniques. Positive control cells were
normal blood mononuclear preparations containing platelets
and negative control cells were from a pre-B lymphoblastic
cell line.

*PF4 = platelet factor 4;
 PDGF = platelet-derived growth factor;
 Gp = platelet membrane glycoprotein.

MATERIALS AND METHODS

Culture initiation

Peripheral blood cells were separated by density gradient centrifugation. The light density cell fraction was washed and resuspended in RPMI (Roswell Park Memorial Institute) nutrient medium containing 10% (v/v) human serum (complete medium). Suspension cultures were initiated by placing cells into culture flasks at a final concentration of 1×10^7 per 10 ml of complete medium supplemented with 10% (v/v) of a 5-fold concentrated human-conditioned medium (HCM) obtained from lectin-stimulated lymphocytes. Control cultures contained no HCM. Culture flasks were incubated at 37^0C in 5% CO_2 and a humidified atmosphere. At weekly intervals, cells were examined closely for any cells in mitosis.

Cell line maintenance

All cell lines emerged between 35-45 days after culture initiation. Any culture which appeared to be a potential cell line by the appearance of proliferating cells with basophilic cytoplasm was allowed to continue growing until sufficient cells density could be achieved (Morgan and Brodsky, 1984). Cells were then pelleted and resuspended in fresh complete medium at a final concentration of 5×10^5 per ml. Cells were sampled every three days for viablity and for morphological evaluation. Repeated subculturing has maintained these cells lines for over a year and 100 passages.

EBV-related studies

The cells were tested for EBV (i.e., EBNA and other a antigens and virus) according to previously published methods. Cells were also tested for early EA induction and rescue of transforming EBV.

RESULTS AND DISCUSSION

Of all the 7 megakaryocyte cell lines tested for EBV or antigen expression, only MOR cells exhibited EBNA (Table 2). In various passages of MOR cells examined for EBNA, >70% were positive, but did not contain EBV-VCA/EA. Table 2 shows that MOR megakaryocytes was from a normal individual who had detectable EA antibody which later disappeared, suggesting that at the time cell line was established, there was a primary EBV infection. The infectivity of megakaryocytes thus could be related to the period during which this individual had been shedding EBV.

Table 2

Human Megakaryocyte Cell Lines and EBV Status
(Antigen/Antibody)[1]

Identification (passage no.)[2]	Patient abnormality	Patient's EBV serology		
		VCA	EA	EBNA
SAL (31)	Acute lymphocytic leukemia	1:40	<1:10	ND
EVA (9)	Autoimmune neutropenia	1:80	<1:10	ND
MOR (11,14,16)[3]	Normal	a<1:160	1:10	1:10
		b 1:80	<1:10	1:10
CAT (5)	Normal	ND		
WAT* (2)	Normal	ND		
OGR (13)	Polycythemia vera	1:40	<1:10	ND
PEN (2)	Acute megakaryo-bastic leukemia	ND		

[1]The cells were first carried in medium 1640 containing 10% human serum and later these cells were carried in NU serum (10%) and FCS (5%). In the mixture of NU and FCS the viability of the cells remained >80%. The cells were changed twice a week and subpassaged as desired.

a, b = Different serum bleedings.

[2&3] Cell lines were examined for EBV-nuclear antigen (EBNA), VCA and EA expressing cells. Only MOR cells expressed >70% EBNA positive cells and no other antigens positive cells were observed in MOR or other megakaryocyte cell lines.

MOR cells were superinfectable with nontransforming P3HR1-EBV. However, in comparison to superinfection of Raji (>45%), the percentage of EA detected was approximately 50% less (>20%). Only 2% of other megakaryocyte cells, i.e., OGR, demonstrated EA at one time. In MOR cells, EA could not be induced with TPA, since these cells were highly sensitive to TPA (<5 ng/ml killed >97% of the cells). We did not use other chemicals that have been known to induce EA in EBV-genome-positive B lymphocytes. Secondly, transforming EBV could be rescued from MOR cells by superinfected cells, suggesting similarities to what has been observed with EBV-genome-positive B lymphocytes. The transformed cord blood lymphocytes did not express EBV-VCA/EA, but did contain EBNA-positive cells in approximately 20% of the cells by 14th day. The other megakaryocytes lines were not tested for rescue of transforming EBV.

The mechanism by which megakaryocytes harbor EBV is yet to be explored. It is probable that these cells may contain receptors for EBV and that the EBV genome may be transferred to these cells through a temporary fusion between EBV-carrying B cells and the megakaryocytes. Besides epithelial cells, megakaryocytes can be added to the list of cells that contain EBV or can be infected with EBV. The function and/or behavior of EBV in megakaryocytes is not known.

REFERENCES

Glaser, R., The infection of epithelial cells with Epstein-Barr virus. In: U. Prasad, D.V. Ablashi, P.H. Levine and G.R. Pearson (eds.), Nasopharyngeal Carcinoma: Current Concepts. Univ. of Malaya, Kuala Lumpur, Malaysia, pp. 239-244 (1983).

Kapff, C.T. and Jandl, J.H. (eds.) Blood: Atlas and Source-book of Hematology. Little, Brown & Co., Boston, pp. 12-13 (1981).

Morgan, D.A. and Brodsky, I., Novel peripheral blood-derived human cell lines with properties of megakaryocytes. J. Cell Biol. 100: 565-573 (1984).

EBV PROTEINS

38

ADVANCES IN THE IDENTIFICATION OF EBV-SPECIFIC PROTEINS:AN
OVERVIEW

Gary R. Pearson

Georgetown University Medical Center

Washington, D.C. 20007

Epstein-Barr virus (EBV) induced antigens can readily
be detected in EBV-infected or transformed cells using a
variety of immunological techniques. Monitoring of the
immune responses to these different antigen has been instrumental in establishing an etiological association
between EBV and different neoplastic diseases (Pearson,
1980). In addition, such antibodies have been shown to
be of diagnostic and prognostic importance for certain
histopathological types of nasopharyngeal carcinoma (NPC)
and African Burkitt's lymphoma. Thus these types of investigations over the past 20 years have served a useful
purpose in studies on EBV and cancer. It is known, however, that most, if not all, of the EBV antigens are complexes composed of multiple determinants. Until recently
little was known about the identity of the various proteins
that composed these antigens or their functions. However,
largely due to the development of monoclonal antibody
technology, there has been good progress over the past
few years in dissecting out the major components of the
different EBV antigens complexes. Many of the components
have been identified, characterized and purified. This
has made it possible to initiate studies on the biological
functions associated with each polypeptide. In addition
new immunological assays have been developed using the
purified proteins. Such assays for monitoring antibodies
directed against specific antigenic determinants might be
more discriminating between individuals with as opposed

to those without EBV-associated diseases than than the
current assays and therefore of greater clinical value for
the diagnosis and clinical management of individuals with
EBV-associated diseases. This is yet to be determined.

As noted above, the production of monoclonal antibodies
to individual proteins has been instrumental in the progress in identifying and categorizing these proteins over
the past several years. Criteria for selecting or cate-
loging a protein into an individual antigenic complex are
shown in Table 1. Most of the recent progress has been
with the EBV-induced membrane antigen (MA) complex and
the EBV-induced nuclear antigen (EBNA). This is largely
because of the interest in developing a subviral vaccine
against this virus and because of the probable involvement
of EBNA in the immortalization process. However, some of
the major components of the other EBV-induced antigen
complexes have also now been identified conclusively.
These are listed in Table 2. The purpose of this article
is to review the recent advances in the identification
and characterization of these proteins.

EBV-induced membrane antigens (MA).

Interest in the development of a subviral vaccine has
focused on the MA complex. This is due to the fact that
viral neutralizing determinants were originally shown to
be associated with MA (Pearson et al., 1970,1971). Through
the employment of biochemical and immunological approaches,
three major EBV-induced glycoproteins have been identified
which compose this complex (Pearson, 1980). These have
recently been designated gp 300/350, gp 200/250 and gp 85/
90 to take into account the molecular weight variations
for each of these glycoproteins as reported by different
laboratories (Thorley-Lawson et al., 1982). Generally
cells will express gp 300/350 or gp 200/250 but not both.
Through the use of monoclonal antibodies, it has been
shown that gp 300/350 and gp 200/250 are closely related
and that the differences in the molecular weight of this
major glycoprotein in different cell lines is probably
due to glycosylation and splicing patterns. Neutralizing
determinants have been shown to be express on all three
glycoproteins (Hoffman et al., 1980; Thorley-Lawson and
Gerlinger, 1980; Strnad et al., 1982; Qualtiere et al.,
1982a). In contrast, the determinants involved in antibody-dependent cellular cytotoxicity (ADCC) are expressed
on gp 300/350 and gp 200/250 but not on gp 85/90

(Qualtiere et al., 1982a). Interestingly, monoclonal antibodies to MA have been produced which recognize a membrane determinant on cells producing transforming EBV but not lytic virus (Mueller-Lantzsch et al., 1981; Qualtiere et al., 1982b). Whether this determinant might be related to transformation still needs to be determined.

The membrane glycoprotein that has attracted the most interest for purposes of developing a subunit vaccine is gp 300/350. This is because of the abundance of this protein in cells infected with EBV and because of the availability of a variety of monoclonal antibodies produced against different determinants expressed on this molecule. This glycoprotein has now been purified and partially characterized (Qualtiere et al., 1982a). More than 50% of this component is composed of a complex carbohydrate rich in fucose plus some mannose (Thorley-Lawson et al., 1982). The protein component of gp 300/350 has been estimated to have a molecular weight of approximately 160 K (Thorley-Lawson et al. 1982).

This major EBV glycoproteins has been biologically characterized. As previously noted, neutralizing and cytotoxic antibody determinants have been identified on gp 300/350. This was accomplished through the use of monoclonal antibodies and immunization experiments (Hoffman et al., 1980; Thorley-Lawson and Gerlinger, 1980; Qualtiere et al. 1982a). Epstein and co-workers have reported that the neutralizing determinants are expressed on the protein portion of this glycoprotein (Morgan et al., 1981). If true, it might be feasible to produce a vaccine containing only the protein component through genetic engineering technique. This would increase the liklihood for success in the production of a vaccine suitable for immunizing large populations. Because of the importance of this observation, this question requires further examination.

Of greater importance has been the observation that immunization of non-human primates with this glycoprotein has resulted in the production of high levels of neutralizing and cytotoxic antibodies (Qualtiere et al., 1982a; North et al. 1982). This has been accomplished in owl monkeys and cottontop marmosets. Immunized owl monkeys were shown to be resistant to challenge with the B-35 strain of EBV as shown by the induction of antibodies to

the EBV-induced early antigens (EA) (G. Pearson, unpublished results). Cottontop marmosets immunized with gp
300/350 in lipsosomes vessicles were resistant to lymphoma induction by B-95 virus (Epstein, personal communication). These results demonstrate that is possible to
prevent the induction of EBV-associated diseases by
immunization with this glycoprotein. As discussed above,
whether it will be possible to produce a suitable vaccine
for human use still needs to be resolved. However, it is
likely that this will eventually be possible using more
modern techniques. Success in the prevention of a human
cancer through the use of a subviral vaccine would be a
milestone in cancer research.

EBV-induced nuclear antigens (EBNA)

EBNA was first demonstrated using the anti-complement
immunofluorescense procedure by Reedman and Klein in 1973.
Subsequent studies established that EBNA, which is expressed in the nucleus of every cell that contains viral DNA,
was identical to the previously described soluble or 's'
antigen detected by complement fixation (Pearson, 1980).
Because of its presence in genome-positive cells, this
antigen complex is considered as a major candidate viral
protein involved in the transformation process. Consequently, there have been numerous attempts to purify and
characterize this antigen. Based on results from early
investigations on this antigen, the molecular weight of
EBNA ranged from 48k to greater than 200k. However, more
recently, using the technique of immunoblotting, it was
established by Strnad et al (1981) that the major component of EBNA was a polypeptide with a molecular weight ranging from 72k-80k depending on the resident genome.
This has now been confirmed by many laboratories. Kieff
and coworkers have cloned the fragment of viral DNA which
encodes for EBNA and have reported that the 72k component
of EBNA was encoded by the Bam HI k fragment of EBV DNA
which contains the IR3 repeat sequence (Hennessy et al.,
1983). This region is of different size in different cell
lines and encodes for a glycine-alanine co-polymer corresponding to a molecular weight between 20-30k (Hennessy
and Kieff, 1983). A rabbit antiserum prepared against
this polypeptide detected the 72-80k polypeptide by immunoblotting. More recently, it was shown that a monoclonal antibody to EBNA also reacted with the glycine-
alanine copolymer (Luka et al., in press). In addition
to the 72k polypeptide, a second polypeptide with a mole-

cular weight of approximately 82k and designated EBNA 2 has also been identified with some but not all anti-EBNA positive sera by immunoblotting (Hennessy and Kieff, 1983).

Differences in the EBNA proteins have also been identified by binding properties to chromatin and ds DNA and by isoelectric focusing. Spelsberg et al (1982) identified a loosely bound EBNA component designated Class I EBNA and a component tightly bound to chromatin designated CLASS II EBNA. EBNA I bound to ds DNA and had an acidic pI. In contrast, partially purified EBNA II did not bind to ds DNA and had a more basic pI. Both forms of EBNA gave identical molecular weights (72k) and identical fragments by peptide mapping. These results suggested that the soluble form of EBNA possibly represented a modified form of the chromatin-associated EBNA component. The nature of the postulated modification still needs to be defined. It is clear therefore from these findings that substantial progress has been made over the past few years in identifying and characterizing proteins associated with the EBNA complex. Future studies will most likely concentrate on defining the biological activities associated with the EBNA proteins.

EBV-induced early (EA) and viral capsid antigens (VCA)

There has been less progress in definitively identifying the polypeptides composing the EA and VCA complexes. This is largely due to the lack of monospecific reagents directed against these proteins. Most of the investigations on these complexes, until recently, employed human sera from EBV infected individuals. Proteins were defined as early or late antigens based on whether they were precipitated from cells cultivated in the presence of inhibitors of viral DNA synthesis. This has resulted in the identification of several proteins with molecular weights ranging from 28k to over 200k (Pearson, 1980) which have been catalogued as early or late proteins. However, none of these studies definitively identified proteins composing the VCA complex as defined by IF or proteins associated with the diffuse and restricted components of the EA complex. More recently this question has been approached through the use of monoclonal reagents. Monoclonal antibodies were categorized as EA or VCA proteins as defined in Table I. EA proteins were further grouped as R or D based on their expression in cells fixed in acetone or methanol. This approach has resulted in the production of

a number of monoclonal antibodies directed against proteins
associated with these antigenic complexes. Such antibodies
have also been used to purify and partially characterize
these proteins and for the development of new assays
capable of measuring human antibodies to individual polypeptides.

Major polypeptides composing the VCA complex have
molecular weights of 125k, 152k and 160k. Monoclonal antibodies have now been produced against the 125k and 160k
polypeptides (Takedo et al., 1983; Kishishita et al. 1984;
Vroman et al., in press). The 125k protein has been shown
to be presence in extracts of purified nucleocapsids as
well as in virusproducing cell lines (Takado et al.,1983).
In infected cells, the protein can be detected both in the
nucleus and cytoplasm by IF. Interestingly, labelling
experiments demonstrated that this 125k VCA component was
a glycoprotein and the carbohydrate moiety was shown to be
complex and rich in mannose (Kishishita et al., 1984). One
other interesting observation from two different laboratories was that this glycoprotein was consistently larger in
P3HRI cells as opposed to B-95-8 cells (Takado et al, 1983;
Kishishita et al., 1984). This appears to be related to
the glycosylation process in the different host cells. An
ELISA assay has been developed for measuring antibodies to
the 125k component (Luka et al., 1984). The results suggest that this glycoprotein is a major immunogen following
a primary infection with EBV.

The second VCA protein definitively identified with
monoclonal antibodies has a molecular weight of 160k
(Vroman et al., in press). This is the molecular weight
of the major polypeptide previously demonstrated in EBV
nucleocapsids by Dolyniuk et al. (1976). The 160k protein
is non-glycosylated and expressed primarily in the nuclei
of infected cells. By immuno-electron microscopy, the
monoclonal antibody directed against the 160k protein was
shown to label virus particles as well as soluble protein
in the nucleus. In the ELISA assay, most human sera from
EBV-infected individuals reacted with this protein with
some exceptions. These exceptions were largely sera from
IM patients suggesting that this protein is not recognized
immunologically as early as the 125k protein following
primary EBV infections.

The major proteins associated with the diffuse and restricted components of the EA complex have molecular weights of 47-60k, 85k and 140k. Monoclonal antibodies have now been produced against the 47-60k complex and the 85k protein (Pearson et al., 1983). These antibodies have been used to partially characterize these antigens. By IF staining on cells fixed in acetone or methanol, it was established that the 47-60k complex is part of the diffuse component of the EA complex while the 85k protein is the major component of the restricted element. An in vitro primary translation product with a molecular weight of approximately 47k was identified with the anti-D mono-clonal and was mapped to the Bam HI M fragment. However, Glaser and coworkers (1983) using micro-injection of DNA fragments into cells mapped this EA-D protein complex to the Charon 4A fragment 7. The reasons for these discrepant findings on the gene location for this antigen have yet to determined although this is likely due to the use of dif-ferent techniques for examining expression of EBV DNA fragments. In addition to the EA-D mapping studies repor-ted by Glaser et al, this group also mapped the 85K EA-R component to the Bam HIH fragment using the same approach.

The major components of the 47-60k EA complex have mol-cular weights of 50/52K. However, the molecular weights of these components shift with time in the infection cycle upward to 60k. This shift upwards has been demonstrated both in activated Raji and P3HR-1 cells (Figure 1). Preli-minary findings suggest that this shift is at least par-tially related to phosphorylation as shown in Figure 2. In addition it has been shown that this complex binds to ds DNA. Less is known about the 85k EA-R protein although it does not appear to bind to ds-DNA. Preliminary results indicated that this protein might be polyribosylated. This however, needs to be substantiated. Further studies including determination of the biological activities asso-ciated with these proteins are currently ongoing.

During the development of ELISA assays with the puri-fied EA proteins, it was discovered that the 85k protein did not work well in this assay with human sera. Similarly it was not possible to routinely detect this protein by immunoblotting using human sera. Further investigations have established that the reason for these findings was due to the fact that antibodies in human sera are largely directed against determinants on the tertiary and not on the primary structure of this protein. This fact would account for the ELISA and immunoblotting observations.

Conclusions

It is evident that substantial progress has been made over the past several years in the identification of proteins associated with the different EBV antigen complexes. However more work is needed on this area. This includes the need to identify other proteins associated with these complexes and the assignment of biological activity to each protein component. It is likely that there will be continued progress in this area over the next few years. Completion of such studies should allow us to establish the role of each protein in the virus infection and transformation cycle and will possibly open up to new avenues of treatment for EBV-associated diseases.

TABLE 1

CLASSIFICATION CRITERIA FOR MONOCLONAL ANTIBODIES TO EBV ANTIGENS

Monoclonal	Cell Lines							Classification
	BJAB	Raji	I-Raji[a]	P$_3$HR-1[b]	B-95-8[b]	P$_3$HR-1+PAA[c]	B-95-8+PAA[c]	
2L10	-	-	-	+d	+d	-	-	MA
L2	-	-	-	+e	+e	-	-	VCA
V3	-	-	-	+e	+e	-	-	VCA
R3	-	-	+	+	+	+	+	EA(D)[f]
R63	-	-	+	+	+	+	+	EA(R)[g]

a Raji cells induced to express EA by incubation in the presence of 20 ng/ml TPA plus 3 mM sodium butyrate.

b Cells activated for 48-72 hours with 20 μg/ml TPA plus 3 mM sodium butyrate to induce VCA and EA synthesis.

c Cells activated in the presence of 150 μg/ml PAA to inhibit synthesis of late antigens.

d Monoclonal antibody was positive against both viable and acetone – methanol – fixed cells.

e Monoclonal was positive only on acetone or methanol – fixed cells.

f Monoclonal reactive with EA-positive cells fixed in either acetone or methanol.

g Monoclonal reactive only with EA-positive cells fixed in acetone.

TABLE 2

EBV POLYPEPTIDES COMPOSING THE MAJOR IMMUNOFLUORESCENCE
ANTIGEN COMPLEXES

Antigen	Polypeptide
VCA	160k*
	152k
	125k*
EA(D)	140k
	47-60k*
EA(R)	85k*
MA	gp 300/350*
	gp 200/250
	gp 90*
EBNA	81k
	72k*
	70k
	65k
	48k

* Confirmed with monoclonal antibodies

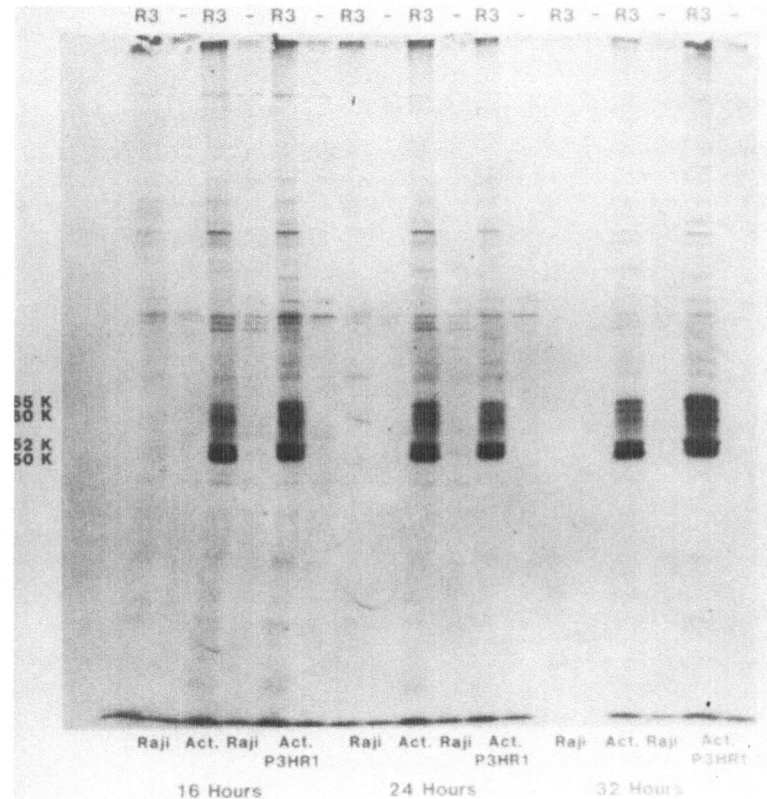

Figure 1. Expression of the major EA(D) polypeptide
complex in TPA and sodium butyrate activated
Raji and P₃HR-l cells as shown by immunoprecipitation at different times after activation.
R3 designates monoclonal antibody reactive
with polypeptides ranging from 47K to 60K that
compose the major EA(D) complex; (-), antibody-
negative serum.

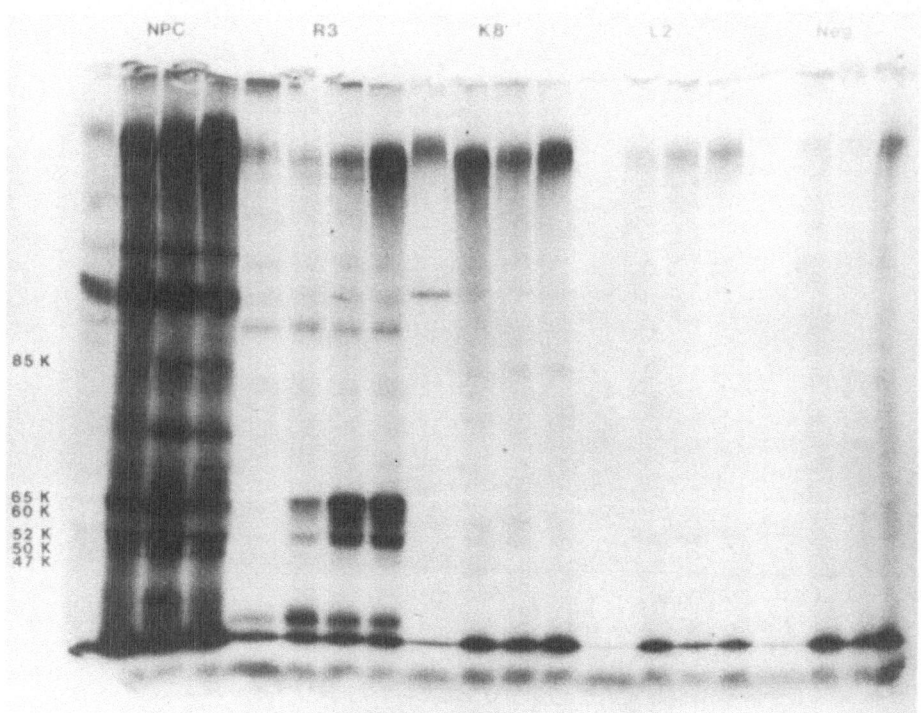

Figure 2.   Immunoprecipitation of ^{32}P-labelled proteins
at different times from TPA and sodium butyrate
activated P$_3$HR-1 cells with human sera and monoclonal antibodies. NPC designates pooled human
antibody-positive serum; R3 is monoclonal antibody to 47-60k polypeptide complex associated
with EA(D); K8 is monoclonal antibody to 85k
component of EA(R) complex; L2 is monoclonal
antibody to 125k VCA protein; neg. is antibody-
negative ascites fluid. Note strong labelling
of polypeptides precipitated at different times
by R3 monoclonal antibody.

References

Dolyniuk, M., Wolff, E. and Kieff, E. Proteins of Epstein-Barr virus. 2. Electrophoretic analysis of the polypeptides of the nucleocapsid and the glucasamine and polysaccharide - containing components of enveloped virus. J. Virol., 18, 289-297 (1976).

Glaser, R., Boyd, A., Stoerker, J., and Holliday, J. Functional mapping of the Epstein-Barr virus genome: identification of sites coding for the restricted early antigen, the diffuse early antigen and the nuclear antigen. Virology, 129, 188-198 (1983).

Hennessy, K., Heller, M., van Santen, V., and Kieff, E. Simple repeat array in Epstein-Barr virus DNA encodes part of the Epstein-Barr nuclear antigen. Science, 220, 1396-1398 (1983).

Hennessy, K., and Kieff, E. One of two Epstein-Barr virus nuclear antigens contains a glycine-alanine copolymer domain. Proc. Nat. Acad. Sci., 80, 5665-5669 (1983).

Hoffman, G.J., Lazarowitz, S.E., and Hayward, S.D. Monoclonal antibody against a 250,000-dalton glyco-protein of Epstein-Barr virus identifies a membrane antigen and a neutralizing antigen. Proc. Nat. Acad. Sci., 77, 2979-2983 (1980).

Kishishita, M., Luka, J., Vroman, B., Poduslo, J.E., and Pearson, G.R. Production of monoclonal antibody to a late intracellular Epstein-Barr virus-induced antigen. Virology, 133, 363-375 (1984).

Luka, J., Chase, R.C., and Pearson, G.R. A sensitive enzyme linked immunosorbent assay (ELISA) against the major EBV-associated antigens. I. Correlation between ELISA and immunofluorescence titers using purified antigens. J. Immunol. Methods, 67, 145-156 (1984).

Luka, J., Kreofsky, T., Pearson, G.R., Hennessy,K., and Kieff, E. Identification and characterization of a cellular protein crossreacting with the Epstein-Barr virus nuclear antigen (EBNA). J. Virol. In press.

424

Morgan, A.J., Smith, A.R., Barker, R.N., and Epstein, M.A. A structural investigation of the Epstein-Barr (EB) virus membrane antigen, gp 340. J. Gen. Virol., 65, 397-404 (1984).

Mueller-Lantzsch, N., Georg-Fries, B., Herbst, H., zur Hausen, H., and Braun, D.G. Epstein-Barr virus strain- and group-specific determinants detected by monoclonal antibodies. Int. J. Cancer, 28, 321-327 (1981).

North, J.R., Morgan, A.J., Thompson, J.L., and Epstein, M.A. Purified Epstein-Barr virus Mr 340,000 glycoprotein induces potent virus-neutralizing antibodies when incorporated in liposomes. Proc. Nat. Acad. Sci., 79, 7504-7508 (1982).

Pearson, G.R., EBV immunology. In: G. Klein (ed.), Viral Oncology, pp. 739-767, Raven Press, N.Y. (1980).

Pearson, G., Dewey, F., Klein, G., Henle, G., and Henle, W. Relation between neutralization of Epstein-Barr virus and antibodies to cell membrane antigens induced by the virus. J. Nat. Cancer Inst., 45, 989-995 (1970).

Pearson, G.R., Henle, G., and Henle, W. Production of antigens associated with Epstein-Barr virus in experimentally infected lymphoblastoid cell lines. J. Nat. Cancer Inst., 46, 1243-1250 (1971).

Pearson, G.R., Vroman, B., Chase, B., Sculley, T., Hummel, M. and Kieff, E. Identification of polypeptide components of the Epstein-Barr virus early antigen complex using monoclonal antibodies. J. Virol., 47, 193-201 (1983).

Qualitiere, L.F., Chase, R., and Pearson, G.R. Purification and biological characterization of a major Epstein-Barr virus-induced membrane glycoprotein. J. Immunol., 129, 814-818 (1982a).

Qualitiere, L.F., Chase, R., Vroman, B., and Pearson, G.R. Identification of Epstein-Barr virus strain differences with monoclonal antibody to a membrane glycoprotein. Proc. Nat. Acad. Sci., 79, 616-620 (1982b).

Spelsberg, T.C., Sculley, T.B., Pikler, G.M., Gilbert, J.A., and Pearson, G.R. Evidence for two classes of chromatin-associated Epstein-Barr virus-determined nuclear antigen. J. Virol., 43, 555-565 (1982).

425

Strnad, B.C., Schuster, T. Klein, R., Hopkins, R.F. III, Witmer, T., Neubauer, R.H., and Rabin, H. Production and characterization of monoclonal antibodies against the Epstein-Barr virus membrane antigen. J. Virol., 41, 258-264 (1982).

Strnad, B.C., Schuster, T.C., Hopkins, R.F., Neubauer, R.H., and Rabin, H. Identification of an Epstein-Barr virus nuclear antigens by fluoroimmunoelectrophoresis and radioimmunoelectrophoresis. J. Virol., 38, 996-1004 (1981).

Takada, K., Fugiwara, S., Yano, S., and Osato, T. Monoclonal antibody specific for capsid antigen of Epstein-Barr virus. Med. Microbiol. Immunol., 171, 225-231 (1983).

Thorley-Lawson, D.A., Edson, C.M., and Gerlinger, K. Epstein-Barr virus antigens - a challenge to modern bio-chemistry. Adv. Cancer Res., 36, 295-348 (1982).

Thorley-Lawson, D.A. and Gerlinger, K. Monoclonal anti-bodies against the major glycoprotein (gp 350/220) of Epstein-Barr virus neutralizes infectivity. Proc. Nat. Acad. Sci., 77, 5307-5311 (1980).

Acknowledgements. This work was supported by United States Public Health Service Grant CA20679 from the National Cancer Institute.

39

BACTERIALLY SYNTHESIZED EBNA AS A REAGENT FOR ENZYME-

LINKED IMMUNOSORBENT ASSAYS

G. Milman+, D.K. Ades+, M.-S. Cho*, S.C. Hartman+,
G.S. Hayward*, A.L. Scott+ and S.D. Hayward*

+Department of Biochemistry, School of Hygiene and
Public Health and *Department of Pharmacology and
Experimental Therapeutics, School of Medicine, The
Johns Hopkins University, Baltimore, Maryland, 21205.

SUMMARY

We have synthesized the carboxy-terminal one third of
the EBNA protein as a fusion polypeptide in bacteria.
When inoculated into rabbits the purified 28K EBNA
polypeptide elicited antibodies which gave the same
immunofluorescence staining patterns on lymphoblastoid
cell lines as EBNA-positive human serum. The 28K EBNA
polypeptide bound tightly to double-stranded DNA
suggesting that the carboxy-terminal portion of EBNA is a
DNA binding domain. The bacterially synthesized EBNA
polypeptide was also employed in an enzyme-linked
immunosorbent assay (ELISA). The pattern of anti-EBNA
antibody titers measured by the ELISA method in sera from
normal individuals and from patients with rheumatoid
arthritis, Burkitt lymphoma, nasopharyngeal carcinoma and
acute infectious mononucleosis was comparable to that
observed in previous serological studies which had
employed the classical anti-complement immunofluorescence
assay (ACIF). However, in contrast to the ACIF assay, the
more sensitive ELISA method was able to detect anti-EBNA
antibody in acute infectious mononucleosis serum.

INTRODUCTION

The Epstein-Barr Virus (EBV) nuclear antigen (EBNA) is expressed in producer and non-producer lymphoblastoid cell lines and in biopsy cells from Burkitt lymphoma and nasopharyngeal carcinomas (Reedman et al., 1974; Huang et al., 1974; Klein et al., 1974). The coding region for EBNA has been mapped to the BamHI-K fragment of the EBV genome by DNA transfection experiments (Summers et al., 1982). Subsequent analyses revealed that BamHI-K contains a 2.0 kb EBNA exon and that 700 bp of this exon is composed of a triplet repeat sequence which is translated in the EBNA polypeptide as a stretch of alternating glycine and alanine residues. The size of the triplet repeat array varies from EBV isolate to isolate and, as a consequence, the EBNA polypeptide also shows size variation between isolates (Heller et al., 1982; Hennessy et al., 1983; Hennessy and Kieff, 1983).

Progress in characterizing the EBNA polypeptide and elucidating its functions in EBV immortalization has been hampered by the low abundance of the protein in lymphoblastoid cells and by the lack of mono-specific antibodies. EBNA is traditionally detected using EBV sero-positive human serum in an anti-complement immunofluorescence assay (ACIF) (Reedman and Klein, 1973). Conversely, EBV carrying lymphoblastoid cells and ACIF constitute the assay for measurement of anti-EBNA antibody titers in human serum. We have taken advantage of the availability of the complete EBV DNA sequence (Baer et al., 1984) and the identification of the reading frame within the BamHI-K EBNA exon (Hennessy and Kieff, 1983) to construct a high expression plasmid which synthesizes a segment of the EBNA protein in bacteria. This partial EBNA polypeptide was used to generate monospecific anti-EBNA anti-serum in rabbits and as the basis of an ELISA method for quantitating EBNA antibody in human serum.

MATERIALS AND METHODS

ELISA for Anti-EBNA Antibody

Microtiter plates adsorbed with the purified 28K EBNA polypeptide (40 ng/well) were incubated with dilutions of patient serum and binding was detected using peroxidase-tagged goat anti-human IgG. The concentration of anti-EBNA IgG in serum was determined by reference to an IgG standard curve.

Serum Samples

Acute infectious mononucleosis serum was provided by Werner Henle, The Children's Hospital, Philadelphia. All samples were anti-IgM-VCA positive. Paul Levine (National Cancer Institute, Bethesda) provided Burkitt lymphoma and nasopharyngeal carcinoma serum. Marc Hoffberg (The Johns Hopkins University, Baltimore) provided samples from rheumatoid arthritis patients. Serum from normal individuals was provided by David Levy (Johns Hopkins University, Baltimore). The normal group consisted of volunteers ranging in age from early 20's to late 30's.

RESULTS

Synthesis of the EBNA Polypeptide in Bacteria

The 2.2 kb SmaI subfragment of BamHI-K indicated in Fig. 1 was inserted in both the sense and anti-sense orientations into a unique SmaI site in an expression vector pHE6. The plasmid pHE6 is a derivative of pGM10 (Waldman et al., 1983) and contains the strong bacteriophage lambda leftward and rightward promoters controlled by the lambda temperature sensitive repressor gene CI857. Polypeptides coded by foreign inserted DNA are synthesized as fusion products of the lambda N-protein. The 2.2 kb SmaI fragment inserted into pHE6 in the correct reading frame directs the synthesis of the carboxy-terminal 191 amino-acids of the BamHI-K EBNA protein linked to 36 amino acids coded by the lambda N gene and linker sequences. A Coomassie stained polyacrylamide gel of a lysate of induced bacteria (Fig. 2A, Track 1) shows a heavily stained band migrating at 28K. This band was not present in lysates of bacteria containing the vector with the EBNA insert in the antisense direction. Immunoblot analysis with EBNA positive human serum identified the over-produced protein as being the EBNA fusion polypeptide. We estimate that the EBNA product represented 8% of the soluble bacterial protein after induction.

Passage of bacterial lysates over a phosphocellulose
column resulted in removal of the bulk of the bacterial
proteins, which did not bind and were present in the
flowthrough fractions (Fig. 2A, Track 2). The 28K EBNA
protein was eluted by high salt (Fig. 2A, Track 3).
Purification to the point of homogeneity was achieved
using hydroxylapatite chromatography (Fig. 2A, Track 4).
The purified 28K EBNA was subsequently used for
monospecific antibody production and for the ELISA method.

Binding of the EBNA Fragment to DNA

The bacterially synthesized 28K EBNA fragment binds to
double stranded DNA. Fig. 2B illustrates the results
obtained when a mixture of supercoiled, nicked circular
and linear pHE6 plasmid DNA was electrophoresed through
agarose after incubation with varying quantities of the
EBNA polypeptide. The migration of each of the DNA forms
through the gel was significantly slowed by the addition
of the EBNA protein and the degree of retardation
increased with the addition of increasing amounts of EBNA.

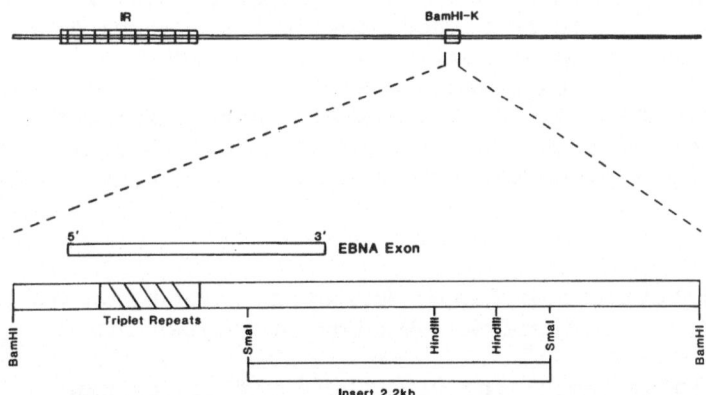

Fig. 1: Schematic representation of the BamHI-K fragment
showing the relative locations of the triplet repeat array,
the 2.0 kb EBNA exon and the 2.2 kb SmaI subfragment which
was inserted into the pHE6 expression plasmid.

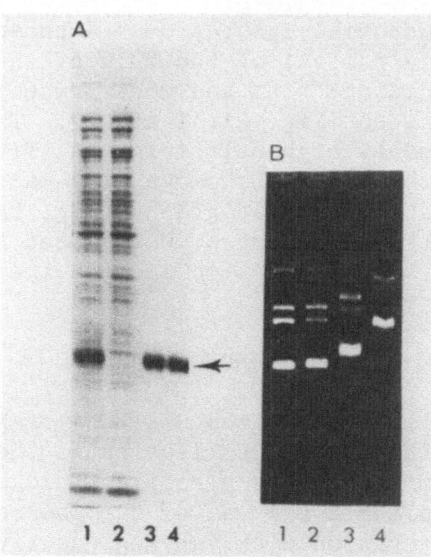

Fig. 2A: Coomassie stained polyacrylamide gel showing
purification of the truncated EBNA protein from a
bacterial lysate. Track 1: Crude extract. Track 2: Flow
through fraction from a phosphocellulose column. Track 3:
High salt eluate from phosphocellulose. Track 4: High
salt eluate from hydroxylapatite. The 28K EBNA protein
band is indicated by an arrow.
Fig. 2B: Ethidium bromide stained agarose gel showing
migration of pHE6 DNA after incubation with 0,2,4 and
6 ug of purified 28K EBNA (Tracks 1,2,3 & 4 respectively).

Reactivity of Rabbit Antiserum Raised Against the
Purified EBNA Fusion Polypeptide

    Rabbits inoculated with the purified 28K EBNA
developed antibodies which reacted by ACIF with EBV-
carrying lymphoblastoid cell lines (Fig. 3) but not with
the genome negative BJAB cell line. The pattern of staining was identical to that obtained by ACIF with EBNA-
positive human serum.

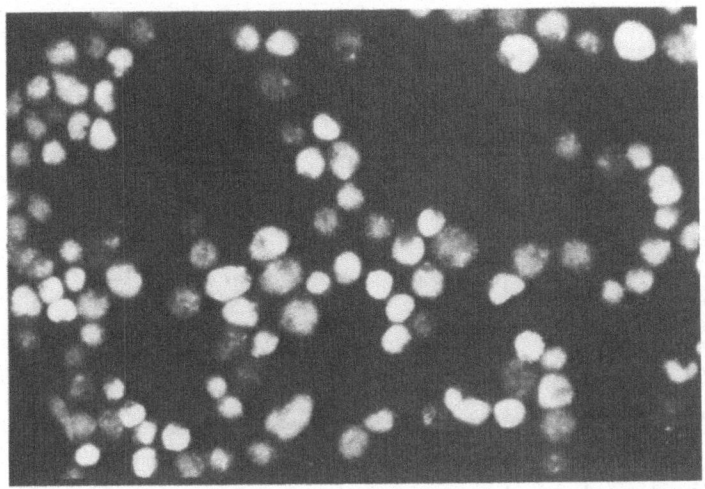

<u>Fig. 3</u>: Methanol fixed P3HR-1 lymphocytes stained using rabbit serum and ACIF.

In immunoblot analyses the rabbit serum recognized the same EBNA protein bands as did EBNA-positive human serum; an 84K protein in extracts of P3HR-1 lymphocytes and a 76K protein in extracts of Raji cells.

ELISA Screening of Human Sera

We have used the purified EBNA polypeptide and an ELISA procedure to screen serum from normal adults and patients with infectious mononucleosis, rheumatoid arthritis, Burkitt lymphoma and nasopharyngeal carcinoma (Table 1). All normal individuals who were EBV sero-positive had detectable levels of anti-EBNA antibody. The titers of the rheumatoid arthritis and Burkitt lymphoma patients spanned the same range as those of the normals but the mean titers of these patients were somewhat elevated relative to the normal group (270 and 372 µg/ml vs. 135 µg/ml anti-EBNA IgG). The 10 nasopharyngeal carcinoma sera all showed elevated EBNA titers. The mean titer was 976 µg/ml and 80% of the samples had greater than 250 µg/ml anti-EBNA IgG, in contrast to the normal

Table 1: ELISA Screening of Serum for Antibody Titers to K-EBNA

Classification	Number of Samples	Anti K-EBNA Titre (µg/ml IgG)				% of Samples with High Titer (>250 µg/ml)
		Mean	Low	Median	High	
Normals						
VCA positive	88	135	2	49	2586	8
VCA negative	12	<1	<1	<1	1°	
Rheumatoid Arthritis						
VCA positive	80	270	2	62	2778	23
VCA negative	3	<1	<1	<1	1	
Burkitt Lymphoma	10	372	4	27	1993	30
Nasopharngeal Carcinoma	10	976	179	382	4713	80
Acute Infectious Mononucleosis	19+	4	<1	2	21	0

° 1 µg/ml IgG marks the effective lower limit of the assay
+ All samples were anti-IgM(VCA) positive with titers between 1:20 and 1:160.

group where only 8% of the samples fell into this
category. The anti-EBNA titers determined by the ELISA
method reflect the same generalized patterns that have
been described in previous studies for these different
disease states (Henle and Henle, 1979). However, the
greater sensitivity of the ELISA assay becomes apparent on
examination of the acute infectious mononucleosis
samples. All 19 sera were EBNA negative in the classical
ACIF assay but 15 of the 19 samples had demonstrable
anti-EBNA antibody titers by ELISA. The acute sera had
very low anti-EBNA titers, the mean titer being 4 μg/ml
compared to 135 μg/ml for the normal group.

DISCUSSION

Characterization of EBV-encoded proteins has, in
general, proven difficult due to the low levels of viral
expression in EBV-immortalized lymphoblastoid cell lines.
One approach which circumvents this problem is to
synthesize specific proteins in bacteria using expression
plasmids. We have constructed an efficient expression
vector, pHE6, and have used pHE6 containing a SmaI subfragment of BamHI-K to obtain gram quantities of 28K EBNA.
The truncated EBNA polypeptide exhibited DNA binding
activity, suggesting that the carboxy-terminal portion of
EBNA is a DNA-binding domain. The binding observed was
non-specific in that the assay used plasmid DNA lacking
EBV sequences. Detection of specific binding may require
more sophisticated assays.

Both the range of the anti-EBNA titers and the mean
titers measured by ELISA in the different patient groups
were consistent with results obtained previously by ACIF.
Such a concordance validates the use of the bacterial 28K
EBNA protein for ELISA. Differences between the ELISA
results and those obtained by ACIF may be due, in part, to
the increased sensitivity of the ELISA method. Infectious
mononucleosis patients are usually EBNA negative by ACIF
until the convalescent phase of their illness. However,
the ELISA method detected anti-EBNA antibody in 15 of 19
acute infectious mononucleosis samples. A further measure
of the sensitivity of the assay is the observation that
high-titer samples which fall into the 1:320 to 1:640
range in the ACIF assay were diluted between 1:50,000 and
1:200,000 for ELISA.

In summary, we have shown that an ELISA method using a bacterially synthesized fragment of EBNA as the antigen provides a specific and sensitive assay, suitable for rapid screening of multiple serum samples for anti-EBNA antibody. The ELISA method offers a viable alternative to the ACIF assay for EBNA and extension of these methods to include EBV early (EA) and late (VCA) antigens is in progress.

ACKNOWLEDGEMENTS

We are grateful to W. Henle, P. Levine, M. Hochberg and D. Levy for serum samples and B. Barrell and colleagues for making available the EBV BamHI-K nucleotide sequence prior to publication. We thank Loretha Myers for excellent technical assistance.

This work was supported by PHS Grant ES03131 (GM) awarded by the National Institute of Environmental Health Sciences and PHS Grants CA37314 (GSH) and CA30356 (SDH) awarded by The National Cancer Institute, DHHS.

REFERENCES

Baer, R., Bankier, A.T., Biggin, M.D., Deininger, P.L., Farrell, P.J., Gibson, T.J., Hatful, G., Hudson, G.S., Satchwell, S.C., Sequin, C., Tuffnell, P.S., and Barrell, B.G., DNA sequence and expression of the B95-8 Epstein-Barr virus genome. Nature, 310, 207-211 (1984).

Heller, M., van Santen, V. and Kieff, E., Simple repeat sequence in Epstein-Barr Virus DNA is transcribed in latent and productive infections. J. Virol., 44, 311-329 (1982).

Henle, W. and Henle, G., Seroepidemiology of the Virus. In: M.A. Epstein and B.G. Achong (ed.), The Epstein-Barr Virus, pp 61-78, Springer Verlag, New York (1979).

Hennessy, K. and Kieff, E., One of two Epstein-Barr virus nuclear antigens contains a glycine-alanine colpolymer domain, Proc. Natl. Acad. Sci. USA, 80, 5665-5669 (1983).

Hennessy, K., Heller, M., van Santen V., and Kieff, E., Simple repeat array in Epstein-Barr virus DNA encodes part of the Epstein Barr virus nuclear antigen, Science, 220, 1396-1398 (1983).

Huang, D.P., Ho, J.H.C., Henle, W. and Henle, G., Demonstration of Epstein-Barr virus associated nuclear antigen in nasopharyngeal carcinoma cells from fresh biopsies, Int. J. Cancer, 14, 580-588 (1974).

Klein, G., Giovanella, B.C., Lindahl, T., Fialkow, P.J., Singh, S. and Stehlin, J., Direct evidence for the presence of Epstein-Barr virus DNA and nuclear antigen in malignant epithelial cells from patients with anaplastic carcinoma of the nasopharynx, Proc. Natl. Acad. Sci. USA, 71, 3737-3741 (1974).

Reedman, B.M. and Klein, G., Cellular localization of an Epstein-Barr Virus (EBV)-associated complement-fixing antigen in producer and non-producer lymphoblastoid cell lines. Int J. Cancer, 2, 499-520 (1973).

Reedman, B.M., Klein, G., Pope, J.H., Walters, M.K., Hilgers, J., Singh, S. and Johansson, B., Epstein-Barr Virus associated complement fixing and nuclear antigens in Burkitt lymphoma biopsies, Int. J. Cancer, 13, 755-763 (1974).

Summers, W.P., Grogan, E.A., Shedd, D., Robert, M., Liu, C.R., and Miller, G., Stable expression in mouse cells of a nuclear neoantigen after transfer of a 3.4 megadalton cloned fragment of Epstein-Barr virus DNA., Proc. Natl. Acad. Sci. USA, 79, 5688-5692 (1982).

Waldman, A.S., Haeusslein, E. and Milman, G., Purification and characterization of herpes simplex virus (type 1) thymidine kinase produced in Escherichia coli by a high efficiency expression plasmid utilizing a lambda PL promoter and CI857 temperature sensitive repressor, J. Biol. Chem., 258, 11571-11575 (1983).

Epstein-Barr virus nuclear antigen (EBNA): antigenicity of
the molecule encoded by the BamH1 K fragment of the EBV
genome

M.J. Allday and A.J. MacGillivray

Biochemistry Laboratory,
School of Biological Sciences,
Falmer, Brighton BN1 9QG, U.K.

Summary

The major antigenic component of the Epstein-Barr virus
nuclear antigen complex (EBNA) is a highly polymorphic (69-
94K mol.wt.) polypeptide. Its molecular weight correlates
with the size of the large Hind III subfragment of the
BamH1 K fragment of the EBV genome which contains the third
internal repeat region, IR3.

The region of the molecule encoded by IR3, plus a small
number of flanking residues, forms a large peptide fragment
which the amino acid sequence predicts is trypsin resistant.
Immunoblotting of trypsin digested, partially purified,
BamH1 K encoded EBNA (BK EBNA) from various cell lines
demonstrates that this region is antigenic and shows it
must be an immunogenic determinant in the native molecule.
It also confirms that it is responsible for the size polymorphism. Immunoblotting other peptide fragments with a
variety of sera show that different anti-EBNA antisera
contain antibodies to different epitopes on BK EBNA. The
possibility that anti-RANA (Rheumatoid Arthritis Associated
Nuclear Antigen) might represent a specific component of
anti-EBNA antisera is considered.

Introduction

In recent years it has become apparent that EBNA, as
demonstrated by anticomplement immunofluorescence, ACIF
(Reedman and Klein, 1973), using polyspecific human antisera, is not a single entity but rather a complex of at
least two and possibly more antigenic components. The major

component has been identified as a polypeptide encoded by
the large Hind III subfragment of the BamH1 K fragment of
the EBV genome (Hennessy and Kieff, 1983). This polypeptide
is subject to a size polymorphism which correlates with the
length of the triplet nucleotide repeat array, IR3, in
different EBV-carrying cell lines (Hennessy et al., 1983).
The region encoded by IR3 in the marmoset lymphoblastoid
line B95-8 is a glycine-alanine copolymer (Hennessy and
Kieff, 1983).

In addition to this polymorphic BamH1 K encoded EBNA
(BK EBNA) two other EBV-associated nuclear antigens have
been described. Grogan et al. (1983) have shown that an
antigen which locates in the nucleus is encoded by the BamH1
M fragment. The molecular weight of this molecule has not
been reported. Hennessy and Kieff (1983) describe a non-
polymorphic 82K polypeptide (they term EBNA 2) which is
serologically distinct from BK EBNA (which they term EBNA 1).

There is also a second antigen-antibody system which,
although it is also associated with the nucleus of EBV-
carrying lymphocytes, was originally described as being
distinct from EBNA-anti-EBNA. This is RANA-anti-RANA
(Rheumatoid Arthritis Associated Nuclear Antigen) (for a
review see Zvaifler and Depper, 1982). Recently Billings et
al. (1983) showed that antibodies to BK EBNA and RANA
identify the same polypeptide on immunoblots. However, this
does not preclude the possibility that the distinction
between anti-RANA and anti-EBNA is real. It may be that the
antibodies recognise different epitopes on the EBNA molecule.
Because of the autoimmune nature of Rheumatoid Arthritis
(RA) we have focused our attention on the peptide sequence
encoded by IR3 as a potential anti-RANA epitope. This is
because DNA hybridization studies have shown that IR3 is
related to DNA sequences found on human and mouse chromosomes (Heller et al., 1982), the peptide is antigenic
(Hennessy and Kieff, 1983) and antibodies raised against it
appear to cross-react with cellular antigens (Dillner et al.,
1984).

Materials and Methods

Cells

All cells were grown in continuous suspension culture
in RPM1 1640 supplemented with 10% foetal calf serum.

Partial purification and concentration of BK-EBNA

Nuclear pellets were produced from MST cells (B95-8
transformed lymphocytes, deKretser et al., 1983). using the
method described by Snary and Crumpton (1974) and stored at

-70°C until required. BK EBNA was extracted from the
nuclei in essentially the same manner used to prepare class
I EBNA (Sculley et al., 1983). The clarified extract was
rapidly heated to 70°C in a preheated glass vessel and
shaken vigorously for 10 minutes, then cooled on ice. The
precipitate formed was pelleted by centrifugation at
10,000g for 20 minutes at 4°C. The supernatant was mixed
with at least two volumes of ethanol and allowed to precipitate overnight at -20°C. This precipitate was pelleted by
centrifuging at 10,000g for 30 minutes at 4°C and resuspended in 1/10 its original volume of extraction buffer. It
was then stored in small aliquots at -70°C until use.

Trypsin digestion of EBNA

100μl aliquots of the partially purified and concentrated EBNA extract were incubated for various periods with
10μl of 0.25% (w/v) trypsin at 37°C. The reaction was
terminated by mixing with 100μl of SDS buffer, boiling for
3 minutes and immediate separation by SDS PAGE.

SDS gel electrophoresis and Immunoblotting

Separation on SDS-10 and 15% polyacrylamide gels was
performed according to the procedure of Laemmli (1970).
Proteins were electrotransferred (overnight) to a nitrocellulose filter (Schleicher and Schull) as described by
Burnette (1981). The nitrocellulose filter was washed for
10 minutes in PBS (pH 7.2) and then incubated for 1 hr at
40°C in PBS containing 3% w/v bovine serum albumin, BSA
(Fraction V, Sigma). This was followed by 1-2 hr incubation
at room temperature with first antibody in PBS containing
3% BSA (human serum, 1/100 or immunoaffinity purified anti-
IR3 antibodies, 1/10).

After three 10 minute washes in PBS the blots were
incubated with second antibody diluted 1/1000 in PBS
containing 3% BSA for 30 minutes (anti-human IgG, γ-chain
specific, peroxidase conjugate, Sigma). Blots were then
washed as above and incubated with 10ml of substrate mix
(0.4 mg/ml 3-3' diaminobenzidine tetrahydrochloride; Grade
II, Sigma) in phosphate/citrate buffer (pH 5.0) + 0.012%
H_2O_2). Colour development was terminated by washing in
distilled water after 30 sec-2 min.

Sera and Antibodies

EBV seropositive and seronegative sera were obtained
from healthy individuals at Sussex University. RANA positive
sera from patients with seropositive rheumatoid arthritis
were a gift from Dr. Patrick Venables (Kennedy Institute,
London). Immunoaffinity purified human antibodies to a
synthetic peptide corresponding to amino acid sequences

within the IR3 encoded region were a gift from Joakim Dillner (see Dillner et al., 1984).

Antibodies were eluted from single bands on immunoblots (i.e. microaffinity purified) and re-used in the immunoblot procedure using the method described by Smith and Fisher (1984).

Results

During extraction and storage at -70°C BK EBNA appears to undergo considerable degradation, producing 4-5 major bands in addition to the 79K native molecule, presumably through proteolysis (see Figure 1). In addition two minor bands can be seen (22 and 27K mol.wt.) which are only recognised by some EBNA-positive sera but not EBV-negative sera. Twenty anti-EBNA-positive antisera were used in immunoblotting experiments (not all shown). Half of these were RF (Rheumtoid Factor)/RANA-positive sera. Nine sera (5 normal, 4 RF/ RANA-positive recognised the two low mol.wt. bands, eleven (5 normal, 6 RF/RANA-positive) did not. This suggests that these antigens are probably associated with EBV but not with Rheumatoid Arthritis or RANA.

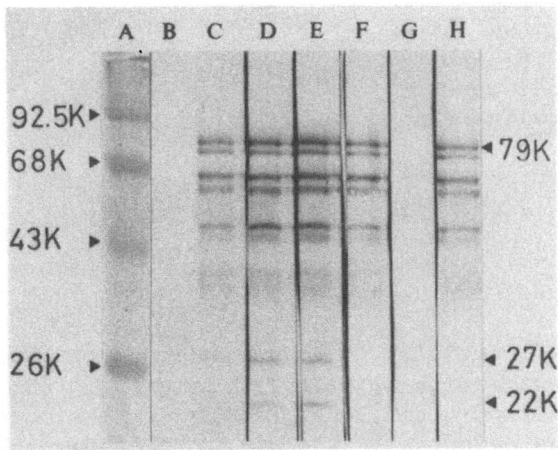

FIGURE 1 - Immunoblot (from SDS-10% polyacrylamide) of partially purified and concentrated BK EBNA. The blot was probed with six different human sera. A: Mol.wt. standards; B: Whole cell extract from EBV-negative Ramos probed with EBNA-positive serum GM; C: Serum GM; D: Serum ST (EBNA-positive); E: Serum 44 (EBNA-positive); F: Serum 74 (EBNA-positive); G: Serum RB (EBNA-negative); H: Serum JC (EBNA-positive).

440

Figure 2 shows the amino acid sequence of BK EBNA predicted from the DNA sequence of the <u>BamH1</u> K fragment (R. Baer, personal communication). The potential cleavage sites for trypsin are indicated. If trypsin cleavage occurs at the available sites the glycine-alanine copolymer encoded by IR3 plus 8 residues should remain as a large (>20K mol. wt.) resistant peptide. Assuming it is antigenic and immunogenic in the native molecule then this fragment should be detectable on immunoblots using EBNA-positive human serum.

FIGURE 2 - Amino acid sequence of BK EBNA predicted from the DNA sequence of the <u>BamH1</u> K restriction fragment from B95-8. The IR3-encoded glycine-alanine co-polymer (underlined) and potential trypsin cleavage sites (●) are indicated. (The cleavage sites are on the carboxyl-terminal side of Arginine (R) and Lysine (K) but bonds involving Proline (P) are unlikely to be cleaved and repetitive

```
                                                                      70
MSDEGPGTGPGNGLGEKGDTSGPEGSGGSGPQRRGGDNHGRGRGRGRGRGGGRPGAPGGSGSGPRHRDGV

                                                                     140
RRPQKRPSCIGCKGTHGGTGAGAGAGGAGGAGGAGAGGGAGAGGGAGGAGGAGGAGAGGGAGAGGGAGGAG

                                                                     210
GAGAGGGAGAGGGAGGAGGAGAGGGAGGAGGAGAGGGAGAGGGAGGAGAGGGAGGAGGAGAGGGAGAGGAGGA

                                                                     280
GGAGAGGAGAGGGAGGAGGAGAGGAGAGGAGAGGAGAGGAGGAGAGGAGGAGAGGAGGAGAGGGAGGAGGA

                                                                     350
GGGAGGAGAGGAGGAGAGGAGGAGAGGAGAGGGAGAGGAGAGGGGRGRGGSGRGRGGSGGRGRGGS

                                                                     420
GGRRGRGRERARGGSRERARGRGRGRGEKRPRSPSSQSSSSGSPPRRPPPGRRPFFHPVGEADYFEYHQE

                                                                     490
GGPDGEPDVPPGAIEQGPADDPGEGPSTGERGQGDGGRRKKGGWFGKHRGQGGSNPKFENIAEGLRALLA

                                                                     560
RSHVERTTDEGTWVAGVFVYGGSKTSLYNLRRGTALAIPGCRLPGLSRLPFGMAPGPGPQTGPLRESIVC

                                                                     630
YFMVFLQTHIFAEVLKDAIKDLVMTKPAPTCNIRVTVCSFDDGVDLPGWFPPMVEGAAAEGDDGDDGDEG

       640
GDGDEGEEGQE
```

sequences of R or K are only cleaved slowly, Croft 1980).

To test this, partially purified and concentrated EBNA from MST was digested with trypsin and immunoblotted. After 40 minutes digestion a single major band with a mol.wt. of about 27K was detectable (Figure 3, track D). This was still stable after a further 60 minutes incubation. In further blotting experiments with different EBNA-positive sera (10 normal and 10 RF/RANA-positive), all recognised this trypsin-resistant band (results not shown). Two of the normal sera also showed a second trypsin-resistant band of mol.wt. approximately 20K (Figure 3, track B). This peptide was degraded and undetectable by immunoblotting after about 100 minutes incubation.

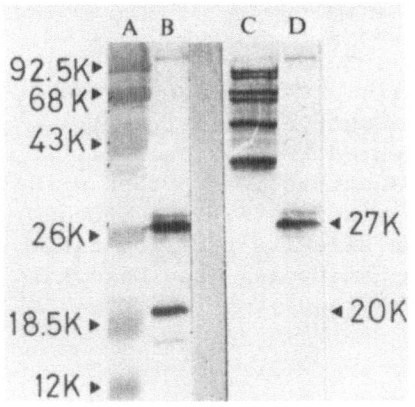

FIGURE 3 - Immunoblot (from SDS-15% polyacrylamide gel of BK EBNA extract before (Track C) and after (Track B & D) 40 minutes digestion with trypsin. The blot was probed with two different human EBNA-positive sera. A: Mol.wt. standards; B: Serum ST; C & D: Serum JC.

Extracts prepared from Raji (see Reedman and Klein, 1973) and BL8 (Harris et al., in preparation) immunoblotted in a similar manner, revealed in each case that the difference between the mol.wt. of the native molecule (69K in Raji and 94K in BL8) and that of the IR3-peptide (19K in Raji and 44K in BL8) is approximately 50K (not shown). This would be expected if only the IR3 encoded region of the molecule is polymorphic.

Figure 4 shows an immunoblot of the BK EBNA extract and a trypsin digested extract probed with both immunoaffinity purified human anti-IR3 antibodies (Dillner et al., 1984) and EBNA-positive human serum. The major bands recognised by the EBNA-positive serum are also demonstrated by the mono-specific anti-IR3 antibodies. This confirms that these peptides contain the IR3 encoded glycine alanine repeat and supports the suggestion that they are degradation products òf the 79K mol.wt. polypeptide. The anti-IR3 antibodies, as expected, also reveal the 27K mol.wt. peptide produced by trypsin cleavage.

These monospecific antibodies do not, however, show the 22K and 27K mol.wt. bands or the 20K mol.wt. trypsin resis-tant band. This suggests that these peptides do not contain the IR3 encoded sequence. To determine whether they are related to the 79K mol.wt. native molecule antibodies were eluted from strips excised from nitrocellulose blots and used to probe another blot of the MST nuclear extract (Figure 5). This shows that the antibodies eluted from 22K and 27K recognise the 79K mol.wt. molecule and all degradation products previously shown. The antibodies eluted from 20K

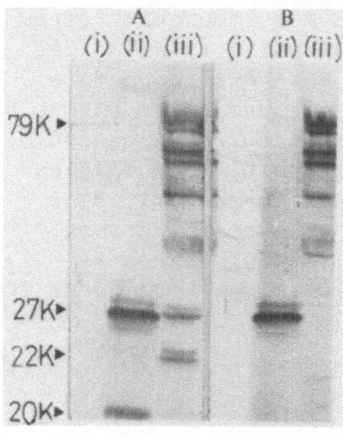

FIGURE 4 - Immunoblot (from SDS-10% polyacrylamide gel) of (i) EBV-negative Molt-4 cell extract (ii) Trypsin digested EBNA extract (iii) EBNA extract. A: probed with human serum (ST); B: probed with affinity purified anti-IR3 antibodies (see Materials and Methods).

recognise the 79K mol.wt. molecule and only one product of degradation. We conclude that these eluted antibodies bind to at least two epitopes on BK EBNA located outside the IR3 encoded region.

FIGURE 5 - Immunoblots (from SDS-10% polyacrylamide gels). A: EBNA extract; B: Trypsin digested (40 minutes) EBNA extract both probed with EBNA-positive serum (ST). Bands arrowed 1,2 & 3 were excised and antibodies were eluted using pH 2.3 glycine-HCl buffer. These antibodies were used to probe EBNA extract on the nitrocellulose strips shown in C.

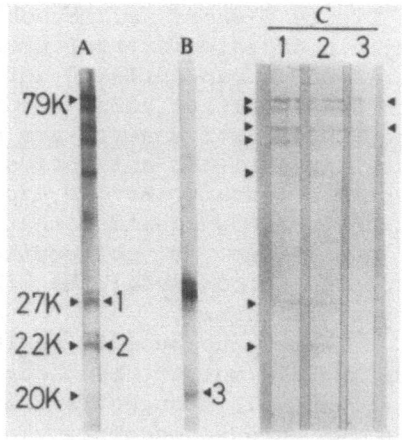

Discussion

The variable region of BK EBNA is encoded by the IR3 repeat and in B95-8 it consists entirely of glycine and alanine residues (Figure 2). The absence of arginine and lysine from this sequence of over 200 residues makes it resistant to a number of proteases including trypsin. We have shown by immunoblotting nuclear extracts after trypsin digestion that the major antigenic product is a peptide which also shows polymorphism. A reasonable assumption based on the mol.wt., polymorphism and amino acid sequence is that this fragment represents the region of BK EBNA encoded by IR3 (together with 8 additional residues). This peptide was recognised by antibodies in all sera examined, suggesting it is a commonly recognised and probably major antigenic and immunogenic determinant of BK EBNA.

Antibodies are found in some EBNA-positive sera which bind to the native 79K mol.wt. BK EBNA and products of proteolytic degradation not apparently containing the IR3-encoded sequence. This indicates that in addition to epitopes within the IR3 region there are antigenic and immunogenic determinants located elsewhere on the BK EBNA molecule. That these are not recognised by all EBNA-positive sera (9/20 and 2/20) suggests they are minor determinants.

Finally, RANA appears to be part of the BK EBNA molecule and anti-RANA antibodies may be directed against a specific determinant located within the IR3 encoded region (see Introduction). It is not possible, however, to establish conclusively from our present data whether RANA is an epitope(s) within this region. The trypsin-resistant peptide is recognised by both the normal and RA sera examined. This indicates that anti-IR3 antibodies are not restricted to RA sera. However, there is general agreement that, while up to 90% of normal sera contain anti-RANA antibodies, in RA anti-RANA titres are higher than normal (see Venables et al., 1984). Therefore, if RANA is located within the trypsin-resistant peptide, one would expect elevated anti-IR3 titres in patients with sero-positive RA. We are currently purifying sufficient of the IR3-peptide to set up a quantitative immunoassay and test this hypothesis.

Acknowledgements

We are grateful to Drs. Anne Harris and Julia Bodmer for providing cell lines BL8 and MST. Financial support for the research was provided by the Nuffield Foundation.

References

Billings, P.B., Hoch, S.O., White, P.J., Carson, D.A. and
Vaughn, J.H. Antibodies to the Epstein-Barr virus nuclear
antigen and to rheumatoid arthritis nuclear antigen identify
the same polypeptide. Proc.Natl.Acad.Sci. USA, 80, 7104-
7108 (1983).
Croft, L.R. Enzymic cleavage of proteins, in Introduction to
Protein Sequence Analysis, published by John Wiley and Sons,
Chichester, U.K. (1980).
Burnette, W.N. 'Western Blotting':electrophoretic transfer
of proteins from SDS-polyacrylamide gels to unmodified
nitrocellulose and radiographic detection with antibody and
radioiodinated protein A. Anal.Biochem. 112, 195-203 (1981).
deKretser, T., Crumpton, M.J., Bodmer, J.G. and Bodmer, W.F.
Biochemical analysis of the polymorphism of three HLA-D
region antigens. Mol.Biol.Med. 1, 59-76 (1983).
Depper, J.M. and Zvaifler, N.J. The association of Epstein-
Barr virus and rheumatoid arthritis, in Gorini, S. (ed),
Advances in Inflammation Research, Vol.3, Raven Press,
New York, pp.29-40 (1982).
Dillner, J., Sternas, L., Kallin, B., Alexander, H., Ehlin-
Henriksson, B., Jornvall, H., Klein, G. and Lerner, R.
Antibodies against a synthetic peptide identify the Epstein-
Barr virus determined nuclear antigen (EBNA). Proc.Natl.Acad.
Sci. USA. 81, 4652-4656 (1984).
Grogan, E.A., Summers, W.P., Dowling, S., Shedd, D.,
Gradoville, L. and Miller, G. Two Epstein-Barr viral nuclear
neoantigens distinguished by gene transfer, serology and
chromosome binding. Proc.Natl.Acad.Sci. USA, 80, 7650-7653
(1983).
Heller, M., Henderson, A. and Kieff, E. Repeat array in
Epstein-Barr virus DNA is related to cell DNA sequences
interspersed on human chromosomes. Proc.Natl.Acad.Sci. USA,
79, 5916-5920 (1982).
Hennessy, K., Heller, M., van Santen, V. and Kieff, E.
Simple repeat array in Epstein-Barr virus DNA encodes part
of the Epstein-Barr nuclear antigen. Science 220, 1396-1398
(1983).
Hennessy, K. and Kieff, E. One of two Epstein-Barr virus
nuclear antigens contains a glycine-alanine copolymer domain.
Proc.Natl.Acad.Sci. USA, 80, 5665-5669 (1983).
Laemmli, U.K. Cleavage of structural proteins during the
assembly of the head of bacteriophage T4. Nature 227, 680-
685 (1970).
Reedman, B.M. and Klein, G. Cellular localization of an

Epstein-Barr virus (EBV)-associated complement-fixing
antigen in producer and non-producer lymphoblastoid cell
lines. Int.J.Cancer 11, 499-520 (1973).
Sculley, T.B., Kreofsky, T., Pearson, G.R. and Spelsberg,
T.C. Partial purification of the Epstein-Barr virus
nuclear antigen(s). J.Biol.Chem. 258, 3974-3982 (1983).
Smith, D.E. and Fisher, P.A. Identification, developmental
regulation and response to heat shock of two antigenically
related forms of a major nuclear envelope protein in
Drosophila embryos: application of an improved method for
affinity purification of antibodies using polypeptides
immobilized on nitrocellulose blots. J.Cell.Biol. 99,
20-28 (1984).
Snary, D. and Crumpton, M.J. Preparation and properties of
lymphocyte plasma membrane. Contemp.Top.Mol.Immunol. 3,
27-56 (1974).
Venables, P.J.W., Smith, P.R. and Maini, R.N. Rheumatoid
arthritis, nuclear antigen and Epstein-Barr virus.
Arthritis Rheum. 27, 476-477 (1984).

41

The Use of Antibodies against Synthetic

Peptides for Studying the EBV Nuclear Antigen.

J.Dillner, L.Eliasson,* L.Sternås, B.Kallin,
G.Klein and R.A. Lerner.*

Department of Tumor Biology, Karolinska
Institute, Stockholm, Sweden; and * Research
Institute of Scripps Clinic, La Jolla,
California.

Five peptides corresponding to amino acid
sequences predicted from the BamHI K fragment of
the EBV genome have been synthesized
(Table 1). The antisera raised against peptide
no. 107, a copolymer of alanine and glycine
deduced from the third internal repeat (IR3)
sequence, gave brilliant nuclear staining in the
anticomplement immunoflourescence assay (ACIF)
on eight EBV-carrying lines (Figure 1a),whereas
five EBV-negative lines were not stained (Table
2). The nuclear staining was competed out by
addition of the synthetic peptide (Figure 1b).
19 out of 21 EBNA-positive sera reacted with the
synthetic peptide in an ELISA test, whereas EBV-
negative sera did not react or gave very weak
reactions (Figure 2). The 19 EBNA-positive sera
that reacted with the peptide were all healthy
donor sera whereas the 3 EBNA-positive sera that
failed to react were Burkitt lymphoma (BL)
patient sera.In a later series of experiments 34
EBNA-positive healthy donor sera were all found
to react with the synthetic peptide, whereas out
of 48 BL sera 19 did not have antibodies to this
peptide (not shown).
The peptide-specific antibodies were purified
from EBNA-positive sera by means of affinity

Table 1. Synthetic peptides and the antibody response against them

Peptide no.	Amino acid sequence	Reading frame	Rabbit antipeptide response: ELISA titer, rabbit 1/rabbit 2	Anti-EBNA response in ACIF, rabbit 1/rabbit 2
107	Ala-Gly-Ala-Gly-Gly-Gly-Ala-Gly-Gly-Ala-Gly-Ala-Gly-Gly-Gly-Ala-Gly-Gly-Ala-Gly-Cys* †	2	1:10,240/1:640	1:32/1:4
106	Gly-Gly-Gly-Ala-Gly-Ala-Gly-Gly-Ala-Gly-Ala-Gly-Gly-Gly-Gly-Arg-Cys*	2	1:5,160/1:10,240	0/0
108	Cys-Arg-Ala-Arg-Gly-Arg-Gly-Arg-Gly-Arg-Gly-Glu-Lys-Arg-Pro-Met*	2	1:5,120/1:5,120	0/0
105	Cys-Lys-Ser-Gln-Gly-Glu-Arg-Ser-Trp-Thr-Trp-Arg-Lys-Glu-Ala-His*	1	1:10,240/1:10,240	0/0
109	Cys-Lys-Glu-Pro-Gly-Gly-Glu-Val-Val-Asp-Val-Glu-Lys-Arg-Gly-Pro*	3	1:40,960/1:20,480	0/0

*These residues were not in the primary amino acid sequence but were added to allow coupling
†The sequence is repeated three times in the IR3 region of the B95-8 virus

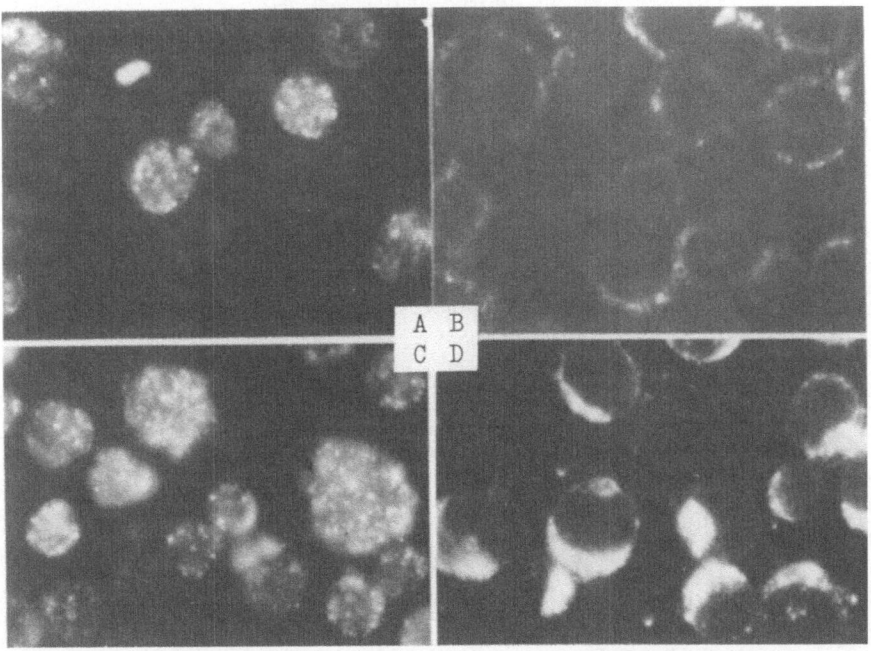

Figure 1.ACIF on P3HR-1 cells with:A:rabbit antipeptide serum,B: rabbit antipeptide serum plus the synthetic peptide, C:human affinity purified antipeptide serum and D:human antipeptide serum plus the synthetic peptide.

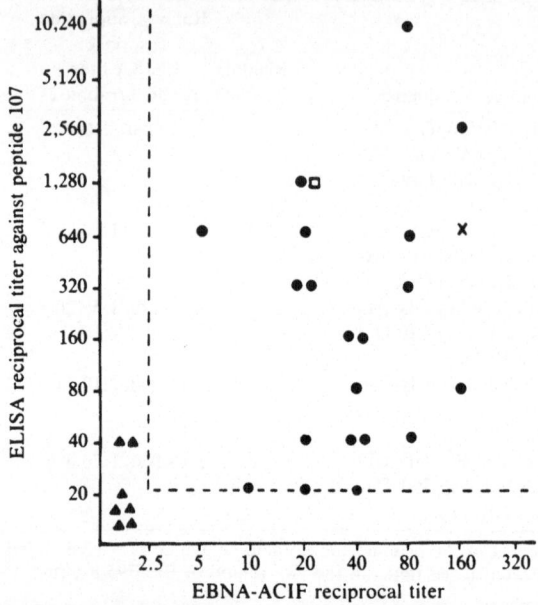

FIG. 2. EBNA-ACIF titers of human EBV-positive and EBV-negative sera against Raji cells and their titers against the synthetic peptide 107 in a direct binding ELISA test. ▲. EBV-negative sera; ●. EBV-positive sera: □ and ×. two EBV-positive sera that were used for affinity purification.

Table 2. Nuclear ACIF reaction with affinity-purified rabbit and human sera

| | | ACIF with | | |
Cell	EBV carrier state	EBNA-positive human sera diluted 1:4	Affinity-purified rabbit antibodies to peptide 107 diluted 1:4	Affinity-purified human antibodies diluted 1:4
BL 30	Negative	−	−	−
BJAB	Negative	−	−	−
Ramos	Negative	−	−	−
Loukes	Negative	−	−	−
BL 2	Negative	−	−	−
Raji	Positive	+++	++	++
LSB-1	Positive	+++	+	+
P3HR-1	Positive	+++	+++	+++
Namalwa	Positive	+++	+	+
Daudi	Positive	+++	+++	+++
B95-8	Positive	+++	+++	+++
Cherry	Positive	+++	+++	+++
Jijoye	Positive	+++	+++	+++

chromatography with the peptide coupled to AH-sepharose. These antibodies gave an EBV-specific nuclear staining similar to that of the rabbit antipeptide antibodies (Figure 1c). This staining was also removed by absorption with the synthetic peptide (Figure 1d).

Immunoblotting

Immunoblottings of DNA-binding proteins from EBV-carrying and EBV-negative cell lines showed that the rabbit antisera against the three peptides predicted from reading frame no. 2, could identify an EBV-specific protein varying in size among different EBV-positive cell lines (Figure 3a,4) in line with the results of Strnad et al (1981) and Hennessy and Kieff (1983). The anti-107 rabbit serum was also used to identify two cellular proteins ,44 and 49 kilodaltons in size (Figure 3a). This reaction was peptide-specific since it was removed by preabsorption with the synthetic peptide (Figure 3b). This cross-reaction is of particular interest in view of the findings of Heller et al (1982) that sequences homologous to the IR3 are present in multiple copies on the human chromosomes.

The affinity-purified human antipeptide antibodies also reacted with the 70- 92 K protein on immunoblottings (Figure 5). The reaction aginst these polypeptides was removed by preabsorption with the synthetic peptide, both in the case of the rabbit anti-peptide (Figure 3b) and the human antipeptide antibodies (not shown). In addition to the peptide-specific reactions, some reactions were seen which were not abolished by addition of an excess of peptide, notably in the 31 to 34 kilodaltons size category (Figures 3,5). These antibodies are likely to have bound to the affinity column non-specifically.

ELISA test

The rabbit anti-107 antibodies could be used as catching antibody for the native EBNA molecule in an ELISA test, where after binding of the antigen, a human EBNA-positive sera is allowed to react and finally an enzyme-linked anti-human antibody. This method has provided a highly

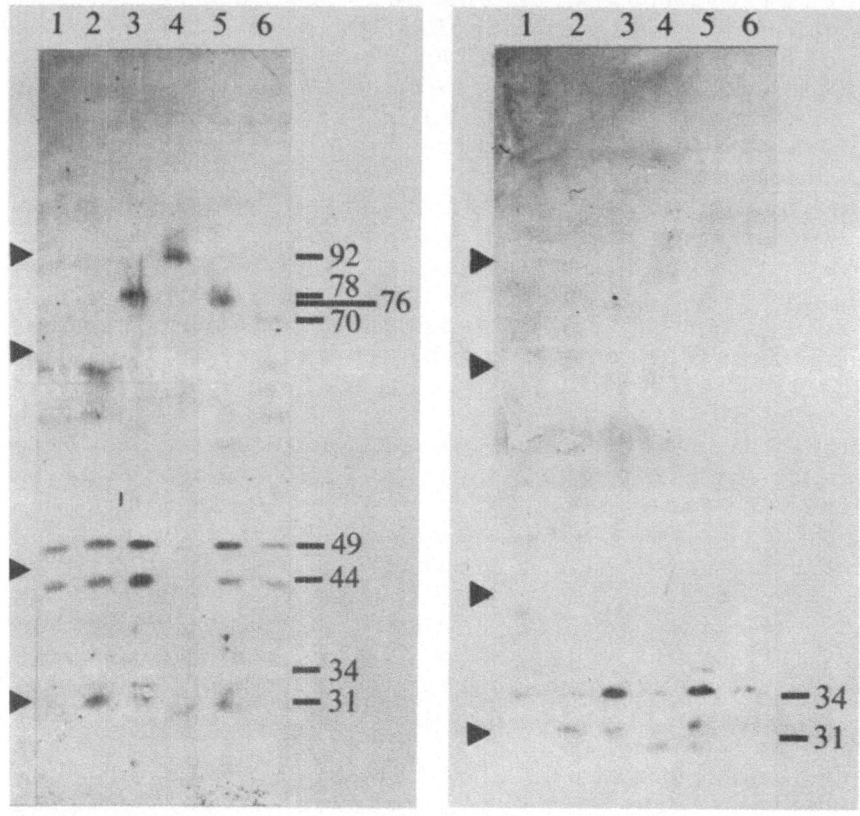

Figure 3a   and   Figure 3b.
Immunoblots on 10 % polyacrylamide gels with
rabbit antipeptide 107 serum,diluted 1:4 (A) and
the same serum after preincubation with the
synthetic peptide (B). The lanes are DNA-binding
proteins from EBV-negative cells ;1:BJAB,
2:Loukes and EBV-positive cells ;3:B95-8,
4:Cherry, 5:P3HR-1 and 6:Raji. Figures indicate
MW in kilodaltons.

---

sensitive quantification method for EBNA, even
in crude cell extracts. Although the rabbit
antipeptide 107 antibodies also react with some
cellular proteins (Figure 3), EBV-negative cells
did not give any reaction in the ELISA test,
presumably due to that EBNA was the only protein
recognized by both the rabbit serum and the
human anti-EBNA positive serum.

Comparison between amount of EBNA and number of
EBV genomes per cell
The EBNA quantification assay was used to
investigate if the amount of BamHI K encoded
EBNA in the cell was correlated to the number of
EBV genomes per cell. Eleven EBV-positive and
three EBV-negative cell lines were used in this
comparative study (Figure 6). The three EBV-
negative lines originate from BL biopsies
(Bjab,Loukes and Ramos). Six of the EBV-positive
cell lines (Bjab/B95-8, Ramos HR1K, Ramos/B95-8,
AW Ramos, EHR-D Ramos, EHR-H Ramos and EHR-O
Ramos) are the in vitro EBV-converted
counterparts of Bjab and Ramos. Four lines are
derived from BL (Raji, Rael, Namalwa, Daudi) and
one is an EBV-carrying lymphoblastoid cell of
non-neoplastic origin (6410). In a previous
study (Ernberg et al,1977) measurement of the
relative EBNA content in cells by
immunofluorescence resulted in a significant
correlation (r=0.86) between the level of EBNA-
fluorescence and the number of EBV-genomes. In
this study the correlation was not found
(r=0.46). This discrepancy can be accounted for
by the higher sensitivity and specificity of the
ELISA assay employed in this study. The main
difference was that the antibodies used were
specific for the BamHI K EBNA protein.

The further use of the synthetic peptide
antisera
The EBNA quantification system has also been
used for monitoring transfection effectivity and
EBNA expression when transfecting an eucaryotic
viral expression vector with a BamHI K insert
into CV1 cells (M-L. Hammarskjöld et al, in
preparation) and also for monitoring EBNA during
purification of the antigen.
A special application of the antipeptide
antibodies presented here is the differential
characterization of the BamHI K-encoded and the
BamHI WYH-encoded EBNAs, where in the case of
the BamHI WYH-encoded EBNA a specific serum is

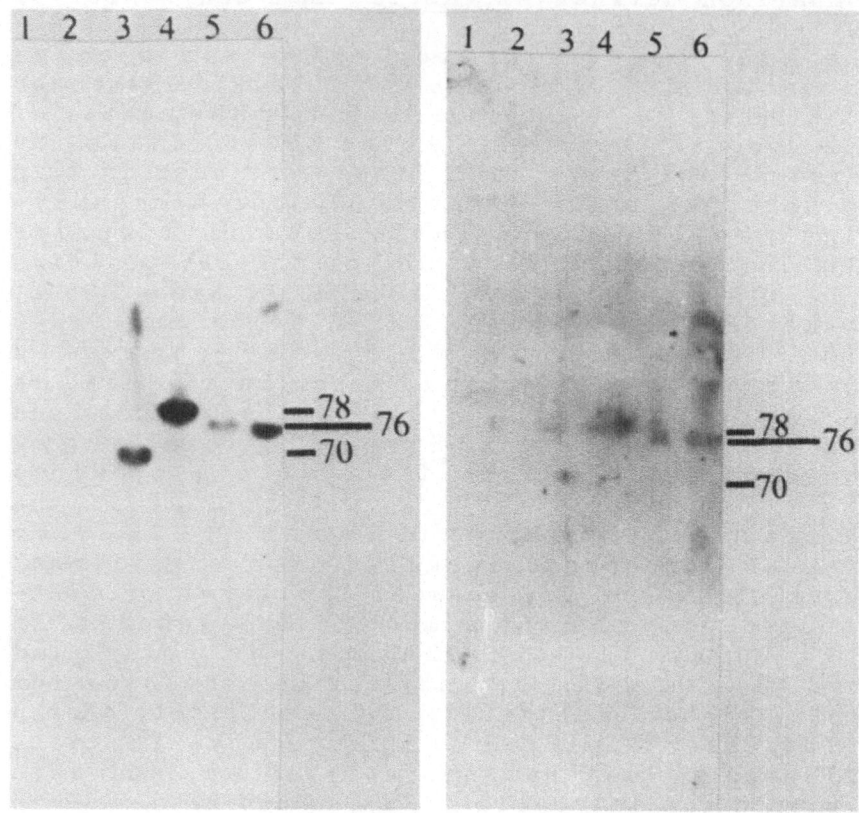

Figure 4a   and   Figure 4b
Immunoblots on 7.5% polyacrylamide gels with
rabbit antipeptide 106 serum,dil.1:2 (A) and
rabbit antipeptide 108 serum,dil.1:2 (B). The
lanes are nucleic extracts from EBV-negative
cells, 1:BJAB, 2:Loukes and EBV-positive cells,
3:Raji, 4:B95-8, 5:Jijoye and 6:P3HR-1.

generated by excessive preabsorption of EBNA-
positive sera with P3HR-1 or Daudi cells. This
has been used for studying the serology of the
two EBNAs, notably the appearance of EBNA
antibodies during the course of infectious
mononucleosis and for studying the binding of
the two EBNAs to metaphase chromosomes (Dillner
et al, in press)

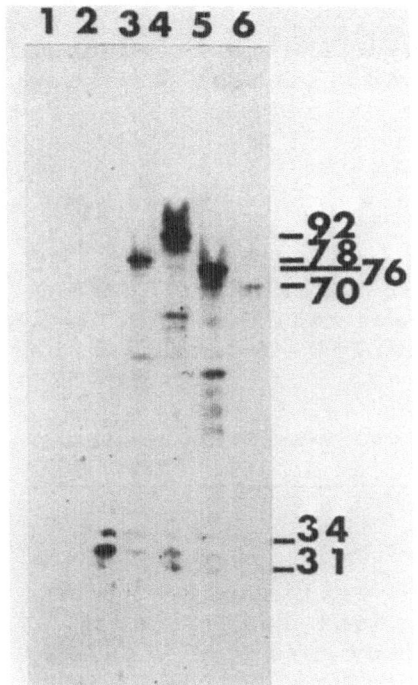

Figure 5.Immunoblot on a
10 % polyacrylamide gel
with human EBNA-positive
serum affinity purified
on a column of peptide
107 bound to AH-
sepharose.Dilution 1:4.
Lanes are DNA-binding
proteins from EBV-
negative cells,1:BJAB,
2:Loukes and EBV-
positive cells,3:B95-
8,4:Cherry,5:P3HR-1 and
6:Raji.

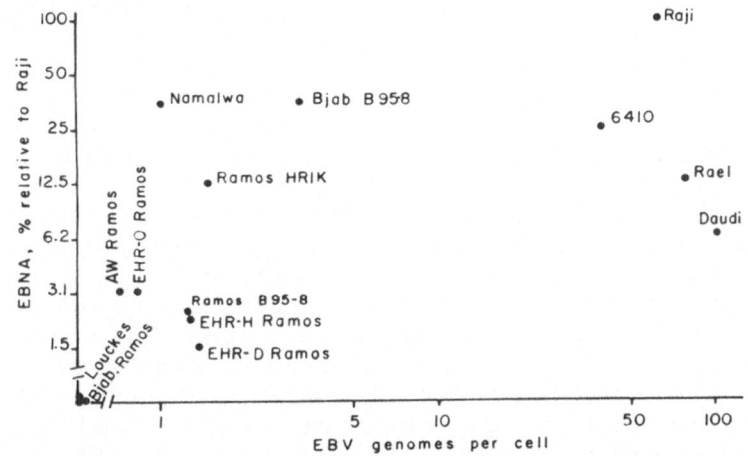

Figure 6
Comparison between EBNA content and number of
EBV genomes. EBNA was measured in an ELISA assay
using affinity-purified rabbit antipeptide 107
serum attached to the bottom of the plate. The
reaction was done with extracts of different
cell lines, followed by a human EBNA-positive
serum and an antihuman antibody alkaline

phosphatase conjugate. Finally the enzyme substrate solution was added. The reaction could be completely removed by addition of an excess of synthetic peptide. The detection level for Raji cells was a dilution of cell extract corresponding to 40.000 cells.

The number of EBV genomes was determined in a dot blot assay. Purified DNA was dotted on a nitrocellulose filter and hybridized to a 32p-labelled probe of the EBV BamHI M fragment dotted on the filter in equivalents of 100-0.5 EBV genomes per cell. The EBV-negative lines Bjab, Loukes and Ramos were negative in both assays.

Concluding remarks

The regular occurrence of antibodies to the synthetic peptide 107, present in 53 out of 53 EBNA-positive healthy donor sera tested, is unique for synthetic peptides and indicates that the peptide corresponds to a regularly immunogenic determinant of the native EBNA protein. The wide usability of these antipeptide antisera can no doubt be accounted for by this fact. For comparative purposes the other four synthetic peptides were also studied with respect to the occurrence of antibodies to these peptides in 34 EBNA-positive and 5 EBV-negative healthy donor sera. No reactivity was found with the peptides 105 and 109 deduced from other reading frames and in no case was reactivity found with the EBV-negative sera. The peptide 106 was however recognized by 3 EBNA-positive sera and the peptide 108 was recognized by 5 EBV-positive sera. This suggests that these peptides contain weak determinants of EBNA.

Although both the 106 and 108 rabbit antipeptide antibodies detect EBNA on immunoblots (Figure 4), they do not stain EBNA in ACIF, nor can they be used as catching antibody in the ELISA test. In spite of this, their antipeptide titers are equal to or better than those of the 107 antipeptide sera (Table 1).

To conclude the approach of using synthetic peptide-generated antibodies has proven to be highly successful for identifying the native EBNA molecule. In this study we succeeded in 3 out of 3 cases. When, as in the case of the peptide 107, one happens to find an important determinant of the native protein, the antipeptide sera have also been shown to be useful for a wide variety of purposes.

Acknowledgements
Tables 1 and 2 and Figures 1,2 and 3 are reprinted from Proc.Natl.Acad.Sci.USA 81,4652 (1984). This work was supported by Public Health Service Grant 5R01, CA 28380-03 awarded by the National Cancer Institute and by the Swedish Cancer Society.

References:
Dillner,J.,Sternås,L.,Kallin,B.,Alexander,H., Ehlin,B.,Jörnvall,H.Klein,G and Lerner,R.A. Antibodies against a synthetic peptide identify the EBV-determined nuclear antigen(EBNA). Proc.Natl.Acad.Sci.USA 81,4652-4656 (1984).

Dillner,J., Kallin,B., Ehlin,B., Timar,L and Klein,G. Int.J.Cancer in press.

Ernberg,I.,Andersson-Anvret,M.,Klein,G., Lundin,L. and Killander,D;Relationship between amount of EBV-determined nuclear antigen per cell and number of EBV DNA copies per cell. Nature;266(5599);269-271,(1977).

Heller,M.,Henderson,A. and Kieff,E. Repeat array in EBV DNA is related to cell DNA sequences interspersed on human chromosomes. Proc.Natl.Acad.Sci.USA 79,5916-5920 (1982).

Hennessy K. and Kieff,E.One of two EBV nuclear antigens contains a glycine-alanine copolymer. Proc.Natl.Acad.Sci.USA;80:5665-5669.(1983).

Strnad,B.,Schuster,T.,Hopkins,R.,Neubauer,R and Rabin,H. Identification of an EBNA by fluoroimmunoelectrophoresis and radioimmuno-electrophoresis. J.Virol. 38,996-1004 (1981).

42

CHARACTERIZATION OF TWO FORMS OF THE 72,000 MW EBNA AND A CROSS-REACTING CELLULAR PROTEIN

LUKA, J.,[1] KREOFSKY, T.,[2] SPELSBERG, T.C.,[2] PEARSON, G.R.,[1] HENNESSEY, K.,[3] and KIEFF, E.[3]

[1]Georgetown University, [2]Mayo Foundation, [3]University of Chicago

INTRODUCTION

Human cells transformed by Epstein-Barr virus (EBV) express virally determined nuclear antigens designated EBNA. The major component of EBNA, identified by immuno-blotting, is a polypeptide with a variable molecular weight ranging from 72,000 to 89,000 daltons depending on the resident EBV genome (Strnad et al. 1981, Spelsberg et al. 1982, Luka et al. 1983). Kieff and co-workers (Heller et al. 1982, Hennessy and Kieff, 1983, Hennessy et al. 1983), showed that the 72K EBNA was encoded from a region on EBV DNA which contained the IR3 repeat sequence. This region encodes for a glycine-alanine polymer. The correct reading frame for translation of this EBV gene is known from the nucleotide sequence of a fusion gene between IR3 and Lacz, since the fusion gene encodes for a protein with epitopes for EBNA (Hennessy and Kieff 1983). Human and mouse cell DNA have interspersed repeat elements related to this EBV triplet, but it is not known, if the cellular sequences encode protein.

The 72K EBNA has been reported to be heterogeneous in isoelectric point and chromatin binding (Spelsberg et al. 1982, Sculley et al. 1983, Luka et al. 1983, Luka et al. submitted). Two classes of EBNA were identified based on their binding properties to chromatin. The loosely bound or class I EBNA (EBNA I) was extracted with 0,4M NaCl, while the tightly bound or class II (EBNA II) was extracted from chromatin with either 5M urea and 2M NaCl or 4M guanidine-HCl. The partially purified EBNA I had a molecular weight of 70,000 while the EBNA II displayed two subunits with 65,000 and 70,000 (Sculley et al. 1983). Whether the 65K polypeptide was a degradation product or a different antigen has not been determined.

In the present work we purified the 72k EBNA from low salt extract, and partially purified the EBNA II from urea extracted chromatin. Two-dimensional gel analysis indicated that EBNA I was a highly modified protein, while EBNA II was a basic protein (Luka et al. submitted). Further analysis also indicated that the basic form of EBNA is expressed in newly infected cells and undergoes a modification after several hours (Luka et al. manuscript in preparation). We also identified a protein with an approximate molecular weight of 62,000 which cross-reacts with the IR3 region of EBNA (Luka et al. in press).

MATERIALS AND METHODS

Cells. The EBV-genome positive Raji, NC37, P3HR-1, Namalva, B95-8, IB4, Jijoye, Ag 876, and the EBV-genome negative BJAB, Loukes, K562, Molt-4 and the mouse NS-1 cell lines were used. All cell lines were grown in RPMl 4640 medium with 5% fetal calf serum.

Sera. A number of anti-EBV positive and negative human sera were used. In addition, a rabbit serum prepared against the polypeptide encoded for by the IR3 unit of EBV DNA was also used.

Cell extraction and purification of proteins.

This was done according to Luka et al., in press and Luka et al., submitted.

Immunoblotting.

The protocol of Luka et al., 1983 was used with slight modifications (Luka et al., in press).

Polyacrylamide gel electrophoresis.

SDS-polyacrylamide gel electrophoresis was performed according to Laemmli (1970).

Two-dimensional gel electrophoresis.

This was carried out as described by O'Farrel (1975) with a pH gradient between 10 and 3,5.

ELISA assay.

The ELISA test was carried out as previously described (Luka et el. 1984).

Monoclonal antibody production.

Monoclonal antibody production was done according to the protocol in Luka et al., in press.

RESULTS AND DISCUSSION

The low salt extract (0.4M NaCl) from Raji cells was chromatographed on Blue Sepharose, Polybuffer exchanger PBE 94 Pharmacia and ds-DNA cellulose columns. The class I EBNA bound to all three columns and after the last step was purified 5,500-fold to homogeneity (Figure 1A). Western blotting indicated that the 72K polypeptide was EBNA (Figure 1B).

The pellet from low-salt extraction was sonicated in 6M urea and 0.3M NaCl to extract the tightly bound EBNA. Following the sonication and centrifugation no residual EBNA activity remained in the pellet as determined by immunoblotting. The urea extract was then directly applied to the PBE 94 column. After this step further purification on a hydroxyapatite column was done. The partially purified EBNA II became soluble in low salt buffer after this step.

The hydroxyapatite-purified EBNA II was then analyzed for binding to ds-DNA cellulose and Blue Sepharose columns. However, this preparation failed to bind to both columns. Based on the possibility that free DNA was present in the fraction, which might have prevented binding to these columns, the fraction was treated with DNase I. After this treatment the antigen still did not bind to the columns.

The EBNA I and II preparations were analyzed by two dimensional gel electrophoresis (2-D PAGE), transferred to nitrocellulose paper and identified by human anti-EBNA positive sera. The soluble class I EBNA showed several modified forms between pH 6 and 8 (Figure 2A), while the class II EBNA gave one form between pH 8,5-9.0 (Figure 2B). However, both heat treatment (70°C, 10 min)

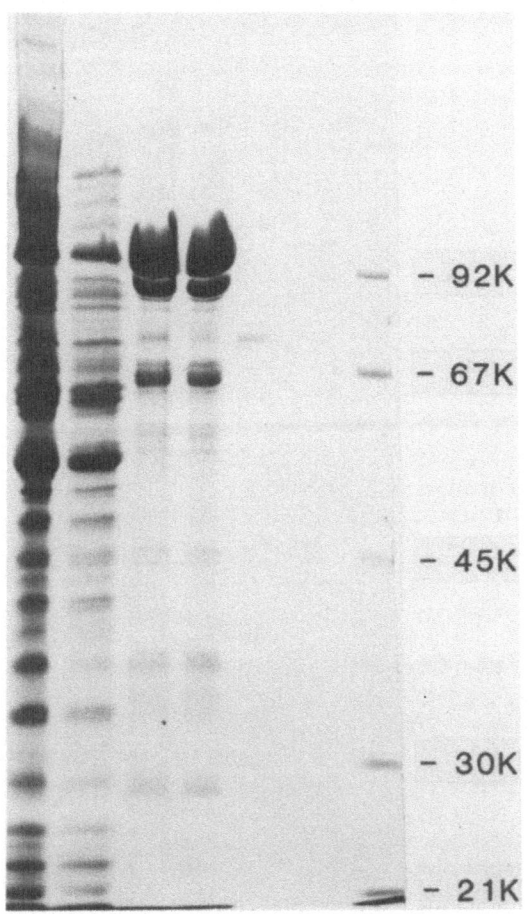
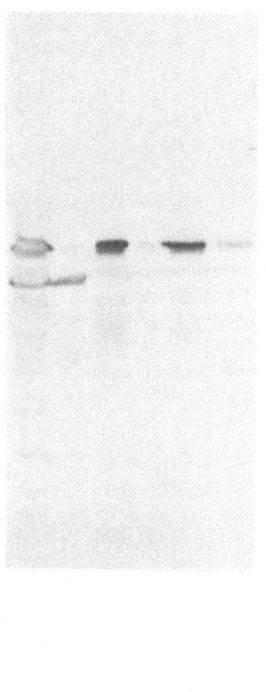

Fig. 1A. SDS-PAGE of EBNA from Raji cells at different
          stages of purification. a) Blue-Sepharose 2M
          eluate, b) PBE 94 flow-through, c) 2M eluate,
          d) ds-DNA cellulose flow-through, e) 0.6M
          eluate, f) 1.5M eluate, g) marker proteins.

      B. Identification of EBNA by immunoblotting at
          different stages of purification. a-f as in
          A.

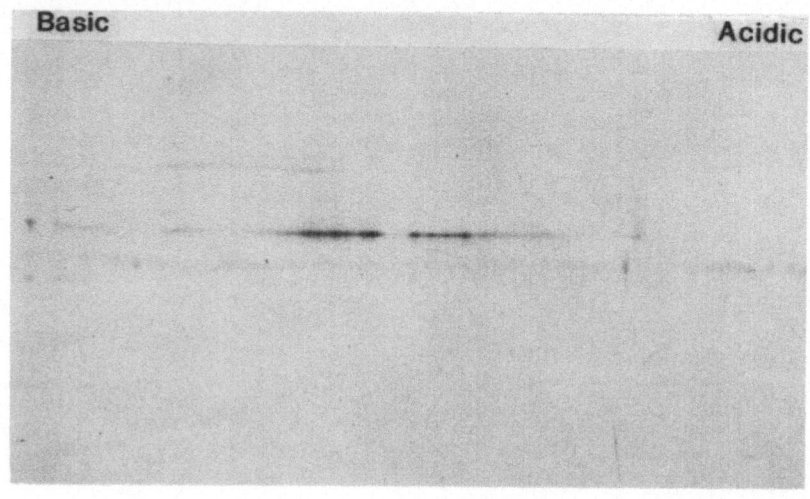

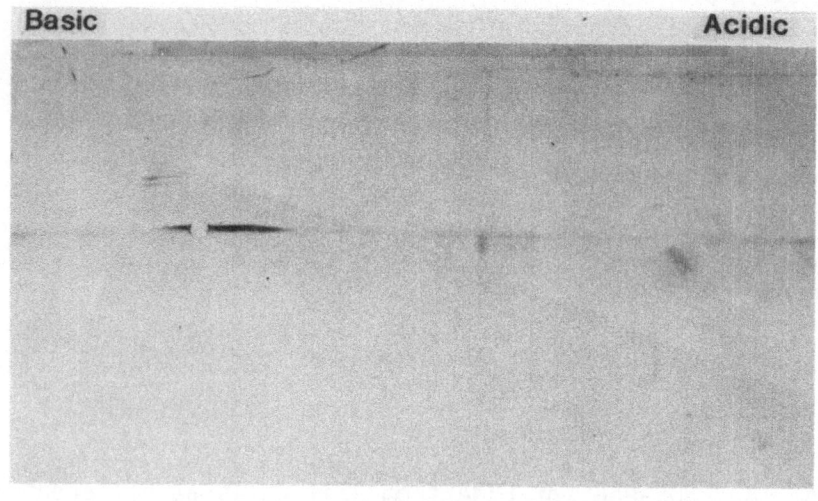

Fig. 2. Identification of the EBNA I (A) and EBNA II (B) on 2-D PAGE by immunoblotting. The basic end represents pH 10, the acidic end pH 3.5.

The EBNA I and II preparations were analyzed by two dimensional gel electrophoresis (2-D PAGE), transferred to nitrocellulose paper and identified by human anti-EBNA positive sera. The soluble class I EBNA showed several modified forms between pH 6 and 8 (Figure 2A), while the class II EBNA gave one form between pH 8,5-9.0 (Figure 2B). However, both heat treatment (70°C, 10 min) and TCA precipitation (10%) converted the EBNA I into the basic form. This probably explains why only the basic form of EBNA was detected in some earlier studies (Luka et al., 1983).

Both EBNA I and EBNA II were also subjected to peptide mapping using partial proteolysis according to Cleveland et al. (1977). The results indicated no major differences between the two forms.

When Raji cells were superinfected with Epstein-Barr virus strains B95-8 and P3HR1, new synthesis of EBNA occurred in the Raji cells only with B95-8. Whereas no P3HR1-specific EBNA (76K) synthesis occurred in the cells following P3HR-1 virus infection (only EA synthesis), the B95-8 virus induced a new EBNA type (78K) in the Raji cells. This EBNA and the original 72K EBNA are still expressed after 5 months of this cell line. The synthesis of this new EBNA was analyzed by 2-D PAGE in culture 12, 24, and 48 hours after infection. The newly synthesized EBNA was basic 12 to 24 hours after infection, and became modified like EBNA I after 24 hours. The original 72K Raji EBNA showed both forms at these times.

During the purification of the 72K EBNA, a 62K protein was also detected with some anti-EBNA positive but not with anti-EBNA negative sera. This protein was also detected in EBV-genome negative cell lines of both human and mouse origin (Figure 3). The protein was partially purified by Blue Sepharose and hydroxyapatite chromatography.

This protein was also detected by a rabbit serum raised against the glycine-alanine copolymer of EBNA. Further characterization by adsorption of human sera with SDS gel-purified 62K cellular protein or 72K EBNA indicated that the cellular protein had a crossreacting epitope with EBNA.

From immunization in vitro, with purified EBNA, a hybridoma clone was selected, which produced monoclonal antibodies against EBNA. Further characterization of

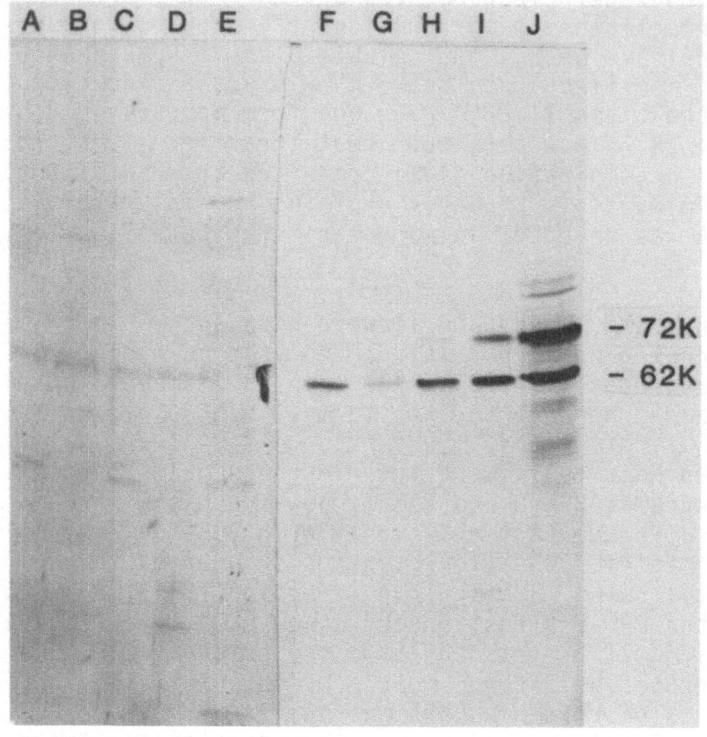

Fig. 3. Identification of EBNA and a 62K protein in different cell lines by immunoblotting. Lanes a-e, anti-EBNA negative serum f-j, anti-EBNA positive serum. a, and f,: Molt-4, b and g,: U562, c and h: NS-1, d and i,: Ramos-B-95-8, e and j,: Raji (Luka et al., 1984). (Reproduced with permission of the American Society of Microbiology, Washington, D.C.).

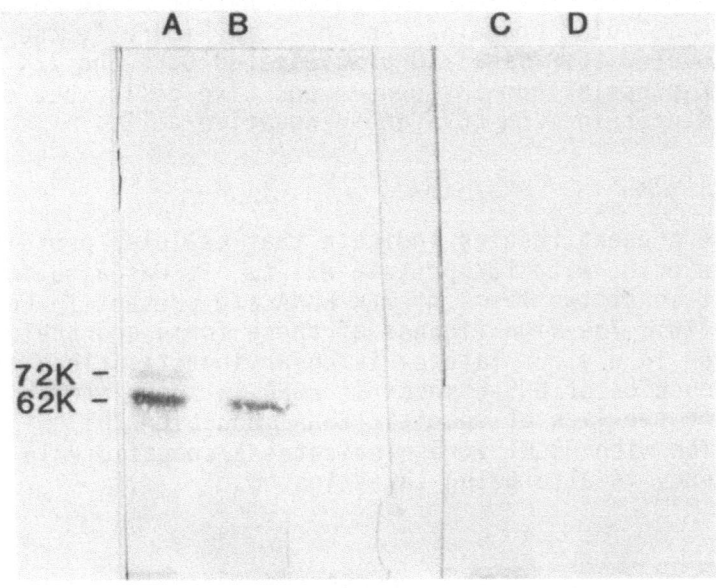

Fig. 4.   Identification of proteins that were affinity
          purified with monoclonal antibody 135 coupled to
          Affigel-10. The eluted proteins were immunoblot-
          ted and stained with an anti-EBNA positive human
          serum (a and c), or an anti-EBNA negative human
          serum (b and d). a and c, affinity-purified pro-
          tein from Raji cell extract. b and d, affinity-
          purified protein from Loukes cell extract (Luka
          et al., 1984). (Reproduced with permission of the
          American Society of Microbiology, Washington, D.C.)

this monoclonal antibody indicated that it could detect
the 62K cellular protein. As shown in Figure 4, the antibody coupled to Affigel-10 precipitated both the 72K EBNA
and 62K protein from EBV genome positive cells, but only
the 62K protein from EBV-genome negative cells.

Conclusions

The present results indicate that cellular protein(s)
crossreacting with 72K protein exist. It was also indicated
that at least two forms of 72K EBNA are present in EBV infec
ted cells. The significance of these forms and their possible
function in a virus latency is under investigation. Whether
the induction of B95-8 specific EBNA in superinfected Raji
cells or the lack of specific EBNA induction following superinfection with P3HR1 virus indicates a specific role of EBNA
in latency is also being investigated.

REFERENCES

Cleveland, D.W., Fischer, S.G., Kirschner, M.V. and
Laemmli, U.K. Peptide mapping by limited proteolysis
in sodium dodecyl sulfate and analysis by gel electrophoresis. J. Biol. Chem. 252:1102-1106. (1977).

Heller, M., Henderson, A. and Kieff, E.   A repeat
sequence in Epstein-Barr virus DNA is related to
interspersed repeated cell DNAs which are at specific
sites on human chromosomes. Proc. Natl. Acad. Sci. USA
79:5916-5920. (1982).

Hennessy, K., Heller, M., vanSanten, V. and Kieff, E.
Simple repeat array in Epstein-Barr virus DNA encodes
part of the Epstein-Barr nuclear antigen. Science
220:1396-1398. (1983).

Hennessy, K. and Kieff, E. One of the two Epstein-Barr
virus nuclear antigens contains a glycine-alanine copolymer domain. Proc. Natl. Acad. Sci. USA 80:5665-5663.
(1983).

Laemmli, U.K. Cleavage of structural proteins during
the assembly of the head of bacteriophage T4. Nature
(London) 227:680-685. (1970).

Luka, J., Jornuall, H. and Klein, G. The Epstein-Barr virus determined nuclear antigen: a previously identified 48k component and higher molecular weight forms of the antigen are structurally related. Intervirology 20:213-222. (1983).

Luka, J., Chase, R.C., and Pearson, G.R. A sensitive enzyme-linked immunosrobent assay (ELISA) against the major EBV-associated antigens. I. Correlation between ELISA and immunofluorescence titers using purified antigens. J. Immunol. Methods 67:145-156. (1984).

Luka, J., Kreofsky, T., Spelsberg, T.C. and Pearson, G.R. Characterization of the Epstein-Barr virus nuclear antigens. I. Purification of two forms of the 72,000 molecular weight EBNA. Submitted for publication.

Luka, J., Kreofsky, T., Pearson, G.R., Hennessy, K., and Kieff, E. Identification and characterization of a cellular protein cross-reacting with the Epstein-Barr virus nuclear antigen. J. Virology. 52:833-838. (1984).

O'Farrell, P.M. High resolution of two dimensional electrophoresis of proteins. J. Biol. Chem. 250:4007-4021. (1975).

Sculley, T.B., Kreofsky, T., Pearson, G.R. and Spelsberg, T.C. Partial purificatin of the Epstein-Barr virus nuclear antigen. J. Biol. Chem. 258:3974-3982. (1983).

Spelsberg, T.C., Sculley, T.B., Pihler, G.M., Gilbert, J.A. and Pearson, G.R. Evidence for two classes of chromatin associated Epstein-Barr virus-determined nuclear antigen. J. Virology 43:555-565. (1982).

Strnad, B.C., Schuster, T.C., Hopkins, R.F., Neubauer, R.H. and Rabin, H. Identification of an Epstein-Barr virus nuclear antigen by fluoroimmunoelectrophoresis and radioimmunoelectrophoresis. J. Virology 38:996-1004. (1981).

Acknowledgement. This work was supported by Public Health Service Grant CA-25340 awarded by NCI, NIH, Bethesda, Maryland 20205.

43

IDENTIFICATION OF EBV-SPECIFIC ANTIGENS FOLLOWING

MICROINJECTION OF SUBGENOMIC DNA FRAGMENTS

Ann Boyd*, Jay Stoerker**, Jane Holliday**,
and Ronald Glaser**
*Department of Biology, Hood College and Program
Resources, Inc., NCI-Frederick Cancer Research
Facility, Frederick, Maryland 21701 and **Department
of Medical Microbiology and Immunology, The Ohio State
University College of Medicine and Comprehensive
Cancer Center, Columbus, Ohio, 43210

SUMMARY

The regions of the Epstein-Barr Virus (EBV) genome which
code for proteins within the early antigen (EA) and viral
capsid antigen (VCA) complex were identified by indirect
immunofluorescence (IF) 2-4 days after microinjection of
subgenomic cloned fragments of EBV DNA. Two new regions
have been identified as part of the early antigen (EA)
complex, namely, the Charon 4A cloned fragments which cover
map units 38-47 and 83-93 respectively. One DNA fragment
from map units 45-54, produces a protein in human cells
after microinjection which reacts with EA$^-$VCA$^+$ human sera.
Attempts to transform human B-lymphocytes from cord blood
with a variety of EBV DNA fragments is described.

INTRODUCTION

Epstein-Barr Virus (EBV) is a human oncogenic virus
member of the Herpesvirus family which has been putatively
associated with nasopharyngeal carcinoma (NPC) (Epstein et
al., 1964; Henle et al., 1971). The restricted host control
for EBV replication and transformation has impaired mapping
of the genome for functional proteins and/or antigens.
Human epithelial cells permissive for EBV replication have
been described (Glaser et al., 1976, 1980; Ben-Basset et
al., 1982; Sixbey et al., 1983) whereas, transformation by
EBV has been reported only with B-lymphocytes (Pope, 1979).

Previous studies to map antigen-coding regions of the EBV genome by transfection (Stoerker and Glaser, 1983) in human epithelial cells suggested that the Bam HI fragment (H) from B95-8 virus DNA or the equivalent region cloned in a Charon 4A vector (EB 26-36) (Buell et al., 1981) is, in part, responsible for early antigen (EA) expression. The use of microinjection to study the expression of the EBV genome was first reported by Graessmann et al. (1980), using P3HR-1 virus DNA. Expression of the P3HR-1 DNA was evidenced by EA synthesis. A coding region for the EA-diffuse (EA-D) and EA-restricted (EA-R) antigens was identified by microinjection of EB 26-36 and EB 61-72 viral DNA respectively (Glaser et al., 1983). Continued efforts to map additional regions of the EBV genome for antigenic expression and preliminary results of attempts to transform human cord B-lymphocytes are reported here.

MATERIALS AND METHODS
Cell Lines

The CNE epithelial tumor cell line used in this study is EBV genome-negative and is derived from a Chinese NPC tumor (Laboratory of Tumor Viruses of the China Cancer Institute, 1978). The cells were maintained in RPMI 1640 medium supplemented with 15% fetal calf serum (FCS).

Cord Cells

Human cord blood was obtained from Dr. Jennifer Niebyl, M.D., Chief, Maternal and Fetal Medicine, Johns Hopkins University School of Medicine, Baltimore, Maryland. Cord blood was diluted 2-fold in serum-free RPMI medium and overlaid onto 4 ml of Ficoll in 15ml conical centrifuge tubes. The cells were centrifuged 20 min at 4C at 1,000 x g and the buffy coat layer removed with a pasteur pipet. The cells were diluted in 50 ml serum-free RPMI medium, centrifuged, and resuspended in RPMI 1640 medium containing 20% FCS.

Preparation of cells for microinjection

Petri dishes were incubated with 2.5 ml 2% glutaraldehyde for 2 hr at room temperature, washed 4 times with distilled water, and coated with 200 µl of Anti-B1 monoclonal antibody, (Coulter Clone, Coulter Electronics Inc, Hialeah, Florida). The antibody was allowed to dry onto

the dishes overnight in the laminar flow hood. The lymphocyte cell suspension was added to the plates, which were
incubated 2 hr at 37°C. The unattached cells and medium
were removed and fresh medium was gently added to the
attached cells. The cells were injected with 1-5 µl of
EBV DNA at a concentration of 100 ng/µl in sterile phosphate buffered saline (PBS). After microinjection, the
cells were returned to the 37°C incubator and refed with
growth medium every 3-4 days.

CNE cells were seeded onto glass coverslips in petri
dishes 24 hr prior to microinjection. The EBV DNA fragments
were microinjected at a concentration of 100 ng/µl as
previously described (Glaser et al., 1983).

Immunofluorescence

CNE cells were fixed in acetone for 10 min at room
temperature 2-4 days after microinjection. Indirect immunofluorescence (IF) was used to detect EBV-specific antigens
in the injected cells as described by Glaser et al. (1983)
using pre-characterized human sera, EA$^+$VCA$^+$ and EA$^-$VCA$^+$.
Epstein-Barr nuclear antigen (EBNA) was detected according
to the procedure of Reedman and Klein (1973).

Preparation of DNA for microinjection

The B95-8 DNA was obtained from Dr. Meihan Nonoyama,
Showa University Research Institute for Biomedicine, St.
Petersburg, Florida and was prepared for microinjection
as described by Glaser et al. (1983). Recombinant DNA used
in these studies from bacteriophage lambda vectors was pre-
pared according to the method of Blattner et al. (1977).
The Charon 4A vectors were obtained from Dr. Bill Sugden
(McArdle Lab., Univ. of Wisconsin, Madison, Wis.) and were
prepared as described earlier (Glaser et al., 1983).
Plasmid pBR322 vector recombinant plasmids were prepared
as described by Stoerker and Glaser (1983) and contained
Bam HI restriction fragments H, HFX, or K which were obtained
from Dr. Elliott Kieff (Kovler Viral Oncol. Lab., U. Chicago,
Chicago, Ill.) DNA was extracted, ethanol precipitated and
resuspended in sterile PBS for microinjection.

RESULTS

Antigen expression of subgenomic fragments of EBV

Continued efforts to map the entire EBV genome for antigen coding regions were performed by microinjection of human cells with cloned fragments of the viral DNA. Antigen detection was determined in IF assays with pre-characterized human sera to detect EA and/or VCA expression. Cells (100-200) were microinjected with single fragments of EBV DNA (100-200 ng/μl) and examined by IF assay for antigen expression. The results of preliminary mapping of about 70% of the genome of EBV are summarized in Table 1.

TABLE 1

MAPPING EBV ANTIGENS BY IMMUNOFLUORESCENCE AFTER

MICROINJECTION OF EBV DNA

DNA:	Immunofluorescence:		
	EA$^+$	VCA$^+$	EBNA$^+$
EB26-36	+	-	-
EB38-47	+	-	-
EB45-54	+	+	-
EB53-61	-	-	+
EB61-72	+	-	-
EB69-79	-	-	-
EB75-84	-	-	-
EB83-93	+	-	-

CNE Cells were seeded on glass coverslips, and microinjected 24 hr later with 100ng/μl DNA cloned fragments. Two to four days later, the cells were fixed in acetone and stained for EBV antigen expression using pre-characterized human sera.

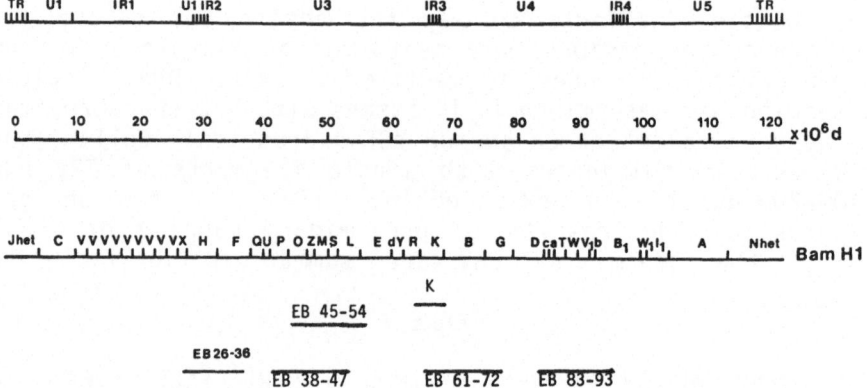

Figure 1. Diagram of the Bam HI restriction map of B95-8 EBV DNA showing the pBR 322 Bam K and Charon 4A EB 26-36, EB 38-47, EB 61-72, EB 83-93 cloned fragments which induce EBV antigens in CNE cells following microinjection, (Glaser et al., 1983). (Reprinted with permission).

Four regions of the EBV genome were found to induce detectable EA: EB 26-36, EB 38-47, EB 61-72, and EB 83-93 (Figure 1). Earlier studies have identified the EB 26-36 region as coding for the EA-R antigen and EB 61-72 as coding for the EA-D antigen based on reactivity with monoclonal antibodies (Glaser et al., 1983). The EB 38-47 and EB 83-93 regions are newly identified regions coding for the EA complex based on reactivity with EA⁺VCA⁺ sera and lack of reactivity with EA⁻VCA⁺ sera. Identification of the EA-D and/ or EA-R specificty of these two regions is in progress.

Sera designated EA⁻VCA⁺ were used to differentiate VCA from EA. Preliminary data suggest that the region EB 45-54 codes for part of the VCA complex (Figure 2). Expression of the EBV nuclear antigen (EBNA) following injection of the BamHI (K) fragment was reported previously (Glaser et al., 1983) in agreement with findings reported by Summers et al. (1982).

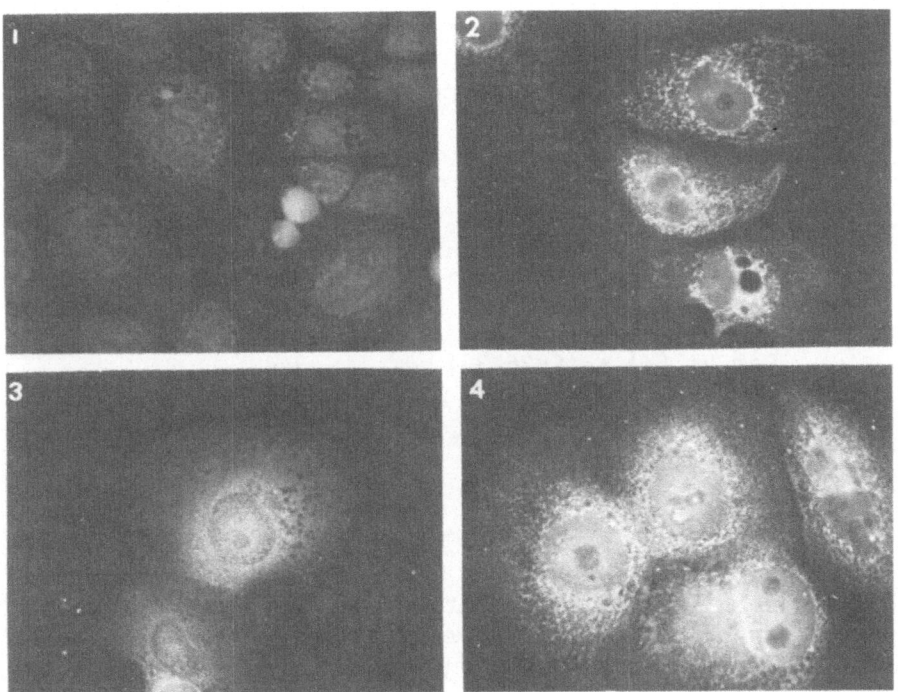

Figure 2. Immunofluorescence photomicrographs of CNE
cells microinjected with (1) PBS control; (2) EB 38-47 (EA);
(3) EB 83-93 (EA); (4) EB 45-54 (VCA).

Transformation of B-lymphocytes

Human B-lymphocytes from cord blood were enriched and
prepared for microinjection as described in Materials and
Methods. The various DNA fragments used to microinject B-
lymphocytes are summarized in Table 2. Cells injected with
pBR322 and several EBV fragments were dead within 2-3 months.
In contrast, B-cells injected with the combination of Bam HI
fragments (HFX and K) were still viable in culture for one
year. Although viable, the cells did not replicate and an
established or tranformed cell line was never obtained.
Representative live cells were stained by IF. The only
EBNA-positive cells were those injected with a mixture of
the Bam HFX and K fragments (Table 2). There were no remaining viable cells left in these cultures by the end of
the 15th month. This experiment has been repeated with the
same results.

TABLE 2

MICROINJECTION OF B-LYMPHOCYTES

DNA	Survival In Culture (months)	EBNA
Bam HI (HFX)	2	-
Bam HI (HFX) + (K)	12	+
Bam HI (K)	2	-
Bam HI (K) + (EB26-36)	3-4	-
(EB26-36) + (EB53-61)	3-4	-
(EB26-36)	2	-
(EB53-61)	2	-
B95-8	2	-
pBR322	2	-

Human Cord B-lymphocytes were prepared and injected as described in Materials and Methods. Viable cells were observed microscopically. Representative cells were stained for EBNA by anti-complementary immunofluorescence (Reedman and Klein, 1973).

DISCUSSION

It is possible to map functional regions of the EBV genome using the precise method of microinjection. In this study, the functional map of EBV relative to antigen expression has been expanded. A combination of precharacterized human sera were used to detect EA and/or VCA antigens in human CNE cells after microinjection. The recently developed monoclonal antibodies to EA-D, EA-R, and VCA (Pearson et al., 1983; Kishishita et al., 1984) have made it possible to identify proteins with EA and/or VCA antigenic determinants. The new EA regions identified in this study are the EB 38-47 and EB 83-93 regions which are part of the EA complex. EB 38-47 overlaps the Bam HI M fragment mapped as EA-D specific by Pearson et al. (1983).

Earlier studies by Graessmann et al. (1980) and ourselves (Glaser et al., 1983) were not successful in detecting VCA. The data here (Table 1) include the first report of VCA reactivity by the EB-45-54 fragment. This DNA fragment codes for several late polypeptides identified by Hummel and Kieff (1982). One monoclonal antibody to VCA antigen precipitates a 125 Kb protein from the EB 45-54 region (Pearson et al., 1983). Since there are several proteins synthesized in the early and late classes and since the early antigen is a complex of polypeptides, it is probable that the VCA antigen may be a complex of proteins coded from more than one region of the genome. It is possible that some of the late proteins are not detected by this experimental approach because early proteins of EBV may have regulatory control over late proteins. To test this, regions coding for EA are being co-microinjected with regions coding for late transcripts (Hummel and Kieff, 1982). Further, restriction enzyme cleavage may inactivate some genes within the promoter, or structural DNA sequence(s). Therefore, EBV framents are being coinjected with the pSV2neo gene (Southern and Berg, 1980) in an effort to enhance expression of the EBV fragment.

Transformation of human B-lymphocytes by intact EBV virus has been described (Miller et al., 1971). The Bam HI K fragment has been shown to code for the EBNA antigen by transfection (Summers et al., 1982) and by microinjection (Glaser et al., 1983). EBNA has been associated with EBV transformation (Klein et al., 1980). Preliminary efforts to transform human cord B-lymphocytes by microinjection of subgenomic fragments of EBV DNA suggest that the Bam HI fragments HFX and K may be cooperatively involved in at least the initiation of transformation. However, it is not yet clear if these regions of the genome are able to confer transformation leading to established cells. Transformation by EBV may follow a multi-step process which is yet to be determined. However, the use of microinjection provides an important tool for studying the functional map of the EBV genome, antigen expression, and the regions required to initiate and maintain transformation.

ACKNOWLEDGMENTS

This work was supported by Grants CA-29066 and CA-36357 from the National Cancer Institute, NIH, and by the Ohio State University Comprehensive Cancer Center Core Grant CA-16058, and in part by Public Health Service Contract NOI-CO-23910 with Program Resources Inc. from NCI.

REFERENCES

Ben-Bassat, H., Mitrani-Rosenbaum, S., and Goldblum, N.,
    Induction of Epstein-Barr virus nuclear antigen and DNA
    synthesis in a human epithelial cell line after Epstein-
    Barr virus infection. J. Virol. 41, 703-708, (1982).

Blattner, F.R., Williams, B.G., Blechl, A.E., Denniston-
    Kiefer, D.O., Moore, D.D., Schumm, J. W., Sheldon, E.L.,
    and Smithies, O., Safer derivatives of bacteriophage
    lambda for DNA cloning. Science 196,161-169, (1977).

Buell, G.N., Reisman, D., Dintner, C., Crouse, G., and
    Sugden, B., Cloning overlapping DNA fragments from
    the B95-8 strain of Epstein-Barr virus reveals a site
    of homology to the internal repetition. J. Virol 40,
    977-982, (1981).

Epstein, M.A., Achong, B.G., and Barr, Y.M., Virus particles
    in cultured lymphoblasts from Burkitt's lymphoma. Lancet
    1,702-703, (1964).

Glaser, R., De-The, G., Lenoir, G., and Ho, J.H.C.,
    Superinfection of epithelial nasopharyngeal carcinoma
    cells with Epstein-Barr virus. Proc. Natl. Acad. Sci. USA
    73, 960-963, (1976).

Glaser, R., Lang, C.M., Lee, K.J., Schuller, D.E., Jacobs,
    D., and McQuattie, C., Attempt to infect normal
    nasopharyngeal epithelial cells with the Epstein-Barr
    virus. J. Nat. Cancer Inst. 64, 1085-1090, (1980).

Glaser, R., Boyd, A., Stoerker, J., and Holliday, J.,
    Functional mapping of the Epstein-Barr virus genome:
    Identification of sites coding for the restricted early
    antigen, the diffuse early antigen and the nuclear
    antigen. Virol. 129,188-198, (1983).

Graessmann, A., Wolf, H., and Bornkamm, G.W., Expression
    of Epstein-Barr virus genes in different cell types
    after microinjection of viral DNA. Proc. Natl. Acad. Sci.
    USA 77,433-436, (1980).

Henle, G., Henle, W., and Klein, G. Demonstration of two
    distinct components on the early antigen complex of
    Epstein-Barr virus-infected cells. Int. J. Cancer
    8:272-278, (1971).

Hummel, M., and Kieff, E., Mapping of polypeptides encoded by the Epstein-Barr virus genome in productive infection. Proc. Natl. Acad. Sci. USA 79, 5698-5702, (1982).

Kishishita, M., Luka, J., Vroman, B., Poduslo, J., and Pearson, G. R., Production of monoclonal antibody to a late intracellular Epstein-Barr virus-induced antigen. Virol. 133, 363-375, (1984).

Klein, G., Luka, J., and Zeuthen, J., Transformation induced by Epstein-Barr virus and the role of the nuclear antigen. Cold Spring Harbor Symp. Quant. Biol. 44, 253-261, (1980).

Laboratory of Tumor Viruses of Cancer Institute and Other Institutions in China. Establishment of an epithelioid cell line and a fusiform cell line from a patient with nasopharyngeal carcinoma. Scientia Sinica 21,127-134, (1978).

Miller, G., Lisco, H., Kohn, H. I., and Stitt, D., Establishment of cell lines from normal adult blood leukocytes by exposure to Epstein-Barr virus and neutralization by human sera with Epstein-Barr virus antibody. Proc. Soc. Exp. Biol. Med. 137, 1459-1465, (1971).

Pearson, G.R., Vroman, B., Chase, B., Sculley, T., Hummel, M., and Kieff, E., Identification of polypeptide components of the Epstein-Barr virus early antigen complex using monoclonal antibodies. J. Virol. 47,193-201, (1983).

Pope, J.H., Transformation of the virus in vitro. In "The Epstein-Barr Virus" (M.A. Epstein and B.G. Achong, eds.) Springer-Verlag, Berlin/New York, (1979).

Reedman, B. and Klein, G., Cellular localization of an Epstein-Barr virus (EBV) associated complement-fixing antigen in producer and nonproducer lymphoblastoid cell lines. Int. J. Cancer 11,499-520, (1973).

Sixbey, J. W., Vesterimen, E. H., Nedrud, J. G., Raab-Traub, N., Walton, L. A., and Pagano, J. S., Replication of Epstein-Barr virus in human epithelial cells infected in vitro. Nature 306,480-483, (1983).

Southern, P. J., and Berg, P., Transformation of mammalian cells to antiobiotic resistance with a bacterial gene under control of the SV40 early region promoter. J. Mol. Appl. Genet. 1, 327-341, (1982).

Stoerker, J., and Glaser, R., Rescue of transforming Epstein-Barr virus (EBV) from EBV genome positive epithelial hybrid cells transfected with subgenomic fragments of EBV DNA. Proc. Natl. Acad. Sci. USA 80,17261729, (1983).

Summers, W. P., Grogan, E.A., Shedd, D., Robert, M., Liu, C.R., and Miller, G., Stable expression in mouse cells of nuclear neoantigen after transfer of a 3.4 megadalton cloned fragment of Epstein-Barr virus DNA. Proc. Natl. Acad. Sci. USA. 79,6588-5692, (1982).

44

LOCALIZATION OF EPSTEIN-BARR VIRUS EARLY ANTIGEN (EA) BY ELECTRON MICROSCOPY

MASAMICHI KISHISHITA, YOHEI ITO, JANOS LUKA, AND

GARY R. PEARSON

Kyoto University, Japan, and Georgtown University,

Washington, D.C.

SUMMARY

The localization of polypeptides (50-52,85K) associated with the diffuse and restricted components of early antigen EA in Epstein-Barr virus (EB virus)- infected cells was demonstrated with immune electron microscopy (EM) using colloidal - gold staining. P3HR-1 cells were activated to express the viral antigens by treatment with 4 mM n-butyrate and 20 ng/ml TPA. After 32-38 hrs, the cells were fixed with 2% glutaraldehyde, dehydrated with alcohol and propylene oxide, and embeded in epon. Ultra-thin section were prepared and incubated first with the appropriate monoclonal antibody, then with Staphylococcal protein A - gold, and finally with uranyl acetate. EM examination revealed that the 50-52 K component of EA-D was localized both in nucleoplasm and cytoplasm of EA-positive cells, while the 85 K component of EA-R was located mainly in the cytoplasm, as indicated by specific colloidal-gold staining.

INTRODUCTION

To identify components of Epstein-Barr virus (EB virus) specific antigens and to purify and characterize these proteins, monoclonal antibodies have been developed against different EB virus proteins (Hoffman et al., 1980; Thorley-Lawson and Geilinger, 1980; Qualtiere et al., 1982; Strnad et al., 1982; Morgan et al., 1983; Pearson et al., 1983; Takada et al., 1983; Kishishita et al., 1984). In this study, using two monoclonal antibodies (Pearson et al., 1983) against proteins associated with different components of the EB virus early antigen complex, their location in infected cells was determined by immune electron microscopy.

MATERIALS AND METHODS

Cell line. The Epstein-Barr virus (EB virus) - producing P3HR-1 was cultured in RPMI-1640 medium supplement with 10% heat-inactivated fetal calf serum. The EB virus-specific antigens were induced with 3 mM n-butyrate and 20 ng/ml TPA at 37°C for 32-38 hrs incubation (Kishishita et al., 1984).

Monoclonal antibodies. The preparation of the two monoclonal antibodies used in this study, designated R3 which is directed against 50-52K polypeptides (EA-D) and R63 which reacts with an 85K polypeptide (EA-R), was described previously (Pearson et al., 1983).

Immunofluorescence assay (IF). EB virus - specific antigen(s) production was examined by immunofluorescence using acetone - fixed smears of activated cells with the monoclonal antibodies or with a standard NPC serum (EA; 1:1280, VCA; 1:2560).

Preparation of cells for immune electron microscopy. Each pellet of activated or non-activated cells (10×10^6 cells) was fixed with 2% glutaraldehyde in 0.1 M phosphate buffer for 2 hrs. The cell pellet was washed twice in PBS and fixed in osmic acid to measure the immunological binding sites for each antigen. The cell pellets were dehydrated with alcohol and with propylene oxide. After dehydration, the pellets were embedded in epon at 45°C and at 60°C for 16 hrs and 8 hrs, respectively.

Colloidal-gold staining. Ultra-thin sections on 300 mesh grid were preincubated with 1% BSA-PBS for 5 min and incubated, first with each monoclonal antibody diluted with 1% BSA-PBS, then with protein A-gold (E. Y Lab. Inc.,) diluted with 5% BSA-PBS at the concentration 1:5, and finally with uranyl acetate. Each incubation period was 30 minutes.

RESULTS AND DISCUSSION

After 32-38 hrs activation, 60% of P3HR-1 cells expressed EB virus-specific antigens by immunofluorescence assay staining with a NPC serum as opposed to less than 1% of the non-activated cells. As shown in Fig. 1, optimal staining was noted with 1:200 or 1:400 dilutions of each antibody. Dilutions of 1:50 and 1:100 produced non-specific binding. In case of colloidal-gold staining, optimal specific staining was observed with the 1:200 or 1:400 dilutions of each antibody but at the 1:50 and 1:100 dilution non-specific binding was observed both in the cells and outside of the cells. In non-activated cells colloidal-gold was not observed at 1:200 or 1:400 dilution of each antibody. As shown in Fig. 2, R3-antigens (50-52K) were localized both in the nucleoplasm and cytoplasm of EA-positive cells, while R63 antigen (85K) was found mainly in the cytoplasm (Fig. 3). No staining was detectable in nuclear and cytoplasmic membranes, or in viral particles with either antibody. Using the post-embedding method of staining with colloidal gold did not clearly reveal specific organelles reacting with these monoclonal antibodies. Some colloidal gold, however, reacted with chromatin of P3HR-1 cells in the R3 antibody stained preparations. Recently we demonstrated that 50-52 K antigens were DNA binding and that the molecular weight of the antigens shifted to 60K with time possibly due to phosphorylation. These results strongly suggest that 50-52K antigens are bound to DNA in vivo. Finally, we have obtained more successful results on EA and VCA localization by using immunoperoxidase staining, and these results will be submitted separately.

REFERENCES

Hoffman, G.F., Lazarowitz, S.G., and Hayward, S.D. Monoclonal antibody against a 250,000 glycoprotein of Epstein-Barr virus identifies a membrane antigen and a neutralizing antigen. Proc. Natl. Acad. Sci. USA 77, 2979-2983. (1980).

Kishishita, M., Luka, J., Vroman, B., Poduslo, J.F., and Pearson, G.R. Production of monoclonal antibody to a late intracellular Epstein-Barr virus-induced antigen. Virology 133, 363-375. (1984).

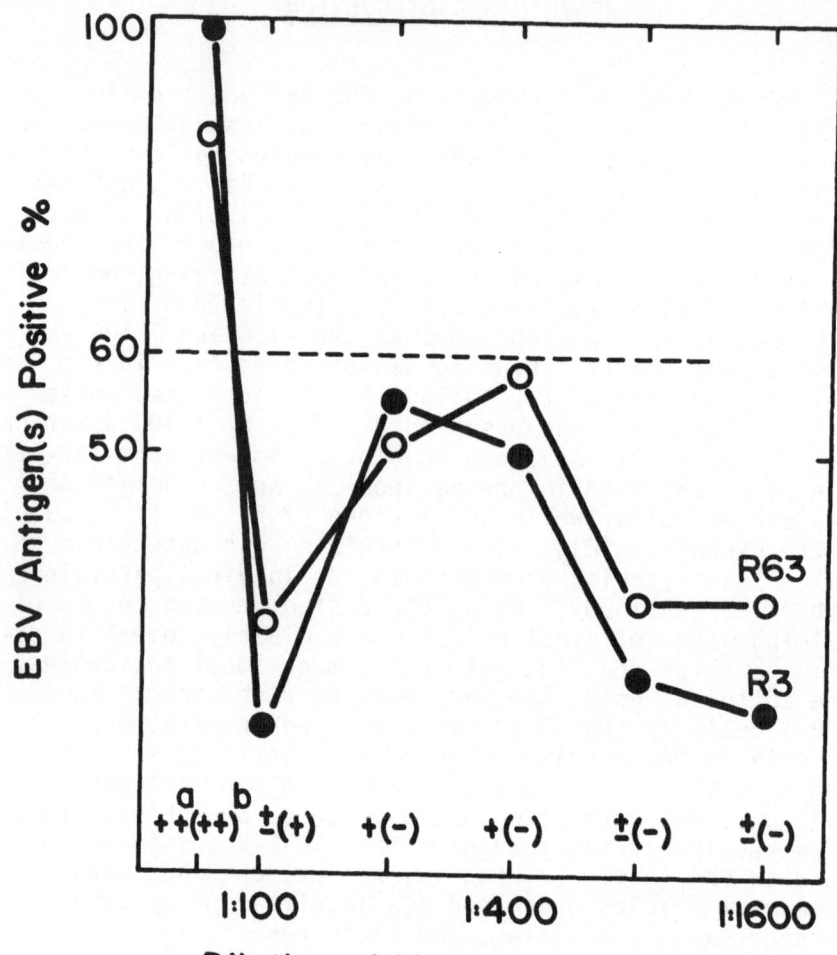

Fig. 1. Epstein-Barr virus - early antigens detected in
activated P3HR-1 cells using two monoclonal
antibodies. Binding of different dilutions of
the colloidal-gold conjugate to activated cells
(a: positive inside cell, b: positive outside
cell).

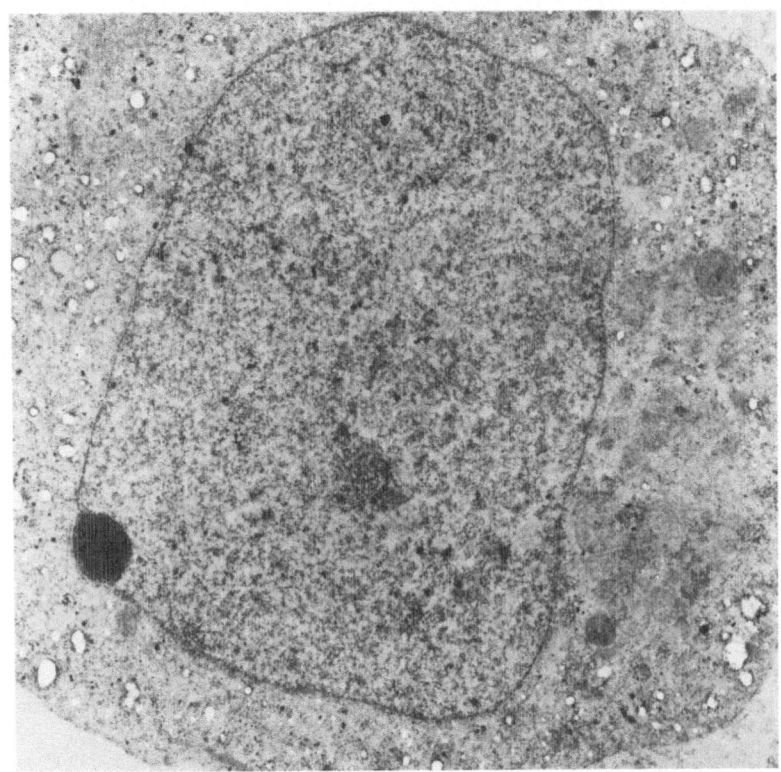

Fig. 2.   Ultrastructural localization of the 50-52 K component of EA-D in P3HR-1 cells by colloidal-gold staining.   X 12500.

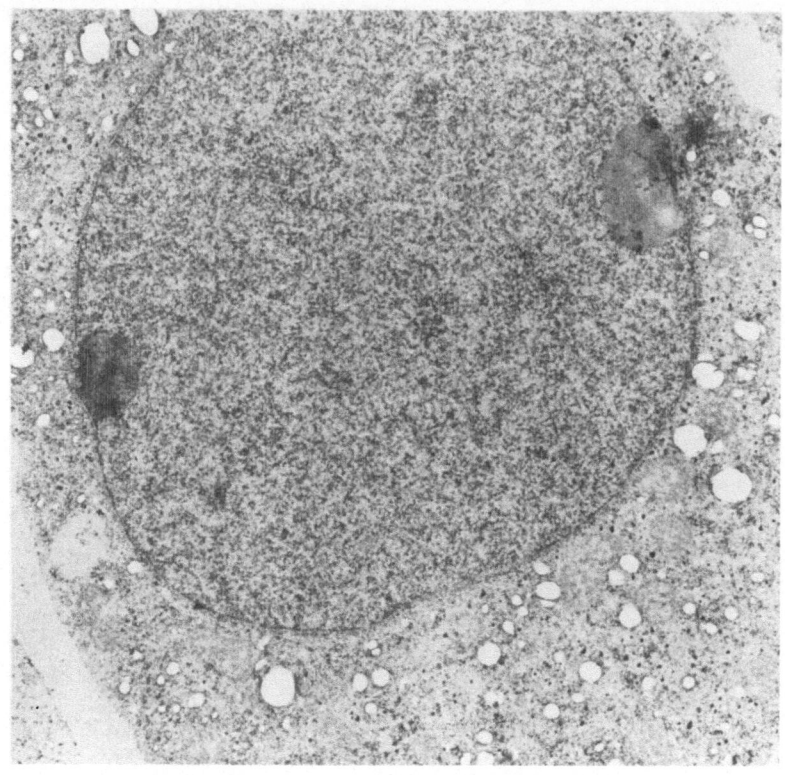

Fig. 3.   Ultrastructural localization of the 85K EA-R
component in P$_3$HR-1 cells by colloidal-gold
staining.   X 12500.

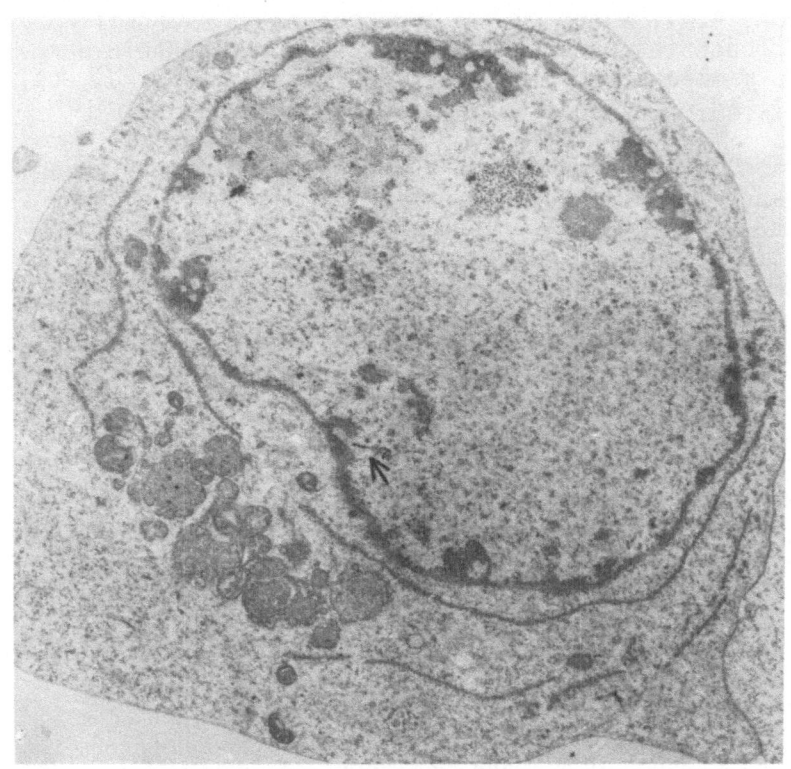

Fig. 4.   Ultrastructural localization of the 50-52 K
protein in osmic acid treated P$_3$HR-1 cells by
colloidal-gold staining.   Arrow shows colloidal-
gold reacting with chromatin.   X 10000.

Morgan, A.J., North, J.R. and Epstein, M.A. Purification and properties of the gp 340 component of Epstein-Barr virus membrane antigen in an immunogenic form. J. Gen. Virol. 64, 455-460. (1983).

Pearson, G.R., Vroman, B., Chase, B., Sculley, T., Hummel, M., and Kieff, E. Identification of polypeptide components of the Epstein-Barr virus early antigen complex using monoclonal antibodies. J. Virol. 47, 193-201. (1983).

Qualtiere, L.F., Chase, R., Vroman, B., and Pearson, G.R. Identification of Epstein-Barr virus strain differences with monoclonal antibody to a membrane glycoprotein. Proc. Natl. Acad. Sci. USA 79, 616-620. (1982).

Strnad, B.C., Schuster, T., Klein, R., Hopkins, R.F., III, Witmer, T., Neubauer, R.H., and Rabin, H. Production and characterization of monoclonal antibodies against the Epstein-Barr virus membrane antigen. J. Virol. 41, 258-264. (1982).

Takada, K., Fujiwara, S., Yano, S., and Osato, T. Mono-clonal antibody specific for capsid antigen of Epstein-Barr virus. Med. Microbiol. Immunol. 171, 225-231. (1983).

Thorley-Lawson, D.A., and Geilinger, K. Monoclonal anti-bodies against the major glycoprotein (gp 350/220) of Epstein-Barr virus neutralize infectivity. Proc. Natl. Acad. Sci. USA 77, 5307-5311. (1980).

45

Selection and production by genetechnological methods of medically relevant EBV-related antigens

Wolf, H., Motz, M., Kühbeck, R., Seibl, R., Bayliss, G.J., Modrow, S., and Fan, J.

Max von Pettenkofer Institute, Munich, FRG.

SUMMARY

Immunoprecipitation of Epstein-Barr virus specified proteins with various sera from normal adults, patients with fresh infectious mononucleosis or nasopharyngeal carcinoma can be used to identify those EBV proteins, antibodies against which are relevant for the characterization of the immune status and diagnosis of a particular disease. Some of these antigens have been localized on the Epstein-Barr virus genome by hybrid selected translation. With the use of sequence data, these genes can be subcloned from EBV-DNA and expressed in procaryotic cells. Data on the expression are presented and the application of the described methods for the production of diagnostic reagents and vaccines is discussed.

INTRODUCTION

The first suggestive evidence that Epstein-Barr Virus might be causally related to nasopharyngeal carcinoma and African Burkitt's Lymphoma was derived from serological data. (For review, see Epstein and Achong 1979). Using mainly indirect immunofluorescence on cells producing virus or at least early viral genes, significantly higher antibody titers to these antigens were found in patients sera. Although helpful for the establishment of a relationship between EBV and these diseases, these first tests (which detected unspecified immunoglobulin classes against a group of proteins named Early Antigen (EA) and

another group of proteins named Virus Capsid Antigens
(VCA)) were of limited value for definite diagnosis of the
malignancies from a single serum. Additionally they could
not be used for the control of therapy. Furthermore, the
preparation of antigens and the evaluation of tests was
not easy and restricted the diagnostic procedure to a
limited number of laboratories.
The introduction of antigen and antibody class specific
tests, specifically the determination of peripheral IgA
antibodies for the two antigen families EA and VCA and
also the first attempts to subdivide at least the
EA-family (EA D or R, Henle et al 1971), achieved
remarkable improvements of the diagnostic and prognostic
value of the tests. However, the test systems did not
allow automatic reading and thus did not favour mass
testing. Attempts to develop ELISA tests from the lysates
of antigen producing cells were made, however these tests
suffer from variable degrees of unspecificity due to
contaminating cellular materials that cannot be eliminated
using economic procedures. Monoclonal antibodies can be
used to prepare highly purified antigens and overcome the
problems of unspecificity and increase the diagnostic
resolution. Because EBV does not effectively replicate in
tissue culture this procedure still is very expensive.
Genetic engineering seems to offer a way out.

MATERIALS AND METHODS

The procedures for tissue culture, labelling of cells,
immune precipitation and hybrid selected translation are
detailed elsewhere (Bayliss and Wolf 1981, Bayliss et al
1983; Seibl and Wolf in press)

Cloning of viral DNA for expression
DNA of a plasmid containing the desired reading frame plus
additional viral sequences in pBR 322 (Skare) was digested
with restriction enzymes and separated on agarose gels.
The desired bands were electroeluted, purified through
Elutip columns (Schleicher & Schuell) and cloned with
standard procedures into the selected vectors: pUC8 and
pUC9, (Messing et al 1982) and pUR228 (Ruether et al
1983). Strains JM83 and BMH 71-18 of E. coli were
transfected with standard procedures and the clones with
inserts were tested in rapid lysis assays, using
appropriate restriction enzymes for the orientation of the
insert relative to the plasmid. Clones with correct

orientation and reading frame were grown, induced with IPTG (1mM) and incubated for another 1 1/2 hours. The bacterial proteins were separated on reducing SDS polyacrylamide gels.

RESULTS

Finding antigens important for diagnosis

Using immunoprecipitation, we have shown that EA and VCA are not one antigen, but that both consist of several polypeptides (Fig 1). We have used this technique to test sera for the presence of antibodies to specific proteins. The results are shown in Fig. 2. Whereas IgG antibodies very early after infection are not very dominant, we were able to show that antibodies to p150 and p143 are invariably present in healthy individuals after primary contact to EBV even after years and in NPC patients. Antibodies to p138 were regularly present with high titers in sera from NPC patients and only in reduced titers present in most of the other sera. For this reason, we labelled this antigen as relevant for primary screening for NPC.

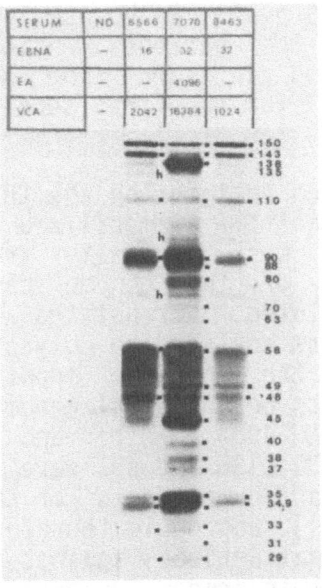

Figure 1:
Raji cells were superinfected with P3HR1 virus. Cells were labeled with ^{35}S-methionine from 12-16 hours post

infection and immuneprecipitated with sera from different
patients. The proteins were analysed on polyacrylamide
gels.

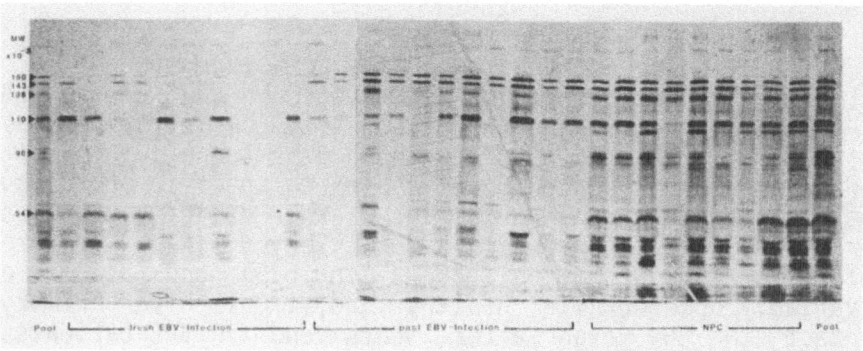

Figure 2:
Different patients sera were tested by immunoprecipitation
with superinfected Raji cells as in Fig. 1. Antibodies
against p 150 and p 143 can be used as indicators of the
immune status of the patients. Antibodies against p 138
are regularly found in sera from NPC patients and are
therefore suitable for the first serological screening of
sera.

Localizing antigens on the EBV genome
Using cloned viral DNA immobilized on nitrocellulose
filter mRNA which was originally transcribed from the
particular segment of the genome can be selected. This
mRNA can be eluted, translated in vitro and the
translation products analysed on polyacrylamide gels. This
procedure allowed the physical mapping of viral primary
translation products on the viral genome (Hummel and Kieff
1982; Cohen et al 1984; Seibl and Wolf in press). The
proteins p150, p143, p138 and p90 were found to correspond
to primary translation products of the same size (Seibl
and Wolf in press) and could therefore be mapped on the
genome. A monoclonal antibody against p150 confirmed this
finding (Jilg et al in prep). The protein p54 was shown to
be processed from a precurser of 47kd using a monoclonal
antibody and mapped to the BamH1 M fragment (Pearson et al
1983). p54 and p90 were identified to be a component of
EA-R and EA-D respectively by monoclonal antibodies

against these proteins (Pearson et al 1983). The unprocessed precursers of the glycoproteins 350kd/250kd were identified with a specific antiserum and could consequently be mapped on the genome (Edson and Thorley-Lawson 1983; Hummel et al 1984). EBNA and possibly LYDMA were also mapped on the genome (Kieff et al this issue). The EBV-specific DNA-polymerase was localized on the genome by sequence comparison with HSV (Baer et al 1984). p138 has been identified as an analog to an immediate early protein of HSV (JCP8, McGeoch, Seibl and Wolf unpublished) with similar approaches.

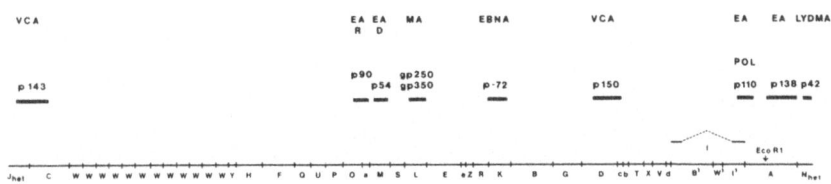

Figure 3:
Map location of important EBV-proteins on the EBV-genome relative to a Bam H 1 restriction map.

Expression of viral information in procaryotes
Initially, we attempted to express a large segment of the gene coding for p138 in several bacterial expression vectors. These attempts were partially successful, however the yield of p138 was small due to its large size and to its degradation by bacterial proteases (data not shown). Therefore, we decided to dissect the gene coding for this antigen. We assumed that there might be a few antigenic epitopes and that these could be identified in the products of the recombinant hosts. Fig. 4 shows computer graphics which give a suggested secondary structure of the protein on the basis of the algorithms derived by Chou and Fasman (1974). Superimposed on the suggested structure are the hydrophilic (dark circles) and hydrophobic (light circles) values of the respective area (average calculation of 5 neighbouring amino acids).

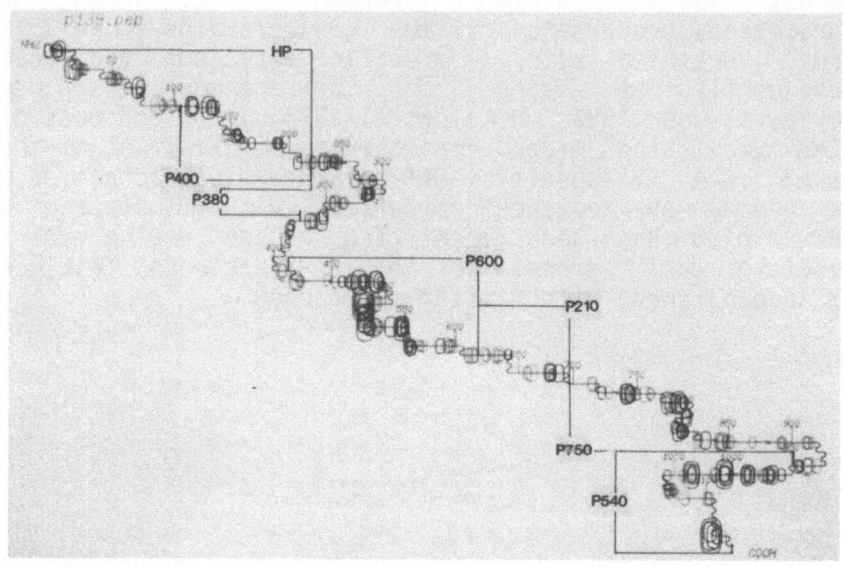

Figure 4:
Computer plot of a Chou-Fasman calculation of the p 138
secondary structure. Additionally, the hydrophobic (light
circles) and hydrophilic (dark circles) regions are
indicated.
Antigenic sites can be expected in hydrophilic regions
with a ß-turn. This situation is given in the p 600 region
and at the COOH-terminus of the protein.

Based on these graphics, amino acid sequences near amino
acid position 520 and at the COOH-terminus of p 138 should
be antigenic and therefore recognizable by NPC sera. To
test this hypothesis, we dissected the viral p 138 coding
region into small segments (200 to 750 bp) and cloned them
into the expression vector pUR288 following the
ß-galactosidase reading frame. The resulting products are
fusion proteins with the large ß-galactosidase (116 kD)
and the respective region of p 138 (fig. 4). These
proteins are stable due to the large bacterial protein
fused to them. Electrophoretic transfer of proteins from
bacterial lysates separated on SDS polyacrylamid gels onto
nitrocellulose membranes (western blot) allowed the
detection of the antigenic fragments of p 138 (fig. 5b).
Only two of them, P600 and P540, are antigenic. This is in
good accord with the prediction based on considerations of
the structural and hydrophobic properties of the primary

amino acid sequence.

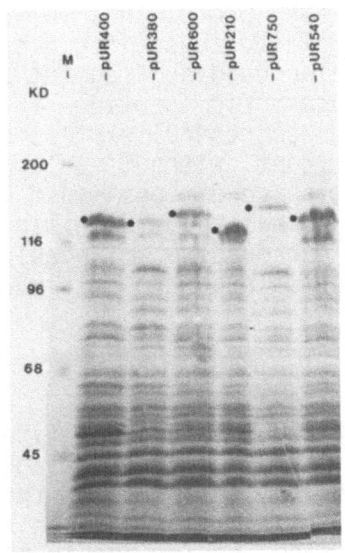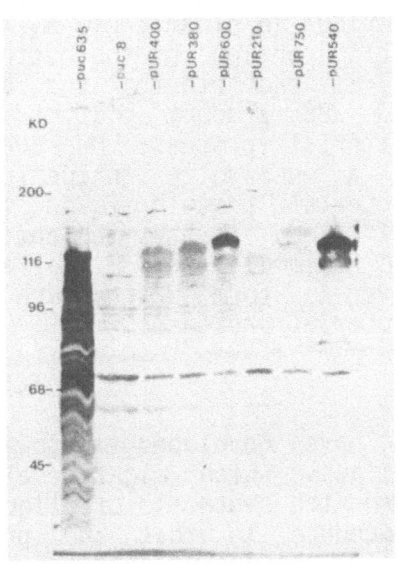

Figure 5:
a. A coomassie brilliant blue-stained SDS polyacrylamide
slab gel analysis of lysates of IPTG induced bacteria
carrying the various plasmids. Fusion proteins with
molecular weights between 120 and 150 kD are indicated
with a closed circle. Track M:molecular weight markers.
Tracks pUR400 -pUR540 lysates of bacteria carrying plasmids
containing the regions of p138 as shown in fig. 3.

b. An enzyme-linked immuno assay of proteins transferred
from a gel (similar to that shown in panel a) onto
nitrocellulose paper. After electrophoretic transfer of
the proteins (Western blot) and saturation of the blot
with BSA a pool of high titer antiserum was applied. After
washing the bound immunoglobulins were visualized by
sequential reaction with peroxidase coupled to antibodies
against human IgG and diaminobenzidine. Only fusion
proteins from bacteria containing pUR600 and pUR540 show
specific reactions. Plasmid pUC 635 contains almost the
entire p138 coding region fused to only 60 amino acids of
the B-galactosidase, however the protein is unstable and
is rapidly degraded. pUC 8 is the negative control for
this track containing the vector plasmid free from EBV

derived sequences.

The p600 but not the p540 region could be expressed after cloning in plasmid pUC 8, the resulting protein has only a small portion of the bacterial ß-galactosidase (fig 5 c). It is probable that the large ß-galactosidase fragment in pUR 288 product protects the eucaryotic peptide from bacterial protease degradation. For successful expression of a eucaryotic antigenic determinant without a large bacterial protein fused to it two conditions must be fulfilled. Firstly the antigenic site has to be determined and secondly this site and its surrounding amino acids have to form a structure which is resistant to attack by bacterial proteases.

DISCUSSION

We have developed a powerful strategy to produce viral antigens which might be almost as fast, and more reliable than the synthesis of oligopeptides. The advantage of this procedure is that the products are less vulnerable to rapid changes of antigenicity with minor variations in the length of the product. The same approaches and computer programs which are used to predict antigenic determinants for peptide synthesis allow us to select those clones which will probably yield antigenic products. Although the construction of the clones is restricted to specialized laboratories, the preparation of antigens from established recombinant bacteria should be very inexpensive and it should be possible to do this in developing countries. It can be expected that highly specific products allow cheaper, more widely usable and standardizeable diagnostic tests, which will also have an increased diagnostic value, especially in conjunction with antibody class specific test protocols. A similar approach to the development of clones expressing antigens suitable for use in vaccines should also be of value. This would involve the identification of regions of proteins with the known potential to induce neutralizing antibodies and their subsequent cloning in the vectors described above. Polypeptides containing not only oligopeptides as antigenic sites but including flanking sequences may have advantages over in vitro synthesized shorter oligopeptides which frequently elicit weak immune responses even when used in combination with strong adjuvants.

REFERENCES

Baer, R., Bankier, A. T., Biggin, M. D., Deininger, P. L., Farrell, P. J., Gibson, T. J, Hatfull, G., Hudson, G. S., Satchwell, S. C., Seguin, C., Tuffnell, P. S., and Barrell, B. G., DNA sequence and expression of the B95-8 Epstein-Barr virus genome. Nature 310,207-211 (1984).

Bayliss, G. J., and Wolf, H., The regulated expression of Epstein - Barr Virus. III. Proteins specified by EBV during the lytic cycle. J. Gen. Virol. 56,105 - 118 (1981).

Bayliss, G. J., Deby, G., and Wolf, H., An immunoprecipitation blocking assay for the analysis of EBV induced antigens. J. Virol. Methods 7,229-239 (1983).

Chou, P., and Fasman, G., Conformational parameters for amino acids in helical, ß-sheet and random coil regions calculated from proteins. Biochemistry 13,211-245 (1974).

Cohen, L. K., Speck, S. H., Roberts, B. E., and Strominge, J. L., Identification and mapping of polypeptides encoded by the P3HR-1 strain of Epstein-Barr virus. Proc. Natl. Acad. Sci. 81,4183-4187 (1984).

Edson, C. M., and Thorley-Lawson, D. A., Synthesis and Processing of the Three Major Envelope Glycoproteins of Epstein-Barr Virus. J. Vir. 46,547-556 (1983).

Epstein, M. A., and Achong, B. G., The Epstein - Barr Virus. Springer Verlag Berlin, Heidelberg, New York (1979).

Henle, G., Henle, W., and Klein, G., Demonstration of 2 distinct components in the early antigen complex of Epstein - Barr Virus infected cells. Int. J. Cancer 8,272-282 (1971).

Hummel, M., and Kieff, E., Mapping of polypeptides encoded by the Epstein-Barr virus genome in productive infection. Proc. Natl. Acad. Sci. 79,5698-5702 (1982).

Hummel, M., Thorley-Lawson, D. A., and Kieff, E., An Epstein-Barr Virus DNA Fragment Encodes Messages for the Two Major Envelope Glycoproteins (gp350/300 and gp 220/200). J. Virol. 49:413-417 (1984).

Messing, J. and Vieira, J. A new pair of M13 vectors
for selecting either DNA strand of double-digest
restriction fragments. Gene 19:269-276 (1982).

Pearson, G.R. Vroman, B. Chase, B., Sculley, T.,
Hummel, M. and Kieff, E. Identification of Polypep-
tide Components of the Epstein-Barr Virus Early
Antigen Complex with Monoclonal Antibodies. J.
Virol. 47:193-201 (1983).

Ruther, U. and Muller-Hill, B. Easy identification
of cDNA clones. EMBO Journal 2:1791-1794 (1983).

Seibl, R. and Wolf, H. Mapping of Epstein-Barr
virus proteins on the genome by translation of
hybrid-selected RNA from induced P3HR1 cells and
induced Raji cells. Virology (in press).

Skare, J. and Strominger, J. Cloning and Mapping of
Bam H1 endonuclease fragments of DNA from the trans-
forming B95-8 strain of Epstein-Barr Virus. Proc.
Natl. Acad. Sci. 77:3860-3864 (1980).

This work was supported by E.M.F.T. 01ZR102.

46

Identification of Multiple Epstein-Barr Virus Nuclear Antigens

T.B.Sculley, P.J.Walker, D.J.Moss, J.H.Pope

Queensland Institute of Medical Research

Herston, Brisbane, 4006. Australia.

SUMMARY

Using the protein immunoblot technique, the Epstein-Barr virus (EBV) nuclear antigen (EBNA) was identified in a variety of EBV genome-positive cell lines. The antigen portrayed a molecular weight of between 70,000 and 75,000 daltons depending on the cell line examined. When a similar immunoblot was performed, employing sera from patients with rheumatoid arthritis, a number of antigens in addition to EBNA were identified. The most prominent of these antigens exhibited molecular weights of 92,000 and 110-115,000 daltons. Unlike EBNA, the 92,000 dalton protein maintained the same molecular weight in each of the cell lines in which it was present. The 92,000 dalton protein was present in all of the EBV genome-positive cell lines except QIMR-GOR and lines carrying the P3HR-1 virus.

INTRODUCTION

During the course of a permissive infection of B-lymphocytes by Epstein-Barr virus (EBV), a number of viral-induced antigens can be detected. These antigens were originally defined by immunofluorescence procedures and have been termed viral capsid antigens (VCA),

membrane antigens (MA), early antigens (EA) and EBV
nuclear antigens (EBNA) (Thorley-Lawson et al., 1982).
However, in the case of a nonpermissive infection of
cells, resulting in transformation, only EBNA can be
directly observed. The presence of EBNA in cells early
after EBV infection and prior to cellular DNA synthesis,
as well as its association with the chromatin, suggest
that it may play an essential role in the transformation
event. Recent reports, however, have indicated that
additional EBV-induced antigens may be expressed in EBV-
transformed lymphocytes. Spelsberg et al. (1982)
presented evidence for the existence of two distinct EBNA
components in NC_{37} cells while Strnad et al. (1981) noted
the presence of an additional 81,000 dalton antigen in
B95-8 and Raji cells. Hennessy and Kieff (1983) also
observed a similar antigen in a variety of EBV-
transformed cell lines. Using the immunoblot technique
and by employing sera from patients with rheumatoid
arthritis (RA) we have identified the presence of at
least three new, EBV-induced antigens in EBV-transformed
cell lines.

METHODS

Cell lines and growth conditions, as well as a
description of the sera used in this study, are described
by Sculley et al. (1984 b).

Immunoblot

The procedure used was essentially as described by
Burnette (1981), with modifications. The SDS gel was
incubated in transfer buffer (20mM Tris base, 150mM
glycine, 20% methanol) for 20 min. A sheet of Whatman
3mm paper was placed under the gel and the nitrocellulose
sheet on top of the gel while it was still under
solution, to prevent air bubbles being trapped next to
the gel. The sandwich' was removed from the buffer and
excess liquid caught between the nitrocellulose and gel
eased off (if left, the liquid can cause streaking of
protein during transfer). A pre-wetted sheet of Whatman
3mm paper was placed on top of the nitrocellulose and the
whole thing placed between two Scotch Brite pads, then
into a commercial transfer apparatus. Electrophoretic

transfer was performed at 18V for 16h at 4^{0}C in transfer buffer containing 0.05% SDS. After transfer the nitrocellulose sheet was incubated in 3% BSA, PBS, 0.05% Tween-20 at RT for 90min, then in sera, diluted 1:20 in 3% BSA, PBS, Tween-20, for 60min at RT. The nitrocellulose was then washed four times for 5min each time in PBS-Tween-20, followed by a 60min incubation in iodinated protein A (1.0 x 10^{5}cpm per ml)at RT, then a final wash, four times for 5min each, in PBS-Tween-20. The nitrocellulose was then blotted dry between paper towels and exposed to X-ray film overnight at -70^{0}C.

All of the solutions contained sodium azide and could be used repeatedly if stored at 4^{0}C. The protein A solution could be reused, as long as fresh protein A was added each time.

RESULTS

Detection of EBNA by Immunoblot

SDS extracts were prepared from each of the cell lines shown in Table 1. Proteins in these extracts were electrophoresed on 10% polyacrylamide-SDS gels, then transferred to nitrocellulose papers and the papers incubated with either anti-EBNA positive or anti-EBNA negative sera obtained from clinically normal individuals. The result obtained with an anti-EBNA positive serum is shown in Figure 1, and is representative of all the anti-EBNA positive sera used. One major antigen was detected in each of the EBV genome-positive cell lines. The antigens were not detected in EBV genome-negative lines, nor were they identified by any EBV seronegative sera. That these antigens were only detected with sera containing antibodies to EBNA, as well as their molecular weights and size variation in different cell lines, are consistent with previous reports (Hennessy and Kieff, 1983; Sculley et al., 1983; 1984a; Strnad et al., 1981) indicating that they represent EBNA.

The molecular weight of EBNA was dependent upon the strain of infecting virus, with the antigen in Raji cells portraying a molecular weight of 70,000 daltons, and cell lines containing either P3HR-1 or QIMR-WIL virus

Table 1.

Cell line*	Infecting virus
1. B95-8	Hawley
2. JT	QIMR-WIL
3. Raji	Endogenous
4. JA	QIMR-WIL
5. BJAB	-
6. BK B95-8	B95-8
7. BK CRUK	IM virus
8. BJAB B95-8	B95-8
9. QIMR-GOR	Endogenous
10. QIMR-WIL	Endogenous
11. P3HR-1	Endogenous
12. AW-Ramos	P3HR-1
13. EHRB-Ramos	P3HR-1
14. Ramos	-

* Cell lines are listed in the order in which they appear in Figures 1 and 2.

expressing 73,000 dalton proteins. CRUK and QIMR-GOR viruses caused expression of 71,000 dalton EBNAs while the B95-8 strain of EBV induced a 75,000 dalton EBNA.

Detection of EBV-induced Antigens with Rheumatoid Arthritis Sera

An identical transfer to the one shown in Figure 1 was incubated with sera from individuals diagnosed as having rheumatoid arthritis (Figure 2). The rheumatoid sera reacted with a number of antigens in addition to EBNA, the most prominent of these antigens having molecular weights of 92,000, 110,000 and 115,000 daltons. None of these additional antigens were present in the two EBV genome-negative cell lines, BJAB or Ramos. Nor were they detected with EBV seronegative sera, regardless of whether they were from rheumatoid patients or controls, indicating that the antigens were EBV-specific. Unlike EBNA, the 92,000 dalton antigen exhibited a consistent molecular weight in each of the cell lines in which it was present. Likewise the 110-115,000 dalton antigens

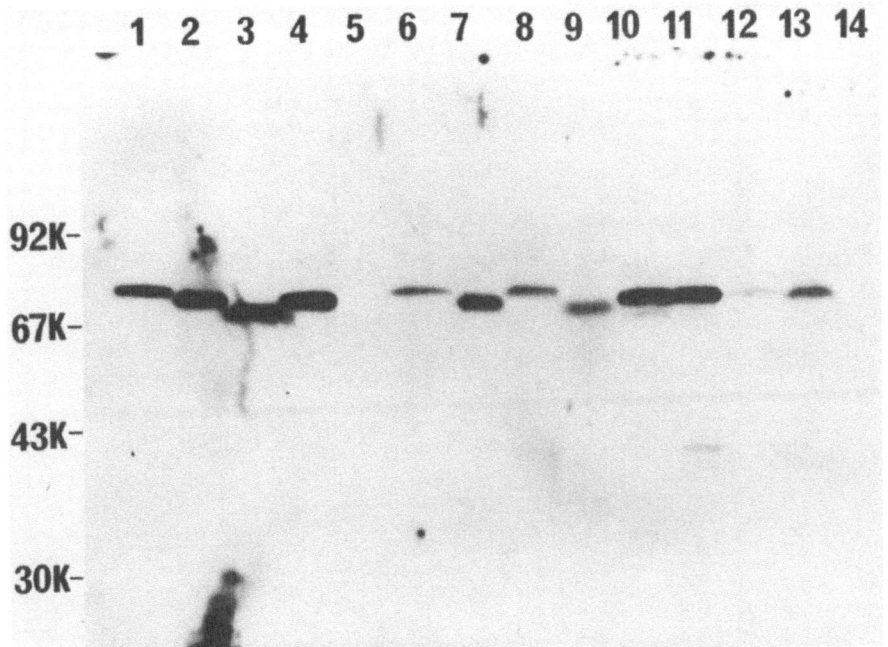

Figure 1. Detection of EBNA in different cell lines by protein immunoblot with sera from clinically normal individuals. Each of the cell lines, as well as their order, are listed in Table 1. (Sculley et al., 1984, reprinted with permission).

appeared to maintain consistent molecular weights in each of the cell lines in which the 92,000 dalton protein was present. The 92,000 dalton protein was absent from QIMR-GOR and cell lines containing the P3HR-1 strain of EBV, and the 110-115,000 dalton antigens were either undetectable in these lines or their molecular weights were altered.

A large number of low molecular weight antigens were also detected by the rheumatoid sera, particularly in QIMR-WIL, QIMR-GOR and P3HR-1 cell lines. As these lines are virus producers and the sera used contained anti-EA and anti-VCA antibodies, these proteins probably represent EA and VCA components. Some, apparently virus strain-specific, antigens were also noted in cell lines containing either QIMR-WIL or CRUK viruses.

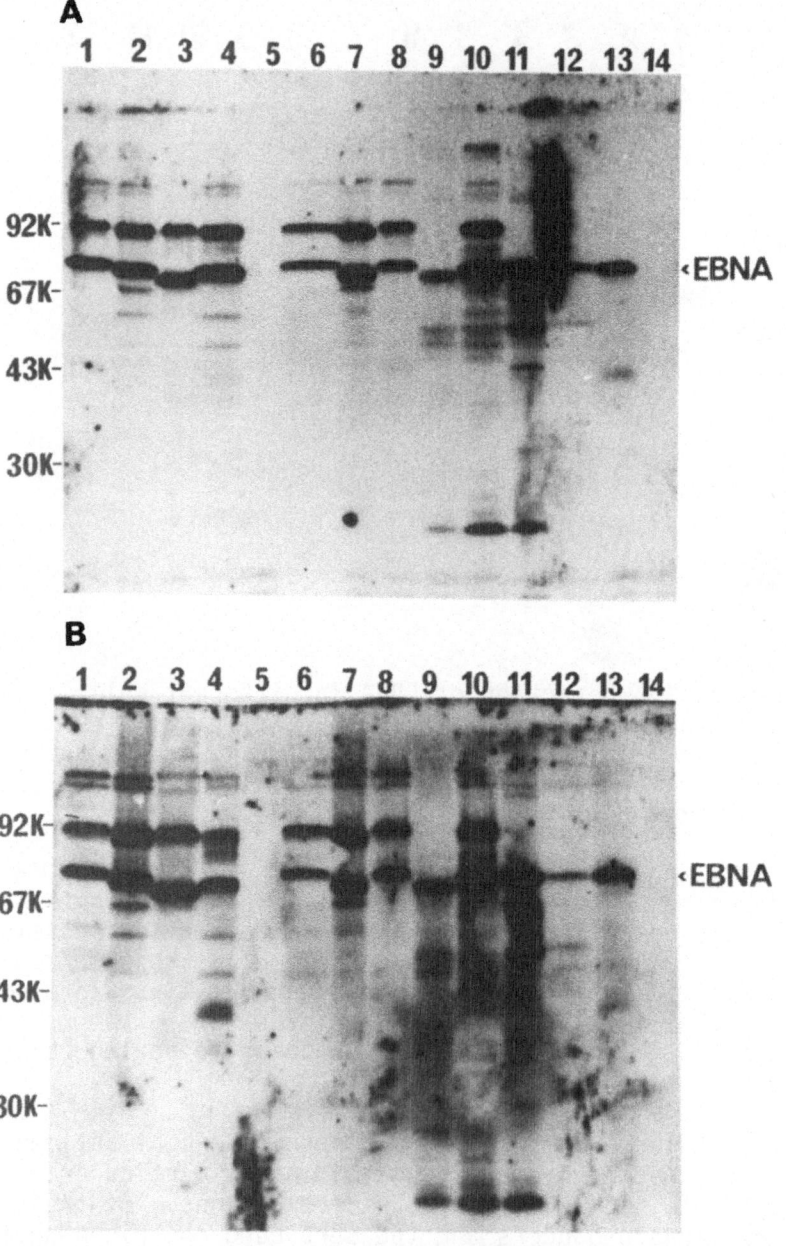

Figure 2. Detection of EBV-induced antigens in different cell lines by protein immunoblot with sera from patients with rheumatoid arthritis (Sculley et al., 1984, reprinted

with permission).

DISCUSSION

During initial attempts to identify additional EBV-induced antigens in EBV genome-positive cell lines, a number of inconsistencies were encountered. Sera would often react with the 92,000 dalton antigen in one immunoblot but not in subsequent immunoblots. These inconsistencies were overcome by modifying the procedure to include 0.05% SDS in the transfer buffer and to block the nitrocellulose sheets with BSA, then incubate with sera, immediately the transfer was completed. The SDS presumably facilitates transfer of the antigen onto the nitrocellulose paper, while immediate blocking and incubation with sera were found necessary when it was discovered that the 92,000 dalton antigen could not be detected on nitrocellulose left in PBS overnight. Similarly, incubation of the nitrocellulose with BSA, antibody and protein A at room temperature rather than 37°C increased reaction with the 92,000 dalton antigen. Apparently the 92,000 dalton antigen does not bind well to the nitrocellulose and prolonged incubation in PBS is sufficient to remove the antigen; this loss is probably enhanced if the PBS contains detergent. Similarly, incubation of the nitrocellulose at elevated temperatures could increase the rate at which the antigen is removed.

The 92,000 and 110-115,000 dalton antigens were detected with 15 of 21 sera from patients with rheumatoid arthritis but only 1 of 8 sera from clinically normal individuals. Reaction with these antigens showed no correlation with the presence of anti-EBNA, anti-VCA or anti-EA antibodies in sera. Further evidence that the 92,000 dalton antigen was not related to VCA or EA components stems from the fact that the 92,000 dalton antigen was not present in the QIMR-GOR or P3HR-1 cell lines, both of which produce VCA and EA.

The 92,000 dalton antigen was present in all of the EBV genome-positive cell lines except QIMR-GOR and all lines containing the P3HR-1 virus. The absence of the antigen in lines carrying the nontransforming P3HR-1 virus suggests that it may be associated with the tranformation process. The QIMR-GOR cell line contained a transforming strain of EBV at one stage (Pope et al., 1969), but recent attempts to transform B-lymphocytes with the virus have failed, suggesting that the virus may

have lost transforming ability (unpublished results). If
expression of the 92,000 dalton protein by EBV is
eventually found to be required to either initiate or
maintain the transformed state, then the present results
suggest that other virus strains similar to P3HR-1 may
exist. The present technique of assaying EBV measures
the transforming ability of the virus, and any strains
that may infect lymphocytes but remain in a latent state
would not be detected.

Antigens similar to the 92,000 dalton protein have
been identified by Strnad et al. (1981) and Hennessy and
Kieff (1983). These antigens, though having reported
molecular weights of 81,000 and 82,000 daltons
respectively, show characteristics analagous to the
92,000 dalton protein. The antigens were found to have a
consistent molecular weight in each of the cell lines in
which they were present, and were only identified by
select sera. The differences in molecular weight between
these antigens and the 92,000 dalton protein may result
from variations in sample preparation, electrophoresis
conditions or molecular weights standards used.

ACKNOWLEDGEMENTS

This work was supported by grants from the
Queensland Cancer Fund and the Australian Arthritis and
Rheumatism Foundation.

REFERENCES

Burnette, W.N. "Western Blotting": electrophoretic
transfer of proteins from SDS-polyacrylamide gels to
unmodified nitrocellulose and radiographic detection with
antibody and radioiodinated protein A. Anal. Biochem.
112, 195-203 (1981).

Hennessy, K. and Kieff, E. One of two Epstein-Barr virus
nuclear antigens contains a glycine alanine copolymer
domain. Proc. Natl. Acad. Sci. U.S.A. 80, 5665-5669
(1983).

Pope, J.H., Horne, M.K. and Scott, W. Identification of

the filtrable leukocyte-transforming factor of QIMR-WIL cells as a herpes-like virus. Int. J. Cancer 4, 255-260 (1969).

Sculley, T.B., Kreofsky, T., Pearson, G.R. and Spelsberg, T.C. Partial purification of the Epstein-Barr virus nuclear antigen(s). J. Biol. Chem. 258, 3974-3982 (1983).

Sculley, T.B., Spelsberg, T.C. and Pearson, G.R. Characterisation of Epstein-Barr virus nuclear antigen(s) in different cell lines by radioimmunoelectrophoresis. Intervirol. (In press) (1984a).

Sculley, T.B., Walker, P.J., Moss, D.J. and Pope, J.H. Identification of multiple Epstein-Barr virus-induced nuclear antigens with sera from patients with rheumatoid arthritis. J. Virol. (In press) (1984b).

Spelsberg, T.C., Sculley, T.B., Pikler, G.M., Gilbert, J.A. and Pearson, G.R. Evidence for two classes of chromatin-associated Epstein-Barr virus-determined nuclear antigen. J. Virol. 43, 555-565 (1982).

Strnad, B.C., Schuster, T.C., Hopkins III, R.F., Neubauer, R.H. and Rabin, H. Identification of an Epstein-Barr virus nuclear antigen by fluoroimmunoelectrophoresis and radioimmuno-electrophoresis. J. Virol. 38, 996-1004 (1981).

Thorley-Lawson, D.A., Edson, C.M. and Geilinger, K. Epstein-Barr virus antigens - a challenge to modern biochemistry. Adv. Cancer Res. 36, 295-348 (1982).

47

BRIEF COMMUNICATION

PRODUCTION OF HUMAN MONOCLONAL ANTIBODIES BY EBV

IMMORTALIZATION

PAIRE, J. and DESGRANGES, C.

Laboratory of Epidemiology and Immunovirology
of Tumors - Faculty of Medicine Alexis Carrel
rue Guillaume Paradin - 69372 LYON CEDEX 2

Many attempts to produce human monoclonal antibodies
have been made since the early mouse hybridoma work of
Kohler and Milstein (1975). Human myeloma cell lines have
been established in culture only rarely, Olsson and Kaplan
(1980) and Croce et al (1980) were the first to get stable
human hybridomas secreting human monoclonal antibodies with
defined specificity. Yet, it has been known for years that
infection of human B cells in vitro with EBV yields lymphoblastoid cell lines producing immunoglobulins (Rosen et al,
1977). Steinitz and co-workers (1977) were the first to
apply this technique to obtain human antibodies of pre-
determined specificity. We report here the establishment
of different cell lines resulting from EBV transformation
that produce monoclonal Ig: anti-Plasmodium falciparum,
anti-herpes simplex virus, anti-Rhesus D and anti-thyro-
globulin.

MATERIALS AND METHODS

Leukocyte donors: Lymphocytes were obtained mostly from
naturally hyperimmune donors (in particular for Plasmodium
falciparum, herpes simplex virus and thyroglobulin).
Nevertheless, in one case, for the antibody against Rhesus
D antigen, the donor was voluntarily immunized and reimmunized with D antigen.

Separation of mononuclear cells: Peripheral blood mononuclear cells were separated on a Ficoll-Hypaque gradient
and washed in Hank's solution.

Enrichment of producing B cells: Some lymphocytes are
directly infected with EBV, but when it is possible,
lymphocytes producing antibodies were enriched by
rosetting. B cells from reimmunized Rhesus D donors, at a
concentration of 10^7 c/ml, were mixed with equal volumes of
papain-treated D red blood cells. The rosetted cells were
recovered before infection. The thyroglobulin antibody-
producing B cells were rosetted with erythrocytes coated
with purified thyroglobulin by the chromid chloride method.

EB virus infection: Cells were pelleted and resuspended in
the supernatant culture medium of B95-8 cell line.

Immortalization: EBV-infected cells were cultured in RPMI
1640 medium containing penicillin, streptomycin (100
IU/ml), glutamine (2 mm/1) and 20% of fetal calf serum.
They were directly subcloned in 384 plates (Greiner),
treated or untreated with Cyclosporin A (0.1 µg/ml).

RESULTS

Plasmodium falciparum: With hyperimmune subjects for
Plasmodium falciparum, but without any disease, we obtained
5 different cell lines from 5 patients: B 38, B 39, B 4,
B 11 and B 18. The antibodies produced by these cell lines
were detected by immunofluorescence on human erythrocytes
infected by Plasmodium falciparum from malaria patients.
The different monoclonal antibodies recognized the trophozoite stage, but showed different patterns of immunofluorescence. B 38 and B 39 gave a particular trophozoite
staining with large patches. B 4 and B 11 reacted strongly
with the ring form of the parasite and B 18 produced a
distinct patching staining pattern at the parasitized red
blood cell (RBC) level. No anti-RBC reactivity was seen on
normal RBC. An immunoprecipitation with ^{35}S-labeled
parasitized human RBC and electrophoresis on SDS polyacrylamide gel showed that B4 recognized a protein with a
molecular weight of 66 K, B 11 one with 240 K m.w., B 18
one with 115 K m.w. and B 39 one with 96 K m.w. For B 38,
we could not identify a protein. All these molecular
weights correspond to know proteins of Plasmodium

falciparum. B 38 and B 39 are IgM λ while B4, B 11 and
B 18 are IgG λ and κ. They all produce 1 to 5 µg/ml of Ig.

Herpes simplex I and II: With hyperimmune subjects for
herpes simplex, we obtained three different cell lines from
three patients: RIP, ABD, Ly 131. Antibodies produced by
these cell lines were tested by ELISA test and by IF. Proteins that they recognized were identified by immunoprecipitation monoclonal antibodies ^{35}S-labeled Hep cell
lines infected with HSV-I or HSV-II and electrophoresis on
SDS polyacrylamide gel. RIP recognized a glycoprotein with
a molecular weight of 53 K, named gD, ABD the same glycoprotein gD and weakly a glycoprotein with a molecular
weight of 110 K, named gB. Ly 131 identified the gB. RIP
and ABD secreted one IgG_1 κ with a production of 15 to 20
µg/ml and Ly 131 one IgG_1 λ with a production of 4 to 5
µg/ml.

Rhesus D: Lymphocytes were obtained from a Rhesus-negative
blood donor who had been immunized and reimmunized with D
antigen. We obtained two different cell lines from the
same donor: CO 8.8 and CO 7.12. These two clones secreted
IgG_1 λ. CO 8.8 produced 15 to 20 µg/ml of Ig and CO 7.12
5 to 10 µg/ml. By immunoprecipitation of ^{125}I-membrane-
labeled erythrocytes and by electrophoresis on SDS polyacrylamide gel, we saw that only the clone CO 7.12 recognized a protein with a molecular weight of 29 K.

Thyroglobulin: Lymphocytes were obtained from patients
with autoimmune diseases (Basedow disease). One cell line
was obtained--BA 10.16. It was tested by a competitive
radioimmunoassay (RIA) with ^{125}I-labeled thyroglobulin.
This clone secreted IgG κ with a production of 5 µg/ml.

DISCUSSION

Human monoclonal antibodies have advantages over the
conventional murine antibodies. For gammaglobulin therapy,
many patients develop antibody response to the mouse Ig
which prevents effective treatment. For parasitic
diseases, such as malaria, in which vaccination is considered as primordial, human monoclonal antibodies against
Plasmodium falciparum may provide a more direct identification than murine monoclonals of different derived epitopes important for immunity and can be used in some cases

for passive protection (Monjour et al, 1983; Lungden et al, 1983). HSV-I and HSV-II monoclonal antibodies may be used for identification of the different glycoproteins of the viruses and in the future for immunization of immuno-deficient patients with suitable neutralizing antibodies (Seigneurin et al, 1983). For Rhesus D, the monoclonal CO 8.8 is now produced in large volume in fermenters with synthetic media and this monoclonal is currently used as red blood cell typing reagent. The antibody reacted strongly with all D positive cells and even with Du cells. One clone recognized a protein band with a molecular weight of 29 K and will permit a clearer understanding of the molecular structure of the Rhesus D. This monoclonal anti-body is being considered for in vivo use in man to prevent Rhesus disease of the newborn as previously described (Crawford et al, 1983). For the monoclonal antibody against thyroglobulin, it will be possible to label this monoclonal and use it for immunoscintigraphy, as a probe in the investigation of autoimmune thyroid disease. This work is now in progress. The usefulness of the EBV transforma-tion technique for the establishment of human thyroid lines secreting monoclonal antibodies, at the same level of production as obtained with murine hybridomas, is now established. It is also possible to produce these mono-clonals in large volume with fermenters and at low cost with synthetic media, to get stable and well defined immunologic reagents.

REFERENCES

Crawford, D.H., Barlow, M.J., Harrison, J.F., Wingler, L. and Huehns, E.R. Production of human monoclonal antibody to Rhesus D antigen. Lancet, i:386-388 (1983).

Croce, C.M., Linnenbach, A., Hall, W., Steplewski, Z. and Koprowski, H. Production of human hybridomas secreting antibodies to measles virus. Nature (London), 288:488-489 (1980).

Kohler, G. and Milstein, C. Continuous cultures of fused cells secreting antibodies at predefined antigenic specificity. Nature (London), 257:495-497 (1975).

Lungden, K., Wahlgren, M., Troye-Blomberg, M., Berzins, K., Perlmann, H. and Perlman, P. Monoclonal anti-parasite and anti-RBC antibodies produced by stable EBV-transformed cell lines from malaria patients. J. Immunol. 131:2000-2003 (1983).

Monjour, L., Desgranges, C., Ploton, I., Alfred, C. and Karabinis, A. Production of human monoclonal antibodies against asexual erythrocytic stages of Plasmodium falciparum. Lancet, 8337:1337-1338 (1983).

Olsson, L. and Kaplan, H.S. Human-human hybridomas producing monoclonal antibodies of predefined antigenic specificity. Proc. Natl. Acad. Sci. (USA), 77:5429-5431 (1980).

Rosen, A., Gergely, P., Jondal, M., Klein, G. and Britton, S. Polyclonal Ig production after Epstein-Barr virus infection of human lymphocytes in vitro. Nature (London), 267:52-54 (1977).

Seigneurin, J.M., Desgranges, C., Seigneurin, D., Paire, J., Renversez, J.C., Jacquemont, B. and Micouin, C. Herpes simplex virus glycoprotein D: Human monoclonal antibody produced by bone-marrow cell line. Science, 221:173-175 (1983).

Steinitz, M., Klein, G., Koskimies, S. and Makela, O. EB virus induced B lymphocyte cell lines producing specific antibody. Nature (London), 269:420-422 (1977).

IMMUNOLOGY

48

CELLULAR IMMUNITY IN EBV INFECTIONS

J.H.POPE

Queensland Institute of Medical Research,

Brisbane, Australia.

INTRODUCTION

Effective study of cellular immunity to EBV had to wait for developments in basic cellular immunology. Once a high level of definition was achieved in this area, particularly in T cell markers, the recognition of the NK system, and the vital role of lymphokines, the stage was set for progress. The present paper considers several aspects of cellular immunity to EBV, concentrating on ADCC and the role of NK cells, while others will consider EBV-specific T cell immunity.

ANTIBODY-DEPENDENT CELLULAR CYTOTOXICITY (ADCC)

Antibody with specificity to surface markers on target cells, though not cytolytic in its own right, may cause lysis in the presence of leucocytes. Since EBV-specific ADCC was first reported (Pearson and Orr, 1976; Jondal,1976), it has been explored with regard to the viral antigens involved and to the clinical significance of the reaction.

The target cells most commonly used in ADCC assays have been LCL or BL lines superinfected with EBV (Pearson and Orr, 1976). Induction of the viral replication cycle was required and the P3HR-1 virus was commonly used. The B95-8 virus was found effective by some (Patel and Menezes,

1982) but not others (Patarroyo et al., 1980). A strong
association of an ADCC reaction with the presence in human
sera of antibody to MA was reported (Pearson and Orr,1976;
Takaki et al.,1980), and only targets expressing MA were
susceptible (Aya et al., 1980). Analysis suggested that of
the several known forms of MA the late form was important
in ADCC (Takaki et al.,1980). Induction of the viral cycle
in EBV genome-positive BL or LCL by butyrate also conferred
similar sensitivity to ADCC (Patarroyo et al.,1980).
Differences in cells transformed by different strains of
EBV have been documented (Katsuki and Hinuma, 1975), and
Sairenji et al.(1982) noted differences in susceptibility
to ADCC. Fine antigenic variations also may be
strain-related (Edson and Thorley-Lawson, 1981;
Mueller-Lantzsch and zur Hausen, 1981; Franklin et
al.,1981).

The MA complex has been intensively studied. Qualtiere
and Pearson (1979) and North et al.(1980) used human sera
to precipitate surface-labeled proteins of super-infected
Raji cells, and detected four major EBV-specific surface
glycoproteins and one non-glycosylated protein.
Thorley-Lawson and Edson (1979) found that an EBV
hyperimmune rabbit antiserum precipitated three major
polypeptides. Hoffman et al.(1980) showed that a 250K
glycoprotein was a MA constituent and a determinant
involved in virus neutralization. MA appears to be a
complex of gp350/300, gp250/200, p140, and gp85, and the
availability of the DNA sequence of the B95-8 strain will
contribute to definitive studies (Baer et al., 1984). The
EBV-related determinant of specific T cell lysis, referred
to as LYDMA (Jondal, 1976), appears distinct from MA
detected serologically.

As ADCC reactions involve targets, effector cells and
antibody, there are multiple variants in each experiment.
Some have approached this by using a single sample of human
serum or by analysing the results of replicate experiments.
A major analysis was attempted by Takasugi et al.(1982)
allowing for targets, effectors and sera. Analyses showed
selective lysis of targets by ADCC and NK effectors, and
variations in the efficiency of different effectors as well
as the expected serum variation in ADCC titre.

Sera from normal sero-positive persons had ADCC
activity (Jondal, 1976), and others have titrated antibody

reacting in ADCC, in sera of patients with various diseases. This will be considered later.

NATURAL KILLER CELLS (NK) AND ACTIVATED T CELLS

It is possible to refer only to aspects of NK cell biology immediately relevant to EBV. The nonspecific lysis shown by NK cells made their study complex. Earlier problems in this field are being reduced by the identification of NK cells as the set of large granular lymphocytes, by the availability of monoclonal antibodies to surface markers, by the recognition of the activation of NK cells by interferon (IFN), and by the progress made towards an understanding of the mechanisms involved in NK cell killing.

An interesting aspect of NK cells in relation to EBV is the increased sensitivity of cell lines super-infected with EBV. Blazar et al.(1983) showed that active cellular metabolism, producing new surface determinants, was required for EBV-infected Raji cells to develop maximum sensitivity to NK cell lysis. The evidence of Patarroya et al.(1982) indicates that these molecules are probably distinct from the classical serologically-defined EBV antigens, and the determinants involved in NK and ADCC lysis were distinct. Recent work suggests that the increased sensitivity of EBV-infected Raji cells was mediated through IFN production during the cytotoxicity test (Blazar et al., 1984).

In short-term cytotoxicity tests, target cell lines show a range of sensitivity to lysis by NK cells, from the sensitive K-562 and HSB-2 through to the resistant LCL and some BL lines. Masucci et al.(1983) found that separated large granular cells (major NK cell population) inhibited outgrowth of the autologous LCL, and concluded that this early effect complements lysis by EBV-specific cytotoxic T cells. Specific T cells effectively eradicate autologous LCL in vitro, while in cultures from seronegative donors NK cells do not.

An exciting discovery is that NK cell killing involves an extracellular cytotoxic factor mediating lysis (Wright et al., 1983). Increased understanding of the biology of NK cells may eventually allow their manipulation in vivo for disease control. Enhancement of NK activity by IFN or by staphylococcus enterotoxin A (Kimber et al.,1983) also point in this direction.

The definition of natural or spontaneous killer cells on the basis of lysis of appropriate targets by effector cells freshly obtained from the peripheral blood may not be as satisfactory as one based on specific cell markers. An important question arises concerning the nature of effector cells generated in culture over a week or so, and E. Klein has strongly advocated the use of the term activated T cells. Masucci et al. (1980) concluded that the cells appearing in culture in response to a variety of stimuli (cells, fetal calf serum (FCS) or mitogen) were related to a blast response and were distinct from classical NK cells in terms of target cell specificity. IFN activates NK cells and Patarroyo et al. (1983) found that brief IFN treatment of lymphocytes enhanced their lysis of allogeneic LCL as well as of a standard NK-sensitive target (Molt-4). The non-restricted suppression of LCL outgrowth recorded by Schooley et al.(1981) also may have involved activated T cells. As activating factors such as IFN operate during stimulation of lymphocytes in vitro, it will be vital to specifically identify the activated T cells by markers, and use clonal analysis for a comparison with NK cells.

OTHER ASPECTS OF CELLULAR IMMUNITY

The value of studies of EBV-specific delayed hypersensitivity may have suffered from the unavailability of pure antigens. The macrophage migration-inhibition test has specificity and allows demonstration of EBV-related antigens. Recent work suggests the detection of a new membrane antigen by this approach (Szigeti et al., 1984).

It is appropriate to briefly mention here the autologous mixed lymphocyte reaction (AMLR). This characteristically involves a proliferative response of T cells to autologous non-T cells, and there is controversy as to whether this response is actually due to foreign antigens in the system. Avoidance of FCS and sheep erythrocytes reduced the strength of the AMLR (Moody et al.,1983), but it is extremely difficult to totally avoid xenoantigens in in vitro systems. The term AMLR has commonly been extended to include stimulation of lymphocytes by the autologous LCL and it has been reported that EBV antigens were not involved in this reaction (Weksler,1976). However, a recent study has emphasized that the response to stimulation by the autologous LCL depends on the EBV serological status of the donor, and therefore may be viewed as an EBV-specific response rather than as a form of AMLR (Misko et al.,1984).

Yet another aspect whose full significance may not yet have been realized concerns inhibition of outgrowth of EBV-infected B cells by various sub-populations of T cells. Several studies have demonstrated such inhibition which may be mediated by NK cells (Shope and Kaplan, 1979) or IFN (Thorley-Lawson,1980; 1981). Another reversible inhibitory effect totally prevented outgrowth of LCL when cells from seropositive adults were infected (Moss et al.,1976). Although these facets of cellular immunity may not be capable of a major role in combatting EBV infection, they may significantly delay the primary infection until more effective mechanisms come into play. In contrast to these results, v.Knebel Doeberitz et al.(1983) found that the presence of T cells enhanced the outgrowth of spontaneous LCL.

THE ROLE OF CELLULAR IMMUNITY IN THE CONTROL OF EBV ACTIVITY

Approximately half of the susceptible individuals contracting primary infection with EBV develop infectious mononucleosis. The factors determining the severity of infection are unknown, and will be difficult to define because one of the important variables is the infectious dose received by each person. Individuals with subclinical or severe forms each develop apparently similar persistent infections, and this indicates that although the former group probably mounts a more effective short-term response this is still not capable of totally eradicating the virus. Several possible explanations of persistence may be considered. One is that infection is maintained in a latent form, and this has been suggested by Epstein and Achong (1973). The idea of latency is perhaps supported by the demonstration of a block in transformation before proliferation (Moss et al.,1976). A second possibility is that a smouldering infection occurs, with enough cells undergoing viral replication to maintain infection. With herpesviruses, even high titre antibody is not capable of preventing cell-cell transmission of virus, and immune functions (such as ADCC) may not be capable of effectively destroying every cell induced into viral replication before virus release. A third possibility is that virus persists in proliferating transformed cells controlled by the immune defenses. There is considerable evidence suggesting that EBV-infected cells in peripheral blood (PB) do not behave in vitro in a manner expected of transformed cells (Rickinson et al.,1974), although there are probably significant differences in the biological behaviour of transformed cells in vitro and in vivo. Given the gaps in our knowledge of viral persistence in vivo, it is difficult to evaluate the role of cellular immunity in normal persons. However, specific diseases provide some basic insights.

It has been well documented that by many criteria cellular immunity in IM is drastically decreased (Mangi et al.,1974), and that suppressor T cells are prominent (Tosato et al., 1979). Although the level of proliferation of EBV-infected B cells reached in IM varies quite widely, unrestrained proliferation is extremely rare. Certainly, some aspects of EBV cellular immunity are defective, and Jondal (1976) found no ADCC activity in sera from patients in the acute phase. Although there were several early reports of lysis of EBV-infected targets by T cells or lymphocytes obtained in the acute phase (Svedmyr and Jondal, 1975: Royston et al., 1975), this was difficult to reproduce (Klein et al.,1981; Patel et al.,1982). A study of severe IM in a family suggested that NK cells might play a significant role in recovery (Fleisher et al., 1982). Furthermore, EBV-specific T cell immunity was markedly depressed in the acute and early convalescent phases of IM but then recovered (Rickinson et al., 1980). Future work on cellular immune functions in acute IM will have to take into account the recent important demonstration that the majority of PB T cells in acute IM are highly prone to death by apoptosis in vitro (Moss et al.,1984).

NPC has been subjected to the most detailed study with regard to ADCC, and important findings have emerged. Pearson et al.(1978) reported an association of high ADCC titres with a good response to treatment and survival for two years in African cases of NPC. A significant finding was the inverse relationship between the titres of ADCC activity and of IgA antibody to VCA. Subsequent analysis of

the ADCC reactivity with sera from NPC patients showed that
the activity resided in the IgG fraction while IgA was
inactive. However, IgA reacted with the major MA
components, explaining its capacity to block ADCC (Mathew
et al., 1981). Presumably, the IgA results in a lower
effective titre of ADCC activity, in keeping with the
observed association between low ADCC titres at diagnosis
and poor prognosis (Neel et al., 1983). IgA was also
associated with inhibition of a specific blast response.
ADCC thus complements the diagnostic usefulness of the
serological tests for antibody to VCA and EA, and for
EBV-specific IgA. Less is known of EBV antigen expression
on the surface of NPC cells, and perhaps this deficiency
should be redressed. Unless late MA antigens are indeed
present, the ADCC findings may suggest that a high level of
EBV replication in other sites enhances progression of the
tumours. Alternatively, it is conceivable that high ADCC
titres simply parallel the titres to some other
hypothetical tumour antigen in NPC.

In African BL cases ADCC activity was present in sera
and the titres correlated with the response to
chemotherapy, suggesting an anti-tumour role for ADCC
(Pearson et al., 1979). However, ADCC titres did not seem
to be related to the stage or extent of the disease. A
comprehensive study of Hodgkin's and non-Hodgkin's lymphoma
cases with high or low titres of antibody to EBV, revealed
the complexity and individuality of the immune status and
showed that EBV infection remained essentially under
control (Masucci et al., 1984).

Some of the most persuasive evidence concerning the
neoplastic potential of EBV-transformed cells and the vital
role of the immune system in their control, comes from
experience with transplant patients. Under
immunosuppression, EBNA-positive lymphomas have proved to
be relatively common (Hanto et al.,1981). It seems somewhat
paradoxical, in view of this, that EBV-induced tumours are
not more widely encountered in other patients who might be
considered compromised immunologically.

Abnormal humoral responses to EBV antigens in a
variety of diseases have long attracted attention.
Recently, Vilmer et al.(1984) reported some correlations
between EBV antibody titres and cellular immune functions
in Wiskott-Aldrich and Chediak-Higashi syndromes and ataxia

telangiectasia. In spite of the occasional absence of anti-EBNA antibody, deficient mixed lymphocyte response or low NK cell activity, none of the patients showed ill effects of the virus. Katz et al.(1984) found that the defective NK activity in C-H disease (characterized by binding to targets but reduced lysis) was augmented to normal levels during IM. Lymphocytes from patients with systemic lupus erythematosus were ineffective in ADCC tests (Aya et al., 1980). In contrast, in the X-linked lymphoproliferative syndrome in which EBV infection is unusually severe, Harada et al.(1982) found that both EBV-specific cytotoxic T cell activity and NK activity were lower than in the mothers and the control groups. In acquired immune deficiency syndrome the cellular immune functions are seriously impaired through T helper cell defects. A variety of viral and other infections are activated under these conditions, and although Burkitt's-like lymphomas have been reported (Ziegler et al., 1982), EBV-transformed cells do not generally show unrestrained proliferation. This is curious as infections with cytomegalovirus are of major importance and the difference may reflect fine differences in immunity to the two herpesviruses.

It is clear that in addition to EBV-specific T cell immunity, other arms of cellular immunity play a potentially important role in control of the infection, and evaluation of their contributions remains an important aim. Increased knowledge of EBV cellular immunity will allow better evaluation of vaccines when they become available.

REFERENCES

AYA, T., MIZUNO, F., and OSATO, T.,Immunologic cytotoxicity against autologous human lymphocytes transformed or infected by Epstein-Barr virus: role of antibody-dependent cellular cytotoxicity in healthy individuals. J. nat. Cancer Inst., 65, 265-271 (1980).
BAER, R., BANKIER, A.T., BIGGIN, M.D., DEININGER, P.L., FARRELL,P.J., GIBSON,T.J., HATFULL, G., HUDSON, G.S., SATCHWELL, S.D., SEGUIN, C., TUFFNELL, P.S., and BARRELL, B.G., DNA sequence and expression of the B95-8 Epstein-Barr virus genome. Nature, 310, 207-211 (1984).
EDSON, C.M., and THORLEY-LAWSON, D.A.Epstein-Barr virus membrane antigens: characterization, distribution, and strain differences. J. Virol., 39, 172-184 (1981).

BLAZAR, B.A., FITZGERALD, J., SUTTON, L., and STROME, M., Increased sensitivity to natural killing in Raji cells is due to effector recognition of molecules appearing on target cell membranes following EBV cycle induction. Clin. exp. Immunol., 54, 31–38 (1983).

BLAZAR, B.A., STROME, M., and SCHOOLEY, R., Interferon and natural killing of human lymphoma cell lines after induction of the Epstein–Barr viral cycle by superinfection. J. Immunol., 132, 816–820 (1984).

EPSTEIN, M.A., and ACHONG, B.G., Various forms of Epstein–Barr virus infection in man: established facts and a general concept. Lancet, 2, 836–839 (1973).

FLEISHER, G., STARR, S., KOVEN, N., KAMIYA, H., DOUGLAS, S.D., and HENLE, W., A non-X-linked syndrome with susceptibility to severe Epstein–Barr virus infections. J. Pediatrics, 100, 727–730 (1982).

FRANKLIN,S.M., NORTH, J.R., MORGAN, A.J., and EPSTEIN, M.A.,Antigenic differences between the membrane antigen polypeptides determined by different EB virus isolates. J. gen. Virol.,53, 371–376 (1981).

HANTO, D.W., FRIZZERA,G., PURTILO,D.T., SAKAMOTO, K., SULLIVAN, J.L., SAEMUNDSEN, A.K., KLEIN,G., SIMMONS, R.L., and NAJARIAN,J.S., Clinical spectrum of lymphoproliferative disorders in renal transplant recipients and evidence for the role of Epstein–Barr virus. Cancer Res., 41, 4253–4261 (1981).

HARADA, S., BECHTOLD, T., SEELEY, J.K., and PURTILO, D.T., Cell-mediated immunity to Epstein–Barr virus (EBV) and natural killer (NK)-cell activity in the X-linked lymphoproliferative syndrome. Int. J. Cancer, 30, 739–744 (1982).

HOFFMAN, G.J., LAZAROWITZ, S.G., and HAYWARD, S.D., Monoclonal antibody against a 250,000–dalton glycoprotein of Epstein–Barr virus identifies a membrane antigen and a neutralizing antigen. Proc. nat. Acad. Sci. USA, 77, 2979–2983 (1980).

JONDAL, M., Antibody-dependent cellular cytotoxicity (ADCC) against Epstein–Barr virus-determined membrane antigens. I. Reactivity in sera from normal persons and from patients with acute infectious mononucleosis. Clin. exp. Immunol., 35, 1–5 (1976).

KATSUKI,T., and HINUMA, Y., Characteristics of cell lines derived from human leukocytes transformed by different strains of Epstein–Barr virus. Int. J. Cancer, 15, 203–210 (1975).

KATZ, P., ZAYTOUN, A.M., LEE, J.H., Jnr., and FAUCI, A.S.,

In vivo Epstein-Barr virus-induced augmentation of natural killer cell activity in the Chediak-Higashi syndrome. J. Immunol., 132, 571-573 (1984).

KIMBER, I., BAKAS, T., and MOORE, M., Regulation of natural and antibody-dependent cellular cytotoxicity by staphylococcal enterotoxin A. Clin. exp. Immunol., 54, 39-48 (1983).

KLEIN,E., ERNBERG, I., MASUCCI, M.G., SZIGETI, R., WU,Y.T., MASUCCI, G., and SVEDMYR, E., T-cell response to B-cells and Epstein-Barr virus antigens in infectious mononucleosis. Cancer Res., 41, 4210-4215 (1981).

von KNEBEL DOEBERITZ,M., BORNKAMM,G.W., and zur HAUSEN, H., Establishment of spontaneously outgrowing lymphoblastoid cell lines with Cyclosporin A. Med. Microbiol. Immunol., 172, 172-199 (1983).

MANGI, R.J., NIEDERMAN, J.C., KELLEHER, J.E., DWYER,J.M., EVANS, A.S., and CANTOR, F.S., Supression of cell-mediated immunity during acute infectious mononucleosis. N. E. J. Med., 291, 1149-1153 (1974).

MASUCCI, M.G., BEJARANO,M.T.,MASUCCI,G., and KLEIN, E, Large granular lymphocytes inhibit the in vitro growth of autologous Epstein-Barr virus-infected B cells. Cellular Immunol., 76, 311-321 (1983).

MASUCCI, M.G., KLEIN, E., and ARGOV,S., Disappearance of the NK effect after explantation of lymphocytes and generation of similar nonspecific cytotoxicity correlated to the level of blastogenesis in activated cultures. J. Immunol., 124, 2458-2463 (1980).

MASUCCI, G., MELLSTEDT, H., MASUCCI,M.G., SZIGETI, R., ERNBERG, I., BJORKHOLM,M., TSUKUDA, K., HENLE, G., HENLE, W., PEARSON,G., HOLM,G., BIBERFIELD, P., JOHANSSON, B., and KLEIN, G. Immunological characterization of Hodgkin's and non-Hodgkin's lymphoma patients with high antibody titres against Epstein-Barr virus-associated antigens. Cancer Res., 44, 1288-1300 (1984).

MATHEW,G.D., QUALTIERE, L.F., NEEL,H.B.III, and PEARSON, G., IgA antibody, antibody-dependent cellular cytotoxicity and prognosis in patients with nasopharyngeal carcinoma. Int. J. Cancer, 27, 175-180 (1981).

MISKO, I.S., SOSZYNSKI,T.D., KANE, R.G., and POPE, J.H., Factors influencing the human cytotoxic T cell response to autologous lymphoblastoid cell lines in vitro. Clin. Immunol. Immunopath., 32 (in press).

MOODY, C.E., GUPTA, S., and WEKSLER, M.E., Lymphocyte transformation induced by autologous cells. XV. Xenoantigens are not required for the proliferative response observed in the autologous mixed lymphocyte reaction. J. clin. Immunol., 3, 100–102 (1983).

MOSS,D.J., BISHOP,C.J., BURROWS, S.R., and RYAN, J.M. (1984). T lymphocytes in infectious mononucleosis I. T cell death in vitro. Clin. exp. Immunol. (in press).

MOSS,D.J., POPE,J.H., and SCOTT, W., Inhibition of EB virus transformation of non-adherent human lymphocytes by co-cultivation with adult fibroblasts. Med. Microbiol. Immunol., 162, 159–167 (1976).

MUELLER–LANTZSCH, N., GEORG–FRIES, B., HERBST, H., zur HAUSEN, H., and BRAUN, D.G., Epstein–Barr virus srain- and group-specific antigenic determinants detected by monoclonal antibodies. Int. J. Cancer, 28, 321–327 (1981).

NEEL,H.B., PEARSON,G.R., WEILAND,L.H., TAYLOR, W.F., GOEPFERT, H.H., PILCH,B.Z., GOODMAN, M., LANIER, A.P., HUANG, A.T., HYAMS, V.J., LEVINE, P.H., HENLE,G., and HENLE, W., Application of Epstein–Barr virus serology to the diagnosis and staging of North American patients with nasopharyngeal carcinoma. Otolaryngol. Head Neck Surg., 91, 255–262 (1983).

NORTH, J.R., MORGAN, A.J., and EPSTEIN, M.A., Observations on the EB virus envelope and virus-determined membrane antigen (MA) polypeptides. Int. J. Cancer, 26, 231–240 (1980).

PATARROYO, M., BLAZAR, B., PEARSON, G., KLEIN, E., and KLEIN, G., Induction of the EBV cycle in B-lymphocyte-derived lines is accompanied by increased natural killer (NK) sensitivity and the expression of EBV-related antigen(s) detected by the ADCC reaction. Int. J. Cancer, 26, 365–371 (1980).

PATARROYO, M., KLEIN,E., and KLEIN, G., The increased natural killer sensitivity of Epstein Barr virus (EBV)-superinfected Raji cells is not due to the recognition of serologically determined EBV antigens by the effectors. Cellular Immunol., 67, 152–159 (1982).

PATARROYO,M., KLEIN,E., and KLEIN,G., Lymphocyte-mediated lysis of autologous and allogeneic B-cell lines in man. Immunol. Lett., 6, 101–105 (1983).

PATEL, P.C., DORVAL,G., and MENEZES, J. Cytotoxic effector cells from infectious mononucleosis patients in the acute phase do not specifically kill Epstein–Barr virus genome-carrying lymphoid cell lines. Infect. and Immunity, 38,251–259 (1982).

PATEL, P.C., and MENEZES, J., Epstein–Barr (EBV)-lymphoid cell interactions. II. The influence of the EBV replication cycle on natural killing and antibody-dependent cellular cytotoxicity against EBV-infected cells. Clin. exp. Immunol., 48, 589–601 (1982).

PEARSON, G., JOHANSSON,B., and KLEIN, G., Antibody-dependent cellular cytotoxicity against Epstein-Barr virus-associated antigens in African patients with nasopharyngeal carcinoma. Int. J. Cancer, 22, 120–125 (1978).

PEARSON, G.R., and ORR, T.W., Antibody-dependent lymphocyte cytotoxicity against cells expressing Epstein-Barr virus antigens. J. nat. Cancer Inst., 56, 485–488 (1976).

PEARSON,G.R., QUALTIERE,L.F., KLEIN, G., NORIN,T., and BAL, I.S., Epstein-Barr virus-specific antibody-dependent cellular cytotoxicity in patients with Burkitt's lymphoma. Int. J. Cancer, 24, 402–406 (1979).

QUALTIERE, L.F., and PEARSON, G., Epstein-Barr virus-induced membrane antigens: immunochemical characterization of Triton X-100 solubilized viral membrane antigens from EBV-superinfected Raji cells. Int. J. Cancer, 23, 808–817 (1979).

RICKINSON, A.B., JARVIS, J.E., CRAWFORD, D,H., and EPSTEIN, M.A., Observations on the type of infection by Epstein-Barr virus in peripheral lymphoid cells of patients with infectious mononucleosis. Int. J. Cancer, 14, 704–715 (1974).

RICKINSON A.B., MOSS, D.J., POPE, J.H., and AHLBERG, N., Long-term T-cell-mediated immunity to Epstein-Barr virus in man. IV. Development of T-cell memory in convalescent infectious mononucleosis patients. Int. J Cancer, 25, 59–65 (1980).

ROYSTON, I., SULLIVAN,J.L., PERIMAN, P.O., and PERLIN,E., Cell-mediated immunity to Epstein-Barr-virus-transformed lymphoblastoid cells in acute infectious mononucleosis. N. E. J. Med., 293, 1159–1163 (1975).

SAIRENJI, T., JONES, W., SPIRO,R.C., REISERT, P.R., and HUMPHREYS, R.E., Epstein-Barr virus strain-specific differences in transformed cell lines demonstrated in growth characteristics, induction of viral antigens and ADCC susceptibility. Int. J. Cancer, 30, 393–401 (1982).

SCHOOLEY, R.T., HAYNES, B.F., GROUSE, J., PAYLING-WRIGHT, C., FAUCI, A.S., and DOLIN, R., Development of suppressor T lymphocytes for Epstein-Barr virus-induced B-lymphocyte outgrowth during acute infectious mononucleosis: Assessment by two quantitative systems. Blood, 57, 510–517 (1981).

SHOPE, T.C., and KAPLAN, J., Inhibition of the in vitro outgrowth of Epstein-Barr virus-infected lymphocytes by Tg lymphocytes. J. Immunol., 123, 2150–2155 (1979).

523

SVEDMYR, E., and JONDAL, M., Cytotoxic effector cells specific for B cell lines transformed by Epstein–Barr virus are present in patients with infectious mononucleosis. Proc. nat. Acad. Sci. USA, 72, 1622–1626 (1975).
SZIGETI, R., SULITZEANU,D., HENLE,G., HENLE,W., and KLEIN, G., Detection of an Epstein–Barr virus–associated membrane antigen in Epstein–Barr virus–transformed nonproducer cells by leukocyte migration inhibition and blocking antibody. Proc. Nat. Acad. Sci. USA, 81, 4178–4182 (1984).
TAKAKI, K., HARADA, M., SAIRENJI, T., and HINUMA,Y., Identification of target antigen for antibody–dependent cellular cytotoxicity on cells carrying Epstein–Barr virus genome. J. Immunol., 125, 2112–2117 (1980).
TAKASUGI, M., MICKEY, M.R., and LEVINE, P.,Natural and antibody–dependent cell–mediated cytotoxicity to cultured target cells superinfected with Epstein–Barr virus. Cancer Res., 42, 1208–1214 (1982).
THORLEY–LAWSON, D.A., The suppression of Epstein–Barr virus infection in vitro occurs after infection but before transformation of the cell. J. Immunol., 124, 745–751 (1980).
THORLEY–LAWSON, D.A., The transformation of adult but not newborn human lymphocytes by Epstein–Barr virus and phytohemagglutinin is inhibited by interferon: the early suppression by T cells of Epstein–Barr infection is mediated by interferon. J. Immunol., 126, 829–833 (1981).
THORLEY–LAWSON, D.A., and EDSON, C.M.,Polypeptides of the Epstein–Barr virus membrane antigen complex. J. Virol., 32, 458–467 (1979).
TOSATO, G., MAGRATH, I., KOSKI, I., DOOLEY, N., and BLAESE, M. Activation of suppressor T cells during Epstein–Barr virus–induced infectious mononucleosis. N. E. J. Med., 301, 1133–1137 (1979).
VILMER, E., LENOIR, G.M., VIRELIZIER, J.L., and GRISCELLI, C., Epstein–Barr serology in immunodeficiencies: an attempt to correlate with immune abnormalities in Wiskott–Aldrich and Chediak–Higashi syndromes and ataxia telangiectasia. Clin. exp. Immunol., 55, 249–256 (1984).
WEKSLER, M.E., Lymphocyte transformation induced by autologous cells III. Lymphoblast–induced lymphocyte stimulation does not correlate with EB viral antigen expression or immunity. J. Immunol., 116, 310–314 (1976).
WRIGHT, S.C., WEITZEN,M.L., KAHLE,R., GRANGER, G.A., and BONAVIDA, B., Studies on the mechanism of natural killer activity II. Coculture of human PBL with NK–sensitive or resistant cell lines stimulates release of natural killer

cytotoxic   factor   (NKCF)   selectively   cytotoxic   to
NK-sensitive   target   cells.   J.   Immunol.,   130, 2479-2483
(1983).
ZIEGLER,   J.L.,   DREW,   W.L.,   MINER,R.C.,   MINTZ,L.,
ROSENBAUM,E.,   GERSHOW,J.,   LENNETTE,   E.T.,   GREENSPAN,J.,
SHILLITOE, E., BECKSTED,J., CASAVANT, C., and YAMAMOTO, K.,
Outbreak of Burkitt's-like lymphoma in homosexual men.
Lancet, 2, 631-633 (1982).

49

T CELL RESPONSES TO EPSTEIN-BARR VIRUS INFECTION

A.B. Rickinson

Department of Cancer Studies, University of
Birmingham, Birmingham, U.K.

The Epstein-Barr (EB) virus is the best-known member
of a particular class of genetically restricted herpesviruses
which are found in several species of ape and of Old World
monkey (Deinhardt and Deinhardt, 1979) and which display a
unique tropism for host cells of the B lymphocyte lineage.
In each case, virus and host appear to have co-evolved such
that the natural infection, both in the primary and in the
persistent phase, remains largely asymptomatic. This is a
remarkable testament to the efficiency of host control
mechanisms since these agents clearly have the potential to
induce uncontrolled proliferation in infected B cells, a
capacity which is manifest in vitro as the virus-induced
transformation of such cells into permanent virus genome-
positive lymphoblastoid cell lines (Pope, 1979; Rabin et al.,
1978). It seems that an analogous sequence of events can
occur in vivo but only in very rare circumstances, the
classic example being the human X-linked lymphoproliferative
syndrome, where primary EB virus infection of boys with a
genetically-determined deficiency of cellular immune
functions leads to a fatal B lymphoproliferative disease
(Purtilo et al., 1982).

Such clinical observations, allied to the disturbance
of the EB virus-host balance which is known to occur in
patients receiving immunosuppressive therapy for the prolongation of allografts (Strauch et al., 1974), strongly
suggest that infections with the B lymphotropic herpesviruses are under cell-mediated immune control. The very

efficiency of this control itself implies the existence of a multiplicity of surveillance mechanisms, and direct evidence for such multiplicity comes from the analysis of cellular responses to EB virus infection in man (see articles by Pope and by Menezes in this volume). The purpose of the present paper is to focus on those aspects of responsiveness which specifically involve the T cell system. This necessarily involves reference to the broad functional sub-division of T cells into:-

(1) helper/inducer T cells which are involved in the initiation and/or amplification of antibody and of effector T cell responses; such helper cells are antigen-specific and MHC class II antigen-restricted, recognising processed antigen on the surface of specialised presenting cells and mediating help by the release of soluble factors (inter-leukins).

(2) suppressor T cells which are involved in the down-regulation of antibody and of effector T cell responses; in many experimental and clinical situations, suppressor functions appear to be neither antigen-specific nor MHC antigen-restricted. The mechanisms of suppression are not understood, although the effect very often appears to be mediated at the level of the helper cell.

(3) cytotoxic T cells are antigenic-specific and predominantly MHC class I antigen-restricted, recognising only those cells whose membranes display the relevant target structure. Functional subdivision into helper and into suppressor/cytotoxic T cell subsets is generally, though not absolutely, reflected by differences in cell surface phenotype.

Primary EB virus infection and the T cell response

In most communities, primary infection occurs naturally during the first few years of life and is almost always sub-clinical. Very interestingly, but for reasons which are still not clear, a delayed primary infection (as happens increasingly in the Western world) leads in up to 50% of cases to the clinical symptoms of infectious mono-nucleosis (IM) (Henle and Henle, 1979). Primary infection occurs, as always, by the oral route and there is growing evidence to suggest that the primary site of virus repli-cation is not in B lymphocytes but in pharyngeal and/or salivary gland epithelium (Morgan et al., 1979; Sixbey et al., 1984), whence infectious virus is shed into the throat. The infection is generalised via non-productively infected B cells (Svedmyr et al., 1984) emanating from the lympho-epithelial site of virus replication in the pharynx and thus EB virus nuclear antigen (EBNA)-positive B cells, many activated to immunoglobulin synthesis, are detectable in the blood of acute IM patients (Klein, G. et al., 1976; Robinson et al., 1981).

Most important in the present context is the origin of the atypical mononuclear cells which appear in large numbers in the blood and in the tissues coincident with the onset of clinical symptoms (Svedmyr et al., 1984). Most of these cells display T cell markers (with a relative pre-dominance of cells with the suppressor/cytotoxic phenotype (Reinhertz et al., 1980)) indicating that the primary infection has induced an unusually vigorous T cell res-ponse. This reactive T cell population is complex, perhaps more than is currently appreciated, and contains at least three functional activities:-

(1) using leukocyte migration inhibition (LMI) as an assay for lymphokine production following the appropriate pre-sentation of antigen to immune T cells (Szigeti et al., 1984a), it can be shown that T cells specific for certain EB viral antigens associated with the productive infection (early antigen, EA; virus capsid antigen, VCA) are present in IM blood, whereas EBNA-specific T cells have not yet developed (Szigeti et al., 1982). These in vitro observa-tions seem most likely to reflect reactivities within the helper T cell compartment in vivo, and in this context it is interesting to note that the anti-viral antibody response in acute IM is itself preferentially directed towards "late" viral antigens (Henle et al., 1974).

(2) IM blood also contains a potent and broad-ranging suppressor T cell activity which can be demonstrated in several mitogen-driven helper T cell-dependent activation systems in vitro (Haynes et al., 1979; Tosato et al., 1979; Reinhertz et al., 1980). The capacity of these same suppressor cells to regulate the EB virus-induced (helper T cell-independent) activation of immunoglobulin synthesis in B cells in vitro remains an important but unresolved question (Tosato et al., 1979; Bird and Britton, 1979). Certainly conventional suppressor T cells from other sources are inactive in the EB virus system (Andersson et al., 1983),

reflecting the unique nature of the virus as an activation
signal.

(3) a range of cytotoxic reactivities are present in IM
blood of which the most interesting is that mediated by
certain Fcγ receptor-negative T cells with apparent
selectivity for EB virus genome-positive target cells in
in vitro chromium release assays (Svedmyr and Jondal,
1975; Royston et al., 1975). Despite much investigation,
the antigenic specificity of these IM effector cells
remains in doubt, but their lack of any obvious MHC restriction (Seeley et al., 1981) strongly suggests that they
represent a particular subset of natural killer (NK) cells
which are activated in response to viral infection (Klein
et al., 1981). It seems unlikely that this unusual cytotoxic response is truly directed towards a virus-coded
lymphocyte-detected membrane antigen (LYDMA) as was originally suggested (Klein, E. et al., 1976), although the
existence of such an antigen is now made clear by other
lines of evidence (vide infra).

Clearly the T cell system does have a role to play in
controlling primary EB virus infection but the polyclonal T
cell response seen in IM patients is so diverse that it may
well contain immunopathological elements, for instance
suppressor cells which could prevent the induction of
effective virus-specific T cell responses in vivo. The
various functional components within IM T cell populations
will best be identified once methods for the in vitro propagation and cloning of such cells have been established.

Recovery from IM is associated with the restoration of
a normal blood picture, the disappearance of both the
suppressor and the cytotoxic T cell activities described
above, and the establishment of a life-long virus carrier
state which is indistinguishable from that seen in individuals whose primary infection was sub-clinical. Once
again the T cell system appears to play a central role in
host control over EB viral persistence.

Persistence EB virus infection and the T cell response

Two lines of evidence indicate the virus carrier status
of previously-infected individuals, the presence of infectious
virus in throat washings of at least some donors (Gerber
et al., 1972) and the occasional "spontaneous" transformation

of cultured leukocytes to EB virus genome-positive lymphoblastoid cell lines (Nilsson et al., 1971). It is now
clear that virus shedding into the throat, perhaps from
productively-infected epithelium, is a much more stable
accompaniment of the virus carrier state than had been
realised (Yao et al, submitted)indeed suggesting that the
site of chronic replication might serve as a reservoir
continually infecting B cells in transit through the area
(Moss et al., 1981).

Whatever the precise mechanism of viral persistence,
it is clear that virus-B cell interactions are subject to
strict T cell surveillance in vivo. Again there are several
elements of the T cell response identifiable in previously-
infected individuals:-

(1) the LMI assay not only demonstrates the persistence of
immune (helper) T cells specific for "late" viral antigens
but also the existence of helper T cells reactive on the
one hand with EBNA (Szigeti et al., 1981a) and on the other
with a virus-induced transformation-associated membrane
change analogous to, perhaps identical with, LYDMA (Szigeti
et al., 1981b). Again it is interesting to note that anti-
EBNA antibodies are a stable feature of the immune response to persistent infection as are EBNA-specific helper
T cells. The recent observation that certain human sera
are able to block the LMI induced by the transformation-
associated membrane antigen suggests that this moiety may
also be serologically defined (Szigeti et al., 1984b).

(2) T cells cultured from previously-infected donors can
affect the virus-induced in vitro transformation of autologous B cells in several ways (Rickinson and Moss, 1983),
some of these effects   described in the literature as
a "suppression" of transformation. This should not be
interpreted as evidence of a role for classical suppressor
T cells in this context, for the mechanisms underlying these
effects are in fact quite different. Thus the "suppression"
of EBNA induction and of cell proliferation which T cells
can bring about in the early phase of the transformation
sequence is a non-immune phenomenon mediated via interferon
release from non-specifically activated T cells (Thorley-
Lawson, 1981). By contrast, the "late suppression" of
immunoglobulin synthesis reported in virus-infected lymphocyte cultures (Tosato et al., 1982) is indeed an immune
phenomenon (i.e. is seen only with virus-immune donors) but

is almost certainly another manifestation of the virus-specific cytotoxic response described below.

(3) The existence of EB virus-specific cytotoxic T cell precursors (memory T cells) in the blood of all healthy previously-infected individuals was first apparent from the regression of B cell outgrowth which occurs exclusively in cultures of EB virus-infected lymphocytes set up from immune donors (Moss et al., 1978). These precursors are reactivated in the presence of autologous virus-infected B cells to yield effector populations which are MHC restricted in their function (Rickinson et al., 1980; Misko et al., 1984) and specific for a virus-induced cell membrane change which is consistently associated with the virus-transformed state; this specific recognition is now taken to define the antigen LYDMA (Rickinson et al., 1981). The molecular identity of LYDMA, in particular its relationship to the three major viral proteins now thought to exist in virus-transformed cells (Kieff, 1982; Fennewald et al., 1984), remains an issue of central importance; the possibility should not be forgotten that membrane-associated forms of all of these proteins might elicit cytotoxic T cell responses.

The maintenance of virus-specific memory T cells at a high frequency in the circulating T cell pool of all previously-infected individuals (Rickinson et al., 1981) strongly implies an important surveillance function such that virus-infected B cells are usually destroyed as soon as LYDMA is expressed on the membrane i.e. soon after the appearance of EBNA in the nucleus and at the very onset of virus-induced B cell proliferation (Moss et al., 1981). The development of T cell memory in convalescent IM patients mirrors the appearance of anti-EBNA antibodies; moreover patients with immune disfunction who do not mount good cellular responses likewise do not generate anti-EBNA reactivity (Vimer et al., 1984). This is exactly what might be expected if target B cells have to be destroyed at an early phase of the infectious cycle in order to release EBNA in an immunogenic form.

It must be remembered that the availability of appropriate LYDMA-positive stimulator cells has allowed the in vitro reactivation and analysis of the above cytotoxic response. By contrast, we know very little if anything about analogous cytotoxic responses which may be directed

towards productively-infected cells via antigens expressed
late in the infectious cycle. The use of cloned viral genes
in DNA transfection studies should provide the relevant
stimulator and target cells with which to pursue these
issues more fully. The realisation that chronic virus
replication, perhaps in some specialised epithelial site,
may be central rather than peripheral to maintenance of
the virus carrier state adds a further, largely unexplored,
dimension to the question of T cell responses and EB virus
infection.

References

Andersson, U., Britton, S., de Ley, M. and Bird, G. Evidence for the ontogenetic precedence of suppressor T cell functions in the human neonate. Eur. J. Immunol., 13, 6-13 (1983).

Bird, A.G. and Britton, S. A new approach to the study of human B lymphocyte function using an indirect plaque assay and a direct B cell activator. Immunol. Rev., 45, 41-67 (1979).

Deinhardt, F. and Deinhardt, J. Comparative aspects: oncogenic animal herpesviruses. In: M.A. Epstein and B.G. Achong (ed.), The Epstein-Barr virus, pp. 373-415, Springer-Verlag, Berlin, Heidelberg, New York (1979).

Fennewald, S., van Santen, V. and Kieff, E. Nucleotide sequence of an mRNA transcribed in latent growth-transforming virus infection indicates that it may encode a membrane protein. J. Virol., 51, 411-419 (1984).

Gerber, P., Nonoyama, M., Lucas, S., Perlin, E. and Goldstein, L.I. Oral excretion of Epstein-Barr virus by healthy subjects and patients with infectious mononucleosis. Lancet, ii, 988-989 (1972).

Haynes, B.F., Schooley, R.T., Payling-Wright, C.R., Grouse, J.E., Dolin, R. and Fauci, A.S. Emergence of suppressor cells of immunoglobulin synthesis during acute Epstein-Barr virus-induced mononucleosis. J. Immunol., 123, 2095-2101 (1979).

Henle, G., Henle, W. and Horwitz, C.A. Antibodies to Epstein-Barr virus-associated nuclear antigen in infectious mononucleosis. J. infect. Dis., 130, 231-239 (1974).

Henle, G. and Henle, W. The virus as the etiologic agent of infectious mononucleosis. In: M.A. Epstein and B.G. Achong (ed.), The Epstein-Barr virus, pp. 297-320, Springer-Verlag, Berlin, Heidelberg, New York (1979).

Kieff, E. The biology and chemistry of Epstein-Barr virus.

J. infect. Dis., 146, 506-517 (1982).

Klein, E., Klein, G. and Levine, P.H. Immunological control of human lymphoma: discussion. Cancer Res., 36, 724-727 (1976).

Klein, E., Ernberg, I., Masucci, M.G., Szigeti, R., Wu, Y.T., Masucci, G. and Svedmyr, E. T-cell response to B cells and Epstein-Barr virus antigens in infectious mononucleosis. Cancer Res., 41, 4210-4215 (1981).

Klein, G., Svedmyr, E., Jondal, M. and Persson, P.O. EBV-determined nuclear antigen (EBNA)-positive cells in the peripheral blood of infectious mononucleosis patients. Int. J. Cancer., 17, 21-26 (1976).

Misko, I.S., Pope, J.H., Hutter, R., Soszynski, T.D. and Kane, R.G. HLA-DR-antigen-associated restriction of EBV-specific cytotoxic T-cell colonies. Int. J. Cancer, 33, 239-243 (1984).

Morgan, D.G., Niederman, J.C., Miller, G., Smith, H.W. and Dowaliby, J.M. Site of Epstein-Barr virus replication in the oropharynx. Lancet, ii, 1154-1157 (1979).

Moss, D.J., Rickinson, A.B. and Pope, J.H. Long-term T cell-mediated immunity to Epstein-Barr virus in man. I. Complete regression of virus-induced transformation in cultures of seropositive donor leukocytes. Int. J. Cancer, 22, 662-668 (1978).

Moss, D.J., Rickinson, A.B., Wallace, L.E. and Epstein, M.A. Sequential appearance of Epstein-Barr virus nuclear and lymphocyte-detected membrane antigens in B cell transformation. Nature, 291, 664-666 (1981).

Nilsson, K., Klein, G., Henle, W. and Henle, G. The establishment of lymphoblastoid cell lines from adult and from foetal human lymphoid tissue and its dependence on EBV. Int. J. Cancer, 8, 443-450 (1971).

Pope, J.H. Transformation by the virus in vitro. In: M.A. Epstein and B.G. Achong (ed.), The Epstein-Barr virus, pp. 205-223, Springer-Verlag, Berlin, Heidelberg, New York (1979).

Purtilo, D.T., Sakamoto, K., Barnabei, V., Seeley, J., Bechtold, T., Rogers, G., Yetz, J. and Harada, S. Epstein-Barr virus-induced disease in boys with the X-linked lymphproliferative syndrome (XLP). Am. J. Med., 73, 49-56 (1982).

Rabin, H., Neubauer, R., Hopkins, F. and Nonoyama, M. Further characterisation of a herpes virus-positive orangutan cell line and comparative aspects of in vitro transformation with lymphotropic Old World primate herpesviruses. Int. J. Cancer, 21, 762-767 (1978).

Reinhertz, E.L., O'Brien, C., Rosenthal, P. and Schlossman, S.F. The cellular basis of viral induced immunodeficiency: analysis by monoclonal antibodies. J. Immunol, 125, 1269-1274 (1980).

Rickinson, A.B., Wallace, L.E. and Epstein, M.A. HLA-restricted T cell recognition of Epstein-Barr virus-infected B cells. Nature, 283, 865-867 (1980).

Rickinson, A.B., Moss, D.J., Wallace, L.E., Rowe, M., Misko, I.S., Epstein, M.A. and Pope, J.H. Long-term T cell-mediated immunity to Epstein-Barr virus. Cancer Res., 41, 4216-4221 (1981).

Rickinson, A.B. and Moss, D.J. Epstein-Barr virus-induced transformation:immunological aspects. In: G. Klein (ed.), Advances in Viral Oncology, Vol. 3, pp. 213-238, Raven Press, New York (1983).

Robinson, J.E., Smith, D. and Niederman, J. Plasmacytic differentiation of circulating Epstein-Barr virus-infected B lymphocytes during acute infectious mononucleosis. J. exp. Med., 153, 235-244 (1981).

Royston, I., Sullivan, J.L., Periman, P.O. and Perlin, E. Cell-mediated immunity to Epstein-Barr virus-transformed lymphoblastoid cells in acute infectious mononucleosis. New Engl. J. Med., 293, 1159-1163 (1975).

Seeley, J., Svedmyr, E., Weiland, O., Klein, G., Moller, E., Eriksson, E., Andersson, K. and van der Waal, L. Epstein-Barr virus selective T cells in infectious mononucleosis are not restricted to HLA-A and B antigens. J. Immunol., 127, 293-300 (1981).

Sixbey, J.W., Nedrud, J.G., Raab-Traub, N., Hanes, R.A. and Pagano, J.S. Epstein-Barr virus replication in oropharyngeal epithelial cells. New Eng. J. Med., 310, 1225-1230 (1984).

Strauch, B., Andrews, L-L., Siegel, N. and Miller, G. Oropharyngeal excretion of Epstein-Barr virus by renal transplant recipients and other patients with immuno-suppressive drugs. Lancet, i, 234-237 (1974).

Svedmyr, E. and Jondal, M. Cytotoxic effector cells specific for B cell lines transformed by Epstein-Barr virus are present in patients with infectious mononucleosis. Proc. Natl. Acad. Sci. U.S.A., 72, 1622-1626 (1975).

Svedmyr, E., Ernberg, I., Seeley, J., Weiland, O., Masucci, G., Tsukuda, K., Szigeti, R., Masucci, M.G., Blomgren, H., Berthold, W., Henle, W. and Klein, G. Virologic, immuno-logic and clinical observations on a patient during the incubation, acute and convalescent phases of infectious mononucleosis. Clin. Immunol. Immunopathol., 30, 437-450

(1984).

Szigeti, R., Luka, J. and Klein, G. Leukocyte migration inhibition studies with Epstein-Barr virus (EBV)-determined nuclear antigen (EBNA) in relation to EBV-carrier status of the donor. Cell. Immunol., 58, 269-277 (1981a).

Szigeti, R., Volsky, D.J., Luka, J. and Klein, G. Membranes of EBV-carrying virus nonproducer cells inhibit leukocyte migration of EBV-seropositive but not seronegative donors. J. Immunol., 126, 1676-1679 (1981b).

Szigeti, R., Masucci, M.G., Henle, G., Henle, W., Purtilo, D.T. and Klein, G. Effect of different EBV-determined antigens (EBNA, EA, VCA) on the leukocyte migration of healthy donors and patients with infectious mononucleosis and certain immunodeficiencies. Clin. Immunol. Immunopathol., 22, 128-138 (1982).

Szigeti, R., Masucci, G., Ehlin-Henriksson, B., Bendtzen, K., Henle, G., Henle, W., Klein, G. and Klein, E. EBNA-specific LIF production of human lymphocyte subsets. Cellular Immunol., 83, 136-141 (1984a).

Szigeti, R., Sulitzeanu, D., Henle, G., Henle, W. and Klein, G. Detection of an EBV-associated membrane antigen in EBV-transformed non-producer cells by leukocyte migration inhibition and blocking antibody. Proc. Natl. Acad. Sci. U.S.A., 81, 4178-4182 (1984b).

Thorley-Lawson, D.A. The transformation of adult but not of newborn human lymphocytes by Epstein-Barr virus and phytohaemagglutinin is inhibited by interferon: The early suppression of T cells by Epstein-Barr virus infection is mediated by interferon. J. Immunol., 102, 829-833 (1981).

Tosato, G., Magrath, I., Koski, I., Dooley, N. and Blaese, M. Activation of suppressor T cells during Epstein-Barr virus-induced infectious mononucleosis. New Engl. J. Med., 301, 1133-1137 (1979).

Tosato, G., Magrath, I. and Blaese, R.M. T cell-mediated immunoregulation of Epstein-Barr virus (EBV)-induced B lymphocyte activation in EBV-seropositive and EBV-seronegative individuals. J. Immunol., 128, 575-579 (1982).

Vimer, E., Lenoir, G.M., Virelizier, J.L. and Griscelli, C. Epstein-Barr serology in immunodeficiencies: an attempt to correlate with immune abnormalities in Wiskott-Aldrich and Chediak-Higashi syndromes and ataxia telangiectasia. Clin. exp. Immunol., 55, 249-256 (1984).

Yao, Q.Y., Rickinson, A.B. and Epstein, M.A. A re-investigation of the Epstein-Barr virus carrier state (submitted for publication).

50

EPSTEIN-BARR VIRUS AND IMMUNOSUPPRESSION

José Menezes and Syam K. Sundar

Laboratory of Immunovirology, University of
Montreal, Montreal, Quebec, Canada   H3T 1C5

INTRODUCTION

Immunosuppression has been increasingly recognized
in clinical medicine for a number of years, at least in
part because of a variety of technological advances:
organ transplantation, cancer chemotherapy, and the
development of potent pharmacologic agents.   However,
immunosuppression is commonly observed in a variety of
viral infections as well as in virus-induced tumors.   For
example, decades ago, lymphopenia was reported in
diseases which now are known to be caused by viruses.
Conversely, viral reactivation occurs frequently in
immunosuppressed or immunocompromized patients.
Interestingly, Epstein-Barr virus-associated tumors have
also been increasingly observed in immunocompromized and
transplant patients (Hanto et al., 1981; Ziegler et al.,
1982).   These observations imply that there are important
virus interactions with the immune system beyond those
involved in controlling acute infection and the development of subsequent immunity.

The Epstein-Barr virus (EBV) is well-known to interact closely with the immune system.   EBV is a polyclonal
B-cell mitogen and can immortalize cells of B-lymphocyte
lineage (Menezes et al., 1976; Rosen et al., 1977; Bird &
Britton, 1979).   The present paper briefly reviews
several aspects of EBV-related immunosuppression and
discusses how EBV may induce this phenomenon.

EVIDENCE FOR IMMUNOSUPPRESSION IN PATIENTS WITH EBV—ASSOCIATED DISEASES

At this point, one may ask whether there is any evidence for immunosuppression in patients with EBV—associated disorders. The answer is yes, and the evidence can be found at the level of both humoral and cellular compartments of the immune system.

Humoral Evidence

To date several investigators have found inhibitors of cellular immunity in the sera of patients with EBV—induced infectious mononucleosis (IM) (Lai et al., 1974; Wainwright et al., 1979). The serum inhibitory activity found in the acute phase appears to be associated exclusively with IgG (Veltri et al., 1981). It was reported originally that this IgG inhibitory activity was specific (Lai et al., 1974); recent data with purified IgG from sera of IM patients, however, indicate that it is nonspecific (Veltri et al., 1981). Table I summarizes these IM inhibitors and their known characteristics. Recent data from our laboratory suggest that the IgG inhibitor has a marked regulatory effect on lymphokine production by T lymphocytes (Sundar, Bergeron & Menezes, submitted for publication).

Similarly, inhibitory activity effective against cellular immunity was also detected in the sera of patients with nasopharyngeal carcinoma (NPC) (Mathew et al., 1981; Sundar et al., 1982). Recent studies have demonstrated that the serum inhibitory activity is, in most cases, associated exclusively with IgA (Sundar et al., 1982; 1983). However, IgA—unrelated activity was also detected in some sera (Sundar et al., 1983). The properties of these inhibitors are summarized in able II.

Cellular Evidence

Emergence of activated suppressor T cells during acute EBV—induced IM was reported several years ago by several groups of investigators. These suppressor cells can inhibit pokeweed mitogen—induced B cell prolifera-

Table I

PROPERTIES OF INHIBITORS OF CELLULAR IMMUNITY DETECTED IN THE SERA OF PATIENTS WITH IM

Method of detection	Nature	Specificity	Effect on targets/lymphocytes	Reference
LMI[a], LST[b]	IgG	Specific	NT[c]	Lai et al., 1974
LMI, LST	IgG	Non-specific	- No effect on targets • Binds to T-lymphocytes • Reduces spontaneous rosette formation by T-lymphocytes • Does not bind to lymphokines	Wainwright et al., 1979 Veltri et al., 1981
LST	IgG	Non-specific	- No effect on targets • Binds to T-lymphocytes • Inhibits interleukin-2 production by lymphocytes induced by mitogen or antigen	Sundar, Bergeron & Menezes (submitted for publication)

[a]LMI: leukocyte migration inhibition test.
[b]LST: lymphocyte stimulation test.
[c]NT: not tested.

Table II

PROPERTIES OF INHIBITORS OF CELLULAR IMMUNITY DETECTED IN THE SERA OF PATIENTS WITH NPC

Method of detection	Nature	Specificity	Effect on targets/ lymphocytes	Reference
ADCC[a]	IgA	Specific	- No effect on lymphocytes • Binds to target cells (likely to EBV-MA) • Can be reversed by EBV-IgG antibodies present in the sera of patients with NPC who are responding well to treatment	Mathew et al., 1980
LST	IgA	Specific	NT	Sundar et al., 1982
LST	IgA	Specific	- No effect on lymphocytes • Binds to antigens (likely masks antigenic determinants eliciting cellular immune responses)	Sundar et al., 1983
LST	Unknown (not IgA)	Unknown	- No effect on targets • Binds to T-lymphocytes	Sundar et al., 1983 Sundar & Menezes (unpublished data)

[a]ADCC: Antibody dependent cellular cytotoxicity.

tion, immunoglobulin synthesis, and autologous T cell proliferation in response to antigens (Haynes et al., 1979; Johnsen et al., 1979; Tosato et al., 1979; Reinherz et al., 1980). Surface characteristics of these cells have also been studied and it is now quite well established, that in addition to their $T3^+$, $T5^+$, $T8^+$ phenotype, these suppressors are also Ia^+ (Johnsen et al., 1978; Reinherz et al., 1980; Crawford et al., 1981; De Waele et al., 1981). The precise role of these suppressor/cytotoxic T cells in vivo during the course of IM is not known. They likely provide a regulatory function by controlling EBV-infected/activated B lymphocyte proliferation as well as immunoglobulin synthesis. Such control could operate through their interaction with helper T lymphocytes and/or may even be mediated by soluble factors/lymphokines. This latter suggestion is indirectly supported by our data showing that purified T-lymphocyte preparations from patients with acute IM do not have EBV-specific killer activity (Patel et al., 1982). In any event, it is possible that the suppressor/cytotoxic T cells which are induced in IM patients contribute to the benign course of the disease (unless they multiply or persist in an activated form beyond the post-acute phase). Such suppressor cells were also found in transient immunodeficiency during asymptomatic EBV infection (Bowen et al., 1983).

It is noteworthy that experimental evidence from animal studies suggests that immunosuppression, at the time of infection, can be beneficial to the host, by controlling the appearance of autoantibodies as well as the development of the clinical syndrome (Onodera et al., 1982; Nash, 1984). Parenthetically, it is interesting that auto-antibodies have been reported in IM (Linder et al., 1979). In addition fatal IM has occurred in patients who failed to mount a characteristic T cell response (Crawford et al., 1979; Robinson et al., 1980).

While there is ample evidence of the emergence of suppressor T cells in IM, we lack information on whether and how such suppressor cells may occur in other EBV-associated diseases. Our preliminary analysis of peripheral blood lymphocytes of a limited number of patients with EBV-associated NPC have revealed, however, that these patients have increased suppressor T cells (unpublished observation).

POSSIBLE MECHANISMS OF EBV-INDUCED IMMUNOSUPPRESSION

The important question at this point is: by what mechanism(s) may EBV be involved in the generation of immunosuppression.

The possible pathways through which EBV may contribute, directly or indirectly, to induce immunosuppression are shown diagramatically in Figure 1. At present, virtually all these pathways and the factors which may modulate them are open to investigation. As indicated in the figure, EBV may have a direct effect on effector cells or may intervene through antibody production or formation of immune complexes or even by generating suppressor cells to induce an immunosuppressive state. EBV may also interact indirectly through non-structural antigens as well as by transforming or immortalizing the appropriate target cells which may then produce immunosuppressive factors such as plasminogen activator, etc. Our laboratory is studying three of the aspects depicted in the figure: suppressor cells, antibodies and plasminogen activator. Some of our observations will now be briefly reviewed.

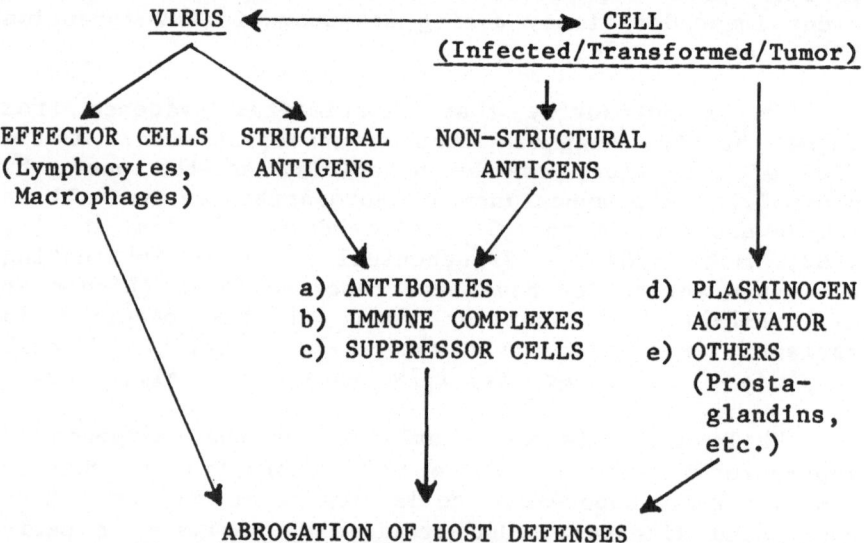

Fig. 1. Possible pathways of virus (EBV)-induced immuno-suppression.

Suppressor Cell Induction In Vitro

Recent in vitro experiments have shown that incubation of sensitized lymphocytes (i.e. from EBV-seropositive healthy individuals) together with an excess of either structural (i.e. soluble) EBV antigens results in an induction of suppressor T cells which express OKT8[+] and Ia[+] phenotype; the suppressor T cells thus generated were found to be antigen-specific since they inhibited the response of sensitized lymphocytes to the inducing antigen only, and not to other antigens or mitogens (Sundar & Menezes, submitted for publication). We infer from these observations that antigen-specific suppressor cells could be generated in vivo. It is noteworthy that the suppressor cells from IM patients discussed above appear to be non-specific. The process by which non-specific suppressor cells arise is also unclear.

Suppressor cells could be at least partly responsible for the decreased cell-mediated immunity reactions observed in patients with various EBV-associated disorders (Fass et al., 1970; Haider et al., 1973; Mangi et al., 1974; Magrath, 1974; Ho et al., 1978; Purtilo et al., 1978) as well as for the occasional syndromes of aplastic anemia and agammaglobulinemia which appear to be a rare sequel of IM in previously normal subjects (Provisor et al., 1975; Lanning et al., 1977; Lazarus & Bachner, 1981).

If EBV antigen-specific suppressor cells can be detected in a patient with EBV-associated disorder, they may prove to be clinically very relevant. In fact, the presence of such specific cells might: (a) imply that EBV or EBV antigen played some role in the etiology/clinical course of the disease, or (b) indicate active EBV replication or production of viral antigen(s) in the host. For example, in the case of persistently active viral infection, the continuous production of virus or its antigens could provide optimal conditions for sustained induction of T suppressors. In this regard, it is interesting that studies with cytomegalovirus infection as well as with animal tumors have indicated that suppressor cells return to normal levels during convalescence as well as following successful removal of tumor by surgery, respectively (Kall & Hellstrom, 1975; Fujimoto et al., 1976; Rubin et al., 1981).

EBV-Specific Antibodies
_____

It is not clear whether in IM or chronic EBV infection EBV-specific antibodies play any inhibitory role in immunity. In NPC, however, EBV-IgA is an outstanding feature of the disease (Henle & Henle, 1976). IgA from NPC patients is able to inhibit antibody-dependent cellular cytotoxicity (ADCC) in vitro (Mathew et al., 1981). Our studies with purified IgA from sera of NPC patients have shown that it specifically abrogates the response of normal sensitized lymphocyte to EBV antigens, but not to phytohemagglutinin (Sundar et al., 1982; 1983). IgA fractions from sera of healthy individuals, from patients with other head and neck cancers and NPC patients in remission do not contain this inhibitory activity (Table III). These and other studies (Henle & Henle, 1976; Mathew et al., 1981; Kamaraju et al., 1983), indicate that EBV-IgA represents a marker of unique clinical significance for NPC, particularly for its prognosis. How this IgA operates as a lymphocyte stimulation inhibitor (LSI) is still unclear; preliminary results indicate that it does not bind to lymphocytes. It will be important to determine whether it acts as IgA-EBV antigen complex.

Whether EBV-antibodies of any other Ig class may play a role in immunosuppression or inhibition of cellular immunity remains to be investigated. EBV-antibody complexes could theoretically be important. It is known that immune complexes can act as blocking factors and impair cellular immunity (Sjogren et al., 1971; Hayami et al., 1973; Hellstrom & Hellstrom, 1974; Hellstrom et al., 1983); they also appear to be of prognostic significance in neoplastic diseases (Carpentier et al., 1981; 1982; Hubbard et al., 1981). Circulating EBV antigen-antibody complexes have been found in patients with EBV-associated malignancies (Oldstone et al., 1975; Heimer & Klein, 1976; Sutherland et al., 1978). The role of such immune complexes in EBV-related disorders is open to investigation. It must be kept in mind that the genetic composition of the host may play a role in the generation of immune complexes, particularly in diseases such as viral infections (Oldstone, 1984).

Table III

LSI ACTIVITY DETECTED IN THE SERA OF PATIENTS WITH NPC

Status	Number of sera positive for LSI/ number tested	% inhibition of lymphocyte stimulation by	
		EB-virus	Soluble antigen
Controls	0/40	0-20	0-21
NPC patients (pretreatment)	31/31	60-98	58-100
NPC patients in remission	0/4	5-15	3-22
NPC patients in relapse	4/5*	79-86	62-92
Patients with other hand and neck cancers	0/30	12-21	5-17

*In the serum of one patient with fatal hepatic relapse, LSI could not be detected.

Plasminogen Activator

It has been suggested that plasminogen activator (PA) produced ty tumor cells may affect host immune response (Newcomb et al., 1978; Wainberg et al., 1982). Our interest in identifying immunomodulating factors which may be produced by EBV-transformed or EBV-genome bearing tumor cells, led us to investigate the production of PA by such cells. We found that EBV-producer lymphoid cells originating from Burkitt's lymphoma (P3HR-1) as well as from experimental EBV-induced marmoset lymphoma (B95-8) synthesized and released large quantities of PA (i.e. 5 x 10^6 cells/ml released 400-800 units of PA into the culture medium during an incubation period of 24 hrs). PA preparations, purified by affinity chromatography using lysine-sepharose columns, abrogated lymphocyte cytotoxicity (Sundar et al., 1984); the data also show that the concentration of PA released by those cells in the native crude form into the medium represents 8 to 16 times the required amount of PA which produced significant inhibition of both NK and ADCC.

Cells of several human lymphoid lines of different origins produce PA, irrespective of the presence or absence of EBV genome in the cells (Sundar, Bergeron & Menezes, submitted for publication). It is thus quite clear that PA synthesis is independent of the presence of EBV genome in the cell, and that it represents simply a by product resulting from cellular transformation by this virus.

Production of PA by EBV-associated tumors in vivo has not been studied. In a preliminary study conducted at the University of Malaysia at Kuala Lumpur, PA was detected in 12 NPC biopsies (Sundar et al., 1984); it is thus possible that PA plays also a role in the pathogenesis of NPC. In general, the role of PA in vivo is not clear. Similarly, how PA operates to abrogate cellular effector immune mechanisms is not known. Our results however show that it affects the effector lymphocytes and not the targets (Sundar et al., 1984). In any event, from the observations described above, it is tempting to speculate that PA released by (EBV-transformed or) tumor cells into the local micro-environment may render ineffective the infiltrating host defenses such as killer lymphocytes and thus help the malignant cells to escape host effector mechanisms. Furthermore, PA may aid malignant cells in the invasion of the surrounding tissue by its ability to activate procolla- genase to collagenase; likewise, plasmin generated by the action of cellular PA may promote the migration of trans- formed cells, as was shown in vitro (Ossowski et al., 1975).

CONCLUSION

Studies describing depression of cell-mediated immunity and features indicative of immunosuppression have been reported in various EBV-associated disorders. In vitro studies have clearly shown that suppressor cells with specific T–lymphocyte phenotype can be detected during EBV-induced acute IM.

Presently, the following important and related questions require investigation. At the disease level: whether the suppressor T cells as well as immunosuppres- sion actually play an important role in the pathogensis of EBV-associated disorders or represent epiphenomena? At a more fundamental, mechanistic level: whether EBV triggers immunosuppression by acting on the immunoregula- tory system; or how EBV interacts with the immunoregula- tory system to trigger immunosuppression?

We have attempted to present in a diagramatic fashion the various pathways by which EBV may trigger, or contribute to the initiation of immunosuppression. Most of these pathways are unexplored at present. Our limited in vitro studies indicate that: (a) both EBV structural and non-structural antigens in excess can induce antigen- specific suppressor T cells; (b) EBV-antibodies of one particular class (IgA) can abrogate virus-directed lymphocyte response, and (c) EBV may also contribute indirectly to immunosuppression as EBV-transformed/tumor cells can release plasminogen activator, a product which has been found to abrogate lymphocyte killer activity.

Finally, it is important to consider that what we refer to as immunosuppression, a phenomenon which is generally detrimental to the host, may in some cases be a natural and useful response in the disease process. In any case, a balance between helper/suppressor T lymphocyte subsets' activity is also an important feature in a normal, healthy individual. The key issue before us is therefore how EBV can disrupt this balance and to what extent. Genetic factors may also play a determining role in the induction of immunosuppression by EBV, and is a major point for future research.

ACKNOWLEDGMENTS

We thank Medical Research Council of Canada, Cancer Research Society, Inc. (Montreal), and "Fondation Justine-Lacoste-Beaubien" for their support. José Menezes is a Senior Scholar from "Fonds de la Recherche en Santé du Québec", and thanks Astra Pharmaceuticals Canada Ltd for the "Astra Research Award in Herpes". We are grateful to Dr. G. Ahronheim for his editorial comments and suggestions.

REFERENCES

BIRD, A.C., and BRITTON, S., A new approach to the study of human B lymphocytes function using an indirect plaque assay and a direct B cell activator. Immunol. Rev., 45, 41-67 (1979).

BOWEN, T.J., WEDGWOOD, R.M., OCHS, H.D., and HENLE, W., Transient immunodeficiency during asymptomatic Epstein-Barr virus infection. Pediatrics, 71, 964-966 (1983).

CARPENTIER, N.J., FIERE, D.M., SCHUH, D., LANGE, G.T., and LAMBERT, P.H., Circulating immune complexes and the prognosis of acute myeloid leukemia, N. Engl. J. Med., 307, 1174-1180 (1982).

CARPENTIER, N.A., LOUIS, J.A., LAMBERT, P.H., and CEROTTINI, J.C., Immune complexes in leukemia and adult malignancies. In: S.B. Rosenfeld (ed.), Immune Complexes and Plasma Exchange in Cancer Patients, pp. 111-134, Elsevier/North Holland, New York (1981).

CRAWFORD, D.H., BUCKELL, P., TEDMAN, N., McCONNELL, I., HOFFBRAND, A.V., and JANASSY, G., Increased numbers of cells with suppressor T cell phenotype in the peripheral blood of patients with infectious mononucleosis. Clin. Exp. Immunol., 43, 291-297 (1981).

CRAWFORD, D.H., EPSTEIN, M.A., ACHONG, B.C., FINERTY, S., NEWMAN, T., LEVERSEDGE, S., TEDDER, R.S., and STEWART, J.W., Virological and immunological studies on a fatal case of infectious mononucleosis. J. Infect., 1, 37–40 (1979).

DeWAELE, M., THEILEMANS, C., and VAN CAMP, B.K.G., Characterization of immunoregulatory T cells in EBV-induced infectious mononucleosis by monoclonal antibodies. N. Engl. J. Med., 304, 460–462 (1981).

FASS, L., HERBERMAN, R., and ZIEGLER, J.L., Delayed cutaneous reactions to autologous extracts of Burkitt's lymphoma clls. N. Engl. J. Med., 292, 776–780 (1970).

FUJIMOTO, S., GREENE, M.I., and SEHON, A.H., Regulation of immune-response to tumor antigens. I. Immunosuppressor cells in tumor bearing mice. J. Immunol., 16, 791–799 (1976).

HAIDER, S., COUTINHO, M.L., EDMOND, R.T.D., and SUTTON, R.N.P., Tuberculin anergy and infectious mononucleosis. Lancet, 2, 74–76 (1973).

HANTO, D.W., FREZZERA, G., PURTILO, D.T., SAKAMOTO, K., SULLIVAN, J.L., SAEMUNDSEN, A.K., KLEIN, G., SIMMONS, R.L., and NAJARIAN, J.S., Clinical spectrum of lymphoproliferative disorders in renal transplant recipients and evidence for the role of Epstein–Barr virus. Cancer Res., 41, 4253–4261 (1981).

HAYAMI, M., HELLSTROM, I., and HELLSTROM, K.E., Serum effects on cell-mediated destruction of Rous sarcomas. Int. J. Cancer, 12, 667–673 (1973).

HAYNES, B.F., SCHOOLEY, R.T., PAYLING-WRIGHT, C.R., GROUSE, J.E., DOLIN, R., and FAUCI, A.S., Emergence of suppressor cells of immunoglobulin synthesis during acute Epstein–Barr virus-induced infectious mononucleosis. J. Immunol., 123, 2095–2101 (1979).

HEIMER, R., and KLEIN, G., Circulating immune complexes in sera of patients with Burkitt's lymphoma and nasopharyngeal carcinoma. Int. J. Cancer, 18, 310–316 (1976).

HELLSTROM, K.E., and HELLSTROM, J., Lymphocyte mediated cytotoxicity and blocking serum activity to tumor antigens. Adv. Immunol., 18, 209-277 (1974).

HELLSTROM, K.E., HELLSTROM, J., and NELSON, K., Antigen specific suppressor ("Blocking") factors in tumor immunity. Biomembranes, 11, 365-388 (1983).

HENLE, G., and HENLE, W., Epstein-Barr virus specific serum antibodies as an outstanding feature of nasopharyngeal carcinoma. Int. J. Cancer, 17, 1-7 (1976).

HO, J.H.C., CHAN, J.C.W., TSE, K.C., NG, M.G., and LEVINE, P.H., In vivo cell-mediated immunity in Chinese patients with nasopharyngeal carcinoma: Etiology and control. In: G. De Thé, Y. Ito (eds), Oncogenesis and Herpesviruses III, pp. 545-553, IARC Publ., Lyon, France (1978).

HUBBARD, R.A., AGGIO, M.C., LOZZIO, B.B., and WUST, C.J., Correlation of circulating immune complexes and disease status in patients with leukaemia. Clin. Exp. Immunol., 43, 46-53 (1981).

JOHNSEN, H.E., MADSEN, M., KRISTENSEN, T., and KISSMEYU-NIELSON, F., Lymphocyte subpopulations in man. Expression of HLA-DR determinants on human T cells in infectious mononucleosis. Acta Pathol. Microbiol. Scand. (C), 86, 307-311 (1978).

JOHNSEN, H.E., MADSEN, M., and KRISTENSEN, T., Lymphocyte subpopulations in man: Suppression of PWM-induced B-cell proliferation by infectious mononucleosis T cells. Scand. J. Immunol., 10, 251-258 (1979).

KALL, M.A., and HELLSTROM, I., Specific stimulatory and cytotoxic effects of lymphocytes sensitized in vitro to either alloantigens or tumor antigens. J. Immunol., 114, 1803-1808 (1975).

KAMARAJU, L.S., LEVINE, P.H., SUNDAR, S.K., ABLASHI, D.V., FAGGIONI, A., ARMSTRONG, G.R., BERTRAM, G., and KRUEGER, G.R.F., Epstein-Barr virus-related lymphocyte stimulation inhibitor: A possible prognostic tool for undifferentiated nasopharyngeal carcinoma. J. Natl. Cancer Inst., 70, 643-647 (1983).

549

LAI, P.K., MACKAY SCOLLARY, E.M., FIMMEL, P.J., ALPERS, M.P., and KEAST, D., Cell-mediated immunity to Epstein-Barr virus and a blocking factor in the patients with infectious mononucleosis. Nature (London), 252, 608-610 (1974).

LANNING, M., KOUVALAINEN, K., SMILA, S., and RAUNEO, V., Agammaglobulinemia with arthritis and celiac disease developing after infectious mononucleosis. Scand. J. Infect. Dis., 9, 144-148 (1977).

LAZARUS, K.H., and BACHNER, R.L., Aplastic anemia complicating infectious mononucleosis: A case report and review of literature. Pediatrics, 67, 907-910 (1981).

LINDER, E., KURKI, P., and ANDERSON, L.C., Autoantibody to "intermediate filaments" in infectious mononucleosis. Clin. Immunol. Immunopathol., 14, 411-417 (1979).

MAGRATH, I.T., Immunosuppression in Burkitt's lymphoma. I. Cutaneous reactivity to recall antigens, alterations induced by tumor burden and by BCG administration. Int. J. Cancer, 13, 839-849 (1974).

MANGI, R.J., NIEDERMAN, J.C., KELLEHER, J.E., DRUYER, J.M., EVANS, A.S., and CANTOR, F.S., Suppression of cell-mediated immunity during acute infectious mononucleosis. N. Engl. J. Med., 291, 1149-1153 (1974).

MATHEW, F.K., QUALTIERE, L.F., NEEL, B.H., and PEARSON, J.R., IgA antibody, antibody dependent cellular cytotoxicity and prognosis in patients with nasopharyngeal carcinoma. Int. J. Cancer, 27, 175-180 (1981).

MENEZES, J., JONDAL, M., LEIBOLD, W., DORVAL, G., Epstein-Barr virus interactions with human lymphocyte subpopulations: virus adsorption, kinetics of expression of Epstein-Barr virus-associated nuclear antigen, and lymphocyte transformation. Infect. Immun., 13, 303-310 (1976).

NASH, A.A., Viruses as regulators of delayed hypersensitivity T-cell and suppressor T-cell function. In: A.L. Notkins, M.B.A. Oldstone (eds), Concepts in Viral Pathogenesis, pp. 225-230, Springer-Verlag, New York (1984).

NEWCOMB, E.W., SILVERSTEIN, S.C., and SILAGI, S., Malignant mouse cells do not form tumors when mixed with cells of non malignant subclone: relationship between plasminogen activator expression by the tumor cells and the host immune response. J. Cell. Physiol., 95, 169 (1978).

OLDSTONE, M.B.A., Virus-induced immune complex formation and disease: definition, regulation, importance. In: A.L. Oldstone, M.B.A. Oldstone (eds), Concepts in Viral Pathogenesis, pp. 201-209, Springer-Verlag, New York (1984).

OLDSTONE, M.B.A., THEOFILOPOULOUS, A.N., GUNVENS, P., and KLEIN, G., Immune complexes associated with neoplasia: Presence of Epstein-Barr virus antigen-antibody complexes in Burkitt's lymphoma. Intervirology, 4, 292-300 (1975).

ONODERA, T., RAY, V.R., MELEZ, K.A., SUZUKI, H., TONOLO, A., and NOTKINS, A.L., Virus-induced diabetes-mellitus: autoimmunity and polyendocrine disease prevented by immunosuppression. Nature, 297, 66-68 (1982).

OSSOWSKI, L., QUIGLEY, J.P., and REICH, E., Plasminogen a necessary factor for cell migration in vitro. In: Proteases and Biological Control, pp. 901-903, Cold Spring Harbor Laboratory, New York (1975).

PATEL, P.C., and MENEZES, J., Epstein-Barr virus (EBV)-lymphoid cell interactions. II. The influence of the EBV replication cycle on natural killing and antibody dependent cellular cytotoxicity against EBV infected cells. Clin. Exp. Immunol., 48, 589-601 (1982).

PROVISOR, A.J., IACRIONE, J.J., CHILCOTE, R.R., NEIBURGER, R.G., CRUSSI, F.C., and BACHER, R., Acquired agammaglobulinemia after life threatening illness with clinical and laboratory features of infectious mononucleosis in three related male children. N. Engl. J. Med., 293, 62-65 (1975).

PURTILO, D.T., HUTT, L., BHAWAN, J., YANG, J.P.S., CASSEL, C., ALLEGRA, S., and ROSEN, F., Immunodeficiency to Epstein-Barr virus in the X-linked recessive lymphoproliferative syndrome. Clin. Immunol. Pathol., 9, 147-156 (1978).

REINHERZ, E.L., O'BRIEN, C., ROSENTHAL, P., and SCHLOSSMAN, S.F., The cellular basis for viral-induced immunodeficiency analysis by monoclonal antibodies. J. Immunol., 125, 1269-1274 (1980).

ROBINSON, J.E., BROWN, N., ARDIMAN, W., HALLIDAY, K., FRANCKE, U., ROBERT, M.F., ANDERSON-ANVRET, M., HOESTMANN, D., and MILLER, G., Diffuse polyclonal B-cell lymphoma during primary infection with Epstein-Barr virus. N. Engl. J. Med., 302, 1293-1295 (1980).

ROSEN, A., GERGLY, P., JONDAL, M., and KLEIN, G., Polyclonal Ig production after Epstein-Barr virus infection of human lymphocytes in vitro. Nature, 267, 52-54 (1977).

RUBIN, H., CARNEY, W.P., SCHOOLEY, R.T., COLVIN, R.B., BURTON, R.C., HOFFMAN, R.A., HANSEN, W.P., COSINI, A.B., RUSSELL, P.S., and HIRSCH, M.S., Effect of infection on T lymphocyte subpopulations: a preliminary report. Int. J. Immunopharmacol., 3, 307-312 (1981).

SJOGREN, H.O., HELLSTROM, I., BANSAL, S.C., and HELLSTROM, K.E., Suggestive evidence that the blocking antibodies of tumor bearing individual may be antigen-antibody complexes. Proc. nat. Acad. Sci. (Wash.), 68, 1372-1375 (1971).

SUNDAR, K.S., ABLASHI, D.V., KAMARAJU, L.S., LEVINE, P.H., FAGGIONI, A., ARMSTRONG, G.R., PEARSON, G.R., KRUEGER, G.R.F., HEWETSON, J.F., BERTRAM, G., SESTERHENN, K., and MENEZES, J., Sera from patients with undifferentiated nasopharyngeal specific Epstein-Barr virus antigen-induced lymphocyte response. Int. J. Cancer, 29, 407-412 (1982).

SUNDAR, S.K., BERGERON, J., and MENEZES, J., Purified plasminogen activating factor produced by malignant lymphoid cells abrogates lymphocyte cytotoxicity. Clin. Exp. Immunol., 56, 701-708 (1984).

SUNDAR, K.S., MENEZES, J., LEVINE, P.H., ABLASHI, D.V., KAMARAJU, L.S., FAGGIONI, A., and PRASAD, U., Relationship of IgA to an EBV-specific lymphocyte stimulation inhibitor (LSI) present in the sera of patients with undifferentiated nasopharyngeal carcinoma. In: Prasad et al. (eds), Nasopharyngeal Carcinoma: Current Concepts, pp. 293-297, University of Malaya, Kuala Lumpur (1983).

SUTHERLAND, J.C., OLWENY, C.L.M., LEVINE, P.H., and MARDINEY, M.R., Epstein-Barr virus-immune complexes in postmortem kidneys from African patients with Burkitt's lymphoma and American patients with and without lymphoma. J. Natl. Cancer Inst., 60, 941-946 (1978).

TOSATO, G., MAGRATH, I., KOSHI, I., DOOLEY, N., and BLEASE, M., Activation of suppressor T cells during Epstein-Barr virus-induced infectious mononucleosis. N. Engl. J. Med., 301, 1133-1137 (1979).

VELTRI, R.W., KIKTA, V.A., WAINWRIGHT, W.H., and SPRINKLE, M., Biologic and molecular characterization of the serum IgG blocking factor (SBF-IgG) isolated from sera of patients with EBV-induced infectious mono-nucleosis. J. Immunol., 127, 320-328 (1981).

WAINBERG, M.A., ISRAEL, E., and MARGOLESE, G.R., Further studies on mitogenic and immune modulating effects of plasminogen activator. Immunology, 45, 715-720 (1982).

WAINWRIGHT, W.H., VELTRI, R.W., and SPRINKLE, M., Abroga-tion of cell-mediated immunity by a serum blocking factor isolated from patients with infectious mononucleosis. J. Inf. Dis., 140, 22-32 (1979).

ZIEGLER, J.L., DREW, W.L., MINER, R.C., MINTZ, L., ROSENBAUM, E., GERSHON, J., LENNETTE, E.T., GREENSPAN, J., SHILLITOE, E., BECKSTED, J., CASAVANT, C., and YAMAMOTO, K., Outbreak of Burkitt's like lymphoma in homosexual men. Lancet, 2, 631-633 (1982).

51

IN VITRO IMMUNOGENICITY OF HUMAN LYMPHOID TUMOUR CELL LINES

D.J. Moss, J.A. Staples, S.R. Burrows, J. Ryan and
T.B. Sculley
Queensland Institute of Medical Research

Herston, 4006, Australia

Numerous studies (e.g. Sample et al., 1971; Field and Caspary, 1972) have detected factors in the serum of tumour-bearing individuals capable of inhibiting immune functions such as PHA stimulation and antigenic stimulation of lymphocytes.

Because of the important implications the existence of such factors have in the surveillance against tumour cells, we have undertaken a comprehensive survey of the existence of these factor(s) in cultured human haemopoietic tumour cell lines. Our approach has been to compare the ability of tumour cell lines and Epstein-Barr virus-transformed B cells (LCL's) to act as a stimulator population in a mixed leucocyte reaction (MLR).

EXPERIMENTAL PROCEDURE

In the first set of experiments, Epstein-Barr (EB) virus-transformed lymphoblastoid cell lines (LCL's) and tumour cell lines listed in Table 1 were tested for their ability to induce an MLR. Responder lymphocytes (5×10^4 cells/well) were mixed with irradiated allogeneic stimulator cells (10^4 cells/well) as triplicate cultures in U-well microtitre plates. Cultures were incubated for 5 days at $37^{\circ}C$ and the level of incorporation of ^3H-thymidine during the last 6 hours of the culture period determined.

In the second set of experiments, the kinetics of stimulation of responder lymphocytes by LCL's and tumour cell lines were investigated. Responder lymphocytes (5×10^4 cells/well) were admixed with irradiated stimulator cells at 10^3, 10^4 or 10^5 cells/well. Proliferation of responder lymphocytes was determined 3, 5, 7 and 9 days later.

In the third set of experiments, responder lymphocytes (5×10^4 cells/well) were cultured in the presence of irradiated LCL stimulator cells (10^4 cells/well) and cell-free supernatant (to 25% final volume in the culture) from tumour cell lines or LCL's. In some experiments, cell-free supernatants from tumour cell lines or LCL's were added to the MLR at dilutions of 10^1 to 10^6. The proliferation of responder cells was determined after 5 days.

In the fourth set of experiments, Balb/c mice were immunized with supernatant culture fluids from the CCRF-CEM cell line and spleen cells fused with the NS1 cell line. Hybridoma cultures were screened for their ability to inhibit the immunosuppressive activity of the CCRF-CEM cell line.

TABLE 1

SUMMARY OF CELL LINES USED IN THIS STUDY

Cell lines	Derivation	Cell type
Lymphoma/leukaemia cell lines		
QIMR-W1 BL	New Guinea Burkitt lymphoma	B
QIMR-W2 BL	New Buinea Burkitt lymphoma	B
QIMR-Agoi BL	New Guinea Burkitt lymphoma	B
QIMR-GOR	New Guinea Burkitt lymphoma	B
Raji	African Burkitt lymphoma	B
BJAB	African Burkitt lymphoma	B
B-95-BJAB	BJAB and EB virus in vitro	B
Ramos	American Burkitt lymphoma	B
AW-Ramos	Ramos and EB virus in vitro	B
QIMR-BM	Australian Burkitt lymphoma	B
QIMR-Joy BL	Australian Burkitt lymphoma	B
HSB2	Acute lymphoblastic leukaemia	T
K562	Chronic myeloid leukaemia	non-B, non-T
HL-60	Acute promyelocytic leukaemia	non-B, non-T
CCRF-CEM	Acute leukaemia	T

RESULTS

Comparison of LCL's and tumour cell lines as stimulators of an MLR

During this investigation, a comparison was made of the ability of 9 haemopoietic tumour cell lines and 20 LCL's to act as the stimulator population in an MLR. Table 2 summarizes the overall results of the 30 experiments performed in which the five-day proliferative response of

---

TABLE 2

SUMMARY OF THE PROLIFERATIVE RESPONSE OF LYMPHOCYTES TO STIMULATION BY A RANGE OF TUMOUR CELLS

Stimulator cell line	Relative Stimulation (%)[1]		Proportion of experiments showing inhibition of MLR[2]
	Mean	Range	
CCRF-CEM	0.3± 3.7	0.07- 1.0	24/25
QIMR-BM	5.5± 5.4	0.4 -15.0	8/10
Ramos	5.4± 8.9	0.13-31.0	10/11
AW-Ramos	3.2± 6.8	0.2 -28.0	5/6
BJAB	12.7±18.0	0.01-52.0	5/7
B-95-BJAB	9.1±12.1	0.6 -62.0	6/8
HSB2	3.0± 2.3	1.1 - 4.0	6/6
QIMR-W1 BL	19.9±32.0	3.5 -78.0	6/8
QIMR-W2 BL	13.8±19.2	3.1 -48.0	6/8

[1] Stimulation expressed as a percentage of the mean stimulation by the LCL's in the same experiment

[2] Proportion of experiments in which the relative stimulation was less than 10%

lymphocytes to a fixed concentration of stimulator cells was assessed. A low level of stimulation of responder lymphocytes was recorded with 13/15 tumour cell lines in the majority of experiments conducted.

Kinetics of stimulation of lymphocytes by tumour cell lines and LCL's

Experiments were conducted to determine whether the difference in the ability of LCL's and tumour cell lines to induce an MLR was independent of stimulator concentration and duration of MLR. The results (not illustrated) indicate that the reduced stimulator capacity of tumour cell lines applies over a broad range of cell concentrations and duration of MLR.

Effect of cell-free supernatants from tumour cell lines on MLR activity

The above results suggest that some tumour cell lines might secrete a factor capable of inhibiting MLR reactivity. To test this hypothesis, cell-free supernatants from tumour cell lines and LCL's were tested for their ability to inhibit MLR reactivity. A summary of the 28 experiments performed is included in Table 3. Immunosuppressive activity was detected in the supernatants of all the tumour cell lines tested. However, the frequency with which this activity was detected showed considerable variation between cell lines. For example, inhibitory activity was nearly always detected (19/20 experiments) in the supernatant from the CCRF-CEM cell line, but considerably less often in the supernatants from the BJAB cell line (4/9).

In titrations of the immunosuppressive factor from the CCRF-CEM supernatant, activity$_3$was detected at a dilution of 10^5 in 5/17 experiments, at 10^3 in 8/17 experiments and at

10^1_3 in 2/17 experiments. Inhibitory activity at dilutions of
10^3 - 10^4 was also detected in supernatants of QIMR-W1 BL,
QIMR-W2 BL, Daudi, QIMR-Joy BL, HSB2, Raji and QIMR-BM cell
lines on several occasions.

Reversal of immunosuppressive activity using monoclonal
antibody directed against proteins in CCRF-CEM supernatant

Hybridomas were screened for their ability to inhibit
the activity of the immunosuppressive activity from the CCRF-
CEM cell line. An IgM monoclonal antibody was isolated that
inhibited the immunosuppressive activity from all the tumour
cell lines listed in Table 1. This antibody (JSD78) was not
mitogenic per se and in no way altered the magnitude of a
normal MLR.

TABLE 3
EFFECTS OF SUPERNATANTS FROM TUMOUR CELL LINES ON MLR

Source of cell supernatant	Proportion of experiments showing inhibition of MLR[1]
QIMR-W1 BL	6/8
QIMR-W2 BL	5/7
QIMR-Joy BL	10/12
BJAB	4/9
B-95-BJAB	5/7
Ramos	3/4
AW-Ramos	5/7
CCRF-CEM	19/20
HSB2	7/8
QIMR-BM	8/10

[1] Experiments in which the proliferative response in the
presence of supernatant from a tumour cell line was less
than 10% of the proliferation in the presence of a LCL
supernatant.

DISCUSSION

In the present study we have compared the capacity of human lymphoid tumour cell lines with that of LCL's to act as the stimulator population in an MLR. The tumour cell lines were derived from lymphomas and leukaemias and are mainly of T cell and B cell origin. While LCL's invariably served as an efficient stimulator population, tumour cell lines frequently failed to stimulate responder lymphocytes. The inability of tumour cell lines to generate an MLR was independent of stimulator dose and duration of MLR.

It was important to determine whether the failure of tumour cell lines to act as the stimulator population in MLR was due to the presence of a soluble immunosuppressive factor secreted by the cell line. The results show that in most experiments such an immunosuppressive factor can indeed be detected in the supernatants of the tumour cell lines but never in the corresponding supernatants from LCL's. It thus seems likely that the failure of tumour cell lines to act as stimulators is due to the release of a soluble immunosuppressive factor, rather than to some aberrant mode of presentation of stimulating antigens.

It should be emphasized that, in a proportion of experiments, tumour cell lines do indeed induce an MLR (Table 1). Furthermore, the results presented in Table 3 show that, on occasions, there is no detectable inhibitory factor in the supernatants from tumour cell lines. While the reason for this experimental variation is not known, the results appear compatible with the fact that the level of inhibitory factor in the CCRF-CEM cell line varies over several orders of magnitude around a mean titre of 10^3. It would thus be reasonable to expect that the level of immunosuppressive factor might occasionally fall below a

detectable level. Results thus far suggest that strict
attention to maintenance of optimal culture conditions is
essential in maintaining a high level of inhibitory factor.

The fact that the JSD78 monoclonal antibody inhibited
the immunosuppressive activity from all of the tumour cell
lines tested is an important and unexpected result. It is
not possible at this stage to determine whether the
monoclonal antibody is directed against the
immunosuppressive factor itself, to a carrier protein or to
the binding site of the immunosuppressive factor.

Of particular importance is the finding that EB virus-
infected Burkitt lymphoma cell lines from endemic regions
(QIMR-W1 BL, QIMR-W2 BL, QIMR-GOR, Raji) produce an
immunosuppressive factor. We (Moss et al., 1978;
Rickinson et al., 1979) and others (Rickinson et al., 1980)
have established techniques for detecting EB virus-specific
T cell-mediated immunity. It will now be important to
determine whether the factor(s) released from these cell
lines can interfere with this specific T cell function
which all normal, healthy, EB virus seropositive
individuals possess, and these studies are in progress.

The in vivo significance of these immunosuppressive
factor(s) remains to be defined. There are numerous
reports indicating that serum from a broad range of tumour-
bearing patients possesses immunosuppressive activity. It
is not at all clear what the relationship is between the
immunosuppressive activity isolated from fresh tumour
biopsies, tumour cell lines, normal cell lines and serum.
It should be possible for us to determine this relationship
using our recently isolated monoclonal antibody.

In conclusion, the present study demonstrates that
soluble immunosuppressive factors are frequently secreted
from haemopoietic tumour cell lines but are rarely, if

ever, secreted by EB virus-transformed cell lines. With
the reagents we now have available it should be possible to
assess whether the factor(s) isolated from tumour cell
lines are present in the serum of cancer patients and in
fresh tumour biopsies. Only then will it be possible to
determine whether such factor(s) play any role in the
initiation and maintenance of human tumours.

REFERENCES

FIELD, E.J. and CASPARY, E.A., Lymphocyte sensitization in
advanced malignant disease: a study of serum lymphocyte
depressive factor. Br. J. Cancer, 26, 164 (1972).

MOSS, D.J., RICKINSON, A.B. and POPE, J.H., Long-term T
cell-mediated immunity to Epstein-Barr virus in man. I.
Complete regression of virus-induced transformation in
cultures of seropositive donor leukocytes. Int. J. Cancer,
22, 622-668 (1978).

RICKINSON, A.B., MOSS, D.J. and POPE, J.H., Long-term T
cell-mediated immunity to Epstein-Barr virus in man. II.
Components necessary for regression in virus-infected
leukocyte cultures. Int. J. Cancer, 23, 610-617 (1979).

RICKINSON, A.B., WALLACE, L.E. and EPSTEIN, M.A., HLA-
restricted T cell recognition of Epstein-Barr virus-
infected B cells. Nature (Lond.), 283, 865-867 (1980).

SAMPLE, W.F., GERTNER, H.R. and CHRETIEN, P.B., Inhibition
of phytohaemagglutinin-induced in vitro lymphocyte
transformation by serum from patients with carcinoma. J.
Natl. Cancer Inst., 46, 1291-1297 (1971).

52

MONOCYTE CONTRASUPPRESSION OF EBV-IMMUNE REGULATORY

T CELLS

G. Tosato[1], S.E. Pike[1], and R.M. Blaese[2]

FDA[1] and NIH[2]

Bethesda, MD 20205

Summary

EBV-immune normal individuals have circulating
regulatory T cells that inhibit the activation of
autologous B cells by EBV. When T cells were fractionated
in subsets enriched for either T8 or T4 expressing
lymphocytes and subsequently cultured with autologous
EBV-infected cells, most of the inhibitory activity
derived from T8-enriched populations. Monocyte depletion
of EBV immune autologous B and T cells was associated with
an enhanced degree of T-cell inhibition. Conversely,
monocyte reconstitution of monocyte-depleted cultures
markedly reduced EBV-immune T-cell suppression. Thus, in
an EBV-specific system of suppression, monocytes act as
regulatory cells with "contrasuppressor" function.

Introduction

Epstein-Barr virus (EBV) is a ubiquitous herpes
virus that infects the majority of adult individuals
worldwide (Henle and Henle, 1979). Following primary
infection, EBV persists in a small proportion of the
circulatory B lymphocytes probably for life (Diehl et al.,
1968; Nilsson, et al., 1971). A number of regulatory
mechanisms have been reported to contribute to the control
of latent EBV infection, including cytotoxicity, natural
killer activity, and suppression (Rickinson et al., 1980;
Tosato et al., 1982, Masucci et al., 1983). These
immunoregulatory mechanisms are believed to prevent an
otherwise uncontrolled expansion of B cells latently

infected with this virus in vivo (Tosato and Blaese, In
Press). We have asked whether monocytes have a role in the
control of infection with EBV and demonstrate that
monocytes have a "contrainhibitory" effect on EBV-immune
T cell suppression.

Material and Methods

Mononuclear cells were obtained from heparinized
peripheral blood of normal individuals seropositive or
seronegative for EBV. These were depleted of monocytes by
a combination of plastic adherence, iron carbonyl
ingestion, and passage through G-10 column to yield
monocyte-depleted mononuclear cells containing less than
4% esterase positive cells (Bianco, 1977). B and T
cell-enriched subsets were obtained by standard
techniques, incubating mononuclear cells and monocyte
depleted mononuclear cells with AET-treated sheep red
blood cells, and separating the rosette-forming cells by
density gradient centrifugation (Tosato, et al., 1982).
Further separation of the T cells in T4 negative and
T8 negative T cell subsets was performed by treatment of
the T cells with OKT4 and OKT8 monoclonal antibodies
followed by complement lysis (Yachie et al., 1982).
Irradiated (3000 R) non-T cells, containing 35-60 per cent
esterase-positive cells were used as a source of
monocytes. The filtered supernatant of the B95-8 cell
line (containing approximately 10^6 transforming units/ml)
was the source of EBV. Monocyte-depleted or non-depleted
B cell-enriched populations were cultured in the presence
of EBV either alone or mixed with autologous T cells,
either unfractionated, or depleted of T4-positive or T8-
positive cells. Monocytes (as autologous irradiated non-T
cells) were added to monocyte-depleted cultures and their
effect on the B-cell response examined. At the end of a
14-day culture period, the immunoglobulin-secreting cell
response was determined by a reverse hemolytic plaque
assay (Tosato, et al., 1982).

Results

It was previously reported that T cells from EBV-
immune normal individuals profoundly inhibit immunoglobulin (Ig) production by autologous EBV-infected
B cells after 12-14 days in culture (Tosato, et al.,

564

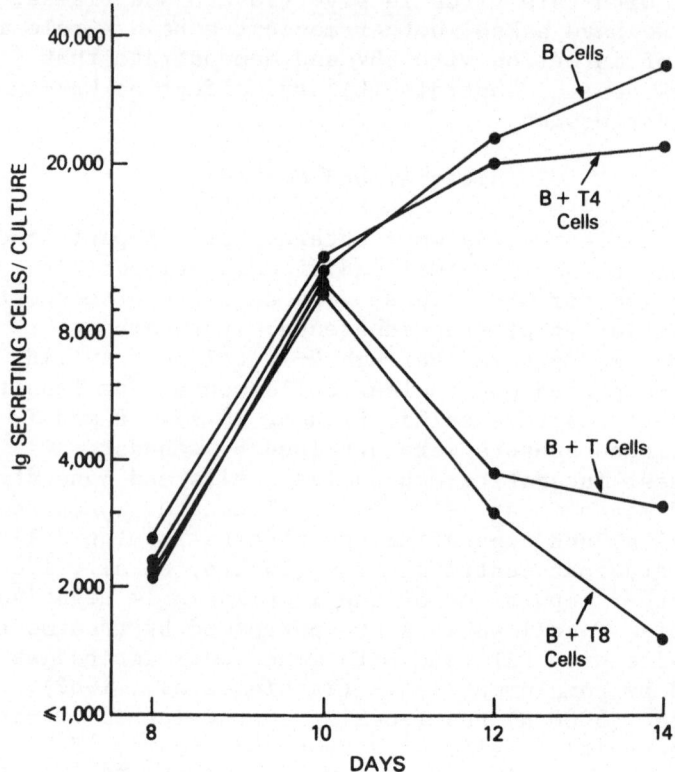

Fig 1.   EBV-immune T cells expressing a T8 phenotype
mediate "late suppression." B cells (20,000)
were cultured in the presence of EBV either alone
or with autologous T, T4-negative or T8-negative
cells (100,000). The Ig secreting cell response
was determined after 8, 10, 12 and 14 days in
culture.

1982). We have called this phenomenon "late suppression."
T cells from EBV non-immune individuals fail to suppress
in this system. To investigate further T-cell regulation
of EBV infection in vitro, we have separated T cells into
subsets depleted of either T4- or T8- bearing lymphocytes
and examined their relative contribution to "late
suppression." In a typical experiment, EBV-immune T cells
and T4-depleted T cells profoundly inhibited Ig production
by the autologous EBV-activated B cells after 14 days in
culture. In contrast, T8-depleted T cells had little
inhibitory effect at this time (Fig. 1). As shown in
Table I, T8-enriched T-cell populations consistently

TABLE I

T Cells Expressing a T8 Phenotype Mediate Late Suppression

Immunoglobulin Secreting Cells/Culture

Exp no.	B cells*	B + T cells	B + T4 neg Cells	B + T8 neg Cells
1	34,173	3,100	1,588	22,372
2	37,400	5,210	7,900	25,800
3	29,700	12,500	16,700	33,200
4	33,948	7,781	3,501	22,114
5	40,500	6,910	6,366	34,754
Mean % suppression		79	78	21

*B cells (2×10^4) were cultured in the presence of EBV
either alone or mixed with autologous T, T4-negative or
T8-negative cells (10×10^4) in microtiter plates. At the
end of a 14-day culture period the number of Ig secreting
cells was determined.

suppressed EBV-induced Ig production by autologous B cells
at 14 days, while T4-enriched T-cell subsets consistently
had little inhibitory effect. These experiments
demonstrate that "late suppression" is mediated by
EBV-immune T cells expressing a T8 phenotype.

Ig production induced by EBV is dependent upon
infection of B cells with the virus and does not require a
cooperative interaction of B cells with T cells (Kirchner
et al., 1979; Bird and Britton, 1979). We asked whether
monocytes have a role in B-cell activation by EBV and in
suppression mediated by EBV-immune T cells. In typical
experiments (Table II), monocytes were not required for
either Ig production by EBV or for suppression by
EBV-immune T cells. These findings demonstrate that
monocytes are not a necessary component in "late
suppression," and suggest that EBV-immune T-cell
regulation derives from a direct interaction of T8-
positive T cells with autologous EBV-infected B cells.

Further analysis revealed that monocytes have a
negative effect on EBV-immune T-cell inhibition. Thus,
many-fold fewer T cells were required to achieve a given
degree of B cell inhibition at 14 days in the absence of
monocytes than in the presence of monocytes. As shown

TABLE II

Monocytes are not Required for EBV-Immune T Cell
Suppression

Immunoglobulin Secreting Cells/Culture

Exp. no.	B cells*		B + auto T cells*	
	with monocytes	without	with monocytes	without
1	15,450	18,238	6,917	2,122
2	29,338	21,899	327	126
3	30,753	28,054	3,772	761
4	26,318	22,063	13,899	513
5	31,837	27,515	6,738	1,279
geo mean	25,911	23,255	3,806	668

*B cells and monocyte-depleted B cells (250,000/culture)
were incubated for 14 days in the presence of EBV either
alone or mixed with monocyte-depleted autologous T cells
(250,000/culture). The Ig-secreting cell response was
determined at the end of the culture period.

in a typical experiment (Fig. 2), 7,800 T cells produced
approximately 50 per cent inhibition of Ig production by
the autologous EBV-infected B cells at 14 days in
monocyte-depleted cultures; in contrast, approximately the
same degree of T-cell inhibition was achieved with 125,000
T cells (8-fold more T cells) in identical cocultures
containing monocytes. Similar experiments repeated in a
group of 16 EBV-immune normal subjects revealed that at
T:B cell ratios ranging from 12.5:1 to 0.78:1, the
degree of T-cell inhibition was, on the average, 40-50 per
cent higher in monocyte-depleted as compared to
monocyte-reconstituted but otherwise identical cocultures
(Fig. 3). This finding could not be attributed to a
"feeder-cell" effect of monocytes, since monocyte-depleted
irradiated B cells or T cells were ineffective in this
system (not shown).

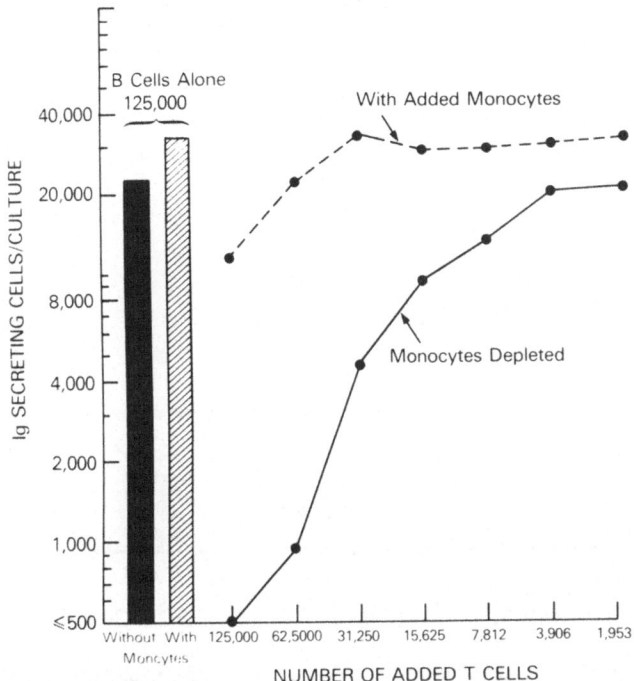

Fig 2. Monocytes have a contrainhibitory effect on EBV-
        immune T-cell suppression. Monocyte-depleted B
        cells (125,000) were cultured in the presence of
        EBV either alone or mixed with 125,000 autologous
        irradiated non-T cells containing 56% monocytes.
        EBV-infected monocyte depleted and monocyte
        reconstituted B cells (125,000) were also
        cocultured with autologous monocyte-depleted T
        cells at varying cell densities. At the end of
        a 14 day culture period the number of Ig-secreting
        cells was determined.

     Monocyte contrasuppression was not restricted by the
major histocompatability complex, since random allogeneic
monocytes were active in this system, nor was dependent
upon immunity to EBV of the monocyte donor (data not
shown).

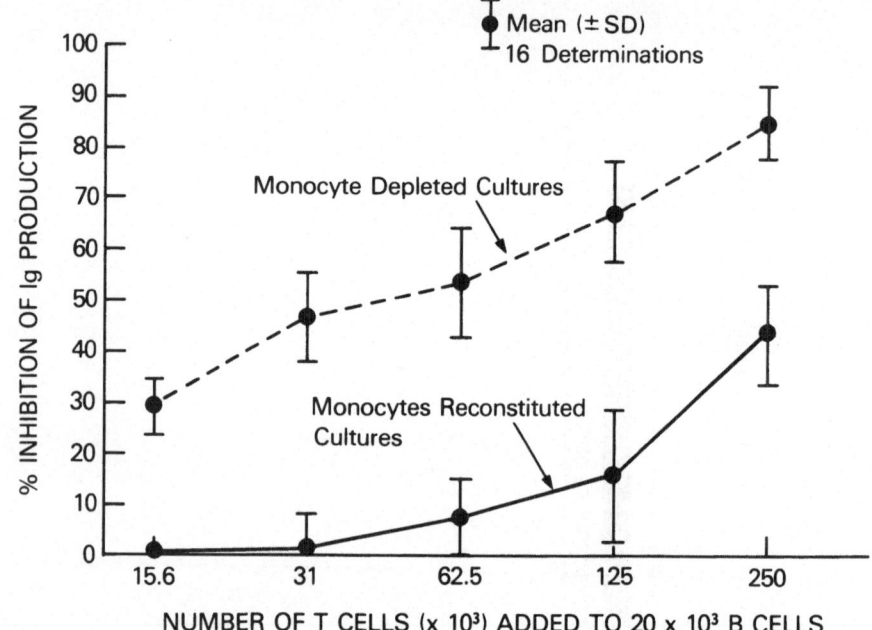

Fig 3. Monocyte contrasuppression of EBV-immune regulatory T cells. Culture conditions are identical
to those described for Fig 2. The results are
expressed as the mean percent suppression (±S.D.)
of 16 determinations.

Discussion

It has been known for some time that T cells from
normal individuals have an ability to inhibit B-cell
"transformation" by EBV (Thorley-Lawson et al., 1977; Moss
et al., 1978). Many immunoregulatory mechanisms have been
reported to contribute to T-cell immunoregulation of EBV
infection, including specific (Rickinson et al., 1980) and
nonspecific (Masucci et al., 1983) cytotoxicity,
suppression (Tosato et al., 1982), and interferon
production (Thorley-Lawson, 1981). Thus, it was reported
that B-cell proliferation and Ig production induced by EBV
are inhibited by T cells, and EBV-activated B cells are
killed by T cells. A number of distinguishing features
characterize individual regulatory processes, including
the requirement of previous immunity to EBV of the T-cell
donor, the stage of B-cell activation on which T cells are

effective, and the necessity of major histocompatibility complex identity between the T cells and the target B cells.

We have used Ig production as a measure of B-cell activation by EBV, and looked at suppression of EBV-induced Ig production as a test for T-cell inhibition of B-cell activation by EBV. The characteristic features of this system have been previously reported (Tosato et al., 1982). These include both a requirement for T-cell immunity to EBV of the lymphocyte donor, and histocompatibility identity between the T cells and the EBV-infected B cells. Suppression by EBV-immune T cells becomes evident 12-14 days after culture initiation, and is reversible, since EBV-infected B cells under suppressor T-cell regulation can be rescued and resume their ability to generate large numbers of Ig-secreting cells if depleted of T cells (Tosato et al., 1982).

In this study, we have further characterized EBV-immune T-cell suppression and attributed the inhibitory function to T cells expressing a T8 phenotype. We cannot exclude the possibility that small numbers of contaminating T4-positive T cells might be required for optimal suppression, but certainly the experiments demonstrate that most of the inhibitory function resides in T cells expressing a T8 phenotype.

In an attempt to further characterize EBV-immune T-cell suppression, we have looked at the effects of monocytes in this system. It was previously reported that monocytes contribute to lymphocyte immortalization by EBV (Pope et al., 1974). Our results suggest that this monocyte function is most likely related to a "contrainhibitory" effect of monocytes on T-cell suppression. Monocyte depletion of purified B cells does not significantly affect Ig production by EBV, suggesting that in these cultures monocytes have little or no direct effect on EBV-activated B cells. However, in the presence of autologous EBV-immune T cells, monocytes reduced significantly the degree of T-cell inhibition. Thus, a given number of added EBV-immune T cells was profoundly suppressive of Ig production by monocyte-depleted autologous B cells, but was only slightly inhibitory if the same target B cells were supplemented with monocytes. This "contrasuppressor" function is not simply due to a "feeder" effect of the monocytes, since irradiated monocyte-depleted B or T cells were not active in this system.

Thus, in addition to other immunoregulatory mechanisms previously reported, monocytes play an important and unique role in the control of EBV infection in man, acting as regulatory cells with "contrasuppressor" function.

References

Bianco, C., Plasma membrane receptors for complement. In Biological Amplification Systems in Immunology. N.K. Day and R.A. Good, eds. Rockefeller University, New York pp. 69-84 (1977).

Bird, A.G., and Britton, S., A new approach to the study of human B lymphocyte function using an indirect plaque assay and a direct B cell activator. Immunol. Rev. 45, 41-66 (1979).

Diehl, V., Henle, G., Henle, W., and Kohn, G., Demonstration of a herpes group virus in cultures of peripheral leukocytes from patients with infectious mononucleosis. J. Virol. 2, 663-669 (1968).

Henle, G. and Henle, W., In "The Epstein-Barr Virus" (M. Epstein and B. Achong, eds) pp. 297-320. Springer-Verlag, Berlin, 1979.

Kirchner, H., Tosato, G., Blaese, R.M., Broder, S., and Magrath, I.T., Polyclonal immunoglobulin secretion by human lymphocytes exposed to Epstein-Barr virus in vitro. J. Immunol. 122, 1310-13 (1979).

Masucci, M.G., Bejarano, M.T., Masucci, G., and Klein, E., Large granular lymphocytes inhibit the in vitro growth of autologous Epstein-Barr virus-infected B cells. Cell Immunol. 76, 311-321 (1983).

Moss, D.J., Rickinson, A.B., and Pope, J.H., Long-term cell mediated immunity to Epstein Barr virus in man. I. Complete regression of virus infected transformation in cultures of seropositive donor leukocytes. Int. J. Cancer 22, 662-668 (1978).

Nilsson, K., Klein, G., Henle, W., and Henle, G., The establishment of lymphoblastoid lines from adult and fetal human lymphoid tissue and its dependence on EBV.   Int. J. Cancer 8, 443-450 (1971).

Pope, J.H., Scott, W., and Moss, D.J., Cell relationships in transformation of human leukocytes by Epstein-Barr virus.   Int. J. Cancer 14, 122-129 (1974).

Rickinson, A.B., Wallace, L.E., and Epstein, M.A., HLA-restricted T-cell recognition of Epstein-Barr virus-infected B cells.   Nature 283, 865-867 (1980).

Thorley-Lawson, D.A., Chess, L., Strominger, J.I., Suppression of in vitro Epstein-Barr virus infection.   A new role for adult human T lymphocytes.   J. Exp. Med. 146, 495-508 (1977).

Thorley-Lawson, D.A., The transformation of the adult but not newborn lymphocytes by Epstein-Barr virus and phytohemagglutinin is inhibited by interferon: the early suppression by T cells of Epstein-Barr virus infection is mediated by interferon.   J. Immunol. 126, 829-833 (1981).

Tosato, G., Magrath, I.T., and Blaese, R.M., T-cell mediated immunoregulation of Epstein-Barr virus (EBV) induced B lymphocyte activation in EBV-seropositive and EBV-seronegative individuals.   J. Immunol. 128, 575-579 (1982).

Tosato, G. and Blaese, R.M., Epstein-Barr virus infection and immunoregulation in man.   Adv. Immunol., in press.

Yachie, A., Miyawaki, T., Yokoi, T., Nagaochi, T., and Taniguchi, N., Ia-positive cells generated by PWM stimulation within OKT4 + subset interact with OKT8 + cells for inducing active suppression on B cell differentiation in vitro.   J. Immunol. 129, 103-106 (1982).

53

ANALYSIS OF INTRATUMORAL LYMPHOCYTE SUBSETS IN PATIENTS
WITH UNDIFFERENTIATED NASOPHARYNGEAL CARCINOMA

T. Tursz, P. Herait, M. Lipinski, G.Ganem, M.C.
Dokhelar, C. Carlu, C. Micheau and G. de The

Groupe d'Immunobiologie des Tumeurs, Institute Gustave-
Roussy, 94805 Villejuif Cedex, France

INTRODUCTION

Typical histopathologic features of undifferentiated
nasopharyngeal carcinoma (NPC) consist of an intrication
of large epithelial tumor cells and numerous lymphocytes
(Shanmugaratnam et al., 1979). The lymphoid cells
within the tumor are cytologically normal, do not
contain EBV and have been shown to belong in majority
to the T-cell lineage (Galili et al., 1980). Whether
these lymphocytes are remnants of the cells present
in the normal nasopharyngeal mucosa or rather indicators
of a local immune reaction to the malignant cells
might be a keypoint in the understanding of this
intrication of epithelial and lymphocytic cells.
     The availability of monoclonal antibodies specific
for subsets of lymphocytes now allows for the analysis
of lymphocytic populations involved in this infiltrate.
Most studies reported so far have been carried out on
lymphocytes isolated from the tumor (Galili et al.,
1980). This method does not allow a direct examination
of the histological distribution of the lymphocytic
subsets within the tumor. Using a panel of monoclonal
antibodies, we have stained frozen and paraffin-
embedded sections of NPC. We present immunological
and immunofluorescence data from 43 different tumors.

METHODS AND MATERIALS

Patients and tissue samples.   Tumor tissues
were from 45 patients with NPC treated at the Institut
Gustave-Roussy. Most of them originated from North
Africa. Diagnoses were made by the same pathologist
(C.M.) and based on histological criteria previously
described.

Twelve biopsy samples were frozen for subsequent immunoperoxidase staining. Five tumors were gently teased and the lymphocytes recovered for immunofluorescence assays. Two specimens were cut in half and treated by both methods. Twenty-nine paraffin-embedded specimens were immunoperoxidase-stained with the monoclonal antibody HNK-1. Peripheral blood was drawn from 31 patients. Monoclonal cells were separated by Ficoll centrifugation and stained by indirect immunofluorescence.

Section immunoperoxidase staining. Indirect immunoperoxidase staining was performed on cryostat sections with all monoclonal antibodies except HNK-1, a reagent which can be used on paraffin sections. Cryostat sections were from tumor specimen fixed in acetone, stored at 70°C, then rehydrated in phosphate-buffered saline (PBS). Paraffin sections were treated with xylene, rehydrated with ethanol and water, and immersed in Tris HCl buffer for 10 min. Endogenous peroxidase was blocked by incubation with 0.3% H_2O_2. Non-specific reactions were prevented by incubation with normal rabbit serum. Sections were then incubated with staining antibody, washed in cold PBS, then incubated with a peroxidase-conjugated goat anti-rabbit Ig antiserum. The reactions were revealed with H_2O_2 and 3-amino-9-ethyl-carbazol or diaminobenzidine. Finally, slides were counter-stained with hematoxilin. The detection of Ig-bearing cells was carried out by the same method using a rabbit anti-human Ig antiserum followed by a peroxidase-conjugated goat anti-rabbit Ig antiserum as developing reagent. Positive cells were semi-quantitatively estimated from + to ++++ on cryostat sections and counted on paraffin sections.

Immunofluorescence assays were performed according to the classical two-step method. Conjugated antisera were from Dako Laboratories, Denmark. Monoclonal antibodies used included OKT3 to the T3 pan-T cell antigen, OKT4 to the helper/inducer T cell subset, OKT8 to the suppressor/cytotoxic subpopulation, OKT6 to the T6 antigen present on mature thymocytes, OKT10 which detects activated T-cells and progenitors

(Reinherz and Schlossman, 1980), all from the Ortho
Diagnostic System. The L1/1/12 hydridoma producing
an anti- HLA-DR monoclonal antibody was supplied by
George Khalil (Hopital St. Louis, Paris).
    HNK-1-producing hybridoma (Abo and Balch, 1981)
was purchased from the American Type Culture Collection
(Rockville, MD). They were used as culture supernatant
or as purified Ig diluted appropriately.

RESULTS

Tumor section indirect immunoperoxidase staining.
Frozen sections of 14 different specimen of NPC were
stained. Tumor cells were easily recognized with
their large nucleus containing several nucleoli and
appearing "chromatin empty" after hematoxin staining.
    Lymphocytes present in the tumor were most often
located around groups of malignant cells, but sometimes
disseminated within the tumor "nests". Most lymphocytes
belonged to the T-cell lineage as defined by their
reactivity with the OKT3 monoclonal antibody (Table 1).

Table 1: Estimated frequency of T-cell associated-
and HLA-DR antigens on lymphoid cells present in NPC
cryostat sections.

Patient	T3	T4	T8	HLA-DR	T10
10	++++	+++	+++	0	ND
20	+	+	+	++	ND
21	+++	+++	++	++	ND
28	+++	+	++	+	+
29	++	+	+	+	++
32	++	++	+	++++	ND
33	++++	+++	++	++++	ND
34	+	+	++	++	ND
35	++++	++++	++	+	ND
36	++++	++++	++	+	ND
37	+++	++	++++	+++	++
38	++++	+++	++	ND	+
39	++++	+++	++	ND	+
40	+	+	+	+	+++

Two patterns of reactivity were observed with respect to the distribution of T-lymphocytes subsets defined by OKT4 and OKT8 antibodies. Most often OKT4+ cells were more numerous (Table 1) and evenly distributed within the lymphoid stroma, whereas OKT8+ cells, more heterogeneously located, appeared to surround the tumor masses. In only 3 patients were OKT8+ cells more numerous, while in 3 patients OKT4+ and OKT8+ cells were present in approximately equal numbers. In six specimens tested with the OKT10 monoclonal antibody, positive cells were constantly detected with a varying number of reactive cells (+ to ++++). HLA-DR-positive lymphocytes were found in all 12 tumors tested but one. The estimated number of positive cells (+++ and ++++) was surprisingly high in 3 tumors.

To estimate the percentage of natural killer cells present in the lymphocytic infiltrates, 29 additional paraffin-embedded specimens were stained with the HNK-1 monoclonal antibody (Table 2).

Table 2: HNK-1-positive cells in paraffin sections of NPC.

% of total lymphoid cells	Number of tumors (total= 29)
0-5	20
6-10	5
11-15	4

In all patients, the percentage of HNK-1+ cells was below 15% of total lymphocytes. Approximately half the sections showed no or very few scattered stained cells. When more numerous, HNK-1-positive cells exhibited the cytological pictures of "large glandular lymphocytes" (LGL), typical of human NK cells.

Indirect immunofluorescence assay on cells isolated from the tumor. Seven surgical specimens were dissociated to get a suspension of cells. After Ficoll centrifugation, mononuclear cells were immuno-fluorescently stained (Table 3).

Table 3: Immunofluorescence staining of mononuclear cells isolated from NPC biopsies.

| Patient | % positive in total mononuclear cells | | | | | |
	T3	T4	T8	HLA-DR	sIg	T3-HLA-DR
28	65	2	55	45	13	29
29	50	30	20	20	10	6
41	29	ND	ND	37	ND	26
42	46	22	22	58	43	28
43	38	30	16	60	ND	19
44	38	20	23	44	35	22
45	43	24	28	40	38	27

On the average, OKT3-positive cells were more numerous than sIg-bearing cells (41.1 ± 10.5% vs. 27.8 ± 6.6%). Within the T lineage, T4 and T8 cells were in roughly equal proportions in three patients, T4+ cells were more numerous in two patients, whereas they were virtually lacking in one patient (Table 3). In all cases HLA-DR-positive cells were more frequent than cells carrying surface Ig (43.4 ± 12.4% vs. 27.8 ± 6.6%). To determine whether the HLA-DR-positive sIg-negative lymphocytes expressed other lymphocyte-associated cell-surface antigens, double staining assays were performed. In every test, a lymphocyte population of 6 to 29% was demonstrated to coexpress the T3 and HLA-DR antigens. In the two patients tested with OKT4 and OKT8 antibodies, the DR-positive T-cell population included both T4 and T8 lymphocytes (data not shown).

DISCUSSION

We have undertaken the phenotypic analysis of lymphocytes present in NPC by a combination of immuno-histological staining of frozen and paraffin sections of tumor biopsies and immunofluorescent staining of lymphoid cells isolated from tumor specimens.

Most of the lymphoid infiltrate was found to be composed of lymphocytes expressing the T3 antigen characteristic of the T-cell lineage (Reinherz and Schlossman, 1980). Among these T lymphocytes, the distribution T helper/inducer and suppressor/cytoxic cells, defined by antibodies OKT4 and OKT8, respectively, varied from one patient to another. However, although less numerous than T4-positive cells in most patients, T8-positive lymphocytes tended to be located predominantly around the tumor cell masses.

Because this phenotypically-defined T8-positive subset is known to include the functional population of cytotoxic cells, it is tempting to postulate that the T8-expressing lymphocytes, found in close relationship with the malignant cells, might play a role in the immune reaction of the tumor. In this regard, it was striking to observe that a relatively high-- although varying from one tumor to another--proportion of T3-positive cells also expressed HLA-DR molecules, as detected in double staining assays. This was in agreement with the observation of more numerous HLA-DR-positive than sIg-bearing cells in lymphoid cells isolated from the tumor. The T10 antigen which, in the periphery, is specifically found on activated T cells, was also detected in all tumors where it was looked for. It is also noteworthy that we studied the reactivity of lymphocytes from NPC tumors with monoclonal antibodies directed against the receptor for interleukin 2 (IL-2) TAC antigen (Uchiyama et al., 1981). This IL-2 receptor is only expressed on activated T cells. Again, we found in all the patients studied, that a high number of T cells (15-25%) were expressing the IL-2 receptor (data not shown). These T cells are are easily grown in IL-2-containing medium. We are presently attempting to clone them in order to look for some specific immune functions.

In conclusion, we report here that the T cells infiltrating NPC tumors demonstrated interesting phenotypic features:

1) Large variations in the T4/T8 ratio, with a usual excess of T4 cells, but with a few striking exceptions;
2) The presence of HNK1+ lymphocytes, often with the morphology of LGL, in high number (6 to 15% of the lymphocytes recovered from the tumor in 9 out of 29 patients), suggesting a role for NK cells in the local defense against NPC;

3) The constant presence of cells with the phenotype of "activated " T lymphocytes, expressing HLA-DR antigens and the IL-2 receptor.

Such data suggest the existence of local immunological reactions in NPC involving both T and NK cells. The lymphocytic infiltration observed in NPC could reflect some important defense mechanism. Similar studies are needed to elucidate their precise nature and their role in the control of the disease. Attempts to correlate the immunological features here described with the clinical grades and/or with the prognosis of NPC are presently being undertaken.

This work was supported by grants from the Centre National de la Recherche Scientifique (CNRS), the Association pour la Recherche sur la Cancer (ARC), and the Institut Gustave-Roussy, Paris and Villejuif, France.

REFERENCES

ABO, T. and BALCH, C.M., A differentiation antigen of human NK and K cells identified by monoclonal antibody (HNK1). J. Immunol., 127, 1024-1029 (1981).

GALILI, U., KLEIN, E., KLEIN, G. and SINGH BAL, I., Activated T lymphocytes in infiltrates and draining lymph nodes of nasopharyngeal carcinoma. Int. J. Cancer, 25, 85-89 (1980).

REINHERZ, E.L., and SCHLOSSMAN, S.F., The differentiation and function of human T lymphocytes: A review. Cell, 19, 821-829 (1980).

SHANMUGARATNAM, K., CHAN, S.H., DE THE, G., GOH, J.E.H., KHOR, T.H., SIMONS, M.J. and TYE, C.Y., Histopathology of nasopharyngeal carcinoma: Correlations with epidemiology, survival rates, and other biological characteristics. Cancer, 44, 1029-1044 (1979).

UCHIYAMA, T., BRODER, S., and WALDMANN, T.A., A
monoclonal antibody (anti-Tac) reactive with activated
and functionally mature human T cells. J. Immunol.,
126, 1393-1397 (1981).

54

POTENTIAL USEFULNESS OF ISOPRINOSINE AS AN IMMUNOSTIMULA-
TING AGENT IN EBV-ASSOCIATED DISORDERS: IN VITRO STUDIES

Syam K. Sundar, Giuseppe Barile, José Menezes

Laboratory of Immunovirology, Ste-Justine

Hospital, Montreal, Quebec, Canada H3T 1C5

SUMMARY

Isoprinosine has been shown to be an immunopotentiating agent on both nonspecific and specific antigenic
stimuli. Here we have investigated whether isoprinosine
enhances in vitro lymphocyte responses to EBV antigens
and phytohemagglutinin. In addition, its effect on
inhibitory action of serum-blocking factors, purified
from the sera of patients with infectious mononucleosis,
was evaluated. The results show that isoprinosine
enhanced significantly in vitro various cellular immune
responses to the above-mentioned stimuli. Moreover,
isoprinosine abrogated the inhibitory action of a known
blocking factor purified from the sera of IM patients.
Therefore, the use of isoprinosine as an immunopotentiating agent in EBV-associated disorders and malignancies
deserves serious consideration.

INTRODUCTION

Immunosuppression is a phenomenon generally observed
in viral infections and malignancies, including in EBV-
associated disorders. Various attempts are being made to
enhance the in vivo cell-mediated immune responses in
patients with EBV-related malignancies. Inhibitory
factors were detected in the sera of patients with infectious mononucleosis (IM) and undifferentiated nasopharyn-

geal carcinoma that could abrogate the responses of sensitised peripheral-blood mononuclear cells (PBL) to specific antigens and recall antigens (Lai et al., 1978; Wainwright et al., 1979; Veltri et al., 1981; Sundar et al., 1982; 1983); it is likely that these inhibitors are, at least partly, responsible for the in vivo immunosuppression observed in these disorders. Because of the well-known effect of isoprinosine, a synthetic inosine-containing complex, as an immunopotentiating agent (Hadden et al., 1981) we have used this novel agent to investigate the responses of PBL to EBV antigens and phytohemagglutinin (PHA). In addition, the effect of isoprinosine on the inhibitory action of purified serum-blocking factors (IgG) was analysed in the lymphocyte stimulation assay. The results presented here indicate that isoprinosine can abrogate the inhibitory effect of serum-blocking factors, thus suggesting that this drug may be useful as an immunopotentiating agent in EBV-associated disorders and malignancies.

MATERIALS AND METHODS

Reagents

Raji cells and P3HR1 virus were purchased from Life Sciences Inc., St.Petersburg, Florida. EBV soluble antigen extraction and titration by complement fixation test and P3HR1 virus inactivation by UV light were carried out as described earlier (Sundar et al., 1982).

Isoprinosine was kindly supplied by New Port Pharmaceuticals, Newport, California. Fresh solutions were prepared in distilled water when needed and sterilized by filtration.

Sera of patients with IM were collected at this hospital and IgG was purified by affinity chromatography on protein A Sepharose columns.

Lymphocyte Stimulation Test

This radiolabelled assay was performed as described earlier (Sundar et al., 1982) using the PBL obtained from

582

EBV-seropositive and seronegative individuals. The effect of sera or IgG fractions was tested against lymphocyte responses to EBV antigens and PHA in the lymphocyte stimulation assay.

Effect of Isoprinosine on the Response of PBL to Various Stimuli

Isoprinosine was added at the beginning of the experiments and its effect on the response of lymphocytes was assayed by the lymphocyte-stimulation test as described above.

Effect of Addition of Isoprinosine at Different Periods of Incubation on the Lymphocyte Responses

Isoprinosine was added at one day intervals and the cultures were further incubated for a total of 6 days at 37°C, after which they were labelled with radioactive thymidine. The incorporated label (counts per minute, CPM) was determined in a liquid scintillation counter.

Calculations

All cultures were carried out in triplicates. The stimulation indices (SI) and % increase due to isoprinosine were calculated as follows:

$$SI = \frac{\text{test cpm} - \text{control cpm}}{\text{control cpm}}$$

$$\% \text{ increase} = \left(\frac{\text{test SI} - \text{control SI}}{\text{control SI}}\right) \times 100$$

RESULTS

Effect of Isoprinosine on the Responses of PBL

The results of experiments dealing with the effect of isoprinosine on the responses of PBL to EBV antigens

Table 1

IMMUNOPOTENTIATING EFFECT OF ISOPRINOSINE ON EBV ANTIGEN- AND PHA-INDUCED LYMPHOPROLIFERATION

Donor		Stimulation index					
		PHA		EBV-particles		Soluble antigen	
No	Antibody titre to EBV-VCA	Without IPN[a]	With IPN	Without IPN	With IPN	Without IPN	With IPN
1	< 5	30	86 (187%)	0.5	0.9 (0%)	0.9	0.8 (0%)
2	< 5	39	208 (433%)	0.7	0.8 (0%)	0.9	0.9 (0%)
3	160	52	198 (280%)	7.6	33.1 (335%)	6.0	28.5 (375%)
4	160	24	91 (279%)	2.3	9.9 (330%)	2.5	14.4 (476%)
5	80	28	153 (446%)	6.1	21.1 (246%)	7.5	31.5 (320%)

[a]IPN = Isoprinosine was used at a concentration of 5 x 10^{-4} M.

and PHA are given in Table 1. Isoprinosine enhanced these responses between 187 and 476%.

Effect of IgG (Purified from the Sera of Patients with IM and from Normal Controls) on Lymphoproliferation and the Effect of Isoprinosine

As shown in Tables 2 and 3, IgG fractions from the sera of IM patients significantly inhibited the lymphocyte responses to EBV antigens and PHA. It is noteworthy that isoprinosine abolished the inhibitory action of these IgG fractions. In contrast to this, the IgG fraction from healthy donor did not show any inhibitory activity.

584

Table 2

INHIBITORY EFFECT OF SERA AND THEIR IgG FRACTIONS ON
LYMPHOCYTE RESPONSES

Source of serum	Stimulating agent	Serum		IgG	
		SI	% inhibition	SI	% inhibition
IM patient	PHA	4.3	88	3.4	90
	SAg	2.3	64	1.4	78
IM patient	PHA	12.9	65	7.7	79
	SAg	1.3	96	1.2	97
IM patient	PHA	9.6	74	7.8	79
	SAg	0	100	0	100
Control autologous serum	PHA	41.9	0	38.5	0
	SAg	5.9	8	5.4	15

The sera were tested on the response of lymphocytes from
a single donor. Responses (SI) to PHA and SAg were 36.6
and 6.4, respectively.

Effect of Addition of Isoprinosine, at Different Times, on Immunopotentiation

Isoprinosine was able to enhance the lymphocyte
responses to EBV antigens by 262, 282, 327, 297, and 342%
when it was added to the cultures on 0-5 days of
incubation, respectively. However, no significant
enhancement (34%) was seen when it was added on day 6,
the day when the cultures were harvested.

DISCUSSION

The results presented above clearly demonstrate that
isoprinosine enhanced the in vitro cellular immune

585

Table 3

EFFECT OF SERUM BLOCKING FACTOR ON MITOGEN- AND EBV
ANTIGEN-INDUCED LYMPHOCYTE RESPONSES AND ITS ABROGATION
BY ISOPRINOSINE

EBV-VCA antibody titers for donors	Treatment	Lymphocyte stimulation test		
		CPM	% inhibition by SBF	% increase due to Isoprinosine (IPN)[a]
160	PHA control	35,746		
	PHA+IPN	136,834		282
	PHA+SBF	4,289	88	
	PHA+SBF+IPN	63,985		79
	SAg[b]	3,286		
	SAg+IPN	15,608		375
	SAg+SBF	262	92	
	SAg+SBF+IPN	7,820		138
< 5	PHA control	45,889		
	PHA+IPN	128,489		181
	PHA+SBF	10,095	78	
	PHA+SBF+IPN	94,168		105
	SAg	1,243[c]		
	SAg+IPN	1,404		13[c]
	SAg+SBF	1,323	0	
	SAg+SBF+IPN	1,350		0

[a]IPN was added at a final concentration 5 x 10^{-4} M.
[b]SAg: EBV soluble antigen.
[c]Statistically insignificant.

responses to EBV-related antigens and PHA. Most significant indeed is the fact that isoprinosine abolished the
inhibitory action of purified serum-blocking factors on
lymphocyte activation by both EBV antigens and mitogen.
It is interesting that this enhancement could be obtained
even when Isoprinosine was added as late as day 5 of the
incubation period.

The effect of isoprinosine has been demonstrated in vivo on both experimental animals and human infections (Hadden et al., 1981). Recently in a double-blind study, this compound was shown to accelerate the restoration of immune responses in patients with solid tumors who received radiotherapy (Fridman et al., 1980), and apparently enhanced the cellular immune responses as well as the production of interleukin-2 in patients with acquired immune deficiency syndrome (AIDS) (Tsang et al., 1984; Grieco et al., 1984).

In conclusion, the results presented here clearly indicate that isoprinosine may be a useful agent to potentiate, as well as to restore, the cellular immune responses in patients with EBV-associated disorders and malignancies.

ACKNOWLEDGEMENTS

This study was supported by grants from Medical Research Council of Canada, "Fondation Justine Lacoste-Beaubien" and the Cancer Research Society (Montreal). JM is a Senior Scholar from "Fonds de la Recherche en Santé du Québec". G.B., on leave from "Instituto di Tecnologie Biomediche" (CNR, Rome, Italy), was supported by a fellowship from "Associazione Italiana per la Ricerca sul Cancro".

REFERENCES

Fridman, H., Calle, R., and Morin, A., Double blind study of isoprinosine influence on immune parameters in solid tumor-bearing patients treated with radiotherapy. Int. J. Immunopharmacol., 2, 194 (1980).

Grieco, M.H., Reddy, M.M., Manvar, D., Ahuja, K.K., and Moriarty, M.L., In vivo immunomodulation by isoprinosine in patients with acquired immunodeficiency syndrome and related complexes. Ann. Int. Med., 101, 206 (1984).

Hadden, J.W., and Giner-Sorolla, A., Isoprinosine and NPT 15392: modulators of lymphocyte and macrophage development function. In: E.M. Hersh, M.A. Chirigos and M.J. Mastrangelo (ed.), Augmenting Agent in Cancer Therapy, pp. 497, Raven Press, New York (1981).

Lai, P.K., Mackay Scollary, E.M., Fimmel, P.J., Alpers, M.P., and Keast, D., Cell mediated immunity to Epstein-Barr virus and a blocking factor in the patients with infectious mononucleosis. Nature (Lond), 252, 608 (1974).

Sundar, K.S., Ablashi, D.V., Kamaraju, L.S., Levine, P.H., Faggioni, A., Armstrong, G.R., Krueger, G.R.F., Hewetson, J.F., Bertram, G., Sesterhenn, K., and Menezes, J., Sera from patients with undifferentiated nasopharyngeal carcinoma contain a factor which abrogates specific Epstein-Barr virus antigen-induced lymphocyte response. Int. J. Cancer, 29, 407 (1982).

Sundar, K.S., Menezes, J., Levine, P.H., Ablashi, D.V., Faggioni, A., Kamaraju, L.S., and Prasad, U., Studies on the relation of IgA to the lymphocyte stimulation inhibitor (LSI) present in the sera of patients with undifferentiated nasopharyngeal carcinoma. In: U. Prasad, D.V. Ablashi, G.R. Pearson and P.H. Levine (ed.), Nasopharyngeal Carcinoma. Current Concepts, pp. 293, University of Malaya, Malaysia (1983).

Tsang, K.Y., Fudenberg, H.H., and Galbraith, G.M., In vitro augmentation of interleukin-2 production and lymphocytes with the TAC antigen marker in patients with AIDS (letter). New Engl. J. Med., 310, 987 (1984).

Veltri, R.W., Kikta, V.A., Wainwright, W.H., and Sprinkle, M., Bologic and molecular characterisation of the serum blocking factor (SBF-IgG) isolated from sera of patients with infectious mononucleosis. J. Immunol., 127, 320 (1981).

Wainwright, W.H., Veltri, R.W., and Sprinkle, M., Abrogation of cell mediated immunity by a serum blocking factor isolated from patients with infectious mononucleosis. J. Inf. Dis., 140, 22 (1979).

55

MECHANISMS OF EXPRESSION OF A BURKITT LYMPHOMA-ASSOCIATED
ANTIGEN (GLOBOTRIAOSYLCERAMIDE) IN BURKITT LYMPHOMA AND
LYMPHOBLASTOID CELL LINES

J. WIELS[1], E.H. HOLMES[2], N. COCHRAN[2], S.I. HAKOMORI[2],
T. TURSZ[1],

[1]Groupe d'Immunobiologie des Tumeurs,

INSTITUT GUSTAVE-ROUSSY, 94805 VILLEJUIF CEDEX,   FRANCE.

[2]Fred Hutchinson Cancer Research Center,
UNIVERSITY OF WHASHINGTON, SEATTLE, WA 98104, U.S.A.

INTRODUCTION

Increasing evidence obtained by chemical analysis and
by examination of monoclonal antibodies directed to tumor
antigens has shown that many tumor -associated antigens in
experimental and human cancers are glycosphingolipids
(Hakomori and Kannagi, 1983). One remarkable example has
been found in characterization of the Burkitt lymphoma-
associated antigen (BLA) defined by a monoclonal antibody,
38.13 (Wiels et al, 1981). The antigens ha been
characterized as globotriaosylceramide (Gb_3) (Nudelman et
al., 1983), and is highly expressed in most Burkitt
lymphoma cell lines, whether the lymphoma cells contain
the Epstein-Barr Virus (EBV) genome (Central- East African
endemic type) or not (European-North American type) (Wiels
et al., 1982). The antigen is not expressed on EBV-
positive lymphoblastoid cell lines, on various other
leukemia and lymphoma-derived cell lines, and on normal
eythrocytes (Klein et al, 1983). This paper describes the
enzymatic basis of Gb_3 antigen synthesis and the
organizational difference in Gb_3 expression in membranes
of Burkitt lymphoma and various lymphoblastoid cell lines.

MATERIALS AND METHODS

Growth of cells.
Various Burkitt lymphoma cell lines (Ramos, Daudi, Put, Namalwa), lymphoblastoid cell lines (Priess, ARH77), the hybrid cell line between a Burkitt lymphoma and a lymphoblastoid cell line (Put/ARH77Cl2), and mouse leukemia L1210 were cultured in RPMI 1640 medium supplemented with 10 % fetal calf serum. Cells were stored frozen at -80°C before use.

Preparation of Golgi membrane-rich fractions from cells.
The preparation of Golgi membrane-rich fractions was carried out at 0-4°C using a modified procedure of Senn et al. (1981) as described elsewhere (Wiels et al, 1984).

Protein determination.
Protein concentrations of cell fractions were determined by the method of Lowry et al. (1951) using bovine serum albumin as a standard.

Enzyme assays.
α-Galactosyltransferase and Galactosidase
The exact procedure as been described elsewhere (Wiels et al., 1984).

Glycolipid purification and characterization.
Cellular glycolipids were extracted and stained as previously described (Nudelman et al, 1983). After cell surface glycolipid labeling with galactose oxidase, followed by treatment with $NaB[^3H]_4$ (Gambergard et. al., 1973), the glycolipids in the labeled cells were extracted.

Cell surface expression of BLA as determined by fluorescence-activated cell sorting.
Classical immunofluorescence tests were undertaken on cells. In order to study the role of membrane organization in Gb_3 expression, cells were treated with either Vibrio cholera sialidase (0.125 international units/ml) for 1 hr at room temperature or 0.125 % final concentration of trypsin for 30 min at room temperature. Cells were then washed twice with PBS before labeling with 38.13 antibody.

RESULTS

$\underline{Gb_3}$ content and α-galactosyltransferase activity in lymphoid cell lines.

The high reactivity of Burkitt lymphoma cell lines Put, Daudi, P3HR1, and Ramos to the antibody 38.13 was correlated with the high chemical quantity of Gb_3 in these cell lines (see Fig. 1). The absence of reactivity of lymphoblastoid cell lines Priess and Remb, and one exceptional Burkitt lymphoma cell line, Namalwa, is consistent with the absence of Gb_3 in these cell lines (lanes 7-9, Fig. 1). The presence or absence of antigen expression and the chemical quantity of Gb_3 in these cell lines, as above, have been correlated with the activity of UDP-Gal : lactosylceramide α-galactosyltransferase, as shown in Table I.

TABLE I. SPECIFIC ACTIVITY OF α-GALACTOSYLTRANSFERASE IN CRUDE HOMOGENATES AND GOLGI MEMBRANE FRACTIONS OF DIFFERENT LYMPHOID CELL LINES

Cell lines	Type	Crude Homogenate	Golgi Membrane pmol/hr/ mg/prot.	BL Antigen Expression pmol/hr/ mg prot.
Ramos	BL°	655 ± 186	3327 ± 587	+
Daudi	BL	113.5 ± 1	627 ± 174	+
Put	BL	72 ± 8	835 ± 2	+
Namalwa	BL	25 ± 6	113 ± 46	-
L1210	Mouse leukemia	14 ± 6	133 ± 51	-
Priess	LCL°°	32.5 ± 6	202 ± 50	-
ARH	LCL	107 ± 12	667 ± 64	-
PUT/ARH77 Cl12	Hybrid	107 ± 58	917 ± 192	+

°BL : Burkitt lymphoma   °°LCL : Lymphoblastoid cell line

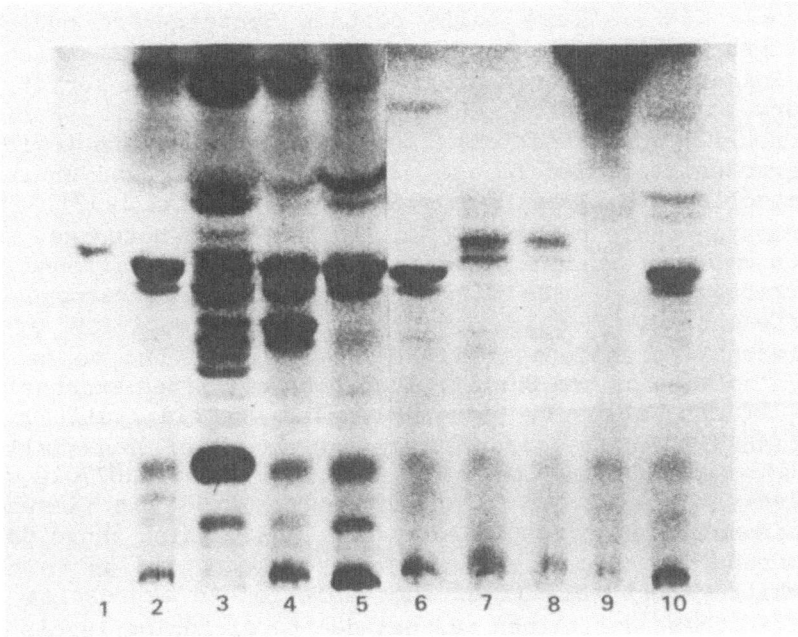

<u>Figure 1</u> : Thin-Layer chromatography (TCL) of neutral
glycolipid fraction of Burkitt lymphoma and lymphoblastoid
leukemia cell lines. Lane 1, reference
globotriaosylceramide (Gb$_3$) isolated from human
erythrocytes ; lane 2, Burkitt lymphoma line Put ; lane 3,
lymphoblastoid cell line ARH77 ; lane 4, a hybrid cell
line between Put and ARH77 (PUT/ARH77 Cl12) ; lane 5,
Burkitt lymphoma line Ramos ; lane 6, Burkitt lymphoma
line Daudi ; lane 7, lymphoblastoid cell line Remb 1 ;
lane 8, lymphoblastoid cell line Priess ; lane 9, Burkitt
lymphoma line Namalwa (unreactive with 38.13 antibody) ;
lane 10, Burkitt lymphoma line Ramos. A slow-migrating
spot with great intensity present in lane 3 was a free
sugar and was eliminated by dialysis. The TLC plate was
reacted with 0.2 % orcinol in 2 M sulfuric acid.

Interestingly, one lymphoblastoid cell line, ARH77, did
not express Gb$_3$ detected by monoclonal antibody 38.13 ;
however, it contained a similar chemical quantity of Gb$_3$
as Ramos, Daudi, and Put (see lane 3 of Fig. 1, as compared
to lanes 2, 5, and 6). It should be noted that ARH77 was

derived from a patient with plasma cell leukemia (Edwards et al, 1982). These cells contain Epstein-Barr nuclear antigen (EBNA), as do most lymphoblastoid cell lines (such as Priess), but other criteria such as aneuploid karyotype (Burk et al, 1978 and R. Berger, Hop. St Louis, Paris, personal communication) and various morphological characteristics led to its classification as an "unusual lymphoblastoid cell line" (Edwards et al, 1982). The chromatographic pattern of Gb_3 in ARH77 was heterogen with several bands, in contrast to the pattern found in Burkitt cells. These differences probably correspond to different fatty acid compositions. The activity of α-galactosyl-transferase in ARH77 was also found to be as high as some of the Burkitt lymphoma cell lines expressing Gb_3 (see Table I, line 7). The hybrid cell line Put/ARH77C12 also expressed a large amount of Gb_3, similar to its two parental lines. Interestingly, Put/ARH77C12 was heavily labeled by the antibody 38.13 in immuno-fluorescence studies (Klein et al, 1983), which shows that membrane Gb_3 was as accessible in this hybrid as in the Burkitt parent cell line. The variation in expression of Gb_3 in these cell lines was not due to differing levels of α-galactosidase. In each case, the activity was very low and nearly identical (results not shown).

Characterization of α-galactosyltransferase in Burkitt Lymphoma cells.

α-Galactosyltransferase activity of Daudi cells was enriched 6.4 fold in a Golgi membrane-rich fraction as compared to the crude homogenate (data not shown). Further characterization of this enzyme indicated that it had maximum activity using cacodylate buffer, pH 5.9, 0.3 % Triton X-100, and 5 mM $MnCl_2$. In order to protect substrates and product from endogenous hydrolytic activity, assays were conducted in the presence of 5 mM CDP-choline and 5 mM galactonolactone. Under these conditions, the α-galactosyltransferase activity was proportional to both time and protein concentration (kinetics not shown). The apparent Km values for LacCer and UDP-galactose were determined to be 0.28 mM and 62 μM, respectively.

Surface exposure of Gb_3 determined by galactose oxidase and Nab $[^3H]_4$.

The degree of cell surface exposure of Gb_3 of Ramos, Daudi, P_3HR1, and Put cells was compared with that of ARH77 cells by galactose oxidase^3-NaB$[^3H]_4$. Gb_3 of Ramos was strongly labeled (Fig. 2B, lane 1), in contrast to that of ARH77 cells which was not labeled (Fig. 2B, lane 2).

Similarly, Gb_3 in Daudi, P3HR1, and Put cells were strongly labeled (data not shown). The presence or absence of Gb_3 in Burkitt and non-Burkitt lymphoblastoid cell lines was assessed by immunostaining of Gb on TLC by the antibody 38.13. The glycolipid fraction isolated from Ramos and ARH77 gave an intense spot (Fig. 2C, lanes 3 and 4, respectively) in contrast to that of mouse leukemia cell L1210 and non-Burkitt lymphoblastoid cells, which did not give a spot by immunostaining (Fig. 2C, lanes 2,5 and 6). Thus, it is clear that ARH77 cells contain as high a level of Gb3 as Burkitt cell lines (Ramos, Daudi, P3HR1, and Put), but Gb3 in ARH77 cells is not exposed at the cell surface.

Antibody reactivity of Gb3 at the cell surface as determined by cytofluorometry.

The remarkable organizational difference between Gb3 in Burkitt cells and ARH77 cells was further substantiated by quantitative immunofluorescence through cytofluorometry. Only the data for P3HR1 and Ramos cell lines in comparison with ARH77 cells are shown in Fig.3. Both P3HR1 and Ramos showed a strong reactivity with the antibody 38.13 (Fig. 3, Ab, Aa, Bb). A similar reactivity was demonstrated by other Burkitt cells (Put, Daudi) (data not shown). In contrast, ARH77 cells showed very little reactivity with the primary antibody 38.13 (Fig. 3, Ca), and the reactivity was not affected by trypsin treatment (Fig. 3, Cb), but was significantly enhanced after sialidase treatment (Fig. 3, Cb). In all these experiments, control cells with or without enzyme treatment showed no significant reactivity if the primary or secondary antibody was omitted (Fig. 3, Aa, Ba, Ca, Cc). A significant intrinsic interaction of Ramos cells with the secondary antibody was observed (Fig. 3 Bc), but it did not influence the high reactivity of Ramos cells with the primary antibody.

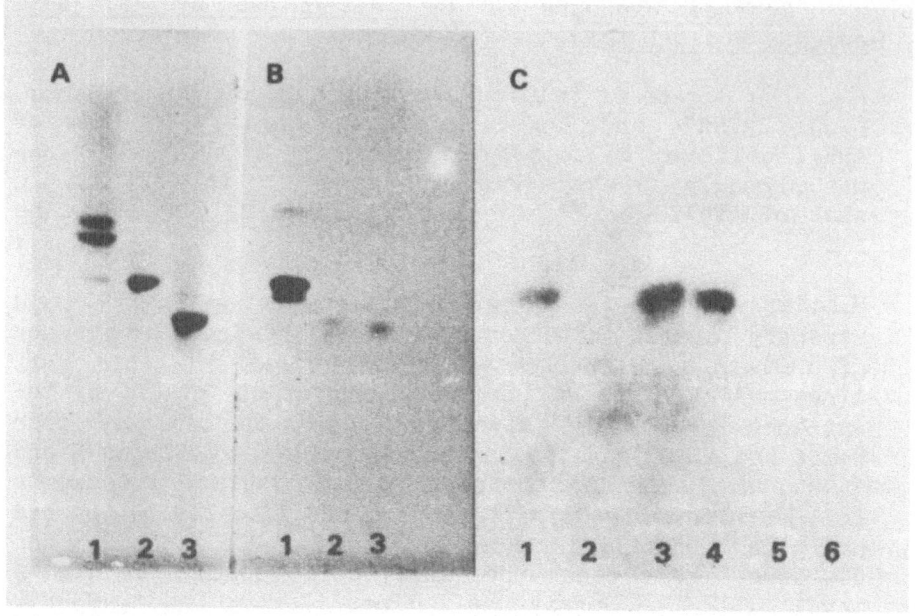

Figure 2 : Surface-labeled globotriaosylceramide and immunostaining pattern of glycolipids of Burkitt lymphoma and lymphoblastoid cells. Panel A, TLC pattern of lactosylceramide (lane 1), globotriaosylceramide (lane 2), and globoside (lane 3). Panel B, lane 1, surface-labeled globotriaosylceramide from Burkitt lymphoma cell line Ramos treated with galactose oxidase and NaB ^{3}H$_4$; lane 2, the neutral glycolipid fraction isolated from ARH77 lymphoblastoid cells treated with galactose oxidase and NaB ^{3}H$_4$; lane 3, the neutral glycolipid fraction of ARH77 cells treated with NaB ^{3}H$_4$ only (without galactose oxidase). Panel C, immunostaining of the neutral glycolipid fraction isolated from various sources. Lane 1, Gb$_3$ isolated from human erythrocytes ; lane 2, neutral glycolipid fraction of L1210 cells ; lane 3, neutral glycolipid fraction of Burkitt lymphoma cell line Ramos ; lane 4, neutral glycolipid fraction of ARH77 cells ; lane 5, neutral glycolipid fraction of lymphoblastoid cell line Remb ; lane 6, neutral glycolipid fraction of lymphoblastoid cell line Priess.

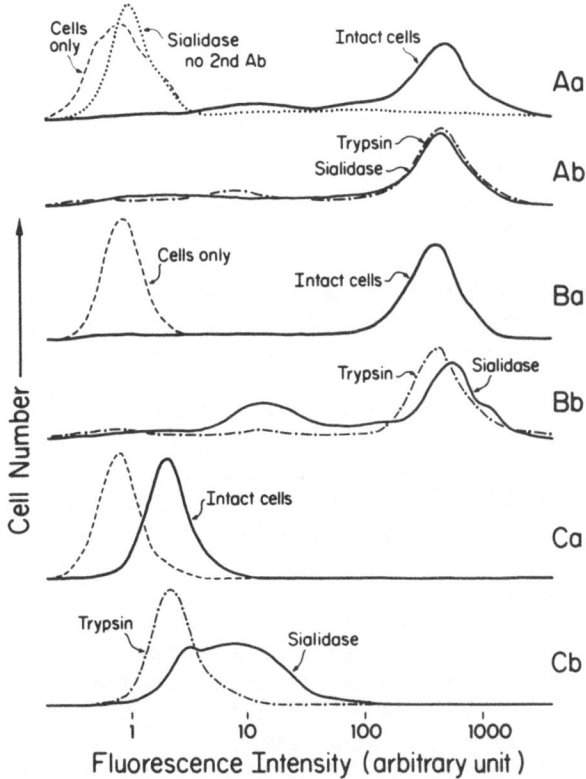

<u>Figure 3</u> : Cytofluorometric pattern of Burkitt lymphoma cell lines (Aa, b, c ; Ba, b, c) and lymphoblastoid cell line ARH77 (Ca, b, c), and the effect of sialidase on exposure of Gb_3. Abscissa, fluorescence intensity ; ordinate, frequency of cells. Panel A, Burkitt lymphoma P3HR1, Panel B, Burkitt lymphoma Ramos, Panel C, lymphoblastoid cell line ARH77.

596

DISCUSSION

A number of Burkitt lymphoma cell lines isolated from both African endemic type (EBV-positive group) or North American type (EBV-negative group) were characterized by the presence of a specific antigen designated BLA, which was defined by the monoclonal antibody 38.13 (Wiels et al, 1981, 1982). Recently, BLA was identified chemically as globotriasylceramide (Gb_3 ; $Gal\alpha 1 \rightarrow 4Gal\ \beta 1 \rightarrow 4Glc\ \beta 1 \rightarrow Cer$), (Nudelman et al. 1983).

The Burkitt lymphoma cell lines Daudi, Ramos, P3HR1, and Put contained a high chemical quantity of Gb3 and had a high reactivity of this antigen at the cell surface. In contrast, non-Burkitt lymphoblastoid cell lines (Priess, Remb, and ARH77) were all characterized by the absence of Gb3 reactivity at the cell surface. Priess and Remb did not contain Gb3, but ARH77 cells contained as high a level of Gb3 as Burkitt lymphoma. Thus, studies have been directed towards two subjects : i) the enzymatic synthesis of Gb3 and characterization of the enzyme (α-galactosyltransferase) responsible for synthesis of Gb3, and ii) surface exposure of Gb3 and its reactivity at the cell surface. The Burkitt cell lines were characterized by high activity of the enzyme responsible for synthesis of Gb3 antigen, i.e. UDP-Gal : lactosylceramide α-galactosyltransferase. Non-Burkitt lymphoblastoid cell lines and one exceptional Burkitt lymphoma cell line, Namalwa, did not contain Gb3 antigen, and the enzyme activity for synthesis of Gb3 was very low. The activation of α-galactosyltransferase thus appears to be the mechanism responsible for Gb3 accumulation in Burkitt lymphoma cells.

The extent of surface exposure and the reactivity of Gb_3 with the antibody 38.13 and with galactose oxidase was much higher in Burkitt lymphoma cells than in the other lines tested. Thus, the chemical quantity, enzyme activity, and immunological reactivity of the antigen expressed at the cell surface have been correlated in those cell lines as described above. In striking contrast, however, one lymphoblastoid cell line, ARH77, contained the same chemical quantity of the glycolipid antigen and the same level of activity of the enzyme for antigen synthesis as Burkitt lymphoma cell lines ; nevertheless, ARH77 did not express the antigen at the cell surface, and the reactivity of the antigen with galactose oxidase-NaB $[^3H]_4$ was negligible. Thus, Gb_3 is chemically present and synthesized in ARH77 cells, but is not exposed at the cell surface. This was further confirmed by cytofluorometric analysis. However, Gb_3 in ARH77 cells became exposed when cells were treated with sialidase, indicating that Gb_3 in ARH77 cells is organized in a cryptic form. A similar change of crypticity of gangliotriaosylceramide (Gg_3) in L5178 lymphoma cells was previously reported (Kannagi et al, 1983). No sialylated or sialidase-sensitive forms of Gg_3 in ARH77 cells or Gg_3 in L5178 lymphoma have been found. The phenomenon is, therefore, ascribable to a change of a second glycoconjugate by sialidase which affects exposure of cryptic Gb_3 and Gg_3. In the case of L5178 lymphoma, the antigen Gg_3 could be masked by co-existing GM_{1b}, which is sialidase-sensitive and is present in "low expressor" cells of Gg_3 in L5178 (Kannagi et al, 1983).

The antigenicity and immunogenicity of carbohydrates at the cell surface are not only defined by the primary chemical structure of the carbohydrate antigens, but also are influenced by the organization of carbohydrates at the cell surface membranes. Even though the antigens are synthesized and are present in plasma membranes, they may not be immunogenic nor recognizable as discrete antigens unless the carbohydrate chains are organized in a distinctive way that makes them reactive with antibodies at the cell surface. Dominant factors affecting expression of the antigens are the density and crypticity of carbohydrate chains at the cell surface. The majority of Burkitt lymphoma cells may be characterized by a high density of Gb_3, except for a few such as Namalwa. They are also characterized by the loss of crypticity which is associated with a deletion of glycolipids with longer carbohydrate chains. The co-presence of longer carbohydrate chains was associated with a remarkable inhibition of the antigenicity of the shorter chain glycolipid antigen (Kannagi et al, 1983). The case of ARH77 is another remarkable example indicating that crypticity is as important as the chemical quantity of glycolipids in the expression of glycolipid antigens.

ACKNOWLEDGEMENTS

This investigation has been supported by NIH grants GM23100, CA20026, and CA19224 (to S.H.), grant 126017 from the Institut National de la Santé et de la Recherche Médicale, grant 80A20 from the Institut Gustave-Roussy, and grant 829 from the UER Kremlin-Bicêtre (to T.T.). Joëlle Wiels has been supported by a fellowship grant through the International Union Against Cancer (ICRETT).

[1]Glycolipids are abbreviated according to the recommendation of the IUPAC Nomenclature Comm. (1977).

REFERENCES

Burk K.H., Drewinko, B., Trujullo, J.M., and Ahearn, M.J. Establishment of a human plasma cell line in vitro. Cancer Res. 38, 2508-2513 (1978).

Edwards, P.A.W., Smith, C.M., Munro Neville, A., and O'Hare, M.J. A human-human hybridoma system based on a fast growing mutant of the ARH 77 plasma cell leukemia derived line. Eur. J. Immunol. 12, 641-648 (1982).

Gahmberg, C.G., and Hakomori, S. External labelling of cell surface galactose and galactosamine in glycolipid and glycoprotein of human erythrocytes. J. Biol. Chem. 248, 4311-4317 (1973).

Hakomori, S., and Kannaji, R. J. Natl. Glycosphingolipids as tumor-associated and differentiation markers. Cancer Inst. 71, 231-251 (1983).

IUPAC-IUB Commission on Biochemical Nomenclature Lipids 12, 453-463 (1977).

Kannagi, R., Stroup, R., Cochran, N.A., Urdal, D.L., Young, W.W., Jr., and Hakomori, S. Glycolipid tumor antigen in cultured murine lymphoma cells and factors affecting its expression at the cell surface. Cancer Res. 43, 4997-5005 (1983).

Klein, G., Manneborg-Sandlund, A., Ehlin-Henriksson, B., Godal, T., Wiels, J., and Tursz, T. Expression of the BLA antigen, defined by the monoclonal 38.13 antibody, on Burkitt lymphoma lines, lymphoblastoid cell lines, their hybrids and other B-cell lymphomas and leukemias. Int. J. Cancer 31, 535-542 (1983).

Lowry, O.H., Rosebrough, N.J., Farr, A.L., and Randall, R.J. J. Biol. Chem. 193, 265-275 (1951).

599

Nudelman, E., Kannagi, R., Hakomori, S., Parsons, M., Lipinski, M., Wiels, J., Fellous, M., and Tursz, T. A glycolipid antigen associated with Burkitt lymphoma defined by a monoclonal antibody. Science 220, 509-511 (1983).
Senn, H.J., Cooper, C., Warnke, P.C., Wagner, M., and Decker, K. Ganglioside biosynthesis in rat liver. Characterization of UDP-N acetylgalactosamine - GM3 acetylgalactosaminytransferase. Eur. J. Biochem. 120, 59-67 (1981).
Wiels, J., Fellous, M., and Tursz, T. Monoclonal antibody against a Burkitt lymphoma associated antigen. Proc. Natl. Acad. Sci. USA 78, 6485-6488 (1981).
Wiels, J., Lenoir, G.M., Fellous, M., Lipinski, M., Salomon, T.C., Tetaud, C., Tursz, T. A monoclonal antibody with anti-Burkitt lymphoma specificity I. Analysis of human haematopoietic and lymphoid cell lines. Int. J. Cancer 29, 653-658 (1982).
Wiels, J., Holmes, E.H., Cochran N., Tursz, T. and Hakomori S.I. Enzymatic and organizational difference in expression of BL antigen (globotriaosylceramide) in Burkitt lymphoma and lymphoblastoid cell lines. J. Biol. Chem. 259 : 14783-14787 (1984).

CONTROL OF EBV-ASSOCIATED MALIGNANT DISEASE

PREVENTION OF EB VIRUS-ASSOCIATED MALIGNANT DISEASES

M A EPSTEIN

Professor of Pathology,   University of Bristol

Department of Pathology
University of Bristol Medical School
University Walk, Bristol BS8 1TD, UK

INTRODUCTION

Epstein-Barr (EB) virus was discovered just over 20 years ago (Epstein et al., 1964) and since that time work from many laboratories has established remarkably close associations between this agent and endemic Burkitt's lymphoma (BL) (Burkitt, 1963), undifferentiated nasopharyngeal carcinoma (NPC) (Shanmugaratnam, 1971), and the lymphomas which occur in immunosuppressed human allograft recipients with an unusually high frequency (Penn, 1978; Kinlen et al., 1979; Weintraub and Warnke, 1982). The basis of these associations is well known (for reviews see Epstein and Achong, 1979; de Thé, 1980; Klein and Purtilo, 1981) and work on cellular oncogene activation in BL now suggests possible explanations. Thus, the virus together with certain co-factors appears to be an essential link in a series of interlocking steps which facilitate characteristic chromosomal translocations (Lenoir et al., 1982) leading to activation of the cellular myc oncogene (Dalla-Favera et al., 1982; Taub et al., 1982) which in turn may cause transformed cells to progress to full malignancy; the HuBlym-1 oncogene also seems to be implicated (Diamond et al., 1983).

But whatever the details of such oncogene activation in BL, and irrespective of whether they operate in the other EB virus-associated tumours, quite recent studies on the experimental induction of lymphomas by the virus

in vivo demonstrate that it can potently, rapidly, and
directly set in motion the chain of events which leads to
the appearance of malignant tumours (Cleary et al., 1984b).
Such a direct role is not entirely surprising in the light
of recent findings on transformation by EB virus in vitro;
this has often been categorized as being merely a form
of "immortalization" (Miller, 1980), yet careful tests
have shown that in addition to the latter phenomenon some
cells are changed in such a way as to possess from the
outset many of the attributes of malignant transformation
(Zerbini and Ernberg, 1983).

The accumulation of information both on the general
biological behaviour of EB virus and on its role in human
malignancies has progressed at an ever increasing pace.
But although this is unquestionably of scientific
importance, it has seemed for some years that the value of
such activities would be considerably enhanced if they
could lead to intervention against infection by the virus
which might in consequence reduce the incidence of the
associated tumours.   It was in this context that proposals
were put forward for a vaccine against EB virus
(Epstein, 1976) and considerable progress in this direction
has subsequently been achieved.

JUSTIFICATION FOR A VACCINE AGAINST EB VIRUS

The striking evidence implicating EB virus, together
with co-factors, in the causation of endemic BL and
undifferentiated NPC has already been alluded to.   BL
occurs frequently only in rather limited areas and even in
these does not involve very large numbers (Burkitt, 1963);
furthermore, the high incidence areas are just those with
many more pressing medical and community health problems.
In contrast, undifferentiated NPC is the most common
tumour of men and the second most common of women amongst
Southern Chinese (Shanmugaratnam, 1971), has a high
incidence amongst Eskimos (Lanier et al., 1980), and there
are moderately high incidence levels in North Africa
(Cammoun et al., 1974), East Africa (Clifford, 1970), and
through most of South East Asia (Shanmugaratnam, 1971).
Thus in world cancer terms undifferentiated NPC is of very
considerable significance and is thought to be responsible
for more than 100,000 deaths a year; this figure alone
justifies efforts to develop a vaccine against EB virus.

The control of a naturally occurring herpesvirus-
induced lymphoma of chickens, Marek's disease (Marek, 1907;
Payne et al., 1976), by inoculation with apathogenic virus

(Churchill et al., 1969; Okazaki et al., 1970)
provided the first example of anti-viral vaccination
affecting the frequency of a cancer. Later work with the
malignant lymphoma which can be induced experimentally by
inoculation of Herpesvirus saimiri in South American sub-
human primates (Meléndez et al., 1969) has shown that
animals given killed virus vaccine were protected against
challenge infection and did not therefore get tumours
(Laufs and Steinke, 1975). Furthermore, in the Marek's
disease system, antigen-containing membranes from cells
infected with Marek's disease herpesvirus markedly
reduced lymphoma incidence when used as an experimental
vaccine in chickens (Kaaden and Dietzschold, 1974) and
even soluble viral antigens extracted from such cells
protected in the same way (Lesnick and Ross, 1975).
Similar approaches with EB virus, one of the five human
herpesviruses, have long appeared worthy of investigation.

A VACCINE BASED ON EB VIRUS MA

It has been known for many years that the virus-
neutralizing antibodies developed by EB virus-infected
individuals are those directed against the virus determined cell surface membrane antigen (MA) (Pearson et al.,
1970; Pearson et al., 1971; Gergely et al., 1971;
de Schryver et al., 1974) and this information prompted
the suggestion that MA be used as an anti-viral vaccine
(Epstein, 1976).

Investigations into the molecular structure of MA have
identified two high molecular weight glycoprotein components of 340,000 and 270,000 daltons (gp340 and gp270)
(Qualtière and Pearson, 1979; Strnad et al., 1979;
Thorley-Lawson and Edson, 1979; North et al., 1980;
Qualtière and Pearson, 1980), and the concordance between
human antibodies to MA and EB virus-neutralization has
been formally explained by the demonstration of these same
glycoproteins in both the viral envelope and the cell
membrane MA (North et al., 1980). Not surprisingly,
therefore, monoclonal antibodies which react with both MA
components neutralize EB virus (Hoffman et al., 1980;
Thorley-Lawson and Geilinger, 1980) and gp340/270 can
themselves elicit virus-neutralizing antibodies (North
et al., 1982a). Most EB virus-producing lymphoid cell
lines synthesize roughly equal amounts of gp340/270.

REQUIREMENTS FOR A VACCINE BASED ON EB VIRUS MA

To elaborate a vaccine against EB virus based on MA
gp340/270, the following essential prerequisites must be

made available or devised:-
Susceptible test animals
    Only two kinds of animal are known to be fully suscep-
tible to experimental infection with EB virus, the owl
monkey (Aotus) (Epstein et al., 1973a and 1973b; 1975)
and the cotton-top tamarin (Saguinus oedipus oedipus)
(Shope et al., 1973; Miller et al., 1977; Miller, 1979).
However, the former "species" has recently been found to
be very heterogeneous with at least nine different
karyotypes (Ma et al., 1976; Ma et al., 1978; Ma 1981),
and shows considerable variation in susceptibility to
certain infections.
    The cotton-top tamarin is therefore the species of
choice for experimental studies in vivo with EB virus
even though it was placed on the endangered species list
some years ago.   For, although there was rather little
information about this animal and the possibility of its
successful propagation in captivity, the necessary
management and husbandry conditions have recently been
defined, and flourishing breeding colonies have been
established (Brand, 1981; Kirkwood et al., 1983;
Kirkwood, 1983; Kirkwood and Epstein, 1984).   Nevertheless, there are severe constraints on the numbers of the
rare and costly tamarins which can be used in each
experiment, similar to those operating in work with
hepatitis B virus where biological tests require the use
of chimpanzees.   Because of these constraints, it is
also necessary to test all methodologies with banal
laboratory animals (which will make antibodies, for
example, even though they cannot be infected with EB
virus), before applying them to the tamarins.
An assay for MA gp340
    In order to work out an efficient and reliable method
for the preparation of antigen, it is essential that the
product can be quantified and monitored at each step to
permit modifications which maximize yields.   Accordingly,
a highly sensitive, quantitative radioimmuno-assay (RIA)
was developed for gp340.   Small amounts of this molecule
were prepared in extremely pure, radioiodinated form,
were shown to be antigenic, and were thereafter used in a
conventional competition RIA to quantify unlabelled samples
of gp340 using a defined system of arbitrary units.   A
full account of the RIA has been published elsewhere
(North et al., 1982b).
A preparation method for MA gp340

As mentioned above, EB virus-producing cell lines usually synthesize equal amounts of gp340/270, but the B95-8 marmoset line (Miller et al., 1972) is anomalous in that it expresses almost exclusively the larger component, thus providing an important advantage for molecular mass-based purification. With the crucial help of the RIA a preparative sodium dodecyl sulphate polyacrylamide gel electrophoresis (SDS-PAGE) procedure was worked out for gp340 from B95-8 cell membranes which included an important new technique for ensuring that the product was renatured and thus in an antigenic form. This was achieved by removing SDS under conditions where protein refolding was prevented; the details have already been given (Morgan et al., 1983).

Enhancement of immunogenicity of MA gp340

gp340 made by the method just described proved only weakly immunogenic in mice and rabbits after repeated injection and the use of Freund's adjuvant. To eliminate the need for these two disadvantageous procedures gp340 was incorporated in liposomes (Morein et al., 1978; Manesis et al., 1979), sometimes with the addition of lipid A (Naylor et al., 1982), and comparative immunogenicity studies were undertaken to determine the best routes and methods of administration. Liposomes containing gp340, with or without lipid A, gave good titres of EB virus-neutralizing antibodies in mice, rabbits, and cotton-top tamarins after rather few inoculations, and all the sera were specific in that they reacted only with MA gp340 and failed to recognise any other molecules from the surface or interior of B95-8 cells. These experiments have been reported in full (North et al., 1982a; Morgan et al., 1984a).

A test for antibodies to MA gp340

In order to exploit immunogenicity studies to the full a sensitive test to quantitate antibody responses to gp340 was essential. A rapid enzyme-linked immunosorbent assay (ELISA) has therefore been developed based on gp340 purified by affinity chromatography using a monoclonal antibody immunoabsorbent (Randle et al., 1984). This ELISA has proved a thousand-fold more sensitive than conventional indirect immunofluorescence tests and has made it possible to follow accurately the sequential production of specific antibodies to gp340 during the immunization of animals. The ELISA is described in a recent publication (Randle and Epstein, 1984).

VACCINATION OF COTTON-TOP TAMARINS

To demonstrate vaccine protection of immunized tamarins
a dose and mode of administration of challenge EB virus
has been worked out which will ensure the induction of
lesions in 100% of unprotected normal animals.   The
lesions have been extensively investigated and both on
histological (Dorfman et al., 1982) and molecular biological (Arnold et al., 1983; Cleary et al., 1984a) grounds
must clearly be regarded as malignant lymphomas with
several interesting features (Cleary et al., 1984b).

Pilot experiment

When this challenge dose of virus was used in a small
scale preliminary experiment, a vaccinated animal whose
serum had been shown to have potent virus-neutralizing
capacity (Pearson et al., 1970; Moss and Pope, 1972) was
found to be totally protected, whereas other animals with
less neutralizing antibody were not (Epstein, 1984).

Confirmatory experiments

The demonstration that purified gp340 in liposomes can
induce virus-neutralizing antibodies in cotton-top
tamarins and the preliminary indication that these protect
against a highly pathogenic dose of challenge virus,
provide a clear mandate for confirmatory tests with larger
numbers of the expensive animals.

In this connection evaluation is under way both of gp340
obtained by the molecular mass-based technique (Morgan
et al., 1983) used from the outset and of gp340 purified
by the more recent monoclonal antibody immunoaffinity
chromatography method (Randle et al., 1984).   Comparison
of the protection induced by inoculation of these two
types of material should give valuable insights into the
biological complexity of gp340, since the first of the
two preparation procedures presumably yields all epitopes
on molecules of the appropriate molecular weight, whereas
the monoclonal antibody is known to bind only about 50 to
60% of the epitopes (Randle et al., 1984).   It will be
interesting to see which immunogen is most efficacious.

DEVELOPMENTS FOR THE FUTURE

Once it has been confirmed that experimentally induced
antibodies to EB virus-determined MA components indeed
protect tamarins against infection by the virus, the
situation will be exactly comparable to that long known for
the Marek's disease herpesvirus and Herpesvirus saimiri
systems (Kaaden and Dietzschold, 1974; Lesnick and Ross,
1975; Laufs and Steinke, 1975).   Planning for a gp340-

based vaccine for man should thus be considered sooner, rather than later.

The most advantageous human context in which to test such a vaccine is in relation to infectious mononucleosis (IM). It is well known that in Western countries groups of young adults can be screened to detect those who have escaped primary EB virus infection in childhood and who are therefore at risk for delayed primary infection which is accompanied by the clinical manifestations of IM in 50% of cases (Niederman et al., 1970; University Health Physicians et al., 1971). Screening could therefore be applied to new students entering Universities or Colleges followed by a double blind vaccine trial amongst informed, consenting volunteers in the "at risk" category. The effectiveness of immunization in preventing infection and reducing the expected incidence of IM would rapidly be evident.

Thereafter, the effect of vaccination and consequential prevention of disease should be assessed in a high incidence region for endemic BL. This tumour has a peak incidence at about the age of seven (Burkitt, 1963) and the influence of vaccination on this should therefore be apparent within a decade. If this were successful there would then be inescapable reasons for tackling the far more difficult, but more important problem of intervention against undifferentiated NPC. Since this is a disease of middle and later life in high incidence areas (Shanmugaratnam, 1971) immunity would have to be maintained over many years.

MA components prepared in the ways discussed here (Morgan et al., 1983; Randle et al., 1984) have never been considered suitable for anything beyond the present experimental prototype vaccine (Epstein, 1984). It was therefore important to know something of the structure of gp340 and of the contribution, if any, of the sugar moiety to antigenicity. Experiments have been undertaken in which gp340 was analysed after treatment with a battery of glycosidases and V8 protease, with or without preliminary exposure during synthesis to tunicamycin. This work, reported by Morgan et al. (1984b), has shown that carbohydrate represents more than 50% of the total mass of gp340, that it is both O- and N-linked, that V8 protease digestion fragments are antigenic, and that specific antibody appears to bind the protein not the sugar.

The seemingly preponderant importance of the protein in

the immunogenicity of gp340 means that for use in man the possibility of exploiting new procedures can be explored. The fragment of EB virus DNA carrying the gene coding for MA has already been identified (Hummel et al., 1984) and the sequence probably relating to this gene is also known (Biggin et al., 1984). The potentiality for cloning the gene and seeking to make the product by expression in suitable pro- or eukaryotic cells is thus very real. In addition, it can be readily envisaged that the practicability of using synthetic gp340 peptides as immunogens will soon be investigated. And however the subunit vaccine molecule is ultimately obtained, yet further possibilities lie in the direction of greatly enhanced immunogenicity using powerful new adjuvants (Morein et al., 1984).

Finally, there is an excellent chance that it may prove feasible to incorporate the EB virus MA gene into the genome of vaccinia virus and thus ensure its direct expression during vaccination in man (Smith et al., 1983). Such an achievement could well solve many of the biological and logistic problems of an EB virus vaccine intended for intervention in relation to undifferentiated NPC.

REFERENCES

Arnold, A., Cossman, J., Bakhshi, A., Jaffe, E.S., Waldmann, T.A. and Korsmeyer, S.J. Immunoglobulin-gene rearrangements as unique clonal markers in human lymphoid neoplasms. New Eng. J. Med., 309, 1593-1599 (1983).

Biggin, M., Farrell, P.J. and Barrell, B.G. Transcription and DNA sequence of the Bam H1 L fragment of B95-8 Epstein-Barr virus. EMBO J., 3, 1083-1090 (1984).

Brand, H.M. Husbandry and breeding of a newly established colony of cotton-topped tamarins (Sanguinus oedipus). Lab. Animals, 15, 7-11 (1981).

Burkitt, D. A lymphoma syndrome in tropical Africa. In: G.W. Richter and M.A. Epstein (ed.), International Review of Experimental Pathology, Vol. 2, pp. 67-138, Academic Press, New York and London (1963).

Cammoun, M., Hoerner, G.V. and Mourali N. Tumors of the nasopharynx in Tunisia: an anatomic and clinical study based on 143 cases. Cancer, 33, 184-192 (1974).

Churchill, A.E., Payne, L.N. and Chubb, R.C. Immunization against Marek's disease using a live attenuated
virus. Nature (Lond.), 221, 744-747 (1969).

Cleary, M.L., Chao, J., Warnke, R. and Sklar, J.
Immunoglobulin gene rearrangement as a diagnostic
criterion of B cell lymphoma. Proc. Nat. Acad. Sci.
(Wash.), 81 593-597 (1984a).

Cleary, M.L., Epstein, M.A., Finerty, S., Dorfman, R.F.,
Bornkamm, G.W., Kirkwood, J.K., Morgan, A.J. and
Sklar, J. Individual tumours of multifocal EB virus-
induced malignant lymphomas in tamarins arise from
different B cell clones. Submitted. (1984b).

Clifford, P. A review: on the epidemiology of nasopharyngeal carcinoma. Int. J. Cancer, 5, 287-309 (1970).

Dalla-Favera, R., Bregni, M., Erikson, J., Patterson, D.,
Gallo, R.C. and Croce, C.M. Human c-myc onc gene is
located on the region of chromosome 8 that is translocated
in Burkitt lymphoma cells. Proc. Nat. Acad. Sci. (Wash.),
79, 7824-7827 (1982).

De Schryver, A., Klein, G., Henle, W. and Henle, G. EB
virus-associated antibodies in Caucasian patients with
carcinoma of the nasopharynx and in long-term survivors
after treatment. Int. J. Cancer, 13, 319-325 (1974).

De-The, G. Role of Epstein-Barr virus in human diseases:
infectious mononucleosis, Burkitt's lymphoma, and naso-
pharingeal carcinoma. In: C. Klein (ed.), Viral Oncology,
pp. 769-797, Raven Press, New York (1980).

Diamond, A., Cooper, G.M., Ritz, J. and Lane, M.-A. Identification and molecular cloning of the Human Blym
transforming gene activated in Burkitt's lymphomas.
Nature (Lond.), 305, 112-116 (1983).

Dorfmann, R.F., Burke, J.S. and Berard, C. A working
formulation of non-Hodgkin's lymphomas: background
recommendation, histological criteria, and relationship
to other classifications. In: S. Rosenberg and H. Kaplan
(ed.), Malignant lymphomas, pp. 351-368, Academic Press,
New York (1982).

Epstein, M.A. Epstein-Barr virus - is it time to develop
a vaccine program? J. Nat. Cancer Inst., 56, 697-700
(1976).

Epstein, M.A. A prototype vaccine to prevent Epstein-Barr
(EB) virus-associated tumours. Proc. Roy. Soc. Lond., B
221, 1-20 (1984).

Epstein, M.A. and Achong, B.G. The relationship of the
virus to Burkitt's lymphoma. In: M.A. Epstein and B.G.
Achong (ed.), The Epstein-Barr Virus, pp. 321-337,
Springer, Berlin, Heidelberg and New York (1979).

Epstein, M.A., Achong, B.G. and Barr, Y.M. Virus particles in cultured lymphoblasts from Burkitt's lymphoma,
Lancet, i, 702-703 (1964).

Epstein, M.A., Hunt, R.D. and Rabin, H. Pilot experiments
with EB virus in owl monkeys (Aotus trivirgatus) I.
Reticuloproliferative disease in an inoculated animal.
Int. J. Cancer, 12, 309-318 (1973a).

Epstein, M.A., Rabin, H., Ball, G., Rickinson, A.B.,
Jarvis, J. and Melendez, L.V. Pilot experiments with EB
virus in owl monkeys (Aotus trivirgatus) II. EB virus in
a cell ine from an animal with reticuloproliferative
disease. Int. J. Cancer, 12, 319-332 (1973b).

Epstein, M., Zur Hausen, H., Ball, G. and Rabin, H.
Pilot experiments with EB virus in owl monkeys (Aotus
trivirgatus) III. Serological and biochemical findings
in an animal with reticuloproliferative disease. Int.
J. Cancer, 15, 17-22 (1975).

Gergely, L., Klein, G. and Ernberg, I. Appearance of
Epstein-Barr virus-associated antigens in infected Raji
cells. Virology, 45, 10-21 (1971).

Hoffmann, G.J., Lazarowitz, S.G. and Hayward, S.D. Monoclonal antibody against a 250,000-dalton glycoprotein of
Epstein-Barr virus identifies a membrane antigen and a
neutralizing antigen. Proc. Nat. Acad. Sci. (Wash.), 77
2979-2983 (1980).

Hummel, M., Thorley-Lawson, D.A. and Kieff, E. An Epstein-
Barr virus DNA fragment encodes messages for the two
major envelope glycoproteins (gp350/300 and gp220/200).
J. Virol., 49, 413-417 (1984).

Kaaden, O.R. and Dietzschold, B. Alterations of the
immunological specificity of plasma membranes of cells
infected with Marek's disease and turkey herpesviruses.
J. Gen. Virol., 25, 1-10 (1974).

Kinlen, L.J., Sheil, A.G.R., Peto, J. and Doll, R.
Collaborative United Kingdom-Australasian study of
cancer in patients treated with immunosuppressive
drugs. Brit. Med. J., 2, 1461-1466 (1979).

Kirkwood, J.K. Effects of diet on health, weight and
litter size in captive cotton-top tamarins (Saguinus
oedipus oedipus. Primates, 24, 515-520 (1983).

Kirkwood, J.K. and Epstein, M.A. Rearing a second generation of cotton-top tamarins (Sanguinus oedipus eodipus).
Submitted (1984).

Kirkwood, J.K., Epstein, M.A. and Terlecki, A.J. Factors
influencing population growth of a colony of cotton-top
tamarins. Lab. Animals, 17, 35-41 (1983).

Klein, G. and Purtilo, D.T. Epstein-Barr virus-induced
lymphoproliferative diseases in immunodeficient patients.
Klein, G. and Purtilo (ed.)., pp. 4209-4303, Cancer
Res., 41 (Supplement) (1981).

Lanier, A., Bener, T., Talbot, M., Wilmeth, S., Tschopp,
C., Henle, W., Henle, G., Ritter, D. and Terasaki, P.
Nasopharyngeal carcinoma in Alaskan Eskimos, Indians and
Aleuts: a review of cases and study of Epstein-Barr
virus, HLA and environmental risk factors. Cancer, 46,
2100-2106 (1980). Laufs, R. and Steinke, H. Vaccination
of non-human primates against malignant lymphoma. Nature
(Lond.), 253, 71-72 (1975).

Lenoir, G.M., Preud'homme, J.L., Bernheim, A. and
Berger, R. Correlation between immunoglobulin light
chain expression and variant translocation in Burkitt's
lymphoma. Nature (Lond.), 298, 474-476 (1982).

Lesnick, F. and Ross, L.J.N. Immunization against Marek's
disease using Marek's disease virus-specific antigens free
from infectious virus. Int. J. Cancer, 16, 153-163 (1975).

Ma, N.S.F. Chromosome evolution in the owl monkey,
Aotus. Amer. J. Phys. Anthropol., 54, 293-303 (1981).

Ma, N.S.F., Jones, T.C., Miller, A.C., Morgan, L.M. and Adams, E.A. Chromosome polymorphism and banding patterns in the owl monkey (Aotus). Lab. Animal Sci., 26, 1022-1036 (1976).

Ma, N.S.F., Rossan, R.N., Kelley, S.T., Harper, J.S., Bedard, M.T. and Jones, T.C. Banding patterns of the chromosomes of two new karyotypes of the owl monkey, Aotus, captured in Panama. J. Med. Primatol., 7, 146-155 (1978).

Manesis, E.K., Cameron, C.H. and Gregoriadis, G. Hepatitis B surface antigen-containing liposomes enhance humoral and cell-mediated immunity to the antigen. FEBS Letters, 102, 107-111 (1979).

Marek, J. Multiple Nerventzundung (Polyneuritis) bei Huhnern. Dtsch. tierarztl. Wschr., 15, 417-421 (1907).

Melendez, L.V., Hunt, R.D., Daniel, M.D., Garcia, F.G. and Fraser, C.E.O. Herpesvirus saimiri. II. An experimentally induced primate disease resembling reticulum cell sarcoma. Lab. Animal Care, 19, 378-386 (1969).

Miller, G. Experimental carcinogenicity by the virus in vivo. In: M.A. Epstein and B.G. Achong (ed.), The Epstein-Barr Virus, pp. 351-372, Springer, Berlin, Heidelberg and New York (1979).

Miller, G. Biology of the Epstein-Barr Virus. In: G. Klein (ed.), Viral Oncology, pp. 713-738, Raven Press, New York (1980).

Miller, G., Shope, T., Coope, D., Waters, L., Pagano, J., Bornkamm, G.W. and Henle, W. Lymphoma in cotton-top marmosets after inoculation with Epstein-Barr virus: tumor incidence, histologic spectrum, antibody responses, demonstration of viral DNA, and characterization of viruses. J. Exp. Med., 145, 948-967 (1977).

Miller, G., Shope, T., Lisco, H., Stitt, D. and Lipman, M. Epstein-Barr virus: transformation, cytopathic changes, and viral antigens in squirrel monkey and marmoset leukocytes. Proc. Nat. Acad. Sci. (Wash.), 69, 383-387 (1972).

Morein, B., Helenius, A., Simons, K., Petterson, R., Kaarianen, L. and Schirrmacher, V. Effective subunit vaccines against an enveloped animal virus. Nature (Lond.), 276, 715-718 (1978).

Morein, B., Sundquist, B., Hoglund, S., Dalsgaard, K. and Osterhaus, A. Iscom, a novel structure for antigenic presentation of membrane proteins from enveloped viruses. Nature (Lond.), 308, 457-460 (1984).

Morgan, A.J., Epstein, M.A. and North, J.R. Comparative immunogenicity studies on Epstein-Barr (EB) virus membrane antigen (MA) with novel adjuvants in mice, rabbits and cotton-top tamarins. J. Med. Virol., 13, 281-292 (1984a).

Morgan, A.J., North, J.R. and Epstein, M.A. Purification and properties of the gp340 component of Epstein-Barr (EB) virus membrane antigen (MA) in an immunogenic form. J. Gen. Virol., 64, 455-460 (1983).

Morgan, A.J., Smith, A.R., Barker, R.N. and Epstein, M.A. A structural investigation of the Epstein-Barr (EB) virus membrane antigen glycoprotein, gp340. J. Gen. Virol., 65, 397-404 (1984b)

Moss, D.J. and Pope, J.H. Assay of the infectivity of Epstein-Barr virus by transformation of human leucocytes in vitro. J. Gen. Virol., 17, 233-236 (1972).

Naylor, P.T., Larsen, H.L., Huang, L. and Rouse, B.T. In vivo induction of anti-Herpes simplex virus immune response by Type 1 antigens and Lipid A incorporated into liposomes. Infect. Immunity, 36, 1209-1216 (1982).

Niederman, J.C., Evans, A.S., Subrahmanyan, L. and McCollum, R.W. Prevalence, incidence and persistence of EB virus antibody in young dults. New Eng. J. Med., 282, 361-365 (1970).

North, J.R., Morgan, A.J. and Epstein, M.A. Observations on the EB virus envelope and virus-determined membrane antigen (MA) polypeptides. Int. J. Cancer, 26, 231-240 (1980).

North, J.R., Morgan, A.J., Thompson, J.L. and Epstein, M.A. Purified EB virus gp340 induces potent virus-neutralizing antibodies when incorporated in liposomes Proc. Nat. Acad. Sci. (Wash.), 79, 7504-7508 (1982a)

North, J.R., Morgan, A.J., Thompson, J.L. and Epstein, M.A. Quantification of an EB virus-associated membrane antigen (MA) component. J. Virol. Methods, 5, 55-65 (1982b).

Okazaki, W., Purchase, H.G. and Burmester, B.R. Protections against Marek's disease by vaccination with a herpesvirus of turkeys. Avian Dis., 14, 413-429 (1970).

Payne, L.N., Frazier, J.A. and Powell, P.C. Pathogenesis of Marek's disease. In: G.W. Richter and M.A. Epstein (ed.), International Review of Experimental Pathology, Vol. 16, pp. 59-154, Academic Press Inc., New York, San Francisco and London (1976).

Pearson, G., Dewey, F., Klein, G., Henle, G. and Henle, W. Relation between neutralization of Epstein-Barr virus and antibodies to cell-membrane antigens induced by the virus. J. Nat. Cancer Inst., 45, 989-995 (1970).

Pearson, G., Henle, G. and Henle, W. Production of antigens associated with Epstein-Barr virus in experimentally infected lymphoblastoid cell lines. J. Nat. Cancer Inst., 46, 1243-1250 (1971).

Penn, I. Malignancies associated with immunosuppressive or cytotoxic therapy. Surgery, 83, 492-502 (1978).

Qualtiere, L.F. and Pearson, G.R. Epstein-Barr virus-induced membrane antigens: immunochemical characterization of Triton X100 solubilized viral membrane antigens from EBV-superinfected Raji cells. Int. J. Cancer, 23, 808-817 (1979).

Qualtiere, L.F. and Pearson, G.R. Radioimmune precipitation study comparing the Epstein-Barr virus membrane antigens expressed on P$_3$HR-1 virus-superinfected Raji cells to those expressed on cells in a B95-8 virus-transformed producer culture activated with tumor-promoting agent (TPA). Virology, 102, 360-369 (1980).

Randle, B.J. and Epstein, M.A. A highly sensitive enzyme-linked immunosorbent assay to quantitate antibodies to Epstein-Barr virus membrane antigen gp340. J. Virol. Methods, 9, 201-208 (1984).

Randle, B.J., Morgan, A.J., Stripp, S.A. and Epstein, M.A. Large-scale purification of Epstein-Barr virus membrane antigen gp340 using a monoclonal immunosorbent. In press.

Shanmugaratnam, K. Studies on the etiology of nasopharyngeal carcinoma. In: G.W. Richter and M.A. Epstein (ed.), International Review of Experimental Pathology, Vol., 10, pp. 361-413, Academic Press Inc., New York and London (1971).

Shope, T., Dechairo, D. and Miller, G. Malignant lymphoma in cotton-top marmosets after inoculation with Epstein-Barr virus. Proc. Nat. Acad. Sci. (Wash.), 70, 2487-2491 (1973).

Smith, G.L., Mackett, M. and Moss, B. Infectious vaccinia virus recombinants that express hepatitis B virus surface antigen. Nature (Lond.), 302, 490-495 (1983).

Strnad, B.C., Neubauer, R.H., Rabin, H. and Mazur, R.A. Correlation between Epstein-Barr virus membrane antigen and three large cell surface glycoproteins. J. Virol., 32, 885-894 (1979).

Taub, R., Kirsch, I., Morton, C., Lenoir, G., Swan, D., Tronick, S., Aaronson, S. and Leder, P. Translocation of the c-myc gene into the immunoglobulin heavy chain locus in human Burkitt lymphoma and murine-plasmacytoma cells. Proc. Nat. Acad. Sci. (Wash.), 79, 7837-7841 (1982).

Thorley-Lawson, D.A. and Edson, C.M. The Polypeptides of the Epstein-Barr virus membrane antigen complex. J. Virol., 32, 458-467 (1979).

Thorley-Lawson, D.A. and Geilinger, K. Monoclonal antibodies against the major glycoprotein (gp350/220) of Epstein-Barr virus neutralize infectivity. Proc. Nat. Acad. Sci. (Wash.), 77, 5307-5311 (1980).

University Health Physicians and PHLS Laboratories. Infectious mononucleosis and its relationship to EB virus antibody. Brit. Med. J., iv, 643-646 (1971).

Weintraub, J. and Warnke, R.A. Lymphoma in cardiac allo-transplant recipients: clinical and histological features and immunological phenotype. <u>Transplantation</u>, 33, 347-351 (1982).

Zerbini, M. and Ernberg, I. Can Epstein-Barr virus infect and transform all the B-lymphocytes of human cord blood? <u>J. Gen. Virol.</u>, 64, 539-547 (1983).

57

A PERSPECTIVE ON TREATMENT OF EBV INFECTION STATES

Joseph S. Pagano, M.D.

Dept. of Medicine and Microbiology and Lineberger Cancer Research Center
Lineberger Cancer Research Center, University of North Carolina School of Medicine, Chapel Hill, NC 27514

The inception of antiviral therapy for Epstein-Barr virus infection, now upon us, is complicated by the different forms and states of EBV infection. EBV can cause at least four and probably five types of infection: 1) acute (both primary and reactivated), 2) persistent, 3) latent, 4) oncogenic and 5) abortive. Fortunately, all of these virologic states are mimicked in cell culture, and thus it is possible to study the effects of antiviral drugs _in vitro_ in ways that may be predictive of the effects of the drugs _in vivo_. Moreover, as we work out the mechanisms of the virus-cell relations in these various states of EBV infection, we can begin to rationalize both effects and predictions. In this talk, I shall review the drugs now known to be effective against EBV _in vitro_ and discuss aspects of the pharmacologic action of some of these drugs about which most is known. I shall then focus on what effect the drugs have on each of the virus-cell relations already cited, and conclude with predictions of whether these drugs are anticipated to be effective in the different human disease states with which EBV is associated.

EBV-Cell Systems. EBV exists in HR1 B-lymphoblastoid cells in a chronic virus-producing state equivalent to persistent infection. Only a minority of cells is producing virus at any given time. If such a cell culture is treated with an inducing agent such as the phorbol ester, TPA, then the major subset of non-virus-producing cells in the culture, which contain latent viral genomes, is induced to produce virus, mimicking virus reactivation. Acute infection is mimicked *in vitro* by infecting Raji cells with EBV harvested from P3HR-1 cells; although exogenous virus is used for infection, and a rapid infectious cycle ensues, this is not a true primary infection inasmuch as the Raji cells already have endogenous viral genomes which in some way contribute to the outcome, perhaps through recombination with exogenous virus. Raji cells provide an intensively studied cellular model of latent EBV infection; the molecular basis for the latent state of the infection seems to be the plasmid or episomal form of the EBV genome. Cell models also exist for the oncogenic relation. Presumably the true oncogenic state is captured in malignant cells explanted from Burkitt's lymphomas, but such cells also contain latent viral genomes in the form of EBV plasmids. Such cell lines, which are monoclonal, may also contain covalently integrated EBV DNA sequences. A B-lymphocytic line exists that contains only integrated viral sequences without episomal forms (Namalwa). Finally, there is a fifth virologic state which seems to exist, both in nature and in *in vitro* models, that is now becoming better defined, namely, abortive infection. *In vitro* this state is produced by exposing Raji cells to TPA which causes activation of EBV gene expression with production of early antigen, but the activation of the genome is aborted and replication of EBV genomes does not ensue. *In vivo*, there is beginning evidence that some kind of corresponding virologic state occurs in nasopharyngeal carcinoma in which activated EBV gene expression appears to begin but stops short of virus replication in the tumor tissue.

Drugs that inhibit EBV replication in vitro. Among the first drugs shown to be effective *in vitro* were phosphonoacetic acid and phosphonoformic acid (Summers, et al., 1976; Datta and Hood, et al., 1981). The action of these two drugs seems to depend on their ability to interact at the pyrophosphate binding site, and thus these drugs directly affect EBV DNA polymerase and do not require prior phosphorylation. Adenosine arabinoside has some inhibitory effects on EBV replication _in vitro_, but the drug has not been studied _in vivo_ (Benz, et al., 1978). Acyclovir is an effective inhibitor with an ED_{50} of 0.3 μmole (Lin, et al, 1984). E-5-(2-bromovinyl)-2'-de oxyuridine (BVDU) has the same relative efficacy _in vitro_ but has not been used _in vivo_ (Lin, et al., 1982). 9-(1,3-dihydroxy-2-propoxymethyl)guanine (BW759U, DHPG) is about six times more active _in vitro_ than Acyclovir with an ED_{50} of 0.05 μmole (Lin, et al., 1984). The halogenated nucleoside analogs, 1-2(deoxy-2-fluro-7-D-ara-binofuranosy l)-5-iodocytosine (FIAC), 1-(2-deoxy-2-fluoro-7 -D-arabino furanosyl)-5-methyl-uracil (FMAU) are approximately ten times more active than DHPG and 50 times more active than ACV _in vitro_ (Lin, et al., 1982).

Mechanism of Acyclovir. In herpes simplex virus infection, Acyclovir is first monophosphorylated by virus-encoded thymidine kinase and then di- and triphosphorylated by cellular enzymes. The triphosphate is the active form of the drug which specifically interacts with EBV DNA polymerase; ACV-triphosphate has at least 100-fold greater affinity for the viral than the normal cellular polymerase (Elion, et al., 1977). In EBV-infected cells, monophosphorylation of ACV seems to be accomplished by cellular kinases, as yet unidentified, rather than by EBV-encoded TK inasmuch as an EBV TK has not been positively identified or isolated, nor is phosphorylation of ACV in EBV-infected cells as efficient as it is in HSV-infected cells (Pagano and Datta, 1982; Colby, et al., 1981). The specificity of the action of ACV and hence its relative nontoxicity thus depends on either selective phosphorylation of the drug in virus-infected cells or preferential affinity of the drug-triphosphate for the viral _versus_ the cellular polymerases, or both. The relative contributions of these various steps in the action of the drug differ with the different herpesviruses (Pagano and Datta, 1982). The nontoxicity of Acyclovir is also promoted by the localization of the active form of the drug in infected tissue inasmuch as ACV-triphosphate cannot permeate normal cell membranes and hence is present in greatest concentration in cells best able to carry out phosphorylation, _i.e._, infected cells, and in cells possessing viral polymerase with high affinity for the triphosphate. However, all these effects are relative, and small amounts of the drug are phosphorylated in normal tissue, the amount depending on cell type and cellular metabolism.

The nature of the interaction with herpesvirus polymerases is as a competitive inhibitor of dGTP. There is evidence for both reversible (Datta, et al., 1980) and nonreversible aspects of this interaction (Furman, et al., 1984). In HSV infection, ACV triphosphate is incorporated into the DNA; the incorporation causes immediate chain termination and binding of the viral polymerase. Incorporation into EBV DNA probably also occurs (work in progress). The likelihood of incorporation into viral DNA is much greater than it is into cellular DNA because of the much greater affinity of the drug for the viral polymerase and the presence of triphosphorylated drug in greater concentrations in virus-infected tissue, at least in HSV infection.

Effects on Protein Synthesis. Recently, Lin, et al., 1985, have obtained evidence indicating that DHPG is much more efficiently phosphorylated in EBV-infected cells than is ACV, which may help to explain why DHPG is the more effective inhibitor. The affinity of EBV DNA polymerase for DHPG-triphosphate is under study. ACV and DHPG also have effects, presumably secondary, on EBV polypeptide synthesis in infected cells. HSV polypeptides have been classified by Roizman, as $\gamma 1$, which are independent of but amplified by viral DNA synthesis, and $\gamma 2$, which are stringently dependent on viral DNA synthesis. Although a similar subdivision of EBV late polypeptides has not as yet been defined, late EBV polypeptides that appear not to be synthesized at all and others which are synthesized in relatively reduced amounts in the presence of ACV have now been pinpointed. DHPG has similar but greater effects than ACV in this regard (Lin, et al., 1984). However, the majority of EBV polypeptides are synthesized normally which suggests that viral polypeptides arising in EBV-cell states that do not involve viral DNA replication will be unaffected by these drugs.

Reversal of Drug Action. Because EBV infection states involve persistent replication or at least persistence of viral genomes, what happens after an effective drug is removed from the infected cells is of prime importance and likely to be relevant to treatment issues. When ACV is applied to the virus-producing HR1 cell line, free viral DNA replication is rapidly abolished and virus production ceases. However, a persistent fraction of viral DNA does remain, detectable by hybridization, in the treated cell cultures. Upon removal of the drug, viral DNA replication and virus production resume rapidly and are soon restored to the levels before treatment. Essentially, virus replication is suppressed so long as the drug is present. This reversibility of inhibition suggests either that the drug triphosphate is reversibly bound to viral DNA polymerase or that new polymerase molecules are generated and become functional upon removal of the drug. There is some preliminary evidence indicating that viral DNA polymerase, which is probably an early polypeptide, is synthesized even in the presence of the drug but remains inactive. Another line of evidence pointing to the possibility that the interaction between ACV triphosphate and EBV DNA polymerase is reversible comes from studies of the kinetics of the interaction of the triphosphate with EBV DNA polymerase in vitro (Datta, et al., 1980). On the other hand, there is increasing evidence that, at least in the case of HSV DNA replication, ACV triphosphate interacts irreversibly with viral DNA polymerase forming a complex with DNA into which the drug incorporates thereby immediately terminating chain elongation (Furman, et al., 1984). New studies are underway to ascertain whether the interaction of the triphosphate with EBV DNA polymerase is also irreversible.

This important issue is made even more interesting by the recent observations that different inhibitors behave differently with respect to duration of drug effect after removal. DHPG is more strongly inhibitory than ACV, and FMAU even more so. In addition, after removal of DHPG from the virus-producing cell line, the kinetics of recovery of virus production follow a slower course, taking approximately 21 days for a full restoration of EBV DNA content to pre-drug exposure levels. The persistent drug effect after removal is even more remarkable in the case of FMAU - more than 58 days (Lin, et al., 1983). These different kinetics of recovery of virus replication probably point to different modes of action of all three drugs in this respect, but the basis is as yet unknown. The differences in degree of persistent effects seem unlikely to be due merely to differences in drug metabolism, but are possibly related to consequences of incorporation of drug into DNA.

Latent infection. All of the drugs so far tested have no effect whatever on replication of the episomal form of the EBV genome in Raji cells, regardless of the differences in potency and mode of action of the various drugs. This lack of effect is quite independent of drug phosphorylation inasmuch as neither Acyclovir, which is poorly phosphorylated in lymphoblastoid cells, nor DHPG, which is better phosphorylated, has any effect on the latent infection. Moreover, phosphonoformate, which does not require phosphorylation, also is without effect on EBV plasmid replication. In HR-1 cells the residual EBV genomes that persist in the presence of high inhibitory concentrations of the drugs are present in the form of EBV episomes or plasmids. Thus, two forms of the EBV genome are replicated in P3HR-1 cells: linear genomes that become encapsidated and circular genomes that remain intracellular. The circular episomal genomes, which are present in a nucleosomal arrangement in the host-cell DNA, are evidently replicated by host DNA polymerase rather than by the viral polymerase, as inferred from the effects of the inhibitors. It is well established that Acyclovir spares host DNA polymerase activity while inhibiting polymerization by the viral enzyme, and it may also be inferred that FMAU has a similar dichotomy of effect on the two classes of polymerase. At this point, it appears that only cytotoxic drugs are liable to have any effect on the maintenance of EBV episomal forms, but such drugs would at the same time produce general, nonselective effects on cellular replication and viability.

Transformation. It is now quite clear from the work of Rickinson and Epstein (1978) and Sixbey and Pagano (1985), that as anticipated by Lemon, et al. (1978), transformation or immortalization of lymphocytes by EBV can be accomplished in the presence of levels of Acyclovir that are greater than the ED_{90} for replication of EBV. In other words, transformation is not mediated by action of the viral DNA polymerase and by a consequent round of viral DNA replication. However, a number of EBV genomes in the form of episomes do in some way become established in the immortalized cells. The number of EBV episomes is limited either in the presence or absence of the drug, and the immortalized cells do not make virus. Cell lines including Burkitt's lymphoma cells that are already transformed are also essentially unaffected by these antiviral drugs in vitro. The only effects so far observed on transformed cells are dose-dependent cytotoxic effects that are probably nonselective.

EBV Pathogenesis and Predictions about Therapy   One view of the pathogenesis of EBV infection is that EBV enters the body through oropharyngeal contact and primarily infects epithelial cells in the oropharynx (Sixbey, et al., 1984; Sixbey and Pagano, 1984). These cells located somewhere in the oropharynx support the active replication of the virus which is excreted in the saliva; the virus also replicates in epithelial cells found in the parotid gland or its duct. A secondary cell target is the B-lymphocyte, and these cells, which are known to bear EBV receptors, are probably infected very early in the course of infection. However, the B-lymphocytes do not appear to be sites of active replication of the virus in normal hosts; rather a small percentage of circulating B-lymphocytes harbor latent EBV genomes and display expression only of EBV antigens not associated with replication of virus. Although these cells do not support virus replication, they are induced into lymphoproliferation which is polyclonal in nature and limited by the normal host immune responses. All of these events occur in acute infectious mononucleois and also presumably in silent EBV infection.

Therefore we might predict that the drugs discussed would interfere with virus that is replicated in epithelial cells and shed in the oropharynx, suppressing replication while drug is being administered. Since infection of B-lymphocytes probably occurs early in infection, the drug would have no effect on the already

latently infected B-lymphocytes. Preliminary results of a
trial of Acyclovir in patients with acute infectious
mononucleosis have confirmed these predictions (Pagano, et
al., 1983). In the patients administration of Acyclovir
transiently suppressed virus excretion but did not abolish
it, nor was there any effect on the ability to establish
EBV infected B-lymphocyte lines from the peripheral blood
of patients being given Acyclovir. However, it is
reasonable to expect that suppression of virus production
would reduce the number of B-lymphocytes that become newly
infected with EBV during the course of infection, perhaps
with favorable effects. Moreover, a respite in virus
replication might tip the balance in favor of the host
immune mechanisms and aid in recovery even though virus
replication and excretion eventually resumed. In fact,
shedding of virus in the oropharynx may continue
asymptomatically for years even in untreated patients.
Finally, some of the manifestations of acute EBV infection
in mononucleosis such as Guillan-Barre syndrome,
hepatitis, and suppression of various aspects of
hemopoiesis are believed to be manifestations of secondary
immune responses to latently infected EBV lymphocytes.
These immunologically based manifestations should be quite
indifferent to the inhibitory effect of the antiviral
drugs except insofar as continuing amplification of the
population of EBV-infected lymphocytes by spread of
infection to additional lymphocytes would be interrupted.
Burkitt's lymphoma seems to be a consequence of one or
more additional steps in this pathogenetic scheme with
chromosomal translocations producing critical activation
of a cellular oncogene (c-myc) which leads to a
monoclonal B-lymphocytic malignancy. None of the
cytogenetic, molecular or cellular changes is likely to be
susceptible to the action of Acyclovir or the other
presently available drugs.

In nasopharyngeal carcinoma tissue, EBV episomes are
found in the epithelial elements of the neoplasm.
Presumably, the transformed epithelial cells arise from
rare epithelial cells infected many years earlier but not
lysed by the virus. The bulk of evidence indicates that
NPC tissue does not contain antigens associated with viral
replication. Recently, evidence has been developed of an
apparent activated transcriptional state in some NPC

tissues (Raab-Traub, et al., 1983), but this activation of
gene expression is believed to stop short of virus
replication, and therefore it is unlikely that Acyclovir
would have any effect on this process.   The drug would
also not be expected to have an effect on the already
transformed epithelial cells bearing EBV episomes.   One
conceivable point of action might be synthesis of EBV
polypeptides equivalent to the $\gamma2$ class of HSV
polypeptides.   If these late polypeptides are among those
that begin to be synthesized during the activated
transcriptional state found in some NPC's, their synthesis
might be limited insofar as it depends upon amplification
of viral DNA templates.   However, it is not yet known
whether any late polypeptides are in fact produced in NPC,
and if they are, what their pathologic consequences might
be.

In the case of NPC, there is, however, some evidence
to suggest that reactivation of EBV may precede or even
trigger the onset of the decrease.   This evidence rests
primarily on IgA responses to EA and VCA that precede (but
do not necessarily lead to) the appearance of a
significant percentage of NPC's (de The, 1982).   It is not
known whether these antibody responses signify
full-fledged virus replication, abortive replication, or
merely an activated transcriptional state, nor is it yet
clear precisely where these antigenic stimuli arise at the
cellular level.   However, if reactivation of viral
replication is a necessary prelude to NPC, then effective
nontoxic antiviral drugs given prophylactically to
high-risk patients in NPC endemic areas might interfere
with appearance of the malignancy.

In the case of invasive B-lymphocytic proliferation
that occurs in immunocompromised hosts, there have been
anecdotal reports of an apparent efficacious effect of
Acyclovir on polyclonal B-cell proliferation (Hanto, et
al., 1982).   However, in two patients with a congenital
immune defect Acyclovir appeared to have no effect
(Sullivan, et al., 1984).   From a theoretical point of
view, we would not expect that either polyclonal or
monoclonal B-lymphocytic proliferation based on already
infected B-lymphocytes would be affected by the antiviral
drugs.   However, if lymphoproliferation in such patients

is based on continuing infection of additional
B-lymphocytes, then antiviral drugs might indirectly curb
polyclonal lymphoproliferation. In all probability, the
intensity and nature of the immune defect would also
determine outcome, with lymphoproliferation in milder
acquired immune defects being more susceptible to
treatment.

In conclusion, we now have at least one drug
available for testing in acute infection states, either
primary or reactivated, that may well have therapeutic
effects in human beings. Other drugs still confined to
the laboratory may point the way to therapeutic agents
useful for treatment of persistent infection states. No
drugs that have been identified until now specifically
inhibit latent EBV infection or growth transformation of
EBV-infected lymphocytes. The treatment of these crucial
EBV infection states presently lies beyond our reach and
will probably require a deeper level of understanding of
the mechanism of establishment and maintenance of the
latent EBV episomes and the transformed state before
further progress. Nevertheless, the prospects for
treatment of EBV infections have brightened substantially
in the short interval since the effect of Acyclovir on EBV
replication was first described in 1980 (Colby, et al.,
1980).

References

Benz, W.C., Siegel, P.J., and Baer, J., Effects of adenine
arabinoside on lymphocytes infected with Epstein-Barr
virus., J. Virol., 27, 475-482 (1978).

Colby, B.M., Furman, P.A., Shaw, J.E., Elion, G.B.,
Pagano, J.S., Phosphorylation of Acyclovir
[9-(2-hydroxyethoxymethyl)guanine] in Epstein-Barr virus
infected lymphoblastoid cell lines., J. Virol., 38,
606-611. (1981)

Colby, B.M., Shaw, J.E., Elion, G.B., Pagano, J.S., Effect
of Acyclovir [9-(2-hydroxyethoxymethyl)guanine] on
Epstein-Barr virus DNA replication., J. Virol., 34,
560-568. (1980)

Datta, A.K., and Hood, R.E., Mechanism of inhibition of Epstein–Barr virus replication by phosphonoformic acid., Virology, 114, 52–59 (1981)

Datta, A.K., Colby, B.M., Shaw, J.E. and Pagano, J.S., Acyclovir inhibition of Epstein–Barr virus replication., Proc. Natl. Acad. Sci., 77, 5163–5166. (1980)

de The, G., Epidemiology of Epstein–Barr virus and associated diseases in man., The Herpesviruses, B. Roizman (ed.). (1982).

Elion, G.B., Furman, P.A., Fyfe, J.A., Selectivity of action of an antiherpetic agent, 9–(2–hydroxyethoxy-methyl)guanine., Proc. Natl. Acad. Sci., 74, 5716–5720 (1977)

Furman, P.A., St. Clair, M.H., and Spector, T., Acyclovir triphosphate is a suicide inactivator of the herpes simplex virus DNA polymerase., J. Biol. Chem., 259, 9575–9579 (1984)

Hanto, D.W., Frizzera, G., Gagi–Peczalska, K.L., Epstein–Barr virus-induced B-cell lymphoma after renal transplantation: acyclovir therapy and transition from polyclonal to monoclonal B-cell proliferation. N. Eng. J. Med., 306, 913–918 (1982)

Lemon, S.M., Hutt, L.M., and Pagano, J.S., Cytofluorometry of lymphocytes infected with Epstein–Barr virus: effect of phosphonoacetic acid on nucleic acid., J. Virol., 25, 138–145 (1978)

Lin, J–C., Nelson, D.J, Lambe, C.U., and Pagano, J.S., Effects of nucleoside analogs in inhibition of Epstein–Barr virus. Proc. International Virology Post–Congress Symposium on Pharmacological and Clinical Approaches to Herpesviruses and Virus Chemotherapy, Oiso, Japan, in press (1985).

Lin, J–C., Smith, M.C., Cheng, Y–C., and Pagano, J.S., Epstein–Barr virus: inhibition of replication by three new drugs., Science, 221, 578–579 (1983)

Lin, J-C., Smith, M.C., and Pagano, J.S., Prolonged inhibitory effect of 9-(1,3-Dinydroxy-2-Propoxy-methyl)guanine against replication of Epstein-Barr virus, J. Virol., 50, 50-55 (1984)

Pagano, J.S. and Datta, A.K., Perspectives on interactions of Acyclovir with Epstein-Barr and other herpes viruses., Amer. J. of Med. (Acyclovir Symposium), 18-26 (1982)

Pagano, J.S., Sixbey, J.W., and Lin, J-C., Acyclovir and Epstein-Barr viraus infection., J. Antimicrob. Chemother., 12, 113-121 (1983)

Raab-Traub, N., Hood, R., Yang, C.S., Henry, B., and Pagano, J.S., Epstein-Barr virus transcription in nasopharyngeal carcinoma, J. Virol. 48, 580-590 (1983)

Rickinson, A.B. and Epstein, M.A., Sensivity of the transforming and replicative functions of Epstein-Barr Virus to inhibition by phosphonoacetate, J. Gen. Virol., 40, 421-431 (1978)

Sixbey, J.W., Nedrud, J.G., Raab-Traub, N., Hanes, R.A., Pagano, J.S., Epstein-Barr virus replication in oropharyngeal epithelial cells. N. Eng. J. Med., 310, 1225-1230 (1984)

Sixbey, J.W. and Pagano, J.S., Epstein-Barr virus transformation of human B-lymphocytes despite inhibition of viral polymerase. J. Virol., in press (Jan. 1985)

Sixbey, J.W. and Pagano, J.S., New perspectives on the Epstein-Barr virus in the pathogenesis of lymphoproliferative disorders. In Current Clinical Topics in Infectious Diseases, J. Remington and M. Schwartz, ed., 146-176 (1984)

Sullivan, J.L., Medreczky, M.D., Forman, S.J., Baker, S.M., Monroe, J.E., Mulder, C., Epstein-Barr virus-induced lymphoproliferation, N. Eng. J. Med., 311, 1163-1167 (1984)

Summers, W.C., and Klein, G., Inhibition of Epstein-Barr virus DNA synthesis and late gene expression by phosphonoacetic acid, J. Virol., 18, 151-155 (1976)

58

CLINICAL AND PATHOBIOLOGICAL FEATURES OF BURKITT'S LYMPHOMA

AND THEIR RELEVANCE TO TREATMENT

Ian T. Magrath, M.B., M.R.C.P.

National Cancer Institute, NIH

Bethesda, Maryland U.S.A

SUMMARY

A consideration of the association of EBV and Burkitt's lymphoma suggests that there is a correlation between the incidence of Burkitt's lymphoma and the proportion of EBV-associated tumors. There appears to be, however, a similar incidence of EBV-negative Burkitt's lymphomas throughout the world. This suggests that there are
at least 2 subtypes of Burkitt's lymphoma, and that the
incidence of the EBV-associated variety is particularly
dependent upon environmental factors, one of which is
early infection with EBV. There are marked clinical differences between tumors in Equatorial Africa and in North
America which probably reflect differences in phenotype,
and the presence of EBV may lead to a different response
to treatment.

Recently, the occurrence of Burkitt's lymphoma in patients with AIDS has been described. This is likely to
provide additional clues to the pathogenesis of the EBV-
associated form of Burkitt's lymphoma. B-cell hyperplasia, whether due to a direct stimulating effect on B-
cells, or a failure of T-cell suppression or cytotoxicity,
appears to be an important prodromal phase of Burkitt's
lymphoma. Novel approaches to intervention, including
both prevention and therapy are likely to be developed as
the precise mechanism of pathogenesis are further worked
out.

OCCURRENCE OF BURKITT'S LYMPHOMA

The observation that Epstein-Barr virus (EBV) is
associated with Burkitt's lymphoma arose directly from
the original observations of Dennis Burkitt and his colleagues that Burkitt's lymphoma in Africa is found in high
frequency only in a geographical region extending approximately 15° north and south of the equator, with a southern
prolongation along the east coast in the region of Mozam-
bique (Burkitt, 1962b). This epidemiological survey was
based on the occurrence of jaw tumors rather than histology and it has subsequently become clear that histologically identical tumors, manifested infrequently as jaw
tumors, occur at a much lower frequency outside this
"lymphoma belt" (Dorfman, 1965; O'Conor, 1965). Although
the geographical distribution in Africa suggested to early
observers the participation of a vectored virus, and led
to Epstein's studies, the nature of the association between Burkitt's lymphoma and EBV remains ill-defined.
EBV is ubiquitous, but the age at which infection occurs
is much earlier in developing countries (Henle and Henle,
1979). Although this could be a factor in the development
of Burkitt's lymphoma, the early age of primary EBV infection is insufficient to account for the existence of the
particularly high incidence in Equatorial Africa, and
there can be little doubt that other environmental factors
are involved in the pathogenesis of this tumor. Based on
epidemiological observations, malaria is a leading candidate, but direct evidence for its participation has not
been obtained.

With the recognition that Burkitt's lymphoma in the
United States and Europe is, in general, not EBV associated, questions regarding the definition of Burkitt's
lymphoma were raised. This issue has been further confounded in recent years by the observation that Burkitt's
lymphoma in certain North African countries has a high
rate of EBV association (Lenoir et al., 1984), and that
Burkitt's lymphoma occurs at high frequency in homosexual
men, particularly if in the prodromal or overt phases of
the aquired immunodeficiency syndrome (AIDS) (Ziegler et
al., 1984). Although the numbers of tumors examined is
still small, a high proportion of the Burkitt's lymphomas
occurring in AIDS appear to be EBV associated (6 of 8)
(Ziegler et al., 1982; Whang-Peng et al., 1984).

THE DEFINITION OF BURKITT'S LYMPHOMA

In 1969 a panel of distinguished pathologists, assembled under the auspices of the World Health Organization, provided a "definition" of Burkitt's lymphoma acceptable to the majority of the panel members (Berard et al., 1969). It should be born in mind that this was a histological definition and therefore could not be precise. The problems of the histological definition have been particularly apparent outside Africa. In the modified Rappaport classification of non-Hodgkin's lymphomas, the subtype "undifferentiated lymphoma" is divided into Burkitt's type and non-Burkitt's type. In order to make this distinction, the pathologist is required to judge the degree of uniformity of the cells and to estimate certain other features such as the proportion of cells containing a single nucleolus. In the absence of numerical descriptors it is not surprising that this distinction has been difficult to make reproducibly, even within a single department of histopathology. The possibility that more objective criteria may be of value should be considered.

Burkitt's lymphoma is unquestionably of B-cell origin, and the vast majority of cases (regardless of the country of origin) express surface IgM and various other B-cell specific or B-cell associated antigens such as B1, B4, BA1 and HLA-DR. These markers do not differ, however, between undifferentiated nonBurkitt's and Burkitt's lymphoma (Sandlund et al., in press). Correlation of histology with the presence of an 8;14, 8;22 or 2;8 translocation, reveals that not only histologically defined Burkitt's lymphoma, whether Equatorial African, North African or North American, but also undifferentiated non-Burkitt's lymphoma, and a proportion of large cell lymphomas, contain the same cytogenetic abnormalities. In our series of undifferentiated lymphomas at the National Cancer Institute, freshly examined tumors and also cell lines derived from lymphomas contain either an 8;14 or an 8;22 translocation regardless of whether they were diagnosed as Burkitt's or non-Burkitt's lymphomas (Sandlund et al., in press). In Bloomfield's series of non-Hodgkin's lymphoma, approximately 50% of the 8;14 translocations detected were present in large cell lymphomas, while the remainder were found in undifferentiated lymphomas (designated as small, noncleaved cell lymphomas, a category which includes undifferentiated lymphomas of both Burkitt's and non-Burkitt's

types) (Bloomfield et al., 1983). These findings indicate
that either the histological categories do not designate
separate entities, or the same cytogenetic abnormalities
can occur in several different diseases.

EBV ASSOCIATION OF BURKITT'S LYMPHOMA

As mentioned above, the majority of Burkitt's lymphomas in equatorial Africa contain EBV DNA. There is
however a small proportion of tumors, 4 or 5%, which are
not associated with EBV (Lenoir et al., 1984). Since patients with EBV-negative tumors possess anti-EBV antibodies (Magrath, 1984), it is difficult to escape the conclusion that African EBV positive and negative tumors differ
phenotypically to at least a small degree; a conclusion
which implies that the clinical features of these subtypes may differ, and also that EBV positive and negative
tumors differ pathogenetically.

The majority of North American Burkitt's lymphomas,
although EBV negative, also arise in patients who have antibodies against EBV. The incidence of Burkitt's lymphoma
in North America is rather difficult to determine because
of a lack of uniformity and reproducibility in histological diagnosis. Assuming that this disease accounts for
between 1/3 and 1/2 of all childhood with non-Hodgkin's
lymphomas, however, the incidence must lie somewhere between 1 and 5 per million children below the age of 16.
In Equatorial Africa, on the other hand, the incidence
of Burkitt's lymphoma is between 50 to 100 per million,
with 2 to 5 cases per million being EBV negative. Clearly,
the incidence of EBV negative Burkitt's lymphoma in North
America and Equatorial Africa is very similar, and the
question must be raised as to whether EBV association does
distinguish two separate variants of the disease. Since
no distinguishing histological features have so far been
detected between EBV positive and EBV negative Burkitt's
lymphoma, confirmation or refutation of this possibility
will be dependent upon careful comparison of the clinical
features of these two diseases, coupled to a detailed
comparison of their respective phenotypes. The possibility that there may also be subtle differences in the
genetic abnormalities, at least at a molecular level,
cannot be excluded.

Recently, it was reported that 85 to 90% of North African Burkitt's lymphomas are associated with EBV (Lenoir et al., 1984). This figure is a little lower than in Equatorial Africa, but markedly higher than in the United States and France. Unfortunately, good incidence figures for North Africa are not yet available. It seems probable, however, that this is an intermediate incidence area. Although there are very few data points, there appears to be a correlation between the incidence of Burkitt's lymphoma in different parts of the world and the proportion of tumors which are EBV-associated (Figure 1). To confirm this, it will be important to study the incidence and EBV association of Burkitt's lymphoma in Southeast Asia, the Middle East, and South America, where information remains scanty. It is entirely possible that in some parts of the world a more even distribution between EBV positivity and negativity of Burkitt's lymphomas will be observed, permitting the collection of more meaningful clinical and epidemiological information relating to these two sub-categories. The collection of information of this kind promises to provide more direct insights into the significance of the EBV association of Burkitt's lymphoma.

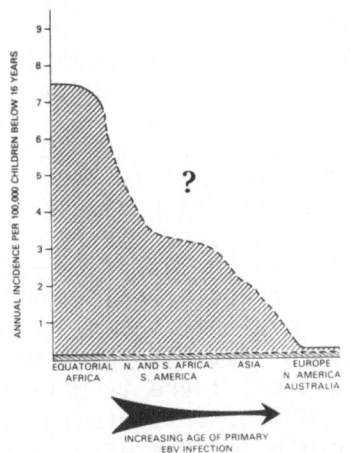

Fig. 1. Graphic depiction of incidence of EBV-associated Burkitt's lymphoma (upper curve) in various parts of the world. Dotted line indicates lack of data. The incidence of EBV-unassociated Burkitt's lymphoma is represented by the lower line, just above the horizontal axis.

CLINICAL DIFFERENCES BETWEEN BURKITT'S LYMPHOMA
IN EQUATORIAL AFRICA AND NORTH AMERICA

Although, ideally, a comparison between EBV-associated and EBV negative Burkitt's lymphoma should take
place within a single geographical area, at present we are
limited, because of numerical considerations, to a comparison of Equatorial African and North American Burkitt's
lymphoma. Such a comparison is, however, of considerable
interest.

Striking differences can be discerned between African
and North American tumors with regard to the frequency of
involved sites both at presentation and relapse, and in
the clinical course of the patients who do not achieve
continuously sustained remission (Magrath, 1984). One
of the most obvious differences is the very low incidence
of jaw tumors in North American patients. Even when jaw
tumors are present, they differ in several respects from
those of African children (Sariban et al., 1984). Firstly,
the frequency is higher in females, as opposed to the male
predominance seen in Africa; secondly, tumors are usually
smaller and involve only a single jaw quadrant in the USA,
as opposed to multiple quadrants in the majority of African children; thirdly, there is no age association, in
contrast to African Burkitt's lymphoma (Burkitt, 1962a)
(Figure 2). Paraplegia and involvement of salivary and
thyroid glands are rare in the United States and orbital
involvement almost never occurs. In contrast, these sites
of involvement are not uncommon in the African child.
Nasopharyngeal involvement which almost never occurs in
Africa, is seen occasionally in the American patient.
Marrow involvement occurs in 20% of American patients at
the time of presentation and cryptic involvement is present in a further 17% (minimum figure) (Benjamin, et al.,
1983), so that 35-40% of patients have marrow involvement
at presentation. Further, there is marrow involvement at
some time in almost 100% of patients in whom treatment
is unsuccessful (Magrath and Sariban, in press). In
African children only 7 to 8% of patients have marrow
involvement at any time in the course of their disease,
although the possibility of cryptic involvement has not
been examined (Magrath, 1984).

637

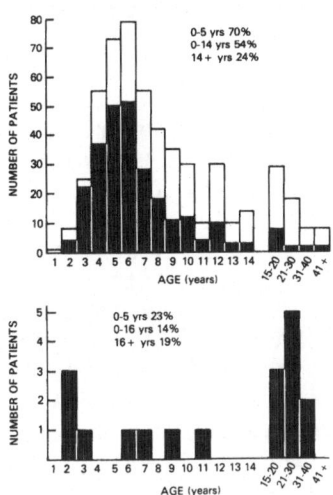

Fig. 2. Frequency of jaw tumors (black histograms) by age in an African series (Burkitt, 1962a) - upper graph, and an American series (Sariban et al., 1984) - lower graph. In the upper graph, the total number of cases at each age is indicated by the unshaded histograms.

Meaningful comparisons between Equatorial African and North American Burkitt's lymphoma with regard to the outcome of treatment cannot be made, since identical therapy has not been administered in each country. A reasonable proportion of patients in both geographical areas (40-70%) appear to be curable by chemotherapy alone (Ziegler et al., 1979; Magrath et al., 1984), but there is a marked difference with regard to the outcome of the treatment of relapse in these countries. Whereas longterm survivors in the United States have been, with a few exceptions, continuously disease-free after the initial induction of remission, in our African series about half of the 72 long term survivors had had one or more recurrences prior to their achievement of sustained remission (of between 4-10 years) (Ziegler et al., 1979). Thus, although the complete response rate and overall survival rate may not differ greatly, the proportion of complete responders who never relapse is quite different in the two diseases. These findings raise questions regarding the meaning of the term "chemotherapy resistance", and suggest that host factors may be particularly important in erradicating tumor in the African child.

In both the African and North American diseases, tumor burden appears to be the predominant predictive factor with regard to prognosis (Magrath, 1984). The titer of antibodies against early antigen seems to correlate very well with tumor burden in the African, and is just as good a predictor of prognosis as tumor burden. This is not the case in American Burkitt's lymphoma in which the frequency of positive antibody titers to early antigen is lower than in African patients. Presumably this reflects the difference in association of the two tumors with EBV, but also implies that antibodies against early antigens (EA) in the African patient may be elevated because of antigen production in tumor cells. Whereas at first this seems improbable because such antigens are difficult to find in preparations of fresh tumor cells, in Burkitt's lymphoma there is a high cell turnover with a spontaneous cell loss rate of 70% (Iverson et al., 1972). Some at least of these dying cells may have undergone a lytic cycle with production of EA and virus capsid antigen (VCA). Rapid removal or death of such cells may prohibit their detection in tumor samples. The association of the early antigen titer with tumor burden is also reflected by the rise in antibodies against early antigens coincident with the initial development of Burkitt's tumor (Magrath et al., 1975). A fall in antibodies to EA frequently occurs after successful therapy of Burkitt's tumor, while the titer often increases at the time of relapse (Henle and Henle, 1979).

Although a number of clinical differences have been discerned between African and American patients, phenotypic differences are less obvious. Our work on cell lines of both origins has demonstrated a higher level of C3 and EBV receptors on African than American cell lines, possibly explaining the difference in EBV association (Magrath et al., 1980). Both receptors are now believed to be carried on a glycoprotein of molecular weight 140 kilodaltons (Fingeroth, et al., 1984). In addition, American cell lines (and tumors) secrete immunoglobulin (IgM), which is uncommonly the case in African cell lines (Benjamin et al., 1982).

SIGNIFICANCE TO THERAPY OF THE EBV ASSOCIATION

The remarkably high salvage rate of African patients with Burkitt's lymphoma who relapse, coupled with the correlation between early antigen titer and tumor burden, suggests that the presence of EBV in the tumor cells and the expression of EBV antigens, including membrane antigens and LYDMA, may permit the mounting of a much greater antitumor response than would be the case in the absence of the viral association. Indeed, it is surprising that EBV-infected cells can escape from the powerful methods of immunoregulation of such cells.

The fact that Burkitt's lymphoma cells appear to be latently infected with EBV provides a potential "Achilles heel" for this tumor, although one which seems unlikely to be exploited. In Lucke's disease, the associated herpes virus proliferates best at low temperature. Thus, during the winter months there is a lytic cycle in the tumor cells with consequent tumor regression. During the summer months, when the virus is in a latent phase, tumor proliferation occurs with consequent death of many of the animals (Magrath, 1983). This phenomenon is provocative in that it implies that stimulation of the EBV replication cycle in Burkitt's lymphoma may be a more useful approach to treatment than the use of drugs which prevent virus replication, such as acyclovir.

BURKITT'S LYMPHOMA IN AIDS

Ziegler and others recently described the clinical features of 90 cases of non-Hodgkin's lymphomas occurring in homosexuals (Ziegler et al., 1984). A high proportion of these individuals had early or late stigmata of AIDS. About 1/3 of the lymphomas were histologically consistent with a diagnosis of Burkitt's lymphoma, or at least undifferentiated non-Burkitt's lymphoma. Very few of these cases have been karyotyped, but a rapid survey of the literature indicates that in all, about ten undifferentiated lymphomas occurring in homosexuals have been studied cytogenetically and all of them bear either an 8;14 or 8;22 translocation (Ziegler et al., 1984; Whang-Peng et al., 1984; Chaganti et al., 1984). It should be born in mind that not all lymphomas occurring in AIDS are necessarily Burkitt's lymphoma. Many of the clinical

features of lymphomas in this patient group resemble those
described in individuals undergoing immunosuppression
for organ transplantation. For example, as in transplant
recipients, about 40% of the non-Hodgkin's lymphomas in
AIDS patients occur in the central nervous system (Ziegler,
et al., 1984). Six of 8 tumors so far examined contain
EBV DNA, but the clinical difference between Burkitt's
lymphoma in patients with AIDS and African Burkitt's
lymphoma appear to be considerable. The absence of jaw
involvement and the very high frequency of central nervous
system and marrow involvement, with the occasional presence of rectal tumors, sharply distinguish the syndrome
from the African disease. It should, of course, be born
in mind that the mean age of the AIDS patients is much
above that of African patients. In AIDS patients it is
very difficult to discern how successful chemotherapy is,
since a large proportion of these individuals die from
infectious complications of AIDS rather than from Burkitt's lymphoma (Ziegler et al., 1984).

The presence in 'AIDS' of Burkitt's lymphoma with
appropriate cytogenetic abnormalities, raises questions
concerning pathogenesis. The striking lymphoid hyperplasia involving both T- and B-cells which occurs prior to
the onset of flagrant AIDS (see this volume Krueger),
coupled with, or caused by, impaired T-cell function, is
almost certain to be important to the pathogenesis of lymphomas in AIDS. These general features are strikingly
similar to those in the African child, where B-cell hyperplasia occurs; possibly because of a combination of
exposure to EBV at an early age, and malaria. Little is
known of the possibility that T-cell impairment occurs in
the African child prior to the onset of disease, but as
seems to be a feature of many human and animal cancers including shistosomal bladder cancer, lymphomas in chickens
and myeloma in mice, Burkitt's tumor is probably preceeded
by a period of marked hyperplasia of the target cell.
Neoplasia presumably occurs when a specific genetic abnor-
mality resulting from juxaposition of a c-myc gene with
a constant region of an immunoglobulin gene occurs
(Magrath, in press).

References

Benjamin, D., Magrath, I.T., Douglass, E.C., and Corash, L.M.: Derivation of lymphoma cell lines from microscopically normal bone marrow in patients with undifferentiated lymphomas: Evidence of occult bone marrow involvement. Blood 61, 1017-1019 (1983).

Benjamin, D., Magrath, I.T., Maguire, R., Janus, C., Todd, H., and Parsons, R.G.: Immunoglobulin secretion by cell lines derived from African and American undifferentiated lymphomas of Burkitt's and non-Burkitt's type. J. Immunol. 129, 1336-1342 (1982).

Berard, C., O'Conor G.T., and Thomas, L.B.: Histopathological definition of Burkitt's tumor. Bull. W. Health Org. 40, 601-607 (1969).

Bloomfield, C.D., Arthur, D.C., Frizzera, G., Levine, E.G., Peterson, B.A., and Gajl-Peczalska, K.J.: Nonrandom chromosome abnormalities in lymphoma. Cancer Res. 43, 2975-2984 (1983).

Burkitt, D.: A lymphoma syndrome in African children. Ann. Royal. Coll. Surg. Eng. 30, 211-219 (1962a).

Burkitt, D.: Determining the climatic limitations of a childrens cancer common in Africa. Brit. Med. J. 2, 1019-1023 (1962b).

Chaganti, R.S.K., Jhanwar, S., Koziner, B., Arlin, Z., Mertelsmann, R., and Clarkson, B.: Specific translocations characterize Burkitt's like lymphoma of homosexual men with the acquired immunodeficiency syndrome. Blood 61, 1269-1272 (1983).

Dorfman, R.F.: Childhood lymphosarcoma in St. Louis, Missouri, clinically and histologically resembling Burkitt's tumor. Cancer 18, 418-430 (1965).

Fingeroth, J.D., Weis, J.J., Tedder, T.F., Strominger, J.L., Biro, F.A., and Feason, D.T.: Epstein-Barr virus receptor of human B lymphocytes is the C3d receptor CR2. Proc. Natl. Acad. Sci. USA 81, 4510-4514 (1984).

Henle, W., and Henle, G.: Seroepidemiology of the virus. In M.A. Epstein and B.G. Achong (eds.), Epstein Barr Virus, pp. 61-78, Springer Verlag, Berlin (1979).

Iverson, U., Iverson, O.H., Ziegler, J.L., and Bluming, A.Z., Kyalwazi, S.K.: Cell kinetics of African cases of Burkitt's lymphoma. A preliminary report. Eur. J. Cancer 8, 305-310 (1972).

Lenoir, G.M., Philip, T., and Sohier, R.: Burkitt-type lymphoma: EBV association and cytogenetic markers in cases from various geographic locations. In I.T. Magrath, G.T. O'Conor, B. Ramot (eds.): Pathogenesis of Leukemias and Lymphomas: Environmental Influences. Raven Press, New York (1984).

Magrath, I.T.: Infectious mononucleosis and malignant neoplasia. In D. Schlossberg (ed.) Infectious Mononucleosis, pp. 225-277, Praeger, New York (1983).

Magrath I.T.: Burkitt's lymphoma: clinical aspects and treatment. In D. Molander (ed.): Diseases of the Lymphatic System: Diagnosis and Therapy, pp. 103-139, Springer Verlag, New York (1984).

Magrath, I.T.: Burkitt's lymphoma: a human tumor model. Am J. Hem Onc., in press.

Magrath, I.T., Freeman, C.B., Pizzo, P., Gadek, J., Jaffe, E., Santaella, M., Hammer, C., Frank, M., Reaman, G., and Novikovs, L.: Characterization of lymphoma derived cell lines: comparison of cell lines positive and negative for Epstein-Barr virus nuclear antigen. II. Surface Markers. J. Natl. Cancer Inst. 64, 477-483 (1980).

Magrath, I.T., Henle, W., Owor, R., and Olweny, C.: Antibodies to Epstein-Barr-virus antigens before and after the development of Burkitt's lymphoma in a patient treated for Hodgkin's disease. N. Engl. J. Med. 292, 621-623 (1975).

Magrath, I.T., Janus, C., Edwards, B.K., et al.: An effective therapy for both undifferentiated (including Burkitt's) lymphomas and lymphoblastic lymphomas in children and young adults. Blood 63, 1102-1111 (1984).

Magrath, I.T., and Sariban, E.: Clinical features of Burkitt's lymphoma in the USA: In G. Lenior, C. Olweny, and G. O'Conor (eds): Burkitt's Lymphoma: A Human Cancer Model. IARC Publications, Lyons, in press.

O'Conor, G.T., Rappaport, H., Smith. E.B.: Childhood lymphoma resembling "Burkitt tumor" in the United States. Cancer 18, 411-417 (1965).

Sandlund, J.T., Kiwanuka, J., Marty, G., Goldschmidts, W., and Magrath, I.T.: Phenotyping of Burkitt's lymphoma cell lines using an ELISA technique. Proceedings of 2nd International Worshop of Human Leucocyte Differentiation Antigens, (in press).

Sariban, E., Donahue, A., and Magrath, I.T.: Jaw involvement in American Burkitt's lymphoma. Cancer 53, 1777-1782 (1984).

Whang-Peng, J., Lee, E.C., Sieverts, H., and Magrath, I.T.: Burkitt's lymphoma in AIDS: cytogenetic study. Blood 63, 818-822 (1984).

Ziegler J.L., Beckstead, J.A., Volberding, P.A., et al.: Non-Hodgkin's lymphoma in 90 homosexual men. Relation to generalized lymphadenopathy and acquired immunodeficiency syndrome. New. Engl. J. Med. 311, 565-570 (1984).

Ziegler, J.L., Magrath, I.T., and Olweny, C.L.M.: Cure of Burkitt's lymphoma: 10 year follow-up of 157 Ugandan patients. Lancet ii, 936-938 (1979).

Ziegler, J.L., Miner, R.C., Rosenbaum, E., et al.: Outbreak of Burkitt's-like lymphoma occurring in homosexual men. Lancet 2, 631-633 (1982).

59

MANAGEMENT OF NASOPHARYNGEAL CARCINOMA

Andrew T. Huang, Ian R. Crocker, Samuel R.
Fisher and Mary Jane Wallman
Duke University Medical Center
Depts. of Radiology, Surgery and Medicine
Durham, North Carolina, U.S.A.

Nasopharyngeal carcinoma is rare in most countries (incidence rate of less than 1 per 100,000) but is a major health problem in some regions (incidence rate as high as 5.8 per 100,000 with a mortality rate of 2.6 per 100,000 reported in 1981 from Taiwan). A search for etiologic factors and development of specific methods of eradication of this cancer remain the focus of investigation by many workers in the field and is the subject of this Symposium. Equally necessary, however, are continued efforts to identify early cases through serologic and cytologic screening, and the development of more effective therapeutic modalities for all stages of illness. The current survival rate of 30-60% reported from various regions of the world is a phase of clinical development which warrants further improvement.

GENERAL CONSIDERATIONS

The clinical management of patients with epithelial carcinoma of the nasopharynx has to a large extent been the responsibility of the radiation oncologist. As in other head and neck sites the successful management of the patient with nasopharyngeal carcinoma relies strongly on an understanding of 1) normal anatomy, 2) usual routes of spread of the primary tumor, 3) primary lymphatic drainage, 4) normal tissue tolerance, 5) dose-time-volume relationships in control of nasopharyngeal carcinoma, and 6) treatment of acute and chronic effects of radiation therapy.

Anatomy and Patterns of Primary Tumor Spread

The nasopharynx is that structure which lies posterior to the nasal cavity joining it to the oropharynx and providing access to the middle ear via the Eustachian tube. It is approximately cuboidal in shape measuring 4x4x3 cm. The roof of the nasopharynx lies chiefly beneath the body of the sphenoid and is irregular due to the presence of the pharyngeal tonsil. The floor is made up of the superior surface of the soft palate. The posterior wall lies anterior to the atlas and axis. The anterior wall is formed by the choanae leading to the nasal cavity. Of great importance are the lateral walls of the nasopharynx. Each are largely occupied by the opening of the Eustachian tube. The fossa of Rosenmüller, where many tumors originate, lies immediately posterior to this. Nasopharyngeal carcinoma spreads continuously in a number of typical patterns. The tumor may spread anteriorly into the nasal cavity quite readily due to the lack of any barrier to its spread in this direction. It may also spread from there to involve the ethmoids, the maxillary antrum and the orbit. When the orbit is involved, paresis of individual muscles of the eye may be seen. As with tumor extension anteriorly there is no barrier to tumor extension inferiorly into the oropharynx. Superior extension into the base of the skull was recognized on polytomograms in 25% of a series of 112 patients published by Fletcher and Million (1965) from the M.D. Anderson Hospital. The tumor may invade the sphenoid sinus directly or enter through the ostium of the nasosphenoid. The tumor may also extend to involve the foramina of the base of the skull leading to a constellation of cranial nerve palsies. Extension intracranially to the cavernous sinus where the 3rd, 4th, 6th, and the 1st and 2nd divisions of the 5th nerves may be involved by the tumor probably occurs through the carotid canal. Laterally a defect in the muscular wall of the nasopharynx (known as the sinus of Morgagni) provides easy access to the parapharyngeal space. Cranial nerves 9, 10, 11, 12 and the cervical sympathetics lie in close proximity and may be involved singly or in combination. The spread of the primary tumor and the associated clinical syndromes are well described by Lederman (1961) in his classic monograph on the subject. In the following table, Fletcher and Million (1965) document contiguous spread on a series of 112 patients seen between 1948 and 1960 (Table 1).

Anatomy of the Lymphatics and Nodal Spread

The nasopharynx is richly supplied with lymphatics and clinical lymph node involvement at presentation occurs in 70 to 90% of cases. Due to its midline location, nodal metastasis is often bilateral. The major lymphatic channels of the nasopharynx pass to the lateral retropharyngeal nodes and from there to the upper deep jugular nodes. There are also direct channels to the upper deep jugular nodes, the mid and lower jugular nodes and the spinal accessory nodes. Rarely are the submental, submaxillary or suboccipital nodes involved other than in the presence of

TABLE 1

THE INCIDENCE OF SPREAD OF NASOPHARYNGEAL CARCINOMA

Site of Spread	# of Patients
Oropharyngeal wall	29
Base of skull	25
Tonsillar Bed	15
Cranial nerves	12
Pterygoid fossa	9
Nasal cavity	5
Maxillary antrum	4
Orbit	3
Soft palate	3
Hard palate	3
Ethmoids	2
Hypopharynx	1

(Reprinted with permission from
Fletcher and Million, 1965)

647

altered lymphatic circulation due to a heavily involved
neck. Interestingly the frequency of lymphatic involvement with nasopharyngeal carcinoma does not correlate with
increasing T stage. Indeed, most authors report a lower
incidence of lymph node involvement with T4 tumors than
with Tl, 2 or 3 tumors.

Clinical Evaluation

     The clinical assessment of the patient with nasophar-
yngeal carcinoma is necessarily supplemented by rontgeno-
graphic assessment of the nasopharynx, oropharynx, nasal
cavity, paranasal sinuses and base of the skull. Conventional tomography has been supplanted by computerized axial tomography. Particularly in patients with advanced
neck disease, chest x-ray, liver function tests and bone
scan are utilized in excluding distant metastases. In patients with anterior extension, ophthalmologic assessment
and follow-up is important. Of greater clinical importance
is attention to the patient's dentition. Xerostomia is almost an inevitable consequence of radiation treatment and
careful attention to dentition should be given at the first
visit. If the teeth are in good repair and the patient is
motivated to follow a carefully prescribed program for
maintaining his dentition, full mouth extraction is not necessary. Any teeth that display severe caries or periodon-
tal disease should be extracted and one to two weeks allowed for the gums to heal before radiation is begun.

IRRADIATION TECHNIQUE

     Even early carcinomas of the nasopharynx should have
generous coverage of the primary tumor volume. This primary tumor volume should include the nasopharynx proper,
the posterior nasal cavity, the posterior maxillary sinus,
the posterior orbit, and the base of the skull including
the entire sphenoid and cavernous sinuses. The importance
of not simply covering the primary tumor as assessed clinically is emphasized by Hoppe et al. (1976) who reported a
correlation between local failure and decreased field size.
Marks et al. (1982) in the historical review of their experience in the Mallinckrodt Institute of Radiology correlated poor local control with marginal coverage of the nasopharynx. The lymphatics of the neck are by custom com-

prehensively irradiated to the level of the clavicles. Ho
(1978) however performed a randomized trial regarding elec-
tive node irradiation in T1, T2, and T3N0 patients and found
no statistically significant improvement in loco-regional
control or survival.

The treatment plan is generally executed with lateral
opposed fields junctioned with an anterior field blocked
in the midline. The level of this junction is governed by
neck node involvement. In the absence of neck node involvement this is usually done above the larynx. Beyond
5000 rad boost doses to the primary should be administered
with either high energy lateral fields (16 to 25 MeV) or
rotational fields. This is to spare the temporomandibular
joints and prevent the development of trismus. In patients
with significant anterior extension of tumor a heavily
weighted anterior field with two wedged lateral fields often becomes necessary. Superior extension necessitates
a more generous superior border.

As far as dosage guidelines are concerned, Million and
Cassisi (1984) recommend 6500 rad (at 180 rad per fraction)
for early T stages and 7000 rad for late T stages. To make
up for under-dosage related to technical factors (bone absorption, etc.) a boost of 500 rad is added to the base of
the skull and nasopharynx. The data justifying doses above
6000 rad for carcinoma of the nasopharynx are scant.
Moench (1972), Hoppe, Marks and Million himself have not
shown any significant correlation between local failure and
increasing doses (above 6000 rad).

The Radiation Therapy Oncology Group (Marcial et al.,
1980) experimented with split course radiation therapy in
the management of patients with carcinoma of the nasopharynx. Patients were randomized to either two courses of
3000 rad in 10 fractions separated by a rest period of
three weeks or 6600 total tumor doses with continuous irradiation. No statistically significant differences were observed in relation to acute toxicity, late toxicity, local
regional control or survival.

649

SURVIVAL

The survival of most modern series of patients with
squamous cell carcinoma of the nasopharynx ranges between
30% and 60%. This variation in survival often reflects
the makeup of the treated patient population. The best
reported 5 year survival (59%) for a series of patients
is from the Stanford group (Hoppe et al., 1976) in which
only 50% of the patients had stage IV disease compared to
80% in most other series. A larger series of 1605 patients
from Taipei reported a 5 year survival of 32% (S.C. Huang,
1980). Although initial studies had reported improved survival for younger patients with carcinoma of the nasopharynx, a report from the Children's Cancer Study Group of
the collective experience of 20 member institutions revealed a five year survival of 51%, well within the range seen
in adult patients (Jenkins et al, 1981). Representative
survivals for stage groupings of nasopharyngeal cancer are
shown in Table 2 (Wang, 1983).

TABLE 2

5 YEAR SURVIVAL RATES FOLLOWING RADIATION THERAPY
(1960-1976)

	N0	N1	N2	N3
T1	18/23 (78%)	3/5	8/16 (50%)	2/8
T2	13/18 (72%)	2/4	9/21 (43%)	4/15
T3	1/8	2/4	4/10	2/12
T4	8/18 (44%)	1/7	1/5	2/12

Total 40/67(60%) 8/19(42%) 22/52(42%) 10/47(21%)

(Reprinted with permission from
Wang, 1983)

FAILURE ANALYSIS

Cooper et al. (1983) in their review of stage IV naso-pharyngeal carcinoma combined the data from M.D. Anderson, Stanford University, the Mallinckrodt Institute and New York University to look at failure rates according to T and N stage. These show local failure to occur in approximate-ly 27.5% of patients. This ranges from 11.5% in Tl cases to almost 50% in T4 cases. Nodal recurrence is less common a problem with only 14% demonstrating recurrent disease in nodes. This ranged from 2.8% for the N0 neck to 21.7% for the N3 neck. Distant failure was found in 21.5% of pa-tients. This seemed to correlate with increasing N stage. As opposed to other sites in the head and neck, recurrences seemed to be manifested over a more prolonged period of time. Below are the tables as compiled by Cooper et al. (1983) (Tables 3, 4, 5).

TABLE 3

LOCAL FAILURES BY "T STAGE"

	T1	T2	T3	T4	T1-T4
NYU	1/4	3/10	0/3	15/28	
MRI	4/42		8/21	26/48	
MDAH	0/6	3/31	4/22	7/23	
SUH	5/38	1/16	6/19	5/9	
Total	8/69 (11.5%)	9/78 (11.5%)	18/65 (27.7%)	53/108 (49.1%)	88/320 (27%)

(Reprinted with permission from
Cooper et al., 1983)

TABLE 4

NODAL FAILURES BY "N STAGE"

	N0	N1	N2	N3	Total
NYU	1/8	0/1	4/14	4/16	
MIR	1/42	0/12	6/21	13/36	
MDAH	0/35	3/30	7/59	21/114	
SUH	1/24	2/25	2/15	2/18	
Total	3/109 (2.8%)	5/68 (7.4%)	19/109 (17.4%)	40/184 (21.7%)	67/470 (14%)

TABLE 5

DISTANT METS BY "N STAGE"

	N0	N1	N2	N3	Total
NYU	0/11	0/2	4/14	2/18	
MIR	7/42	3/12	8/21	20/36	
MDAH	2/35	3/30	19/59	46/114	
Total	9/88 (13.2%)	6/44 (13.6%)	30/94 (31.9%)	68/168 (40.4%)	113/394 (28.7%)

Reprinted with permission from
Cooper et al., 1983)

RETREATMENT

The general consensus from the literature is that retreatment with radiation for carcinoma of the nasopharynx is appropriate and provides some patients with excellent palliation and prolonged survival without undue complication (McNeese and Fletcher, 1981). The best survivals for treatment of recurrent disease are reported from the San Francisco group (Fu et al., 1975) with 41% of their patients alive at 5 years after first recurrence. The Taiwan group (Hsu and Tu, 1983) have had a much larger experience in retreatment and report a 5 year survival of only 5% in the absence of distant metastases. This less optimistic figure for disease-free survival after primary site recurrence is amplified by a report from the Radiation Oncology Cancer Institute in Peking in which there was only 14% disease-free survival following retreatment (Yan et al., 1983).

COMPLICATIONS

Common acute complications during radiation treatment include xerostomia, dysgeusia and mucositis. Although the mucositis and dysgeusia are temporary the xerostomia is relatively permanent. Dryness and crusting in the nasopharynx and external ears are relatively common although minor side effects of treatment. Hearing loss due to a serous otitis, neck fibrosis and varying degrees of trismus are also commonly seen. Endocrine assessment will often reveal subtle biochemical disturbances of the hypothalmic pituitary axis, but clinical hypopituitarism is rarely seen. With the lower neck being irradiated, hypothyroidism is occasionally reported. Serious complications including osteonecrosis, soft tissue ulceration, and myelopathy are usually seen in less than 5% of patients. Although cranial nerve palsies are most often related to recurrent disease, involvement of the 9th, 10th, 11th and 12th cranial nerves by radiation-induced fibrosis leading to impairment of function is well known.

SURGICAL MANAGEMENT

The nasopharynx is a surgically inaccessible area which is difficult to examine by the non-experienced physician. Coupled with its meager sensory supply, carcinomas within this area can occur unheralded. This explains why one of the most common presenting signs of nasopharyngeal carcinomas is a painlessly enlarging neck mass in the posterior digastric nodes or in the posterior triangle along the course of the spinal accessory nerve. Other presenting signs and symptoms of nasopharyngeal carcinoma are a unilateral serous otitis media, epistaxis, nasal obstruction, cranial nerve deficit, headaches, and hyposmia. Because of the history, location of the node, and other physical findings, the surgeon should be alerted to the possibility of a lesion occurring within the nasopharynx. Prudence should be exercised prior to the biopsy of any neck node without an appropriate history and physical examination of the head and neck, since the identification of a primary within the nasopharynx would make this surgical procedure unnecessary.

In diagnosing nasopharyngeal carcinoma, modern fiberoptic nasopharyngoscopes are an excellent means to fully assess the superior, lateral and inferior extent of the lesion. CT scans and recently NMR's are excellent methods to delineate soft tissue extension and/or bone erosion of the base of the skull or lateral vertibral bodies, and complement endoscopy with controlled biopsies from specific sites within the nasopharynx to accurately stage the lesion. If a nasopharyngeal lesion is highly suspect and nasopharyngeal biopsies are negative, needle aspiration of lateral neck masses can be helpful in determining if malignant cells are present, but delineating the specific type of tumor is often quite difficult using this method. Open biopsy of a neck mass should be performed only after all diagnostic steps have been exhausted and are negative. An open biopsy should be performed in appropriate skin creases which could be enlarged if necessary to perform a radical lymphadenectomy. Damage to the spinal accessory nerve should be avoided.

The mainstay for treatment of nasopharyngeal carcinoma remains radiation therapy secondary to the relative surgical inaccessibility of the nasopharyngeal lesion and the primary efferent retropharyngeal lymphatics from the nasopharynx. Myringotomy with tympanostomy tube insertion should be performed prior to radiation therapy if serous otitis media is present or develops during treatment. Surgical extirpation of recurrent or persistent disease in the

neck with a radical neck dissection remains controversial,
but generally, it is not felt to be effective in controlling the disease or prolonging survival. Following radiation therapy, isolated disease, which is not fixed to the
base of the skull or deep neck structures, should be considered for resection.

As with all carcinomas of the head and neck, appropriate long term follow-up is necessary to ascertain the
effectiveness of treatment and the possibility of recurrence. Fiberoptic nasopharyngoscopy is an extremely important aspect in the follow-up of these patients. The
occurrence of new neck masses, cranial nerve dysfunction,
epistaxis, increased headaches, all herald the possibility
of recurrent disease. The possibility of a secondary primary in the upper aerodigestive tract is possible and
should always be considered in the follow-up evaluation
of these patients.

ADJUVANT TREATMENT

Many treatment centers have begun to modify their
treatment modality for patients with advanced loco-regional nasopharyngeal cancer because of their higher rates of
local failure and distant dissemination during early phases
of follow-up after radiotherapy. The currently chosen ad-
junctive therapy is generally in two forms: combination
chemotherapy given before or after radiation, and administration of interferon before and also concurrent with
radiation. Many of the combination chemotherapy regimens
are chosen on the basis of their proven effectiveness in
recurrent nasopharyngeal cancer (A.T. Huang et al., 1983).
Interferon, because of its antiviral property has also
been tested in recurrent nasopharyngeal cancer in Germany
(Treuner et al., 1981) and the United States (Connors et
al., 1982).

In regions where nasopharyngeal cancer is prevalent,
adjuvant chemotherapy is applied infrequently. Many physicians in these regions believe that chemotherapy should
be reserved for recurrent disease when additional radiation
cannot be tolerated or distant metastasis occurs. Many
other clinicians have used chemotherapy as an adjunct to
radiation but no specific data is available. Table 6 summarizes 5 different trials of adjuvant therapy in nasophar-

yngeal carcinoma. The Institute of Radiotherapy in Kuala
Lumpur is currently treating patients with advanced disease
with chemotherapy (cyclophosphamide/Oncovin/Methotrexate/
Adriamycin) and radiation. Dharmalingam et al., (1983) representing the Institute reported 63% objective tumor re-
gressions (19% complete and 44% > 50% partial) in 54 patients after 3 cycles of drugs. Patients were subsequently
treated with radiation. In Italy, in a similar trial using
a different regimen of Adriamycin/Bleomycin/Vinblastin/Da-
carbazine, 8/11 patients responded favorably before they
were given definitive radiation (Galligioni et al., 1982).
The survival rates from these two studies are not yet a-
vailable. Goepfert and his associates in M.D. Anderson
Hospital treated 16 patients with stage III and IV disease
with chemotherapy (Vincristine/Adriamycin/Cyclophosphamide,
Bleomycin/Vincristine/Methotrexate and Bleomycin/Cyclophos-
phamide/Methotrexate/Fluorouracil) combined with radiation
and noted only two recurrences in a median follow-up of 36
months (Goepfert et al., 1981). The M.D. Anderson group observed in these and other patients with squamous cell carcinoma of the head and neck increased fibrotic tissue changes associated with the combined approach. The authors and
their colleagues at Duke University Medical Center (A.T.
Huang et al., 1983) have adopted a different approach by
placing adjuvant chemotherapy after definitive radiation
to avoid compromises in total radiation dosage and also
fortuitously to reduce the possibility of drug-radiation
sensitization and resultant higher incidence of tissue fibrosis. In their 10 patients with stage III and IV disease,
6 cycles of chemotherapy (Bleomycin/Methotrexate/Vinblast-
ine/Lomustine) were initiated 1 month after completion of
radiation. In a median follow-up of 28 months, 8/10 were
free of disease. Increased tissue induration was not observed beyond what is expected from the curative doses of
radiation. These trials indicate that chemotherapy may
have an important and necessary role for the future design
of therapy for advanced disease of the nasopharynx.

A trial of interferon in combination with radiation
for advanced nasopharyngeal cancer is in progress at the
Institute of Radiotherapy in Kuala Lumpur in collaboration
with Burroughs-Wellcome Laboratories (personal communication). The trial was begun in the fall of 1982. A significantly superior survival is said to have been observed
in the group of patients randomly allocated to receive radiation and interferon over the group receiving radiation

TABLE 6

ADJUVANT THERAPY PROGRAMS IN PROGRESS

Investigator (yr. of study)	Regimen	No. Treated (median followup)	Dis.-free Survival	Overall Survival
Goepfert 1976–82	RT+VAC BCMF–RT–BCMF RT+BVM	16 (36 mos)	14	14
A.T. Huang 1977–84	RT–BMVL	10 (28 mos)	8	9
Galligioni 1982	ABVD–RT	11 (27 mos)	4 complete regression 4 part. (>50% regression)	
Dharmalingam 1982	COMA–RT	54	10 complete regression 24 part. regression	
M.K. Tan 1982–present	Interferon–RT	in progress		

alone. The final outcome of this trial will not be known
for three to five years. This combination, if proven efficacious, may lead to trials using many other antiviral
drugs which are in the stage of active development at
present.

REFERENCES

Connors, J.M., Andiman, W.A., Merigan, T.C., Treatment of
nasopharyngeal carcinoma with interferon:Epstein-Barr virus
serology and clinical results of a pilot study. Proc. Amer.
Soc. Clin. Oncol. abstract C-771 (1982).

Cooper, J.S., Del Rowe, J., and Newell, J., Regional stage
IV carcinoma of the nasopharynx treated by aggressive
radiotherapy. Int. J. Radiat. Oncol. Biol. Phys. 9:1737-
1745 (1983).

Dharmalingam, S.K., Singh, P., Tan, M.K., Singaram, S.P.,
and Prasa, U., Cyclophosphamide, vincristine, methotrexate,
and adriamycin in untreated nasopharyngeal carcinoma; a report from Malaysian trial on chemotherapy in nasopharyngeal
carcinoma. In: U. Prasad, D.V. Ablashi, P.H. Levine, and G.R.
Pearson (ed.), Nasopharyngeal Carcinoma, Current Concepts,
p. 402 (1983).

Fletcher, G.H., and Million, R.R., Malignant tumors of the
nasopharynx. Am. J. of Roentgenol Radium Ther. Nucl. Med.
93:44-55 (1965).

Fu, K.K., Newman, H., and Phillips, T.L., Treatment of lo-
cally recurrent carcinoma of the nasopharynx. Radiology
117:425-431 (1975).

Galligioni, E., Carbone, A., Tirelli, U., Veronesi, A.,
Trovo, M.G., Donatella, M.M., Crivellari, D., Roncadin, M.,
Frustaci, S., Tumdo, S., and Grigoletto, E., Combined chemotherapy with doxorubicin, bleomycin, vinblastine, dacarba-
zine and radiotherapy for advanced lymphoepithelioma. Cancer Treatment Reports 66:1207-1209 (1982).

Goepfert, H., Moran, M.E., and Lindberg, R.D., Chemotherapy of advanced nasopharyngeal carcinoma. Proc. Amer. Clin.
Oncol. 22:401 (abstract) (1981).

Ho, J.H.C., An epidemiologic and clinical study of nasopharyngeal cancer. Int. J. Radiat. Oncol. Biol. Phys. 4:
183-198 (1978).

Hoppe, R.T., Gothenet, D.R., and Bagshaw, M.A., Carcinoma
of the nasopharynx; eighteen years experience with mega-
voltage radiation therapy. Cancer 37:2605-2612 (1976).

Huang, A.T., Cole, R.B., and Jelovsek, S.B., Adjuvant
treatment of nasopharyngeal carcinoma. In: Nasopharyngeal
Carcinoma, Current Concepts. University of Malaya Press,
Kuala Lumpur, p. 397-401 (1983).

Huang, S.C., Nasopharyngeal cancer; a review of 1,605 patients treated radically with cobalt 60. Int. J. Radiat.
Oncol. Biol. Phys. 6:401-407 (1980).

Hsu, M.M., and Tu, S.M., Nasopharyngeal carcinoma in Taiwan. Cancer 52:361-368 (1983).

Jenkin, R.D.T., Anderson, J.R., Jereb, B., Thompson, J.C.,
Pyesmany, A., Wara, W.M., and Hammond. D., Nasopharyngeal
carcinoma - a retrospective review of patients less than
30 years of age. Cancer 47:360-366 (1981).

Lederman, M., Cancer of the nasopharynx. Charles C. Thomas
(1961).

Marcial, V.A., Hanley, J.A., Chang, C., Davis, L.W., and
Moscol, J.A., Split-course radiation therapy of carcinoma
of the nasopharynx; results of a national collaborative
clinical trial of the radiation therapy oncology group.
Int. J. Radiat. Oncol. Biol. Phys. 6:409-414 (1980).

Marks, J.E., Bedwineck, J.M., Lee, F., Purdy, J.A., and
Perez, C.A., Dose-response analysis for nasopharyngeal
carcinoma. Cancer 50:1042-1050 (1982).

McNeese, M.D., and Fletcher, G.H., Retreatment of recurrent
nasopharyngeal carcinoma. Radiology 138:191-193 (1981).

Meyer, J.E., and Huang, C.C., Carcinoma of the nasopharynx.
Radiology 100:385-388 (1971).

Million, R.R., and Cassisi, N.J., Management of head and neck cancer. J.B. Lippincott (1984).

Moench, H.C., Phillips, T.L., Carcinoma of the nasopharynx: review of 146 patients with emphasis on radiation dose and time factors. Am. J. Surg. 124:515-518 (1972).

Treuner, J., Niethammer, D., Dannecker, G., Jobke, A., Aldenhoff, P., Kremens, B., Nessler, G., and Bommer, H., Treatment of nasopharyngeal carcinoma in children with fibroblast interferon. In: E. Grundmann, G.R.F. Krueger, and D.V. Ablashi (ed.), Nasopharyngeal Carcinoma: Cancer Campaign, Vol 5, pp 309-316, Gustav Fischer Verlag, Stuttgart, New York (1981).

Wang, C.C., Radiation therapy for head and neck neoplasms. John Wright (1983).

Yan, J.H., Hu, Y.H., and Gu, X.Z., Radiation therapy of recurrent nasopharyngeal carcinoma; report on 219 patients. Acta Radiol. (Oncol.) 22:23-28 (1983).

60

TREATMENT OF NASOPHARYNGEAL CARCINOMA WITH THE ANTIVIRAL

DRUG 9-[(2-HYDROXYETHOXYMETHYL)] GUANINE: A CASE REPORT

J.W. Sixbey, E. Thompson and E.C. Douglass

Division of Infectious Diseases, Department
of Hematology-Oncology
St. Jude Children's Research Hospital
322 N. Lauderdale
Memphis, TN   38101   USA

SUMMARY

A patient with advanced nasopharyngeal carcinoma was treated for 7 weeks with 9-[(2-hydroxyethoxymethyl)] guanine (acyclovir). Although progression of the disease was not interrupted in this patient, further trials of acyclovir are warranted.

INTRODUCTION

The regular association of the Epstein-Barr virus with undifferentiated nasopharyngeal carcinoma raises the possibility for specific antiviral intervention at some stage in the disease. Acyclovir, a synthetic acyclic nucleoside, has known activity against the Epstein-Barr virus both in vitro (Colby et al., 1980; Datta et al., 1980; Lin et al. 1984) and in vivo (Sixbey et al., 1983a; Pagano et al., 1983). However, therapeutic efficacy of the drug in nasopharyngeal carcinoma would seem to require an activated viral state. In lymphocytes, acyclovir interrupts active viral replication but has no apparent effect on latent infection (Colby et al., 1980; Lin et al., 1984) or EBV-induced lymphocyte transformation (Sixbey and Pagano, 1985). Epithelial elements in naso-pharyngeal carcinoma bear only the marker of viral

latency, EBNA (Huang et al., 1974; Klein et al., 1974).
Ongoing viral replication, however, is implied by the distinctive serologic pattern accompanying onset of disease:
high antibody titers to Epstein-Barr early antigen and
viral capsid antigen (Ringborg et al., 1983). Moreover,
recent evidence from our laboratory (Sixbey et al.,
1983b;1984) and previous studies by Trumper and associates
(1977) demonstrate that both normal and malignant epithelia are capable of supporting active EBV replication.

To evaluate the safety and potential efficacy of this
DNA polymerase inhibitor in patients with nasopharyngeal
carcinoma, we treated a 20-year-old woman with metastatic
NPC in second relapse and high antibody titers to antigens
of the viral replicative cycle for 7 weeks with acyclovir.

CASE REPORT

T.T. was diagnosed to have nasopharyngeal carcinoma
in June 1981 after a 2-year history of right jaw pain,
headache, and decreased hearing. She presented with right
ocular proptosis, a large nasopharyngeal mass which limited jaw motion, and bilateral cervical adenopathy. Computerized tomography of her head showed a large nasopharyngeal mass extending through the greater wing of the
sphenoid, into the brain 2-3 cm and forward into the right
orbit. Biopsy revealed nasopharyngeal carcinoma of the
poorly differentiated type. She was classified an IARC
Stage C. She received 2 courses of vinblastine (6 mg/M^2),
bleomycin (15 mg/M^2), and cis-platinum (90 mg/M^2) with
good response and 5 additional courses following radiation
therapy. From August to October 1981, she received electron beam and cobalt irradiation, 5200 rads to a right
temporal port, 5000 rads to the nasal cavity and 3500 rads
to the lower neck and supraclavicular area. Therapy was
associated with approximately a 50% reduction in tumor
mass, and repeat biopsy showed only residual fibrosis.
EBV-specific antibody titers, elevated on admission,
declined immediately post-therapy (Figure 1).

In March 1982, the patient experienced epigastric
pain. In August 1982, a single nodule in the left lobe of
the liver and left upper lobe of the lung appeared on

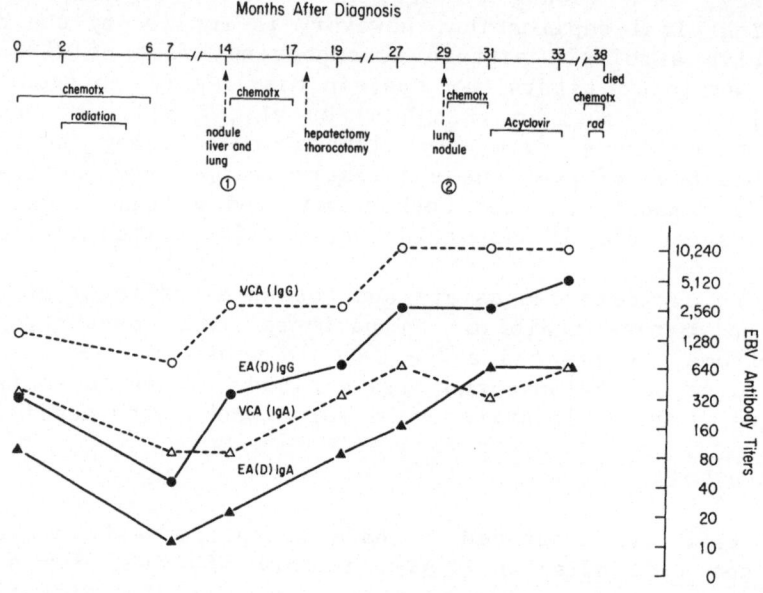

Figure 1.
EBV antibody titers in the months subsequent to diagnosis. An antibody titer rise preceded each relapse.

radiographic studies. She received methotrexate (250 mg/M^2) and 5-fluorouracil (600 mg/M^2) to which she partially responded. In December 1982, the nodules in both lung and liver were surgically resected.

Except for rising antibody titers to EBV (Figure 1), T.T. remained well until November 1983, when a left hilar mass was appreciated on chest radiograph. With this second relapse, the patient received methotrexate and 5-fluorouracil with no response. At this time, acyclovir was administered, 500 mg/M^2 intravenously every eight hours for 3 weeks followed by oral acyclovir 400 mg every 4 hours.

Acyclovir therapy was to be continued as long as there was no evidence of disease progression or unacceptable toxicity. Acyclovir was well tolerated with no adverse effects despite T.T.'s previous history of mild renal impairment secondary to nephrotoxic chemotherapy. Salivary IgA to early antigen and viral capsid antigen, present before therapy, persisted throughout treatment. Serum antibody titers to EBV did not change significantly throughout therapy or for the subsequent 5 months of follow up (Figure 1). Oropharyngeal viral shedding did not occur at any time during study. Seven weeks after institution of drug, tumor progression was evident on chest radiograph. Tumor tissue obtained at autopsy 5 months post-therapy was EBNA positive by the anticomplement immunofluorescence technique (Reedman and Klein 1973). Early antigen and viral capsid antigen could not be demonstrated by indirect immunofluorescence. EBV DNA was detected in approximately 50 percent of cells on a touch preparation of tumor by in situ cytohybridization using a cloned, biotin-labeled EBV DNA probe (Brigati et al., 1983).

DISCUSSION

Prolonged antiviral therapy with acyclovir was well tolerated in this patient with advanced nasopharyngeal carcinoma but failed to produce a discernable clinical response. Factors to be considered with regard to treatment failure include first, the advanced stage of this multi-drug resistant tumor; second, inadequate phosphorylation of acyclovir in malignant tissue to its active triphosphorylated derivative; and third, possible irrelevance of viral replication to the progression of nasopharyngeal carcinoma. In view of the drug's relative safety, further studies promoting intervention with acyclovir at earlier stages of the disease are warranted, particularly in patients where rising antibody titers signal relapse but in whom negative diagnostic workups do not support further cytotoxic chemotherapy.

REFERENCES

Brigati, D.J., Myerson, D., Leary, J.J., Spalholz, B., Travis, S.Z., Fong, C.K.Y., Hsiung, G.D., and Ward, D.C., Detection of viral genomes in cultured cells and paraffin-embedded tissue sections using biotin-labeled hybridization probes. Virology, 126,32-50 (1983).

Colby, B.M., Shaw, J.E., Elion, G.B., and Pagano, J.S., Effect of acyclovir [9-(2-hydroxyethoxymethyl)guanine] on Epstein-Barr virus DNA replication. J. Virol., 34,560-568 (1980).

Datta, A.K., Colby, B.M., Shaw, J.E., and Pagano, J.S., Acyclovir inhibition of Epstein-Barr virus replication. Proc. Natl. Acad. Sci. USA, 77,5163-5166 (1980).

Huang, D., Ho, J.H.C., Henle, W., and Henle, G., Demonstration of Epstein-Barr virus-associated nuclear antigen in nasopharyngeal carcinoma cells from fresh biopsies. Int. J. Cancer, 14,580-588 (1974).

Klein, G., Giovanella, B.C., Lindahl, T., Fialkow, P.J., Singh, S., and Stehlin, J.S., Direct evidence for the presence of Epstein-Barr virus DNA and nuclear antigen in malignant epithelial cells from patients with poorly differentiated carcinoma of the nasopharynx. Proc. Natl. Acad. Sci. USA, 71, 4737-4741 (1974).

Lin, J.-C., Smith, M.C., and Pagano, J.S., Prolonged inhibitory effect of 9-(1,3-dihydroxy-2-propoxymethyl)-guanine against replication of Epstein-Barr virus. J. Virol., 50,50-55 (1984).

Pagano, J.S., Sixbey, J.W., and Lin, J.-C., Acyclovir and Epstein-Barr virus infection. J. Antimicrob. Agents Chemother., 12:Suppl. B, 113-121 (1983).

Reedman, B.M., and Klein, G., Cellular localization of an Epstein-Barr virus (EBV)-associated complement-fixing antigen in producer and non-producer lymphoblastoid cell lines. Int. J. Cancer, 11,499-520 (1973).

Ringborg, U., Henle, W., Henle, G., Ingimarsson, S., Klein, G., Silfversward, C., and Strander, H., Epstein-Barr virus-specific serodiagnostic tests in carcinomas of the head and neck. Cancer, 52,1237-1243 (1983).

Sixbey, J.W., and Pagano, J.S., Epstein-Barr virus transformation of human B-lymphocytes despite inhibition of viral polymerase. J. Virol., 53,299-301 (1985).

Sixbey, J.W., Nedrud, J.G., Raab-Traub, N., Hanes, R.A., and Pagano, J.S.: Epstein-Barr virus replication in oropharyngeal epithelial cells. N. Engl. J. Med., 310,1225-1230 (1984).

Sixbey, J.W., Pagano, J.S., Sullivan, J.L., Gurwith, M., Fleisher, G., and Clemons, R.H., Treatment of infectious mononucleosis with intravenous acyclovir. Clin. Res., 31,542A, (1983a) (Abstract).

Sixbey, J.W., Vesterinen, E.H., Nedrud, J.G., Raab-Traub, N., Walton, L.A., and Pagano, J.S., Replication of Epstein-Barr virus in human epithelial cells infected in vitro. Nature, 306,480-483 (1983b).

Trumper, P.A., Epstein, M.A., Giovanella, B.C., and Finerty, S.: Isolation of infectious EB virus from the epithelial tumor cells of nasopharyngeal carcinoma. Int. J. Cancer, 20,655-662 (1977).

61

EBV-SPECIFIC TRANSFER FACTOR IN THE TREATMENT OF

AFRICAN BURKITT'S LYMPHOMA: A PILOT STUDY

F.K. NKRUMAH[1], G. PIZZA[2], D. VIZA[3],
J. NEEQUAYE[1], C. DE VINCI[2], P.H. LEVINE[4]
[1]Burkitt Tumor Project, Univ. of Ghana
Medical School, Accra, Ghana; [2]Ospedale
Marcello Malpighi, Bologna, Italy; [3]Faculte
de Medecine Broussais, Hotel Dieu, Paris,
France; [4]National Cancer Institute,
Bethesda, Maryland, U.S.A

SUMMARY

**Eleven African patients with Burkitt's lymphoma
entered a study to evaluate the efficacy of a transfer
factor with specific activity against Epstein-Barr virus.
The disease-free interval in the group of treated patients
was significantly longer than in the control group. The
absence of toxicity and the potential efficacy of the
transfer factor indicates the need for more extensive
clinical trials in African Burkitt's lymphoma.**

INTRODUCTION

In previous reports (Pizza et al, 1981; Levine et al,
1983), we described the rationale for using a transfer
factor with specific activity against the Epstein-Barr
virus (EBV) in a combined therapeutic approach to the
control of African Burkitt's lymphoma (BL). Based
primarily on the evidence that late relapse (greater than
one year after completion of chemotherapy) is actually the
result of disease re-induction, possibly mediated through
persistent EBV activity, we proposed that EBV-specific
transfer factor be administered to patients who had
achieved remission following conventional chemotherapy
with the intention of preventing these late relapses.

Supported in part by the Regione Emilia Romagna

This report describes the initial results obtained in this clinical trial.

MATERIALS AND METHODS

Patients with Stage III BL newly diagnosed by the University of Ghana's Burkitt Tumor Project were alternately assigned to receive either transfer factor or no additional treatment after completing a course of conventional chemotherapy (oral cyclophosphamide 40 mg/kg x 3 and intrathecal methotrexate 15 mg x 3). For those receiving transfer factor, one dose of five units (extracted from $5x10^8$ lymphoid cells) was given monthly.

The transfer factor had initially been obtained from the lymphocytes of a patient with nasopharyngeal carcinoma who had been shown to have strong cell-mediated immunity against EBV membrane antigens as measured by the leukocyte migration test (for details, see Pizza et al, 1981). This transfer factor was shown by in vivo and in vitro methods to be able to transfer immunity to Raji cells super-infected with EBV (R-EBV). After replication in the LDV-7 cell line (Viza et al, 1982), this transfer factor was aliquoted and stored at -70°C in Ghana until used in the trial.

In vitro assays monitoring the cell-mediated immunity of patients and controls included the direct leukocyte migration inhibition (LMI) test, which was performed in Ghana as previously described (Levine et al, 1981), and the indirect LMI test, which was performed on coded supernatants in Bologna.

RESULTS

As shown in Table I, the transfer factor used in this study had activity against the B95-8 strain of EBV, cytomegalovirus (CMV), herpes simplex Type II (HVS-2) and R-EBV.

Between May 1981 and August 1984, eleven patients entered the study. As shown in Table II, one of five treated patients had a relapse whereas three relapses occurred in the six non-transfer factor controls. The disease-free interval in the group of treated patients (57.8 + 30.2) was significantly longer than in the untreated controls (24.0 + 15.7, $p < 0.05$). The one patient in the transfer factor group who relapsed had a clinical remission of only 7 weeks.

TABLE I IN VITRO ASSAYS FOR TRANSFER FACTOR ACTIVITY		Supernatant Dilutions			
Donor	Antigen	10^{-4}	10^{-3}	10^{-2}	10^{-1}
#1	B95-8	1.00	0.98	0.93	0.80*
	CMV	1.19	0.93	0.88	0.96
	HVS-2	1.17	1.04	0.91	1.03
	R-EBV	NT	0.90	1.03	0.92
#2	B95-8	NT	0.89	0.83*	1.03
	CMV	0.98	0.80*	NT	0.84*
	HVS-2	1.03	1.02	1.28	1.16
	R-EBV	0.93	1.46	1.01	0.80*
#3	B95-8	0.67*	0.61*	NT	0.57*
	CMV	0.72*	0.70*	0.61*	0.56*
	HVS-2	0.80*	0.71*	0.58*	0.93
	R-EBV	1.12	0.99	1.02	1.62

NT = Not Tested

* = Significant Migration Inhibition (M.I.<0.85)

$$M.I.=\frac{\text{Leukocyte Migration with Antigen + Transfer Factor}}{\text{Leukocyte Migration with Antigen Alone}}$$

10^6 lymphocytes from each normal donor were incubated overnight with transfer factor (TF) at a ratio L:TF=5:1 in 1 ml of RPMI 1640 and 10% fetal calf serum (FCS). The cells were washed twice and exposed 3 hours to the antigens at optimal dosage [viruses 100 transforming units/ml and 5:1 for Raji cells superinfected with EBV (R-EBV)]. Afterwards the cells were washed again twice; 2×10^5 in 0.2 ml of RPMI 1640 + 5% FCS were resuspended and incubated overnight in a CO_2 incubator to produce LIF in the supernatant. The presence of LIF activity was then evaluated on normal human neutrophil cells at various dilutions of the supernatant with the agarose microdroplet leucocyte migration inhibition technique (see Levine et al., 1981).

TABLE II

BURKITT'S LYMPHOMA: CLINICAL RESULTS

Patient	Age	Sex	Treatment[a]	Relapse[a]	Disease-Free Interval[a]	Follow-up[a]	Status
1) M.A.	9	M	36	7	7	77	Dead (77)
2) A.A.	10	M	32	No	60	60	Alive
3) F.I.	11	F	52	No	80	80	Alive
4) L.G.	7	F	8	No	82	82	Alive
5) E.A.	9	F	28	No	60	60	Alive
MEAN					57.8 ± 30.2^b	71.8 ± 10.9	
6) F.A.	6	M	No	19,26	19	170	Alive
7) M.A.	6	M	No	No	20	20	Alive
8) K.A.	9	M	No	12	12	78	Dead (78)
9) K.A.	10	M	No	No	50	50	Alive
10) A.K.	6	F	No	35	35	66	Alive
11) A.Y.	9	F	No	No	8	8	Alive
MEAN					24.0 ± 15.7^b	65.3 ± 57.7	

a All data are in weeks.

b $p < 0.05$

The LMI tests, which could only be performed on the transfer factor group because of transportation problems preventing frequent laboratory evaluation of the untreated controls, showed an improvement of in vitro cellular immunity in the patients treated with transfer factor.

DISCUSSION

A number of reports (Viza et al, 1983, 1984; Rosenfeld et al, 1984; Pizza et al, 1979) have documented the effectiveness of specific TF in the control of diseases associated with potentially oncogenic viruses. In addition, recent reports indicate that TF may increase the length of remission in certain tumors (Kirsh et al, 1984). This pilot study investigating the potential value of EBV-specific transfer factor in the treatment of Burkitt's lymphoma indicates that additional patient evaluation is warranted. While the difference in disease-free interval being statistically significant, the trend is promising and the absence of toxicity clearly improves the benefit:risk proportion. It is of interest that the only patient to relapse in the transfer factor group had a brief remission of only 7 weeks, thus suggesting that the tumor was not completely controlled by the initial chemotherapy.

With regard to the in vitro assays, the results at present can only be used to support the feasibility of our current approach. Because of the absence of data from the control group, it is not possible to state whether the improved CMI is due to the TF or is solely the recovery of normal immunity following chemotherapy-induced remission.

Although patients with Stage III disease provide the best study group for such trials because of the eventual treatment failure of approximately 50% using the current conventional chemotherapy regimen, these results suggest that the study should perhaps also be extended to Stage IV patients (virtually all stage I and II patients do not relapse, thus making a clinical trial in these groups of patients impractical).

REFERENCES

KIRSH, M.M., et al, Transfer factor in the treatment of carcinoma of the lung. Ann. Thorac. Surg., 38, 140-145 (1984).

LEVINE, P.H., NKRUMAH, F.K., ABLASHI, D.V., PEARSON, G.R., FAGGIONI, A., VIZA, D., LVOVSKY, E., DE VINCI, C., and PIZZA, G., Clinical and experimental data on the effect of antiviral agents against oncogenic herpesviruses: Implications for the treatment of nasopharyngeal carcinoma. In: U. Prasad, D.V. Ablashi, P.H. Levine, and G.R. Pearson (eds.), Nasopharyngeal Carcinoma: Current Concepts, pp. 415-422, University of Malaya Press, Kuala Lumpur (1983).

LEVINE, P.H., PIZZA, G., CANNON, G., ABLASHI, D.V., ARMSTRONG, G., and VIZA, D., Cell-mediated immunity to Epstein-Barr virus associated membrane antigens in patients with nasopharyngeal carcinoma. In: A. Grundmann, G. Krueger, and D.V. Ablashi (eds.), Nasopharyngeal Carcinoma: Basic Research as Applied to Diagnosis and Therapy, pp. 137-144, Gustav Fischer Verlag, Stuttgart (1981).

PIZZA, G., VIZA, D., ABLASHI, D.V., JEROME, L., ARMSTRONG, G., and LEVINE, P.H., The possible use of specific transfer factor in the treatment of patients with naso-pharyngeal carcinoma. In: A. Grundmann, G. Krueger, and D.V. Ablashi (eds.), Nasopharyngeal Carcinoma: Basic Research as Applied to Diagnosis and Therapy, pp. 301-308, Gustav Fischer Verlag, Stuttgart (1981).

PIZZA, G., VIZA, D., RODA, A., ALDINI, R., RODA, E., and BARBARA, In vitro produced transfer factor for the treatment of chronic active hepatitis. N. Engl. J. Med., 300, 1332 (1979).

ROSENFELD, F., VIZA, D., PHILLIPS, J., VICH, J.M., BINET, O., and ARON-BRUNETIERE, R., Traitement des infections herpetiques par le facteur de transfert. Presse Med., 13, 537-539 (1984).

VIZA, D., BOUCHEIX, C., CESARINE, J.P., ABLASHI, D.V., ARMSTRONG, G., LEVINE, P.H., and PIZZA, G., Characterization of human lymphoblastoid cell line, LDV/7, used to replicate transfer factor and immune RNA. Biol. Cell, 8, 1721-1726 (1982).

VIZA, D., ROSENFELD, F., and PHILLIPS, J., Specific bovine transfer factor for the treatment of herpes infections. In: C.H. Kirkpatrick (ed.), Immunobiology of Transfer Factor, pp. 245-259, Academic Press, New York (1983).

VIZA, D., VICH, J.M, PHILLIPS, J., and ROSENFELD, F., Orally administered specific transfer factor for the treatment of herpes infections. Lymphok. Res., **4**, 27-30 (1984).

62

BRIEF COMMUNICATION

THE TREATMENT OF NASOPHARYNGEAL CARCINOMA (NPC)

J.H.C. Ho

Medical and Health Department

Institute of Radiology and Oncology

Queen Elizabeth Hospital

Kowloon, Hong Kong

There is an urgent need for a general agreement on a system for the stage classification of NPC, without which the evaluation of the effectiveness of different techniques employed in the treatment and the comparison of treatment results between centres will not be meaningful. If a comparison is to be made between different systems, it has to be a prospective one. A good system should have clearcut definitions of the criteria used in classifying the stages and allowing them to be identified without ambiguity. In this respect T1 (tumour confined to one wall of the nasopharynx) and T2 (confined to two walls) in the UICC and the AJC systems could well be combined under T1, because one cannot be certain from clinico-radiological examination that the tumour is confined to just one wall of the nasopharynx, and it has yet to be shown that the subclassification has anything to offer in guiding treatment or prognosis. Most NPC are eccentric in origin and a small tumour in the fossa of Rosenmuller (lateral pharyngeal recess) is in fact astride two walls. Classification of cervical nodal metastases by size is an arbitrary decision. There will always be some variation in personal judgement when the measurement is done by palpation, and yet a difference of 1 mm in the measured diameter of the node separates N1 from N2, and N2 and N3. Furthermore, Ho (1978) has shown that the laterality or mobility of the cervical nodal metastases are far less important prognostic factors compared with the level of the nodal involvement, and yet this factor was ignored in either the UICC or AJC classification.

The two main causes of treatment failure are an un-controlled primary tumour and distant metastases. Analysis of 1139 NPC patients with stages I-IV disease treated in 1976-78 at Queen Elizabeth Hospital, Hong Kong (Ho, 1978) showed that approximately one-quarter of the patients had primary tumour recurrence by the end of the fifth year after the commencement of treatment. Probably the failure in some of them was due to a geographic miss in the treatment. This could be reduced or avoided when computerized tomography (CT) was used in the demonstration of the extent of the primary tumour and a CT radiotherapy planner in the treatment planning. In the same analysis it was found that the most common sites of clinically detected distant metastases were bone, lung and liver in the following incidence ratio: 1.93: 1.08: 1.00. This is unfortunate because in our experience bone metastases respond very poorly to chemotherapy. They often continue to grow or appear during a course of chemotherapy which may cause regression in some visceral metastases. Most primary tumours also failed to respond even if they had no previous radiation therapy. This means that adjuvant chemotherapy has little to offer in the treatment of NPC until more effective chemotherapeutic agents are found.

Reference

Ho, J.H.C.: Stage classification of nasopharyngeal carcinoma. In: Nasopharyngeal Carcinoma: Etiology and Control, eds. G. de The and Y. Ito. Lyon, IARC Scientific Publ. No. 20 (1978), pp. 99-113.

SPECIAL LECTURE

63

SPECIAL LECTURE

EPSTEIN-BARR VIRUS: PAST, PRESENT AND FUTURE

Gertrude and Werner Henle

Children's Hospital of Philadelphia

Philadelphia, PA

At an occasion like this, it may be permissible to discuss the past, the history of the Epstein-Barr virus, from a rather personal and perhaps slightly biased perspective. This may be less than exciting for those who were among the pioneers in this field and present at the creation, but those of you who joined the steadily growing crowd at later stages might not be aware of how much of the inital research depended on being at the right place at the right moment, on recognizing and grasping an opportunity as it arose, or on just good old plain luck.

Our interest in human cancer viruses preceded the discovery of the Epstein-Barr virus by several years. We had started to work with polyoma virus, a newly discovered mouse tumor virus, to become familiar with essential techniques, among them assessment of the role of interference and interferon in the establishment and maintenance of persistent viral carrier cultures and the use of immuno-fluorescence for monitoring primary and persistent infections.

We thus were not unprepared for the telephone call we and many other virologists received in the late 1950s from Carl Baker, then Assistant Director of the National Cancer Institute, to invite grant applications for support of a search for human cancer viruses. They were thought to exist in analogy to the animal tumor viruses which were extensively studied as potential models for their as yet unknown human counterparts. However, the experiences gained were rarely if ever, applied to human cancers.

No detailed research proposals were required because there were no leads, after all, on which to base them; merely a brief outline of the proposed general approach was requested and, of course, a not too heavily padded budget. As an additional inducement, 10 years of committed support were promised. Can you imagine 10 years without writing a competing grant application?

Carl Baker's telephone call translated our vague plans into action, and we received a 10 year grant entitled "Interference phenomena in the detection of human cancer viruses". Accordingly, we cultured numerous specimens from childhood cancers, supplied by the surgeon-in-chief of the Children's Hospital of Philadelphia, C. Everett Koop, who is now the U.S. Surgeon-General, and we tested the out-growing cells for resistance to vesicular stomatitis virus as an indication of interference or interferon production, induced by an indigenous cancer virus that might be present. We have still to find a single cell culture of man or beast, fish or fowl, reptile or insect that was normally resistant to this omnipotent virus. The cultured childhood tumor cells, unfortunately, proved no exception; they failed to resist infection by vesicular stomatitis virus.

When Chick Koop returned in 1963 from a conference in Africa, he told us about Burkitt's lymphoma and urged us to work on it because the epidemiology of this most frequent tumor of African children strongly suggested that it was caused by a virus. We immediately wrote to Denis Burkitt and other physicians on the African scene only to learn that everyone was already committed to other investigators. We apparently had missed the boat.

A year later, Tony Epstein and his colleagues at the Middlesex Hospital in London had established continuous cultures of Burkitt's lymphoma cells and found, on electron microscopic examination, herpes-like virus particles in a small proportion of the cultured cells. This discovery aroused few ripples of excitement because most virologists assumed that the virus could only be herpes simplex, cyto-megalo or chicken pox virus; none of these, nor any animal herpes virus was as yet suspected of causing cancers. The virus indigenous to the Burkitt cells was therefore thought to be a harmless passenger of no particular interest.

As a last resort, Tony Epstein sent the Burkitt cell cultures for identification of the virus to Klaus Hummeler who recently had spent a sabbatical in his laboratory and had been in charge of our virus-diagnostic service at the Children's Hospital of Philadelphia which had just been dismantled, because the Pennsylvania Health Department withdrew its support for financial reasons. Klaus Hummeler came to our office, waving the bottles, to ask what should be done with them. Whereupon my wife promptly responded, "give them to me", and thereby came our chance to work on Burkitt's lymphoma.

The title of our grant was not a mere figment of our imagination. My spouse showed that the EB-1 and EB-2 Burkitt cutures were highly resistant to VSV, that the cells, following implantation, conferred resistance to monolayer cultures of human, but not rabbit or mouse cells, suggesting production of an interferon which indeed was found in the culture media. Another Burkitt cell line serves now as a potent source of one type of human interferon.

My spouse was unable, however, to transmit the indigenous herpes virus to routine cell cultures or to chick embryos, hamsters or mice of all seasons, using all possible routes of inoculation. This failure provided the first clue that the virus in the Burkitt tumor cultures was previously unknown but for proof, immunologic procedures were required. Only immunofluorescence appeared to fit the situation at hand but where to get sera from Burkitt patients which we believed, alas mistakenly, were needed.

Our chance came when we learned that a Nigerian Burkitt patient had been flown to the Clinical Center at the National Institutes of Health in Bethesda for plasmapheresis and attempts to sediment the expected C-type Burkitt's lymphoma virus from the plasma by highspeed centrifugation. It is truly amazing how much debris, how many "virus-like particles", can be sedimented from anybody's plasma, but C-type virus particles have not been among them. When we called Ray Bryan, then the Director of the National Cancer Institute, to request the supernates from the centrifuge runs, I distinctly heard "My God, they pour them down the drain" - together with all the antibodies we needed. Luckily, one more run was planned and its supernate, labeled P-91, yielded brilliant immunofluorescence in about 10% of the EB-1 and EB-2 cells. A serum from an American child with leukemia, chosen as control, failed to react. The resulting euphoria was of only short duration, however, because it was soon found that many American sera as well as commercial human gamma globulin, also gave positive reactions.

We had to prove that the immunofluorescence was referable to the indigenous herpes virus. For one thing, the percentage of immunofluorescent cells and the percentages of virus-producing cells in given Burkitt cell lines, and soon other lymphoblast cultures, were closely similar. When we presented a lantern slide which listed a dozen cell lines in descending order of the percentage of immunofluorescent cells at a conference of the American Cancer Society at Rye, N.Y., in 1967, George Klein was the first to rise to the discussion to exclaim that the same order applied to the cell membrane immunofluorescence which he had been studying, indicating that the cell membrane antigen was virus-induced. This conference was the most fruitful in our career because it led to a collaboration with George Klein and his associates which has continued to the present day and resulted by now in over 75 joint publications, with more to come we hope.

Soon thereafter, it was shown by Gary Pearson, while a postdoctorate fellow, first in George Klein's and subsequently our laboratory, that the membrane-reactive antibodies were, in part, identical with EBV neutralizing antibodies.

With Klaus Hummeler we showed by negative contrast electron microscopy that viral nucleocapsids, extracted from Burkitt cells, acquired a fringe of antibodies, resembling the head of Medusa, after exposure to immunofluorescence-positive, but not negative sera. The antigen so detected was therefore named later viral capsid antigen to differentiate it from the diffuse and restricted early antigens discovered by us, and the EBV-associated nuclear antigen (EBNA) discovered by Beverly Reedman and George Klein.

Irrefutable proof of the identity of the immunofluorescent and virus-producing cells was obtained when Harald zur Hausen, as a postdoctoral fellow in our laboratory, picked individual fluorescent cells for embedding, thin sectioning, and electron microscopic examination. All were loaded with virus particles, whereas non-fluorescent cells, similarly prepared, showed none.

Soon after his return to Germany, Harald zur Hausen showed that Burkitt cells which did not produce virus, nevertheless contained EBV DNA and that Burkitt's lymphomas and nasopharyngeal carcinomas, harbored EBV DNA in amounts equivalent to multiple viral genomes per cell. These observations have been amply confirmed by others, with ever more refined techniques. Furthermore, any cell harboring EBV DNA was shown by the Stockholm group to express the nuclear antigen EBNA.

To return to immunofluorescence, we made, of course, all along efforts to properly identify the virus with the aid of acute and convalescent sera from patients with primary herpes simplex, cytomegalo or varicella virus infections, as well as with hyperimmune sera to various animal herpes viruses. To compress several months of work into one sentence, the virus indigenous to the Burkitt and other lymphomablastoid cell lines was antigenically un-related to any known herpes virus. It was, indeed, new and we named it-provisionally we thought-EB virus after the EB-1 culture in which it was first observed in order to retire the awkward designations "herpes-like", "herpes-type" or even "leuco" virus. A year or two later, we would have baptized it infectious mononucleosis virus, since all herpes viruses are named after the principal disease they cause, but EBV was already too well entrenched to make a change.

Why was EBV production limited to only a small fraction of the cultured cells? Was the autogenous interferon production the limiting factor or was an enzyme in short supply that another herpes virus could contribute? Helper viruses had just become fashionable, but those we tested, that is herpes simplex, mumps and reo-3 viruses, provided no help. However, help came unexpectedly when a batch of completed culture medium was kept inadvertantly in the warm room for more than a week instead of the 48 hours we used in the prelaminar flow hood days to eliminate bacterially contaminated batches.

Frugal as we were, we used this batch of medium and were surprised by 5- to 10-fold increases in virus-producing cells. We traced this enhancement back to a loss of arginine by action of a fetal calf serum component. We still use arginine-free medium today to raise the number of virus-producing EB-3 cells. Subsequently, various anti-metabolites were shown by others to induce cycles of viral replication in latently infected cells by as yet obscure mechanisms.

The immunofluorescence test was soon applied to seroepidemiologic surveys which showed that antibodies to EBV
were detectable anywhere in the world, even in such remote
regions as the Amazon jungle, South Pacific atolls, Aleutian
islands and, I am sure, Siberia if anyone should care to go
there to collect sera. Depending on the state of hygiene,
the degree of crowding, and other socioeconomic conditions,
nearly everybody was found sooner or later to acquire
antibodies to the virus. However, all sera from African
Burkitt patients, kindly supplied now by Eva Klein and
others, had substantially greater concentrations of antibodies than sera from African control children so that EBV
remained a candidate for the etiology of Burkitt's lymphoma.

The association of EBV with nasopharyngeal carcinoma
and with infectious mononucleosis were both discovered by
chance. When Lloyd Old and his associates needed, and
received sera from African patients with tumors other than
Burkitt's lymphoma as controls for double diffusion precipitation tests with extracts of virus-producing Burkitt
cells, they included sera from several patients with nasopharyngeal carcinomas. To everyone's surprise, these sera
reacted like Burkitt sera, producing up to 5 lines of
precipitation. When we tested the same and additional sera
from nasopharyngeal carcinoma patients we found that they
reacted like Burkitt sera also in the immunofluorescence
test, yielding much higher antibody levels than sera from
patients with other head or neck tumors.

Because EBV had turned out to be one of the most widespread human viruses, we suspected that it might induce a
common illness as its primary clinical activity. In our
search for this disease we were assisted beyond the call of
duty by one of our young female technicians who was seronegative, developed infectious mononucleosis and serocon
verted in its course. This was not a laboratory infection
as proclaimed by the Division of Biological Hazards of
NIH. Our technician was a very pretty girl and thus was
exposed to the kissing disease virus by the natural route.

Before her illness, our technician had often donated
leukocytes for experiments carried out by Volker Diehl,
then a postdoctoral fellow in our laboratory. Her cells
had never grown by themselves in culture; they grew only
when co-cultivated with lethally x-irradiated Burkitt cells
from producer cultures. Leukocytes donated by her during

her illness, however, now grew readily by themselves, yielding permanent EBV-positive cultures. These were the initial clues for the causal relationship of EBV to infectious mononucleosis and for the lymphoproliferative effects of the virus.

We then remembered that our friends at Yale, James Niederman and Robert McCollum, had collected sera from freshman at entry into college and again from those who developed infectious mononucleosis during the ensuing four years in anticipation of future candidate viruses for the etiology of the disease. They sent us coded sera and all pre-illness samples were readily identified by us; they were negative in the immunofluorescence test, whereas all acute convalescent phase sera were strongly positive.

As mentioned, four distinct groups of EBV-associated antigens had been identified in time, and differentiation of the immunoglobulin class of the corresponding antibodies added another dimension to the EBV-specific serology. It was found that each of the three EBV-associated diseases (IM, BL, NPC) evokes at its height a characteristic spectrum of antibodies which differs from the other two and the pattern seen in healthly persons after long-past primary infections. IgM antibodies to the viral capsid antigens are limited to infectious mononucleosis; high IgG antibody titers to the viral capsid and restricted early antigen are characteristic of Burkitt's lymphoma; high titers of IgA and IgG antibodies to the viral capsid and the diffuse early antigen are an outstanding feature of nasopharyngeal carcinoma.

The EBV-specific serology thus became an important diagnostic tool. In the tumor patients, it provides, furthermore, information on their prognosis and serves to monitor them during remissions for evidence of imminent tumor relapses. Most recently surveys for IgA antibodies to viral capsid antigen and IgG antibodies to the diffuse early antigen have been used successfully in China and recently also Alaska for early detection of nasopharyngeal carcinoma patients.

All these studies depended upon efficient, unstinting, harmonious collaborations on a world-wide basis with clinicians, epidemiologists and scientists covering many disciplines. It has been our pleasure and a rewarding experience to work closely with numerous individuals on all continents, some of whom we have yet to meet in person. Through the combined efforts, EBV has become a highly respected and fascinating virus which has attracted the attention of literally hundreds of molecular virologists, immunologists, biochemists, cytogeneticists and others.

The present, therefore, is as exciting as was the past. The current symposium bears witness to that and has left little for us to add except to ask, perhaps, a few questions.

We still need to know more about events during the incubation period of infectious mononucleosis. If the primary targets of EBV are pharyngeal epithelial cells, as seems likely now, why is the incubation period of infectious mononucleosis 4 to 7 weeks instead of the few days observed with the usual respiratory viruses?

We have learned that 13 days after exposure, or 25 days before onset of illness, EBV-carrying lymphocytes were already circulating in the blood of a most cooperative, susceptible Swedish boy, who in the course of a party had kissed a girl who two days later developed signs of infectious mononucleosis. We need to enlist other susceptible and cooperative contacts of cases, as hard as they are to find, and study them within the first few days after exposure, including a search for virus in the oropharyngeal secretions.

What is the actual incidence of chronic active infectious mononucleosis? Are all cases due to EBV or are some due to other viruses that have been aided by the immunosuppressive effect of a primary EBV infection?

How many of the polyclonal lymphoproliferations observed in immunologically compromised patients are EBV-induced? If observed in organ transplant patients, how often do the lymphomas arise from the recipient's cells and how often from the cells of the organ or blood donors?

Regarding Burkitt's lymphoma, we still do not know many details of its genesis. Is EBV one of several initiators of this malignancy or, in other words, are EBV-associated and non-associated tumors the same disease or two distinct entities?

Are the characteristic chromosomal translocations in Burkitt cells truly mere chance events due to enhanced lymphoproliferation induced by malaria, or the acquired immunodeficiency syndrome, or any other excessive antigenic stimulation? Is the translocation of the c-myc oncogene from chromosome 8 to the immunoglobulin gene regions of chromosome 2, 14 or 22 the final step or does the oncogene product set off yet another step in the chain of events leading to the development of the tumor?

The course of events leading to nasopharyngeal carcinoma is even less clear. What is the cell or origin? Does the carcinoma arise from fetal thymic remnants in Waldeyer's ring, as has been suggested many years ago, and supported

now by reports of an association of EBV with some thymic, salivary glands and possibly also some tonsil and larynx carcinomas? All the anatomical sites involved are derived embryologically from the same region, the third and fourth primitive pharyngeal pouches.

While certain epithelial cells have been productively infected with EBV in vitro, there is as yet no evidence of immortalization of epithelial cells by EBV. This should not be too surprising because nasopharyngeal carcinoma cells have not been grown permanently in culture. Once the appropriate growth factors for the carcinoma cells are identified, they presumably will support also the growth of experimentally EBV-transformed epithelial cells.

Despite the deficiencies in our knowlege, one wonders why a key role of EBV in the genesis of Burkitt's lymphoma and nasopharyngeal carcinoma is still questioned, whereas a causative role of the fourth "first human cancer virus", the human T-cell leukemia virus, tacitly accepted as the cause of human adult T-cell leukemia, even though it should raise similar doubts. In fact, EBV has served as model for the HTLV studies; every seroepidemiologic observation made with HTLV is matched by earlier obserations made with EBV. This includes the existence of many seropositive healthy viral carriers which clearly implies that in addition to HTLV other factors are needed to induce a T-cell leukemia or lymphoma.

Truly amazing strides have been made regarding the biochemistry and molecular biology of EBV. The viral genome has been deciphered, fragments of it have been cloned, genes have been transfected to various types of cells, new EBV-specific antigens have been identified, and a number of viral polypeptides and glycoproteins are produced now in

quantity for study of their biological functions, their
usefulness in the specific serodiagnosis and in monitoring
of EBV-associated diseases, and last not least, for production
of a vaccine.

We have no doubt that the glycoprotein vaccine now in
the offering will protect susceptible college students, at
least transiently, against infectious mononucleosis, perhaps
even permanently, if a subsequent exposure to EBV evokes a
rapid recall of antibodies to intercept disease, although
not necessarily inapparent infection. If thus a persistent
viral carrier state ensues it will convert an initially
transient into an ultimately permanent immunity. However,
for prevention of EBV-associated malignant diseases presumably
also latent persistent EBV infections have to be prevented,
which presents a vastly more difficult task.

It is clear that one cannot discuss the present without
at least a few glimpses into the future. We have learned
in the late 1950s, however, that it is hazardous to go
further. At that time, Sir MacFarlane Burnet, the eminent
Australian virologist-immunologist and Nobel laureate,
passed through Philadelphia and told us that he was
abandoning virology to return to his first love, immunology,
because all important things that can be done with viruses
had been done. It showed that the crystal ball of even
the great can at times be awfully cloudy. One thing we
can say today with assurance; not all important things
that can be done with EBV have been done. There is plenty
of work ahead so that EBV will remain an efficient travel
agent for all of you for some time to come. EBV has taken
many of us all over the world, to study EBV-associated
diseases in endemic regions of Africa or the Far East, or
to attend EBV-devoted conferences, symposia or workshops
in many parts of the world, including now this lovely spot
in Greece at the shores of the wine blue sea. There are
still further years of further EBV-induced travels ahead
before the glycoprotein vaccine will put an end to it.

LIST OF PARTICIPANTS

AARONSON, S.
National Cancer Institute
Bldg. 37, Rm. 1E18
Bethesda, MD 20205
USA

ABLASHI, D.
National Cancer Institute
Building 37 Rm. 1E24
Bethesda, MD 20205
USA

ALFIERI, C.
Hospital Sainte-Justine
Pediatric Research Center
3175 Cote Ste. Catherine Rd.
Montreal, Quebec
Canada H3T 1C5

ALLDAY, M.
Biochemistry Lab
School of Biological Sciences
University of Sussex
Falmer, Sussex BN 19 Q9
England

ANDERSEN, R.P.
Viral Genetics Lab.
Dept. Molecular Biology
Abbott Laboratories, North Chicago
USA

ARCHARD, G.
P.O. Box T-29 The Welcome Bldg.
183 Eastor Road
London
England

BARILE, G.
Institute of Technologic Biomedichine
C.N.R. Via Morgagni 30/E
00141 Roma
Italy

BERTRAM, G.
Oberarzt University, HNO-Klinik
Joseph-Stelzmann Str. 9
5000 Koln 410
West Germany

BLAESE, R.M.
National Institutes of Health
Building 10, RM 4N117
Bethesda, MD 20205
USA

BODESCOT, M.
Institute de Res, Sci. sur le Cancer
Villejuif
France

BOYD, A.
8821 Indian Springs Rd.
Frederick, MD 21701
USA

CHEN, J.Y.
College of Medicine
National Taiwan University
Taipei
Republic of China

CLIFT, B.
Chronic Disease Branch
Center Inf. Diseases
Anchorage, AL 99501
USA

DAHLBERG, J.
National Cancer Institute
Building 37, RM 1C15
Bethesda, MD   20205
USA

DESGRANGES, C.
Laboratoire EBV
Faculte de Medicine
A. Carrel 69008
Lyon, France

DILLNER, S.J.
Dept. Tumor Biology
Karolinska Institute
Stockholm, S-10401
Sweden

EPSTEIN, M.A.
Dept. of Pathology
University of Bristol, Medical School
Bristol BS8 ITD
England

FARRELL, P.
Ludwig Institute for Cancer Research
Hills Road
Cambridge CB2 2 QH
England

FOWLER, E.
Dept. Microbiol. and Immunology
2096 MSB-Univ. South Alabama
Mobil, AL   36688
USA

GILDEN, R.
Frederick Cancer Research Facility
P.O. Box B
Frederick, MD   21701
USA

GLASER, R.
Ohio State University
College of Medicine
333 West 10th Avenue
Columbus, OH   43210
USA

GRANLUND, D.
Biotech Research
1600 East Guide Dr.
Rockville, MD   20850
USA

GRIFFIN, B.
Imperial Cancer Research Fund
Lincoln's Inn Fields
London WC 2
England

HAYWARD, D.S.
Department of Pharmacology
Johns Hopkins School of Medicine
725 N. Wolfe St.
Baltimore, MD   21205
USA

HO, J.H.C.
M. and H.D. Institute of Radiology
Queen Elizabeth Hospital
Wylie Road
Kowloon, Hong Kong

HUANG, A.T.
P.O. Box 3942
Duke University Medical Center
Durham NC   27710
USA

HUANG, D.
Medical and Health Dept.
Queen Elizabeth Hospital
Kowloon
Hong Kong

HUTT-FLETCHER, L.
Box J-145 JHMHC
University of Florida
Gainesville, FL 32610
USA

ITO, Y.
Department of Microbiology
Faculty of Medicine
Kyoto University
Kyoto 606
Japan

JILG, W.
Max Von Pettebkofer Institute
Pettenkofer Str 9a
D-8000 Munchen 2
West Germany

JONCAS, I.
Hospital Sainte-Justine
3175 Chemin Sainte-Catherine
Montreal
Canada

JONES, J.F.
NJH/NAC, Dept. Pediatrics
3800 East Colfax Avenue
Denver, CO 80206
USA

KALOGEROPOULOS, N.
Dept. Pathology
Univ. Athens
Medical School
Athens, Greece

KLEIN, E.
Virology Lab.
St. Luke's Roosevelt Hospital
428W 58th Street
New York, N.Y. 10019
USA

KLEIN, R.
Dept. Microbiology
New York Univ. Medical Center
550 First Avenue
New York, N.Y. 10016
USA

KLEIN, G.
Dept. Tumor Biology
Karolinska Institute
S-104 01 Stockholm
Sweden

KIEFF, E.
Dept. Medicine and Microbiology
The University of Chicago
The Pritzker School of Medicine
910 East 58th Street
Chicago, ILL 60637
USA

KOTTARIDIS, S.D.
Papanikolaou Res. Center of Oncology
171 Alexandras Ave.
Athens, Greece

KRUEGER, G.R.F.
Immunopathology Laboratories
University of Koln
Joseph-Stelzmann Str. 9
5000 Koln 41
West Germany

KURSTAK, E.
Director-CVRG-ICVO
Faculty of Medicine
University de Montreal
C.P. 6128, succ. A.
Montreal
Canada H3C 3J7

LANIER, A.P.
Chronic Disease Branch
Center for Inf. Diseases
Anchorage, AL 99501
USA

LAU, H.W.
M. and H.D. Institute of
Radiology and Oncology
Wylie Road
Kowloon, Hong Kong

LEVINE, P.H.
National Institutes of Health
Landow Building, Rm. 8C41
Bethesda, MD 20205
USA

LI, M.T.
Biotech Research Lab
1600 E. Gude Drive
Rockville, MD 20850
USA

LINDAHL, T.
Imperial Cancer Research Fund
Lincoln's Inn Fields
London WC2,
England

LUKA, J.
Georgetown University
Dept. Microbiology
Washingto, D.C. 20007
USA

MAGRATH, I.
National Cancer Institute
Building 10, RM 13N 240
Bethesda, MD 20205
USA

MC CANN, J.
Physicians Radio Network
2980 Berkshire Rd.
Cheveland Hts., Ohio 44118
USA

MENEZES, J.
Lab. of Immunovirology
Ste-Justine Hospital
3175 Ste-Catherine Rd.
Montreal (Quebec)
Canada H3T IC5

MILMAN, G.
Dept. Biochemistry
Johns Hopkins University
615 N. Wolfe St.
Baltimore, MD 21205
USA

MODROW, S.
Max Von Pettenkofer Institut
Pettenkofer Str. 9a
8000 Muchen 2
West Germany

MOSS, J.D.
Queensland Institute of Medical Research
Bramston Ice
Brisbane 4006
Australia

NEEL, H.B.
Dept. Otorhinolaryngology
Mayo Clinic
Rochester, Minnesota 55905
USA

NONOYAMA, M.
10900 Roosevelt Blv.
St. Petersburg, FL 33702
USA

PAGANO, J.S.
Lineberger Cancer Res. Center
Chapel Hill, NC   27514
USA

PAPAS, T.
Lab. Molecular Oncology
NCI-Frederick Cancer Res. Facility
Bldg. 469 Rm 204
Frederick, MD   21701
USA

PAPADAKIS, T.
Evagelismos Hospital
Athens, Greece

PEARSON, G.
Dept. Microbiology
Georgetown University
School of Medicine
Washington, DC   20007
USA

PERRICAUDET, M.
IRSC-CNRS
7 Rue Guy Mocquet
94800 Villejuif
France

PIZZA, G.
Divisione Urologica
Ospedale Provinciale Specializzato
Bologna 11
Italy

POON, F.Y.
M. and H.D. Institute of
Radiology and Oncology
Queen Elizabeth Hospital
Kowloon, Hong Kong

POPE, J.H.
Queensland Institute of Med. Research
Bramston Ice
Herston, Q4006
Australia

PRASAD, U.
Dept. Otorhinolaryngology
University of Malaya
Kuala Lumbur 22-11
Malaysia

PURTILO, D.T.
Dept. Pathology
University of Nebraska Medical Center
42nd and Dewey Ave.
Omaha, NE   68105
USA

RAAB-TRAUB, R.N.
Lineberger Cancer Res. Center 237H
University of North Carolina
Chapel Hill, NC   27514
USA

RAMANDANIS, G.
Hellenic Anticancer Institute
St. Savas Hospital
171 Alexandras Ave.
Athens, Greece

RICKINSON, A.B.
Dept. of Cancer Studies
The University of Birmingham
The Medical School
Birmingham B15 2TJ
England

RICKSTEN, K.A.
Dept. of Med. Biochemistry
Univ. of Goteborg
Box 33031, S-40033
Gotenborg, Sweden

ROUBAL, J.
Institute of Sera and Vaccines
Prague
Czechoslovakia

RYMO, L.
Dept. of Clinical Chemistry
Sahlgren's Hospital
S-413 45 Gothenburg
Sweden

SAEMUNDSEN, K.A.
Dept. of Microbiology
Univ. Iceland-Virus Lab.
Eiviksgata-P.O. Box 855
15-121 Reykzavik
Iceland

SAW, D.
Inst. of Pathology
Queen Elizabeth Hospital
Kowloon, Hong Kong

SCHUBACH, H.W.
Univ. of Minnesota
Section of Med. Oncology
School of Medicine
Box 286 Mayo Memorial Building
420 Delaware Street
Minneapolis, Minnesota   55455
USA

SCULLEY, T.
Queensland Institute of Med. Res.
Bramston Ice, Herston
Brisbane, Australia

SEKERIS, C.
National Research Foundation
Lab. of Biological Research
Athens, Greece

SHANMUGHAM, M.S.
Dept. of Otorhinolaryngology
Singapore General Hospital
Singapore

SHEN, Y.Y.
Fujian Medical College
Fuzhon, Fujian
People's Republic of China

SIMONS, M.J.
Immunogene Typing
P.O. Box 218
Toorak 3142
Australia

SIXBEY, W.J.
Division of Infectious Diseases
St. Jude Children's Res. Hospital
Memphis, Tennessee
USA

SMITH, R.W.
Institute for Postgraduate
Interdisciplinary Studies
P.O. Box 60846
Palo Alto, CA   94306

STRAUS, S.E.
National Institutes of Health
Building 10, Rm. 11N-113
Bethesda, MD   20205
USA

SUNDAR, S.K.
Ste. Justine Hospital
3175 Ste-Catherine Rd.
Montreal (Quebec)
Canada H3T 1CS

693

SWAN, D.
National Cancer Institute
Bethesda, MD 20205
USA

TOSATO, G.T.
Bldg. 29, Rm. 502
NIH/FDA
Bethesda, MD 20205

TRONICK, R.S.
National Cancer Institute
Building 37, Rm. IA07
Bethesda, MD 20205
USA

TURSZ, T.
Laboratoire d' immunogies des tumeurs
Institut Gustav Roussy
94805 Villejuif Cedex 1
France

VOURNAKIS, N.J.
Dept. Biology
Syracuse University
130 College Place
Syracuse, NY 13210
USA

VOLSKY, J.D.
Dept. Pathology and Lab. Medicine
Univ. Nebraska Medical Center
42nd and Dewey Ave.
Omaha, NE
USA

WOLF, H.
Molecular and Tumor Virology
Max von Pettenkofer Institut
Pettenkofer Str. 9a
D-8000 Munchen 2
West Germany

WIELS, J.
Laboratoire d' Immunologie des tumerus
Institute Gustave Roussy
94805 Villejuiff Cedex
France

YANG, J.
Dept. of Hygiene and Tropical Med.
Keppel Street
London, WC1
England

YADAV, M.
Dept. of Genetics and Cellular Biology
University of Malaya
Kuala Lumpur 22-11
Malaysia

ZENG, Y.
Institute of Virology
China National Centre
for Preventive Medicine
China

ZONGZA, V.
Dept. of Biochemistry and Mol. Biology
University of Athens
Atehns, Greece